Hydro-Environmental Analysis

Freshwater Environments

Hydro-Environmental Analysis

Freshwater Environments

James L. Martin

CRC Press
Taylor & Francis Group
Boca Raton London New York

CRC Press is an imprint of the
Taylor & Francis Group, an **informa** business

CRC Press
Taylor & Francis Group
6000 Broken Sound Parkway NW, Suite 300
Boca Raton, FL 33487-2742

First issued in paperback 2017

© 2014 by Taylor & Francis Group, LLC
CRC Press is an imprint of Taylor & Francis Group, an Informa business

No claim to original U.S. Government works

Version Date: 20130531

ISBN 13: 978-1-4822-0607-4 (hbk)
ISBN 13: 978-1-138-07172-8 (pbk)

Library of Congress Cataloging-in-Publication Data

Martin, James Lenial, 1947-
   Hydro-environmental analysis : freshwater environments / James L. Martin.
      pages cm
   Includes bibliographical references and index.
   ISBN 978-1-4822-0607-4
   1. Watershed management. 2. Water resources development--Environmental aspects. 3. Freshwater habitat conservation. I. Title.

TC409.M36 2014
628.1--dc23   2013021198

Visit the Taylor & Francis Web site at
http://www.taylorandfrancis.com

and the CRC Press Web site at
http://www.crcpress.com

Dedication

To my family for their love and support, and
to my students for whom this is written

Contents

PART II   Lakes and Reservoirs

Preface

Adequate supplies of water are a crucial resource for any human activity. The water resource management paradigm, as implemented in both practice and law, for much of our history focused on the management of fresh waterbodies, such as rivers and streams, lakes and reservoirs, to meet the supply needs of water for human use for drinking, recreation, navigation, and often just for conveying wastes. Equally important has been the management of those systems for excess water, either to prevent or control flooding or to remove excess water to convert lands for some human benefit. It has only been in recent decades that there has been a paradigm shift such that consideration of environmental needs is now a required component of all water resource projects. The Clean Water Act (CWA) in 1972 set a new national goal "to restore and maintain the chemical, physical, and biological integrity of the Nation's waters." Many environmental resource and engineering activities today are focused on either the maintenance of environmental quality in, or remediating historical damage and restoration of, our aquatic systems.

Factors impacting the chemical, physical, and biological integrity of our nation's waters are extremely complex and factors affecting those processes and their interactions are often poorly understood. That makes water resource management extremely interesting and also a challenge, for example, determining how and what waterbodies should be restored "to" (per the CWA) and ensuring that this generation's solution does not become the next generation's problem.

This book was developed in part for a graduate course of the same name that the author teaches at Mississippi State University. The course is offered by the Department of Civil and Environmental Engineering, but students populating the course are typically from a wide variety of academic disciplines, both in science and in engineering. That course is intended to introduce the physical, chemical, and biological characteristics of rivers and streams, lakes and reservoirs, estuaries and coastal waters from an engineering perspective. This volume will focus on freshwater environments.

The "design" audience of the book is environmental and water resource engineers and environmental scientists. The book is not a traditional engineering book in that it concentrates on broad and general concepts, rather than specific design criteria. By design, there are few equations in this book and the format is relatively informal. Rather, the book focuses on an introduction to the characteristics of freshwater environments as those characteristics may affect, or be affected by, water resource management and engineering projects. The book is intended to introduce students of environmental science to engineered structures and students of engineering to the aquatic and limnological sciences. The book focuses on regulated freshwater environments, as there are relatively few waterbodies that are not controlled or modified to some degree. Because of the complexity of these aquatic systems, often precise definitions and classification become difficult. However, these definitions and systems of classification are important for a variety of reasons, such as for scientific study, for the survey of these systems and how they change with time, and for regulation such as under the CWA. Often it is the legal or regulatory definition that controls the management of these systems, so this book focuses in part on those regulatory definitions. Finally, the book will introduce some of the basic considerations and principles involved in the restoration or management of aquatic systems.

This volume (freshwater environments) is divided into two parts: Rivers and Streams, and Lakes and Reservoirs. Each part begins by discussing the characteristics of those systems and methods of classification, followed by a discussion of physical, chemical, and biological characteristics. In the section on lakes and reservoirs, the characteristics and operations of regulatory structures are also introduced. Methods commonly used to assess the environmental health or integrity of these waterbodies are then presented, followed by an introduction to considerations for restoration.

The last two chapters deal with an introduction to two unique aquatic environments: wetlands and reservoir tailwaters.

I gratefully acknowledge all of the contributions of my students and colleagues that have helped shape this book. I particularly thank Sandra L. Ortega-Achury for her contributions, advice, help, and the correction of many of my blunders. I also would like to thank my wife for her patience and support. Finally, I would like to thank my major advisor for my PhD, Dr. Steven C. Chapra, who many years ago provided me not only an education but also a profession and a philosophy to live by: "Do what you love, and love what you do, and you'll never work again."

James Lenial Martin

Author

James Lenial Martin is professor and Kelly Gene Cook, Sr. chair in civil engineering in the Department of Civil and Environmental Engineering at Mississippi State University. Previously, he was a research environmental scientist with the U.S. Environmental Protection Agency at its Large Lakes Research Station, a research civil engineer with the U.S. Army Corps of Engineers Waterways Experiment Station, and vice president and director of engineering with AScI Corporation. His degrees include a bachelor of science in wildlife science from Texas A&M, a bachelor of science in civil engineering from Texas A&M, a master of science in biology from Southwest Texas State University, and a PhD in Civil and Environmental Engineering from Texas A&M. He is a registered professional engineer in Mississippi, a founding diplomate in water resources engineering with the American Academy of Water Resources Engineers, and a fellow of the American Society of Civil Engineers. He has over 30 years of experience conducting and managing water quality modeling projects and developing and applying models of hydrodynamics and water quality. He is senior author of *Hydrodynamics and Transport for Water Quality Modeling* and senior editor and author of *Energy Production and Reservoir Water Quality* and has authored/coauthored over 100 technical reports and publications, including Environmental Protection Agency guidance documents and model user documentation.

1 Introduction

1.1 HYDRO-ENVIRONMENTAL ANALYSIS, OR WHAT IS IN A NAME?

The title of this text is *Hydro-Environmental Analysis*. So, the first obvious question that the reader may have is what exactly is "hydro-environmental analysis?" What is the topic area and is this a new field of study? The phrase is not new to this book and its usage is becoming increasingly common. Other books and journals are devoted to hydro-environmental studies. We are sure that we could provide some concise, valid, and very esoteric definitions. But we will not do so. Instead, we will provide a discussion of the purpose of this book, which may hopefully indicate why *Hydro-Environmental Analysis* is an apt title for the content.

> All a man needs is Confidence and Ignorance, and he will be sure to succeed in life.
>
> **Mark Twain (Figure 1.1)**

The book is intended primarily for undergraduate and graduate students of engineering, while it is hoped that students in other disciplines will also find it useful. Engineering is, in large part, based upon the premise that we can predict things. For example, much of engineering is based upon Newton's laws of motion, or laws of conservation, or laws of thermodynamics, etc. Engineers have successfully used these laws, principles, or theories in order to make predictions about what happens to a system, or to relationships between cause and effect. In order to design engineering structures, some idea of the relationships between cause and effect is necessary for predictions, whether they be predictions of the impact of wind loads on a building or the impact of high flows on a dam.

> There is something fascinating about science. One gets such wholesale returns of conjecture out of such a trifling investment of fact.
>
> **Mark Twain**

A basic question is the degree to which we need to understand a system in order to make predictions about it. Engineers, by definition, deal with the art of applying scientific knowledge and theories to practical problems, with the emphasis on the "a" word—applied. So, to what degree do engineers need to understand these theories or that knowledge in order to apply them correctly? Or, what are the consequences if the scientific theories or "conjectures" are wrong? For example, undergraduate engineering students often learn specific computational techniques, and then often only as graduate students do they find out how limited the knowledge base is on which those computations are based. But, if the limitations are not known and methods are applied inappropriately, there may be consequences in terms of human or environmental health.

A quote often attributed to George Box (one of the most influential statisticians of the twentieth century) is that while "all theories are wrong, some are useful." One could argue then that at a minimum, we need to know enough about a system to know when the laws, theories, or knowledge that we use to establish cause and effect relationships and make engineering predictions are wrong. For example, Newton's laws on are not universally applicable. There are natural phenomena that Newton's laws cannot explain, particularly as one approaches the speed of light. For most engineering structures though, where the design is based on Newton's equations, hopefully that is not a major problem. There are many similar theories and laws in other disciplines, which are useful but

1

FIGURE 1.1   Mark Twain. (Image from Library of Congress Ref. No. LC-USZ62-5513.)

are not "perfectly conceived," such as the Bohr atom and the Schrödinger equations for chemical reactions used in chemistry and chemical engineering. These laws and theories, while not universally applicable, are still extremely useful. Newton's laws, for example, still form the basis for much of engineering mechanics.

It may not be necessary to have a complete understanding of general relativity in order to design a rocket to send a person to the moon. A knowledge of the physical relationships encompassed in Newton's laws, which all engineering students are exposed to, may be perfectly adequate. But, how about predicting the impact of heat on the flows in a lake? Or, how about predicting the impact of nutrients on phytoplankton concentrations in a lake? These environmental relationships are not rocket science. They are much more complex! That is, environmental relationships are much more variable and uncertain than some physical relationships are and the factors impacting those relationships are not nearly as well known. In addition, while many if not most engineering projects have some impact on the environment, most engineering students are not exposed to environmental relationships with nearly the same degree of rigor as they are to physical relationships for them to estimate the environmental impact of those engineering projects. However, it is often engineers who either make decisions or provide input to decision makers related to environmental problems. That may be partly due to one common attribute among engineers, in that they are good at making decisions and solving problems, and in environmental management usually someone has to come up with a solution and make a decision. Unfortunately, as will be a recurring theme in this book, one generation's solution is often the next generation's problem. Perhaps as a result of incomplete knowledge of the environmental impacts in the design of past engineering projects (e.g., watershed or channel alterations), many of today's engineering projects deal with the remediation or mitigation of those environmental impacts, or the restoration of impacted waterbodies.

This book is not intended as a substitute for a rigorous curriculum in the environmental sciences, including limnology, and other "ologies" and it is not intended to provide "everything an engineer should know." First, the knowledge base in many areas is enormous, and many professionals spend their entire careers dealing with only one or a few aspects of the topics discussed. Secondly, in many cases an adequate knowledge base simply does not exist.

This text is intended to provide an overview of the factors affecting the quantity and quality of water (hydro), and considerations for the analysis of that quantity and quality in a variety of environmental systems, including rivers, streams, lakes, and reservoirs. The material covered will include that in engineering studies of hydrology, hydraulics, groundwater engineering, atmospheric studies and air pollution, water quality, water resource engineering, and environmental studies. These studies are generally highly computational in nature. While the text is written from an engineering perspective, the reader is cautioned that computational aspects will not be emphasized in this book; the emphasis will be on more broad and general concepts and factors that may affect the use, or misuse, of those computations. That structure will make this book and any courses on which it may be based difficult for some engineers. However, equations will be sprinkled sparingly throughout the text so that those engineers do not go into complete withdrawal.

While the book is organized into a series of sections dealing with rivers and streams, wetlands, and lakes and reservoirs, this separation and some of the distinctions between these waterbodies are arbitrary and artificial. These water systems are inseparable components of the water environment, thus *Hydro-Environmental Analysis* is considered an apt title. An organizing principle for hydro-environmental analysis is presented in Section 1.2.

The artificial separation of these waterbodies has resulted in considerable confusion in the literature, such as just trying to define and distinguish each type of waterbody (river, stream, lake, reservoir, etc.). A common recurring theme of this book will also be in pointing out that confusion where it occurs and in providing a discussion as to how it may impact our methods for analysis and environmental management. *So, let the confusion begin!*

1.2 HYDROLOGIC CYCLE

Consider that the earth's total water supply is, for all practical purposes, constant. However, if we examine each reservoir of that water supply, we find that it is in a continuous state of flux or motion, such as between the oceans, atmosphere, land surfaces, and groundwaters. This cycle of water is referred to as the hydrologic cycle and is the organizing principle upon which much of this book is based.

Assuming that the total quantity of water on the planet is essentially fixed, and that mass is conserved, we should be able to essentially "bookkeep" or quantify the water transfer between each reservoir and component of the hydrologic cycle (Figure 1.2). The components or processes include:

- Atmospheric movement of air masses
- Precipitation
- Evaporation
- Transpiration
- Infiltration
- Percolation
- Groundwater flow
- Surface runoff
- Streamflow

Figure 1.3 illustrates the relative magnitude of each of these processes in the global water balance. As would be expected, the greatest fluxes are between the atmosphere and the oceans, followed by the fluxes between the land surface and the atmosphere. As discussed in the following section, the quantity of water in motion is an extremely small fraction of the earth's water supply. While small on a global scale, these fluxes have enormous impacts on the use and availability of the water supply.

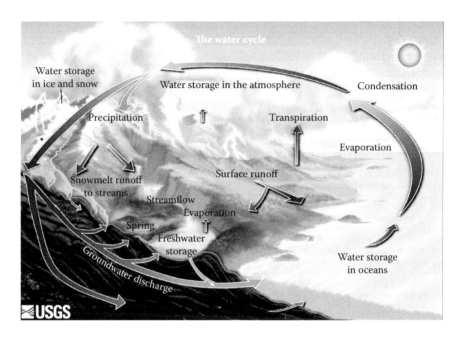

FIGURE 1.2 The hydrologic cycle. (From USGS, The hydrologic cycle (pamphlet), U.S. Government Printing Office, Washington, DC, 1984a.)

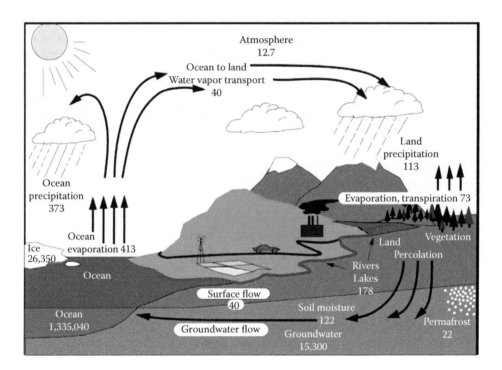

FIGURE 1.3 Global water budget (units for reservoirs in 10^3 km^3, and flow in 10^3 km^3 yr^{-1}). (From Trenberth, K.E., Smith, L., Qian, T., Dai, A., and Fasullo, J., *Journal of Hydrometeorology*, 8, 758–769, 2007. With permission.)

1.2.1 SO, HOW MUCH WATER IS THERE AND WHERE IS IT?

The total water supply of the world is 326 million cubic miles (about 360 trillion gallons or 1.36 trillion cubic kilometers) of which about 0.02 mi³ (0.02 trillion gallons or 0.08 km³) is in motion at any one time. Most of the water is stored in the oceans, frozen in glaciers, held in lakes, or detained underground (Table 1.1). The 317 million cubic miles of water held by the oceans constitutes 97.3% of the earth's supply.

According to the U.S. Geological Survey (USGS 1984a):

- Only about 3,100 mi³ of water (12,921 km³), chiefly in the form of invisible vapor, is contained in the atmosphere at any given time.
- If the water were to fall all at once, the earth would be covered with only about 1 in. of water.
- Of the 102,000 mi³ (425,000 km³) of water that passes into the atmosphere annually, 78,000 mi³ (325,000 km³) falls directly back into the oceans.
- Streams and rivers collect and return to the oceans some 9,000 mi³ (37,000 km³) of water, including a large quantity of water that, as "groundwater," has moved slowly to natural outlets in the beds and banks of streams.

While water is abundant, only about 0.3% is in a form that is usable to us, and most of that water is inaccessible. So, of the 326 million cubic miles or 1.36 trillion cubic kilometers of water, only about 15,000 mi³ (62,500 km³) maintains life processes, principally as soil moisture, which provides water necessary for vegetation. This water reaches the atmosphere again by the process of evapotranspiration.

1.2.2 WHERE DOES IT GO?

In comparison with the earth's total water supply, the total amount of water available is small. However, both the amount and the distribution of this available water have large ecological and human use impacts.

The quantity of water available and the water demands also vary with time and location around the United States, and the world. In 2000, approximately 408 billion gallons per day (1.5 trillion L day^{-1}) was withdrawn in the United States for all uses (Hutson et al. 2000), of which about 220 billion gallons per day (833 billion L day^{-1}) was returned to streams after use. As illustrated

---

TABLE 1.1
Global Water Source and Volume

Water Source	Water Volume (km³)	Percentage of Total
Oceans	1,230,000,000	97.20
Icecaps and glaciers	28,600,000	2.15
Groundwater	8,300,000	0.61
Freshwater lakes	123,000	0.01
Inland seas	104,000	0.01
Soil moisture	67,000	0.01
Atmosphere	12,700	0.001
Rivers	1,200	0.0001
Total water volume	1,360,000,000	100

Source: USGS, The hydrologic cycle (pamphlet), U.S. Government Printing Office, Washington, DC, 1984a.

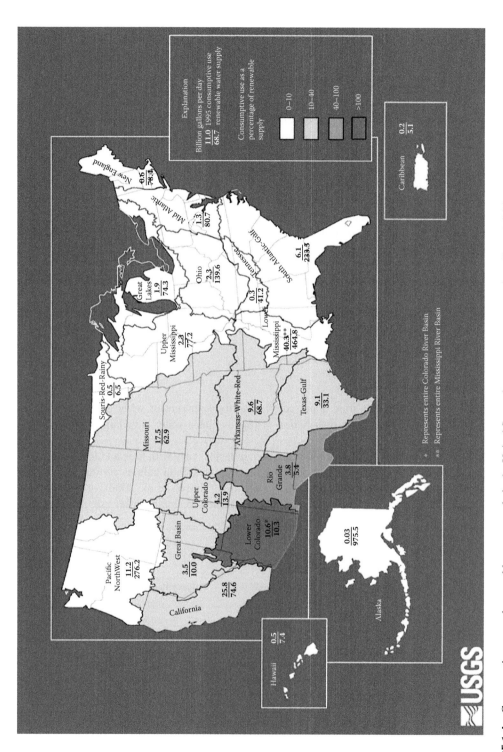

FIGURE 1.4 Consumptive use and renewable water supply in the United States, by water-resources region. (From USGS, National Water Summary 1983—Hydrologic events and issues, U.S. Government Printing Office, Washington, DC, 1984b.)

by Figure 1.4, consumptive demands vary widely across the United States, partly as a function of water-use patterns, and they also vary widely in comparison with the available supply. For example, in the eastern United States, consumptive use is a relatively small fraction of the available supply, while water use in the arid West often approaches or exceeds the available supply.

The demands also vary between specific water uses, including the following:

- Livestock
- Irrigation
- Domestic
- Public water supply
- Thermoelectric cooling
- Mining
- Aquaculture
- Industrial uses

As illustrated in Figure 1.5, of these uses in 2005, thermoelectric power (49%), irrigation (31%), public water supply (11%), and industrial demands (4%) combined to make up 95% of the total U.S. water demands. In 2000, irrigation in California accounted for 20% of all irrigation in the United States (Hutson et al. 2000).

The demands also vary between surface water and groundwater supplies. Of the total demands, approximately 79% were surface water withdrawals in 2000. Of the demands for groundwater, 68% were for irrigation (Hutson et al. 2000). Total irrigation withdrawals in the United States are illustrated by Figure 1.6.

The total demands increased from approximately 180 to 408 billion gallons per day (681 billion to 1.5 trillion L day^{-1}) during the period between 1950 and 2000 (Hutson et al. 2000). Changes also occurred in the distribution of these demands during that period, as illustrated by changes in the distribution of irrigated lands (Figure 1.7).

Total demands have also changed with the growing U.S. population, as illustrated by a comparison of changes in the population and public supply withdrawals between 1950 and 2000 (Figure 1.8).

1.3   PATTERNS IN WATER MANAGEMENT IN THE UNITED STATES

The earth's total water supply is large and it is usually the distribution of water in time and space that is the problem. The quantity and quality of water are rarely "just right" at any specific time or location. Water is commonly in excess or in short supply and/or its quality limits or precludes its use.

The pattern of water management and use has varied with time in the United States. Historic federal laws focused on protecting waters for navigation and other human usages, while state and local laws focused on "who owns the water." Conventional water management in the United States has focused on the control or manipulation of water supplies. Traditional management has included the construction of dams and levees to control excess water; the construction of dams to contain water for water supply, power generation, and other uses; and the construction of conduits of various kinds to redistribute water or provide for commercial uses. Often, traditional water resource engineering projects have also focused on design to withstand the impacts of excess water.

Often, historical water management projects or practices did not take into full consideration the environmental impact of those projects or practices. One consequence is that many of the past generations' water management solutions have become today's or the future's water management problems. For example, many engineering projects today are attempting to reverse the impact of past engineering projects. Examples include projects for reducing or removing flow regulation structures such as dams and canals; removing legacy contaminates (DDT, dioxins, PCBs, etc.) from waterbodies; restoring the natural flow and biota of waterbodies, which were impacted by past projects; and mitigating the impact of excess use of groundwater, which has resulted in depletion and contamination of aquifers, land subsidence, and decreased flows in surface waters.

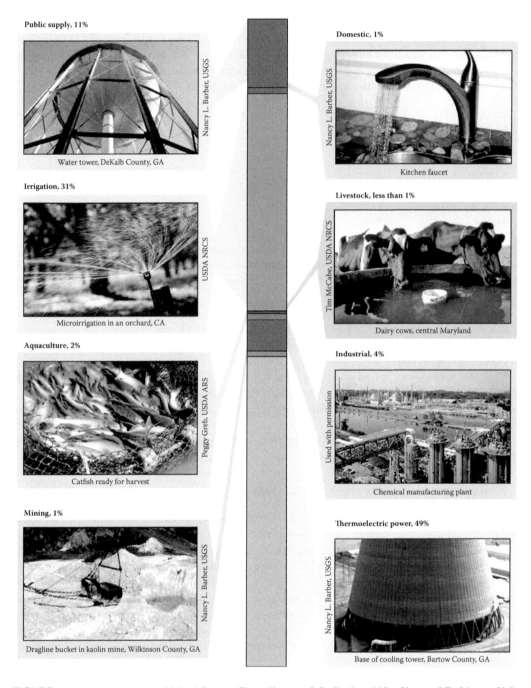

FIGURE 1.5 Water use in the United States. (From Hutson, S.S., Barber, N.L., Kenny, J.F., Linsey, K.S., Lumia, D.S., and Maupin, M.A., USGS Circular 1268, U.S. Geological Survey, Washington, DC, 2000.)

The changes in water management and water use have resulted from (Hirsch 2006):

- Market forces—energy costs and agricultural prices
- Demographics
- Public values (ecosystem services)

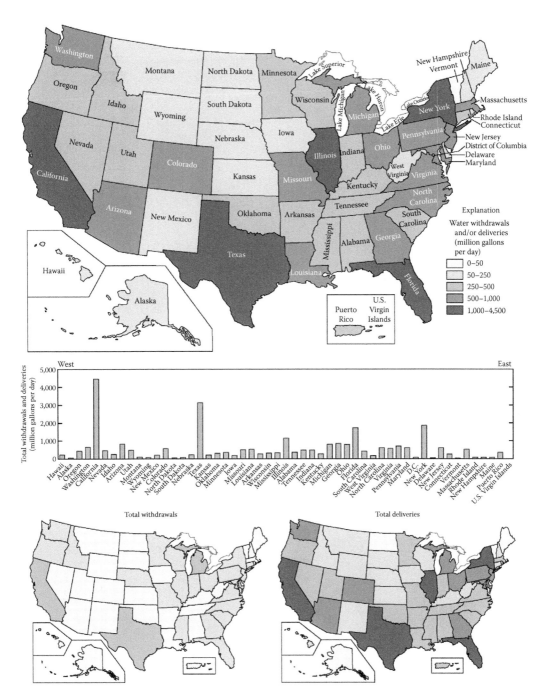

FIGURE 1.6 Total irrigation withdrawals in the United States. (From Hutson, S.S., Barber, N.L., Kenny, J.F., Linsey, K.S., Lumia, D.S., and Maupin, M.A., USGS Circular 1344, U.S. Geological Survey, Washington, DC, 2005.)

- State water laws
- Federal law (Clean Water Act, Endangered Species Act, etc.)
- Negotiations (federal, state, tribal, and users)
- Technology: water reuse, conservation, and aquifer storage and recovery (ASR)

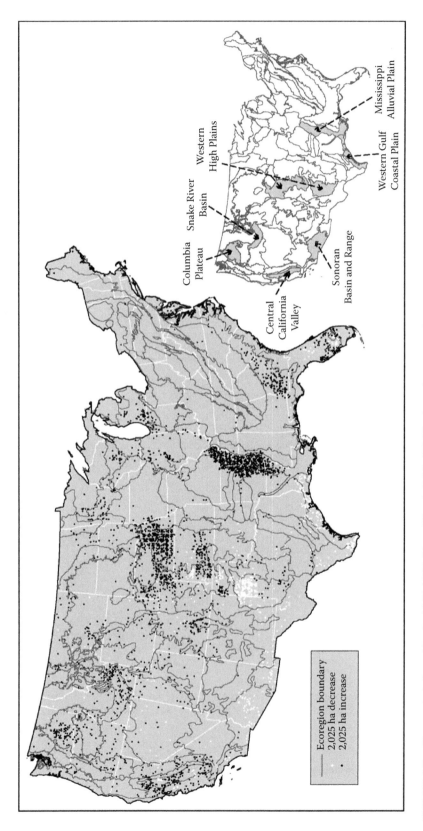

FIGURE 1.7 Changes in irrigated lands, 1947–1997. (From Hirsch, R.M., USGS science in support of water resources management, Presented at the Mississippi Water Resources Conference, April 25–26, Jackson, MS, 2006.)

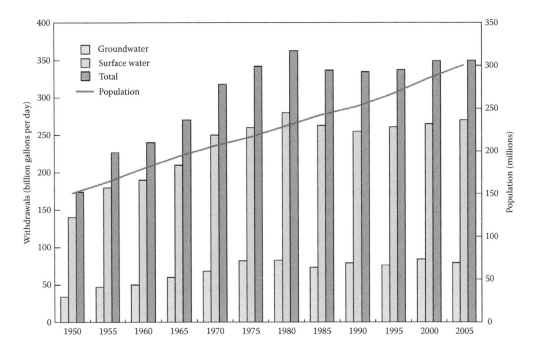

FIGURE 1.8 Changes in public supply withdrawals in the United States. (From Hutson, S.S., Barber, N.L., Kenny, J.F., Linsey, K.S., Lumia, D.S., and Maupin, M.A., USGS Circular 1344, U.S. Geological Survey, Washington, DC, 2005.)

Past water management has focused more on the quantity of water, rather than the quality, with the exception of the impact of that quality on human health and use. Only in recent decades has the quality of the nation's waters for fish and wildlife become of major national concern. The paradigm for water management has, to some degree, shifted in the United States from "how much water can we take from the water supplies" to "how much do we need to leave behind" (paraphrased from Hirsch 2006).

Shifts have also occurred from the management of high and low flows to attempting to manage the flow hydrograph; from management at timescales of days to weeks, to years and decades; and, from the management of individual waterbodies, to watershed scales or larger, including global impacts, and including interactions between groundwaters, surface waters, and the atmosphere. Often though, the knowledge base to accommodate those shifts in water management practices is limited or does not exist. The result in some cases may be that we are learning just enough from our past mistakes to go on and make entirely new ones.

REFERENCES

Hirsch, R.M. 2006. USGS science in support of water resources management. Presented at the Mississippi Water Resources Conference, Jackson, MS.

Hutson, S.S., N.L. Barber, J.F. Kenny, K.S. Linsey, D.S. Lumia, and M.A. Maupin. 2000. Estimated use of water in the United States in 2000. USGS Circular 1268, U.S. Geological Survey, Washington, DC.

Hutson, S.S., N.L. Barber, J.F. Kenny, K.S. Linsey, D.S. Lumia, and M.A. Maupin. 2004. Estimated use of water in the United States in 2000. USGS Circular 1268, U.S. Geological Survey, Washington, DC.

Hutson, S.S., N.L. Barber, J.F. Kenny, K.S. Linsey, D.S. Lumia, and M.A. Maupin. 2005. Estimated use of water in the United States in 2000. USGS Circular 1344, U.S. Geological Survey, Washington, DC.

Trenberth, K.E., L. Smith, T. Qian, A. Dai, and J. Fasullo. 2007. Estimates of the global water budget and its annual cycle using observational and model data. *Journal of Hydrometeorology* 8, 758–769.

USGS. 1984a. The hydrologic cycle (pamphlet). USGS, U.S. Department of the Interior, U.S. Government Printing Office, Washington, DC.

USGS. 1984b. National water summary 1983—Hydrologic events and issues. USGS Water Supply Paper 2250, U.S. Government Printing Office, Washington, DC.

Part I

Rivers and Streams

2 Rivers and Streams, Characteristics

2.1 LET THE CONFUSION BEGIN

First of all, what exactly is a river? And, what is the difference between a river and a stream? Or a brook? A beck? A creek? A crick? A slough? A bayou? Definitions of a river will vary but include:

- A large natural stream of water (larger than a creek).
- A large natural waterway. This is a specific term in the vernacular for large streams, *stream* being the umbrella term used in the scientific community for all flowing natural waterways. In the vernacular, stream may be used to refer to smaller streams, as may creek, run, fork, etc.
- A natural stream of water of considerable volume.
- A large stream.
- A large natural body of water that flows into another body of water.

So, how big is "larger than a natural stream?" For example, compared to the Mississippi River, when does a river become a stream? What is a "natural" stream? For example, if a river or a stream receives most of its flows from industries or waste discharge, is it a river? Are point source discharges "natural?" Are regulated rivers or ephemeral rivers (that only flow for short periods following rainfall events) in the desert southwest really rivers (e.g., Figure 2.1)?

For that matter, what is the difference between a "creek" and a "crick?" Perhaps Patrick F. McManus in his book *A Fine and Pleasant Misery* (1981) provided a fitting description (Figures 2.2 and 2.3):

"There is much in the world today concerning creeks and cricks. Many otherwise well-informed people live out their lives under the impression that a crick is a creek mispronounced. Nothing could be farther from the truth. A crick is a distinctly separate entity from a creek, and it should be recognized as such.

First of all a creek has none of the raucous, vulgar, freewheeling character of a crick. If they were people creeks would wear tuxedos and amuse themselves with the ballet, opera and witty conversation; cricks would go around in their undershirts and amuse themselves with the Saturday-night fights, taverns, and humorous belching. Creeks would perspire and cricks sweat. Creeks would smoke pipes; cricks chew and spit.

Creeks tend to be pristine. They meander regally through high mountain meadows, cascade down dainty waterfalls, pause in placid pools, ripple over beds of gleaming gravel and polished rock. They sparkle in the sunlight. Deer and poets sip from creeks, and images of eagles wheel upon the surface of their mirrored depths.

Cricks, on the other hand, shuffle through cow pastures, slog through beaver dams, gurgle through culverts, ooze through barnyards, sprawl under sagging bridges, and when not otherwise occupied, thrash fitfully on their beds of quicksand and clay.

Cows should perhaps be credited with giving cricks their most pronounced characteristic. In deference to the young and the few ladies left in the world whose sensitivities might be offended, I forgo a detailed description of this characteristic. Let me say only that to a cow the whole universe is a bathroom, and it makes no exception of cricks. A single cow equipped only with determination and fairly good aim can in a matter of hours transform a perfectly good creek into a crick." (McManus, 1981)

FIGURE 2.1   Are these really rivers? (a) Los Angeles storm system. (From Wikimedia.) (b) The Mojave River. (USGS photo.) (c) Historical image of an industrial pollution site, Calumet River. (From Great Lakes Image Gallery, U.S. EPA.)

FIGURE 2.2   (a) A creek? (Photo by James Martin.); (b) definitely a crick. (Photo from NRCS Image Gallery.)

While not intending to overemphasize the influence that grazing and pasture management may have, in some cases that management may prevent turning a "creek" into a "crick" and may be the focus of land management efforts, regulation, or litigation. Examples include numerous regulations designed to protect riparian habitats and rangeland health. Additionally, restoration efforts may be required to turn a "crick" back into a "creek" (see Chapter 8, e.g., livestock exclusion).

FIGURE 2.3 The cow patrol. (Reprinted from a cartoon by Bill Suddick, Toronto; text modified from original. With permission.)

The foregoing description also demonstrates that it is the characteristics of a river or a stream that affect our perception of its quality as well as the approaches that are used to manage and regulate that river or stream. In 1972, the Federal Water Pollution Control Act Amendments (also known as the Clean Water Act [CWA]) were enacted "to restore and maintain the chemical, physical, and biological integrity" of all "waters of the United States," which of course includes rivers and streams. The CWA requires that each state establish the beneficial uses for all waters within the state and the allowable concentrations of specific pollutants in order to protect those beneficial uses (numeric standards). But, the CWA also allows the establishment of narrative standards, which are statements of unacceptable conditions in and on the water. Thus, conditions perceived as unacceptable may result in a river or a stream not meeting the standards established by the CWA.

Differences in the management or regulation of a river may also occur based on whether or not that river is considered navigable. For example, Section 10 of the Rivers and Harbors Act of 1899 stipulates that the U.S. Army Corps of Engineers (COE) regulate all structures that work in, or affect, the navigable waters of the United States. The determination of whether a waterbody is a navigable water of the United States is also made by the COE. That determination may also impact the management of rivers and streams or other waterbodies that are connected to, and impact on, the navigable waterways. As will be discussed later, the connectivity to navigable waterways may impact whether or not wetlands are protected under Section 404 of the CWA. Under this section, the COE controls and permits the discharge of dredged and fill material in navigable and other waters.

Rivers are not static systems. The water levels, which impact the characteristics of rivers and streams, may continually vary in response to storms, runoffs, and groundwater inflows and may impact how rivers and streams are managed and regulated. For example, the protection of navigable waterways under the Rivers and Harbors Act extends to structures below the mean high-water line in tidal waters, or the ordinary high water level in nontidal waters. The high water level is typically estimated based on a high flow that would be expected to occur only once in every 100 years. The 100-year return flow is also often used to establish, for flood insurance purposes, 100-year flood inundation area, or Special Flood Hazard Area. In addition to high flows and water levels, low flows are often critical such as for water quality or environmental impacts. Many agencies are involved in establishing low-flow criteria for rivers and streams, protective, for example, of aquatic life or the water supply.

What is the difference between a river and a reservoir, or a wetland? What are the differences between "natural" and "regulated" rivers, and how are they regulated?

It is clear that there is no precise or exact definition of a river or stream. Whether a waterbody is a river, stream, brook, or beck may be solely due to local custom or the whim of the cartographer who put in on a map. However, the characteristics of rivers and streams do impact how they are managed and regulated and are often used to classify such waterbodies. Some of their general characteristics are described in the following sections.

2.2 CHARACTERISTICS OF RIVERS AND STREAMS

Rivers are *lotic* systems (lotus, from *lavo*, to wash), and are generally dominated by flows, which are usually unidirectional in response to gravity, as opposed to lakes and reservoirs, which are *lentic* systems (*lenis*, to make calm). The characteristics of rivers and streams are strongly related to the watershed (climate, geology, vegetative cover, etc.) and human influences. The differences in how rivers are organized and function have led to the development of a variety of classification schemes, of which some of the more common ones are discussed in this section. While the separation is somewhat artificial, the variations in riverine systems and their classifications are often based on changes occurring with respect to distance, longitudinal, lateral, and vertical, and with time (Figure 2.4). The changes are not confined to the variations within the river channel, but are on a much broader scale, since riverine systems strongly impact the surrounding landscape. Another major factor impacting the characteristics of rivers and their landscape is human activity. Relatively few "natural" landscapes or unregulated or "natural" riverine systems remain.

2.2.1 STREAM CORRIDOR CONCEPT: A LONGITUDINAL VIEW

Consider a river as a "large stream," which originates at some elevation and ultimately flows downstream into some other waterbody, such as an estuary or an ocean. The river may also receive,

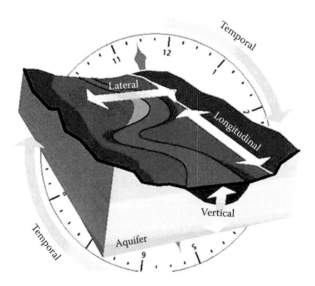

FIGURE 2.4 The four dimensions of a lotic system. (From O'Keefe, T.C., Elliott, S.R., and Naiman, R.J., Introduction to watershed ecology, U.S. Environmental Protection Agency, Watershed Academy Web, 2007, Available at http://cfpub.epa.gov/watertrain/pdf/modules/watershedecology.pdf; Based on Ward, J.V., *Journal of the North American Benthological Society*, 8, 2–8, 1989. With permission.)

and be influenced by, tributary inflows along its path to its outlet. The physical, chemical, and biological characteristics of the river would be expected to vary longitudinally as the channel and the floodplain vary from their origin to the outlet. For example, the channel width and depth would be expected to increase downstream as the drainage area and discharge increase.

A simplified longitudinal model captures these observed changes by disaggregating the river into three zones (Figure 2.5):

- Headwaters zone
- Transfer zone
- Depositional zone

The headwaters zone generally has the steepest slope, and the relatively high current velocities often cut deep channels resulting in V-shaped valleys with rapids and waterfalls being common (Miller 1990). The bed material is usually rocks, boulders, or particles of a relatively large size.

The river slope generally decreases as the river moves into the transition zone and eroded material from the headwaters zone moves into this zone. Typically, the river will become broader and the flow will increase as the rivers merge. The river may begin to meander within its floodplain.

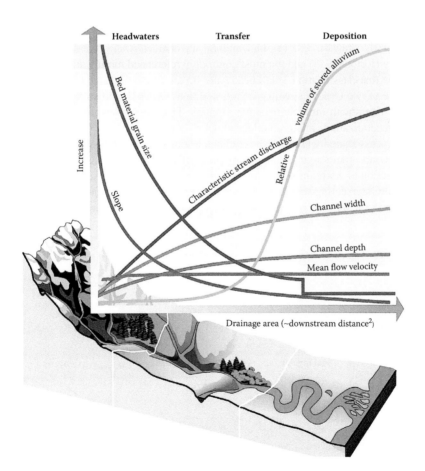

FIGURE 2.5 Changes in the channel in the three zones. (From FISRWG, *Stream Corridor Restoration: Principles, Processes, and Practices*, Federal Interagency Stream Restoration Working Group, 1998.)

In the depositional zone, the gradient flattens from a buildup of sediment over time. The river widens further and meanders toward its mouth. Near the mouth, the meanders and the separate channels that are formed may make the determination of the main channel problematic.

2.2.1.1 Drainage Basins and Networks

These same three zones (headwater, transfer, and depositional zones) can also be seen on a smaller scale within the watersheds of contributing streams. One definition of a watershed is the "area of land that drains water, sediment, and dissolved materials to a common outlet at some point along a stream channel" (Dunne and Leopold 1978). The size and structure of watersheds as well as their topographic and geographic structures vary significantly due to geologic, morphologic, vegetative, soil, and climatic differences. Their form varies greatly due to the climatic regime and the underlying geology, morphology, soils, vegetation, etc. Drainage patterns are primarily controlled by the overall topography and underlying geologic structure. Figure 2.6 illustrates some of the common patterns.

2.2.1.2 Stream Order

The characteristics of rivers and streams may also be influenced, and illustrated, by their connectivity. As a stream or river forms in a watershed, flowing along the path of least resistance, it eventually meets and merges with other rivers or streams to form yet larger rivers or streams, resulting in the planform view, their connectivity like the branches of a tree.

A method of classifying the hierarchy of natural channels according to their position in the drainage system, referred to as *stream order*, (Figure 2.7) permits a comparison of the behavior of a river with others similarly situated. The original method for characterizing rivers by their connectivity was developed by Horton (1945) and the most commonly referenced modification is that proposed by Strahler (1957). In that classification, small headwater streams are designated Order I. Streams formed by the confluence of two Order I streams are referred to as Order II, and so on, with larger numbers indicating larger rivers with multiple tributary streams. The scheme has proven useful for developing and testing generalizations and predictions about river processes. For example, as previously discussed, a stream is associated with the river gradient, drainage area, channel width, and discharge.

Note that a second-order stream is formed by the confluence of two first-order streams and so on, while the intersection of a stream with a lower-order stream does not raise the stream order (e.g., a third-order stream intersecting with a first-order stream is still a third-order stream below the intersection). In general, the number of rivers or streams decreases nearly exponentially, and the stream

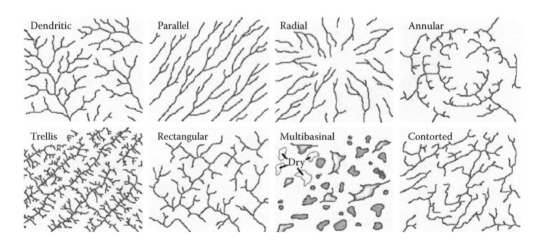

FIGURE 2.6 Watershed drainage patterns. (From Howard, A.D., *AAPG Bulletin Series*, 51, 2246–2259, 1967. Reprinted from American Association of Petroleum Geologists and Datapage, Inc. With permission.)

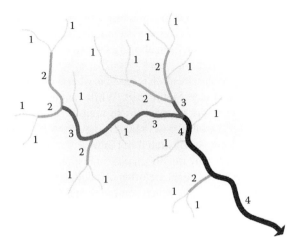

FIGURE 2.7 Stream order classification. (From FISRWG, *Stream Corridor Restoration: Principles, Processes, and Practices*, Federal Interagency Stream Restoration Working Group, 1998.)

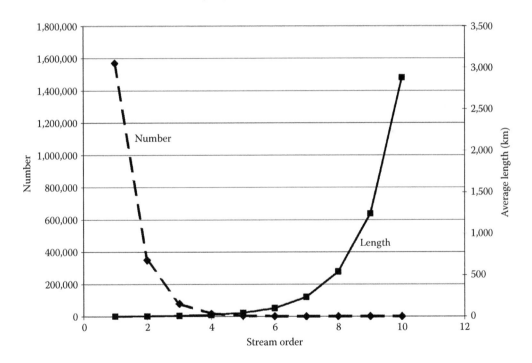

FIGURE 2.8 Number and length of river channels of various sizes in the United States. (Based on data from Leopold, L.B., Wolman, M.G., and Miller, J.P., *Fluvial Processes in Geomorphology*, W.H. Freeman, San Francisco, CA, 1964.)

length increases, as the stream order increases (Figure 2.8). Similarly, the mean drainage area of the contributing watershed increases nonlinearly with the stream order.

There are usually three to four times as many streams of Order $n-1$ as there are of Order n, each of which is generally half as long and drains alittle more than one-fifth of the land area (Cushing and Allan 2001). Most rivers and the great majority of river miles are in the lower order, which should influence how we manage our river systems.

2.2.1.3   Planform Classification

Perhaps one of the more common classification systems of channels is the *planform* classification of Leopold and Wolman (1957). This classification system reflects changes in the shape or pattern of the river or stream in a plan view (as seen from above) as it moves through the three zones (headwater, transition, and deposition; Figure 2.5).

One characteristic on which the classification is based is whether the river has one (*single thread*) or multiple channels (*multithread*) (see Figure 2.9). While single-channel rivers are more common (Figure 2.9a), multichannel rivers often occur in areas with erodible banks, an abundance of course sediments, and highly variable flows (Interagency Stream Restoration Working Group 1998). Multithread *braided* streams, as illustrated in Figure 2.9b, typically have wide and shallow channels, and the channel structure may rapidly change in response to flows, impacting vegetation. Hydraulic analyses of these braided systems are often difficult. Less common than the multithread braided systems are *anastomosed* streams, which commonly have narrow, deep channels and result from rising downstream water levels resulting in sediment deposition.

Another planform characteristic of streams is *sinuosity*, or how crooked or *meandering* the stream is. Sinuosity is a measure of the amount of curvature in a river, as reflected, for example, by the length of the river versus the length of the river valley or by other metrics such as the channel wavelength or the meander radius or curvature. The length of the river is measured along the river channel, or *thalweg*. Sinuosity generally increases in rivers from the headwater to the depositional zones (Figure 2.10).

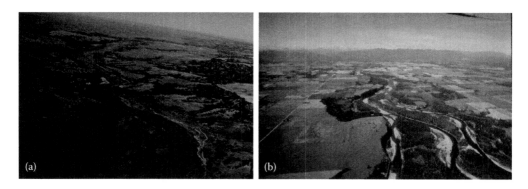

FIGURE 2.9   Single (a) versus multithread (b) streams. (From FISRWG, *Stream Corridor Restoration: Principles, Processes, and Practices*, Federal Interagency Stream Restoration Working Group, 1998.)

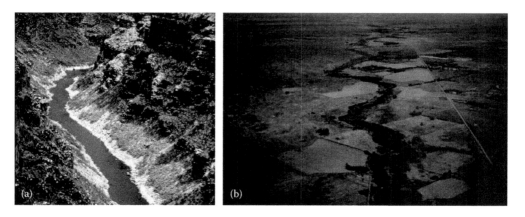

FIGURE 2.10   Sinuosity of rivers: (a) high and (b) low. (From FISRWG, *Stream Corridor Restoration: Principles, Processes, and Practices*, Federal Interagency Stream Restoration Working Group, 1998.)

For example, small meanders are typically associated with high gradients and coarse substrates, while large meanders are associated with low gradients and fine substrates, such as in the depositional zone. The channel sinuosity has large impacts on the channel characteristics and has often been reduced in regulated rivers. A frequent goal of restoration efforts is to restore natural sinuosity.

Streams may allow differences in the presence and frequency, or absence, of *pool* versus *riffle* areas. Riffles are areas of relatively shallow swifter-flowing waters, where the surface is usually turbulent, which usually alternate with, or are separated by, deeper more slowly moving pools. The grain size distribution also varies, with the riffle areas more armored with larger rocks or pebbles, while the pools are more depositional, with smaller grain sizes. The formation and stability of pools and riffles are largely a function of the streambed makeup (Interagency Stream Restoration Working Group, 1998):

- Gravel and cobble streams typically have regularly spaced pools and riffles and are stable in rapidly fluctuating streamflows.
- Sandbed systems typically do not have pools and riffles, since the grain size is the same in the shallow and in the deeper areas, but they typically do have alternating deep pools.
- High-gradient and high-velocity streams typically have pools but not riffles, with the water moving from pool to pool in a stair-step fashion.

The location of the pools and riffles also varies with the sinuosity of the stream (Figure 2.11). For example, in sinuous streams, riffles are found at the entrance and exit of meanders, and control the streambed elevation, while pools are located at the outside bend of the meander.

Low-gradient streams typically form pool–riffle systems while high-gradient and high-velocity streams typically have *step*–riffle systems, as plunge pools are formed as the stream cascades over a step (Figure 2.12). Pools may also form downstream of boulders or logs, as the falling water scours a pool, and aid in dissipating energy. Backwater impacts due to logs, root wads, and debris blocking or partially blocking a channel may also result in pools. Pools may also be of biological origin, such as due to beaver dams. The upward-sloping area of the pool from the bed to the head of a riffle is known as a *glide*, while the transitional feature between a riffle and a pool is known as a *run*. Glides and runs typically have a flatter water surface than riffles.

Pools and riffles, along with other channel features such as runs, steps, and undercut banks, provide a necessarily diverse instream habitat for different organisms. Riffles provide cover for

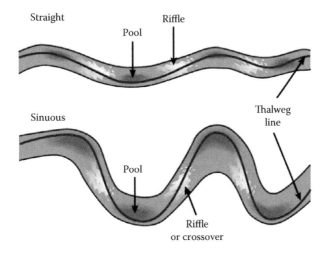

FIGURE 2.11 Sequence of pools and riffles in straight and sinuous streams. (From FISRWG, *Stream Corridor Restoration: Principles, Processes, and Practices*, Federal Interagency Stream Restoration Working Group, 1998.)

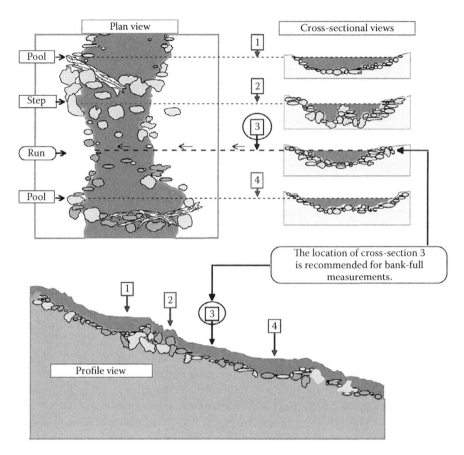

FIGURE 2.12   Location of features in a step-pool system. (From Rosgen, D.L., *Applied River Morphology*, Wildland Hydrology Books, Pagosa Springs, CO, 1996. With permission.)

macroinvertebrates and deliver food for fish in microhabitats within the riffle or downstream pool. For example, certain caddisflies occur typically only in riffles. The coho salmon generally rear in pools formed by large woody debris. The turbulence associated with riffles also aids in maintaining high concentrations of dissolved oxygen. Often, as sedimentation increases due to the cumulative impacts of human or natural influences on watershed changes, the number of glides and runs increases in relation to pools and riffles. The re-establishment of pools and riffles is a common goal of stream restoration.

2.2.1.4   Channel Processes and Stream Morphology

2.2.1.4.1   Gradient

The slope or the rate of change in a bottom elevation with respect to distance (e.g., expressed as meter per kilometer [m km^{-1}], feet per mile [ft. mile^{-1}], or percent) is one longitudinal morphological characteristic of streams impacting other characteristics. The stream slope typically decreases as streams move from the headwater to the depositional zones (Figure 2.5).

An increasing slope, or gradient, is generally associated with higher stream velocities. In hydraulics, stream velocities are separated into two types of flow based on a comparison of the stream velocity to the celerity of a gravity wave ($c = \sqrt{gY}$, where c is the celerity, g is the gravitational acceleration [32.17 ft. s^{-2} or 9.81 m s^{-2}], and Y is the stream depth [feet or meters]). If the stream velocity is equal to the celerity of a gravity wave (c), then the flow is referred to as a *critical* flow. If the velocity is greater than c, then the flow is referred to as *supercritical*, and if smaller, *subcritical*.

One of the reasons this distinction is important is that for a supercritical flow, disturbances due to an obstruction in the stream can only propagate downstream. So, if you change the structure of the stream, for example, it cannot have an impact upstream. However, if the flow is subcritical, then an obstruction may have an impact upstream. For example, a boulder or debris flow may create a backwater effect or increase the water-surface elevation upstream. Channel slopes or gradients that result in supercritical flows are called *steep* slopes in hydraulics, while those producing subcritical slopes are called *mild*. A natural flow in these regions also *gradually varies*, but in a zone where the flow changes from steep to mild, or supercritical to subcritical, there is generally a rapid and large increase in water-surface elevations, and a large loss in energy. These transitions are referred to as *hydraulic jumps* (Figure 2.13), and the flow is categorized as *rapidly varied*. Hydraulic jumps may occur naturally or they may be induced downstream of man-made structures such as spillways; they are important because of the energy that is dissipated and the increased mixing that results from the highly turbulent flows.

The slope or gradient, and the resulting stream velocities, also has a large impact on whether the flow will erode the channel, carry or transport materials suspended in the flow, or allow those materials to be settled out of the flow and deposited. Lunetta et al. (1997) categorized stream reaches, based on the stream slope, as:

>12%: *source reaches* since any material falling into a channel of that steepness in a storm event will immediately move downstream.

4%–12%: *transport reaches* since materials still tend to move through these areas because of high hydraulic energy.

<4%: *response reaches* where the energy of the stream drops and sediment tends to remain in residence for longer periods.

2.2.1.4.2 Valley Width and Confinement

The transport and transport characteristics of rivers will vary as a function of the balance between a particular gradient flow, the sediment supply, and the valley width and confinement (Figure 2.14). Take, for example, the wetted perimeter of a river. The wetted perimeter is the perimeter of the cross section that is wet, or the perimeter over which the bed is in contact with water. So, a wide and shallow river would have a greater wetted perimeter, more contact with the bed and therefore more friction. Entrenched rivers would have less friction because they have a lower wetted perimeter.

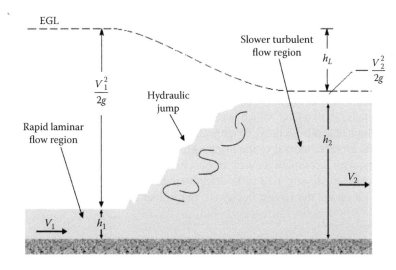

FIGURE 2.13 Illustration of a hydraulic jump.

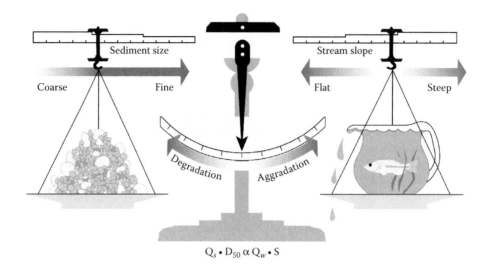

$$Q_s \cdot D_{50} \, \alpha \, Q_w \cdot S$$

FIGURE 2.14 Factors affecting channel degradation and aggradation. (Reproduced from the American Society of Civil Engineers from Lane, E.W., *Proceedings from the American Society of Civil Engineers*, 81, 1–17, 1955. With permission.)

Streams will increase in width when the sediment supply exceeds the transport capacity, and often, as a result, they lose their defined riparian zone.

2.2.1.4.3 Roughness

The flow velocity in a channel is inversely proportional to the channel roughness and directly proportional to the stream slope (usually to the 1.2 power). These concepts are incorporated into all hydraulic models using roughness coefficients such as Manning's number (Manning's *n*). For example, a higher velocity would be expected in a stream with a smooth bottom than a stream with boulders, cobbles, or vegetation.

2.2.1.4.4 Dominant Discharge

The dominant discharge is that discharge that carries the majority of the sediments and is responsible for creating or maintaining the size and shape of the channel (Leopold et al. 1964; Knighton 1984). The dominant discharge is also known as the "channel-forming" discharge and occurs fairly frequently (it is not a rare event). For rivers and streams that are at equilibrium with the sediment loads (not degrading or aggrading), the dominant discharge is about equal to or less (for incised streams) than the bank-full discharge, or the discharge that fills the banks before spilling onto the floodplain. When a streambed fills in (supply exceeds transport out) it is said to be aggraded, while conversely, a stream with a reduced supply may have a capacity greater than its load and will downcut in order to increase its bed load, which is called degrading.

Determining the bank-full discharge (Figure 2.15) is often subjective so a common alternative is to specify the recurrence interval of the discharge, which may vary between once every year to once every 5 years, with a common recurrence interval being once every 1.5 years (Leopold et al. 1964). Another metric commonly used is the effective discharge, which is defined as the increment of discharge that transports the largest fraction of the annual sediment load over a period of years (Andrews 1980).

2.2.1.4.5 Sediment Transport Mechanisms

The term *sediment* covers a wide range of particle types and particle-size classes, ranging from very fine clays to boulders. The factors impacting transport will also vary between some of the sands that

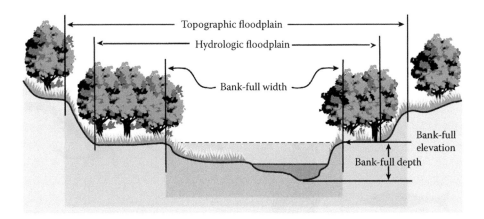

FIGURE 2.15 Lateral variations in the channel and floodplain. (From FISRWG, *Stream Corridor Restoration: Principles, Processes, and Practices*, Federal Interagency Stream Restoration Working Group, 1998.)

are transported as individual particles and some of the fine materials that may form flocs, so their transport is also determined by the other particles around them.

In rivers with moderate slopes, there are basically two types of transport (MacArthur et al. 2007):

- Bed load plus suspended load
- Bed load plus wash load

In the first, the suspended load is the particles that are small enough to be kept in suspension by the turbulence in the river and the bed load is the coarser particles that are transported along the bed intermittently by rolling, sliding, or saltating (bouncing along the bottom). In the second, the bed-load material includes all types of solids found in the bed whether they are transported in the bed or in suspension, and the wash load is the fine particles traveling in suspension that are not commonly found in the sediments. The suspended load is carried by the flow, so it travels at the same velocity, while the bed-load transport may differ from the dominant flow velocity and may only occur during high-flow events.

2.2.1.5 River Continuum Concept

The best-known longitudinal model for rivers is the river continuum concept (RCC) (Figure 2.16), first proposed by Vannote et al. (1980), which attempts to generalize and explain the observed longitudinal changes in stream ecosystems.

The RCC model proposes that (FISRWG 1998):

- Rivers exhibit continuous longitudinal changes in their physical characteristics from their headwaters to their mouth
- The longitudinal gradient in their physical characteristics controls the biotic response

Therefore, the RCC model identifies relationships between the progressive changes in a stream's structure, such as the channel size and streamflow, and the distribution of species. So, the biotic response is then not just a function of some physical characteristic, such as temperature, but the position (location) along the length of the river.

Longitudinal connectivity refers to physical changes along the entire stream and the chemical and biological responses to those changes. For example, a headwater woodland stream (Orders I–III)

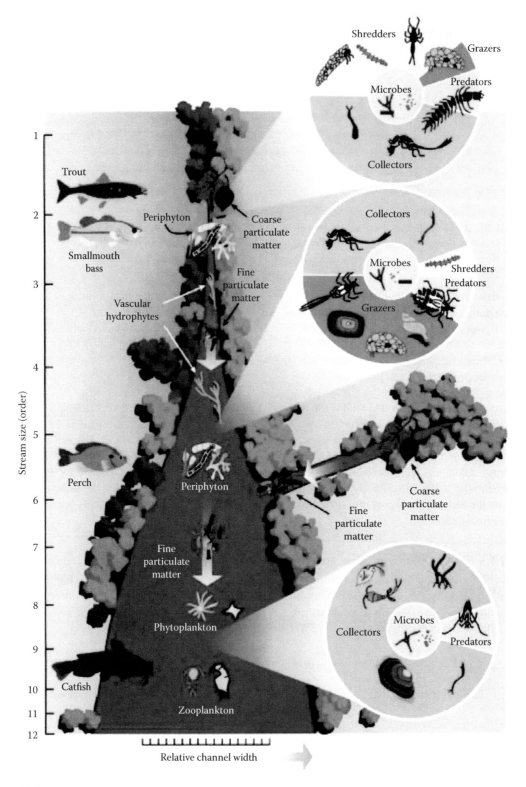

FIGURE 2.16 The river continuum concept. (From Vannote, R.L., Minshall, G.W., Cummins, K.W., Sedell, J.R., and Cushing, C.E., *Canadian Journal of Fisheries and Aquatic Sciences*, 37, 130–137, 1980. With permission.)

would be expected to have a steep gradient with riffles, rapids, and falls. The sunlight reaching these streams is limited due to the shading of the riparian forest, so photosynthesis is limited and *heterotrophs* (the consumers) as opposed to *autotrophs* (the primary producers) dominate. The primary source of carbon and energy then comes primarily from outside of the system (*allocthonous* sources, or from outside the stream), such as from falling leaves. Insects are dominated by those that can shred the course carbonaceous materials, collectors, or predators. Water is cooled by springs and often supports trout.

In the mid-reaches, the river gradient decreases, there are fewer rapids and falls, and the stream is wider. The sunlight reaches the water, the stream is warmer, and primary production increases, so that much of the energy is produced by the autotrophs (*autochthonous* sources, or from inside the stream). Insects feed on the living (or dead) plants. The larger stream also supports a greater diversity of invertebrates and fish.

As the stream approaches the lower regions, it becomes larger (Orders VII–XII), the gradient lessens, the temperature increases, and the flow increases (due to the larger contributing watershed). The stream is generally more turbid, reducing light penetration, and energy is supplied by upstream sources. The primary producers are often dominated by drifting phytoplankton. Backwaters may exist where turbidity has settled and aquatic plants are abundant. Fish species are omnivores and plankton feeders such as carp, buffalo, suckers, and paddlefish (FISRWG 1998).

The RCC is not universally applicable. However, it does serve as a useful conceptual model and has stimulated a considerable amount of research since it was first introduced in 1980 (FISRWG 1998).

2.2.2 STREAM CORRIDOR: LATERAL VIEW

Stream characteristics vary laterally (and vertically) as well as longitudinally. This section will discuss some of those characteristics in the stream channel, in the floodplain, and in the upland areas, as bordered by the upland fringe.

2.2.2.1 Stream Channel

The stream channel is that portion of the stream that normally conveys water. Moderate flows are contained within the stream channel and a bank-full discharge is one that fills the entire channel's cross-sectional area, above which the flow spills onto the floodplain.

The stream channel is typically characterized by lateral variations in the cross-sectional shape, width, depth, bank, and bottom characteristics, as well as longitudinally by its slope, sinuosity, pool and riffle spacings, etc. These physical characteristics impact and reflect the stream's hydraulics and sediment characteristics, as well as its ecological characteristics. For example, wide and shallow systems will have different characteristics from those that are narrow and deep. The width to depth ratio (the bank-full width divided by the mean bank-full depth) is used as one of the four morphological characteristics employed to delineate stream types, as discussed in Section 2.2.3 on the Rosgen classification scheme (others are the slope, sinuosity, and entrenchment ratio). Another channel characteristic used in the Rosgen classification is the median particle size.

The channel characteristics are rarely constant, and are most commonly in a state of flux. This could be natural or due to anthropogenic impacts. For example, destabilizing streams such as in a typical incising channel where the streambed degrades until the critical bank height is exceeded and the bank fails, which increases the channel width and sediment load (Fischenich and Morrow 2000).

2.2.2.1.1 Streamflow and Groundwater Interactions

Not all rivers continuously flow and streams and rivers can be classified based on their duration and frequency of flow. Based on their flows, rivers may be classified as *perennial* or permanent, intermittent, or ephemeral. Even in perennial rivers, flows vary over time, such as within and between seasons. The determination of whether a stream is perennial or not has regulatory implications, and various states may define perennial as part of their statutes. For example, North Carolina (2003) defines a perennial stream as "a well-defined channel that contains water year round during a year of normal rainfall with the aquatic bed located below the water table for most of the year" and "biological, hydrological, and physical characteristics commonly associated with the continuous conveyance of water."

Whether a stream or a river is perennial, intermittent, or ephemeral is most commonly a function of the relationship of the water level to that of the groundwater (Figure 2.17), and whether the stream is a *losing* or a *gaining* system. A gaining stream (effluent) is below the water table and receives discharges from the aquifer. An influent or "losing" reach loses stream water to the aquifer. An ephemeral stream is a losing stream and flows only during or immediately after periods of precipitation. Intermittent streams are also averaged over time as losing streams and flow only happens during certain times of the year. During periods of dry weather, they may dry up but leave a series of disconnected pools that can sustain aquatic life. Perennial streams are gaining systems that flow continuously during both wet and dry periods. Perennial streams can also result from flows from origins other than groundwater. For example, in arid regions, some perennial streams exist due to industrial or wastewater discharges.

The flow in streams may be considered to consist of two components, the base flow (for perennial streams) and the storm flow as illustrated in a graph of flow versus time (Figure 2.18). The storm flow results primarily from runoff from the watershed and the peak flow occurs some time after the peak rainfall (a lag time). The hydrograph shows a rising and falling limb as runoff begins and ends.

The flow in perennial rivers not only occurs when it rains, but also because there is a base flow component that is primarily a function of groundwater inflow from shallow aquifers or discharges from industries or municipal wastewater discharges. As a result, the quantity (and often the quality) of rivers during periods between storms is integrally linked with the groundwater table. The groundwater table can vary as a function of the recharge (from the infiltration of rain) and withdrawals, such as from wells. Groundwater withdrawals for agriculture or other uses have resulted in lowering the water table in many areas, such as the Mississippi Delta, with the consequent impact that the average flow (base flow) in many streams has declined over the years and many formerly perennial streams are now ephemeral or intermittent. The interrelationship between groundwater withdrawals and surface water supplies is critical to many water use and water management programs.

2.2.2.1.2 *Hyporheic Zone*

A *rheic* flow refers to the visible free-running water we normally think of as a stream, river, or other moving flow of water, while *hypo* refers to "under, beneath, or below." So, the hyporheic zone is

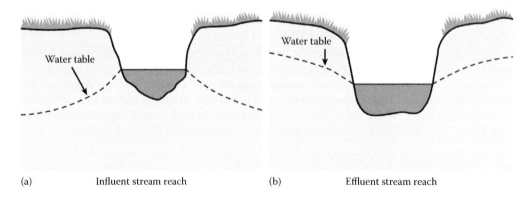

(a)   Influent stream reach   (b)   Effluent stream reach

FIGURE 2.17 (a,b) Groundwater and surface water interactions. (From FISRWG, *Stream Corridor Restoration: Principles, Processes, and Practices*, Federal Interagency Stream Restoration Working Group, 1998.)

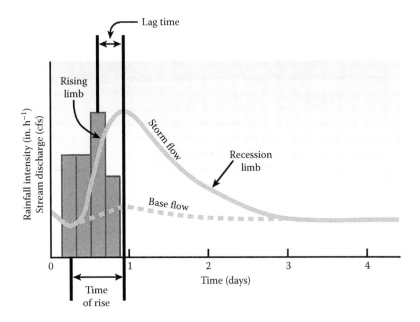

FIGURE 2.18   Flow hydrograph. (From FISRWG, *Stream Corridor Restoration: Principles, Processes, and Practices*, Federal Interagency Stream Restoration Working Group, 1998.)

the area below (and beside) the streambed where water percolates through spaces between the sediments, rocks, and cobbles and other permeable soils (saturated interstitial space, Figures 2.19 and 2.20). Hyporheic zones are often referred to as ecotones or bridges between two different types of habitats or environments, such as surface water and groundwater. Boulton et al. (1998) defined the hyporheic zone as a "spatially fluctuating ecotone between the surface stream and the deep groundwater where important ecological processes and their requirements and products are influenced at a number of scales by water movement, permeability, substrate particle size, resident biota, and the physiochemical features of the overlying stream and adjacent aquifers."

The hyporheic zone is an important component of river hydrology and hydraulics. Depending on the hydraulic gradient, flows may be into or out of this zone (Figure 2.19), adding to or removing flows from the stream or to groundwater, and a hyporheic flow may include much or all of the total river flow. The recognized functions of the hyporheic zone (NRCS 2007) include:

- Regulation of the stream temperature by groundwater upwelling
- Water retention and storage, which can reduce peak flows during floods and sustain base flows during dry periods
- Habitat creation, especially for aquatic invertebrates such as crustaceans, and vertebrates such as larval fishes
- Buffering and filtering nutrients from streamflows and groundwater
- Aquifer recharge
- Nutrient enrichment

2.2.2.2   Floodplain

The floodplain is the area above the streambank, the extent of which may vary. Some streams and rivers flow through valleys, where it is relatively easy to differentiate between the channel, the floodplain, and upland areas. Other rivers flow through deltas, fans, or broad plains (Wetzel 2001).

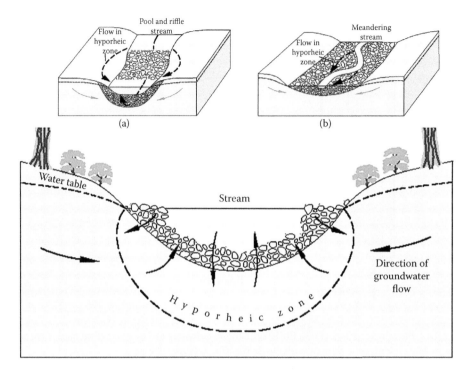

FIGURE 2.19 Hyporheic zone. Water exchange in the hyporheic zone associated with (a) abrupt changes in the streambed slope and (b) stream meanders. (From Winter, T.C., Harvey, J.W., Franke, O.L., and Alley, W.M., *Ground Water and Surface Water: A Single Resource*, U.S. Geological Service, Denver, CO, 1998.)

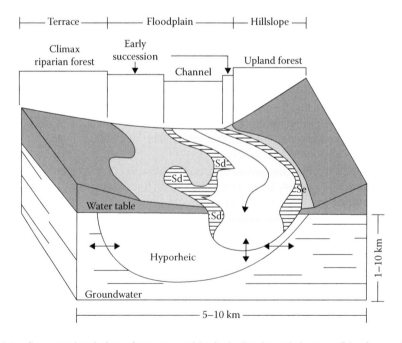

FIGURE 2.20 Cross-sectional view of stream corridor including hyporheic zone (Sd refers to the sediment deposition site and Se refers to the sediment erosion). (From NRCS, *National Engineering Handbook, Stream Restoration Design*, Department of Agriculture, Natural Resources Conservation Service, Washington, DC, 2007.)

2.2.2.2.1 Riparian Zone

The riparian zone is the interface between land and a stream or river. The riparian is the transitional area of the river corridor between the river and upland areas and is defined by Webster's as "relating to or living or located on the bank of a natural watercourse (as a river) or sometimes of a lake or a tidewater." But, definitions vary and include (Fischer et al. 2001, 2002):

- "Associated with water courses. Riparian may refer to vegetation associated with large rivers or with small, even intermittent drainages such as arroyos" (Dick-Peddie and Hubbard 1977).
- "Riparian areas are three-dimensional ecotones of interaction that include terrestrial and aquatic ecosystems, that extend down into the groundwater, up above the canopy, outward across the floodplain, up the near-slopes that drain to the water, laterally into the terrestrial ecosystem, and along the water course at a variable width" (Ilhardt et al. 2000).
- "A distinct ecological site, or combination of sites, in which soil moisture is sufficiently in excess of that otherwise available locally, due to run-off and subsurface seepage, so as to result in an existing or potential soil vegetation complex that depicts the influence of that extra soil moisture" (Anderson 1987).
- "As ecotones, they encompass sharp gradients of environmental factors, ecological processes and vegetative communities. Riparian areas are not easily delineated but are composed of mosaics of landforms, communities and environments within the larger land-scape" (Gregory et al. 1991).
- "Environs of freshwater bodies, watercourses, and surface-emergent aquifers (springs, seeps, and oases) whose transported waters provide soil moisture in excess of that otherwise available through local precipitation to potentially support the growth of mesic vegetation" (Warner and Hendrix 1984).
- "Land inclusive of hydrophytes and/or with soil that is saturated by ground water for at least part of the growing season within the rooting depth of potential native vegetation" (Brosofske 1996).
- "Riparian areas are transitional between terrestrial and aquatic ecosystems and are distinguished by gradients in biophysical conditions, ecological processes, and biota (Figure 2.21). They are areas through which surface and subsurface hydrology connect waterbodies with their adjacent uplands. They include those portions of terrestrial ecosystems that significantly influence exchanges of energy and matter with aquatic ecosystems (i.e., a zone of influence). Riparian areas are adjacent to perennial, intermittent, and ephemeral streams, lakes, and estuarine-marine shorelines" (National Research Council 2002).

So, why is the definition of a riparian zone important? The definition may impact management decisions and it also has regulatory implications. For example, is a riparian zone a wetland? If so, and particularly if the contiguous river is navigable, there is a body of law intended to protect the riparian zone, such as under Section 404 of the CWA. Under this section, anyone wanting to discharge fill material into "waters of the United States" must obtain a permit from the COE. However, the National Research Council (2002) indicated that "riparian areas generally do not satisfy regulatory and other definitions of 'wetland,' and thus are not encompassed by regulatory programs for wetland protection."

In general, no federal statute provides direct protection for riparian zones. Certain federal laws do require an evaluation of the adverse effects that would be caused by federal actions, along with the consideration of less environmentally damaging alternatives, and although not specifically focused on riparian zones, they may include them as part of the evaluation. Other federal laws, such as the Farm Bill, provide incentives for moving intensive agricultural practices away from streams by installing riparian buffers. Other programs promote the development, restoration, or conservation of riparian buffer zones, such as the National Buffer Conservation Initiative, which is intended to encourage the use of conservation buffer strips by agricultural producers and other landowners in both urban and rural settings. The COE considers that although it does not have the authority to directly regulate

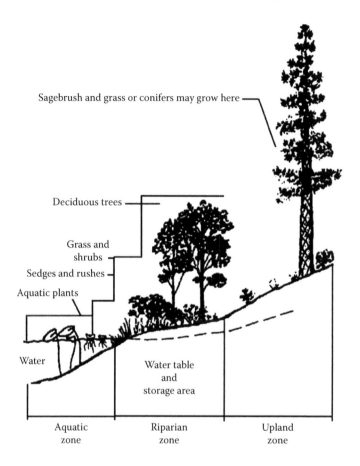

FIGURE 2.21   Riparian zone. (From U.S. Department of the Interior, *The Impact of Federal Programs on Wetlands*, Vol. 2, Report to Congress, 1994. Available at http://www.doi.gov/oepc/wetlands2/. With permission.)

upland areas, it does have the regulatory authority over riparian zones as part of Section 404 permit decisions (ERDC 2002). As such, the COE may require vegetated buffer strips as part of the mitigation for filling wetlands. Other efforts to protect riparian zones often come from state and local programs.

Riparian zones are a critical component of river management, by protecting the water quality and providing wildlife habitats. They are "a critical element of the overall aquatic ecosystem in virtually all watersheds" (Federal Register 2002).

Among the benefits of healthy riparian buffer zones (Figure 2.22) are the dissipation of the stream energy for overbank flows, the reduction of erosion and bank maintenance, increased flood storage and groundwater recharge, and the provision of habitats and organic matter, and wildlife travel corridors.

2.2.2.2.2   Landforms and Deposits

As a result of the lateral migration of a channel, a variety of structures form on the floodplain (FISRWG 1998). These structures include (Figure 2.23):

- Chute: a new channel formed across the base of a meander. As it grows in size, it carries more of the flow.
- *Oxbow*: a term used to describe the severed meander after a chute is formed.
- Clay plug: a soil deposit developed at the intersection of the oxbow and the new main channel.

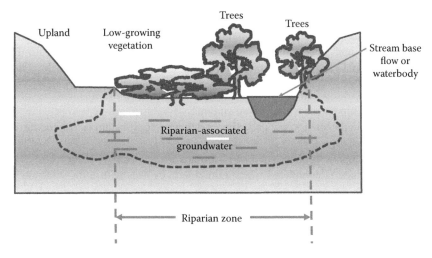

FIGURE 2.22 Schematic of a generic riparian zone showing zone of influence. (Based on NRCS, Riparian areas environmental uniqueness, functions, and values, RCA Issue Brief #11, United States Department of Agriculture, Natural Resources Conservation Service, Washington, DC, 1996.)

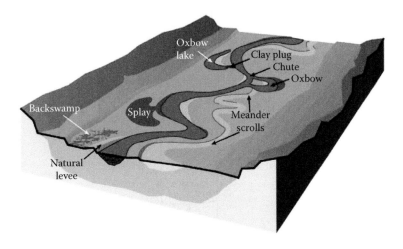

FIGURE 2.23 Landforms and deposits. (From FISRWG, *Stream Corridor Restoration: Principles, Processes, and Practices*, Federal Interagency Stream Restoration Working Group, 1998.)

- Oxbow lake: a body of water created after clay plugs the oxbow from the main channel.
- Natural levees: the formations that build up along the bank of streams that flood. As sediment-laden water spills over the bank, the sudden loss of depth and velocity causes coarser-sized sediment to drop out of suspension and collect along the edge of the stream.
- Splays: delta-shaped deposits of coarser sediments that occur when a natural levee is breached. Natural levees and splays can prevent floodwaters from returning to the channel when floodwaters recede.
- *Backswamps*: a term used to describe floodplain wetlands formed by natural levees.

2.2.2.3 Upland Areas

The upland areas are those above the existing floodplain and are connected to the floodplain by an upland fringe. The fringes have no typically defining shapes or features. One example of a fringe that has resulted from changes in the floodplains is terraces (FISRWG 1998).

2.2.2.3.1 Terrace Formation

Terraces are the flat-topped (benches) remnants of abandoned floodplains, formed through the interplay of incising and floodplain widening. The Federal Interagency Stream Restoration Working Group (FISRWG 1998) provided an example of terrace formation by channel incision (see Figure 2.24):

- Cross-section A represents a nonincised channel.
- Due to changes in the streamflow or sediment delivery, equilibrium is lost and the channel degrades and widens.
- The original floodplain is abandoned and becomes a terrace (cross-section B).
- The widening phase is completed when a floodplain evolves within the widened channel (cross-section C).

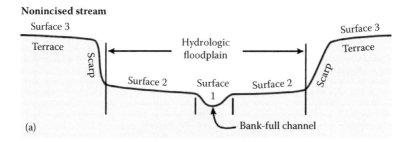

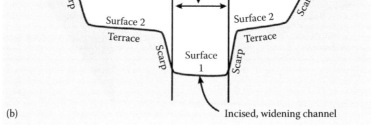

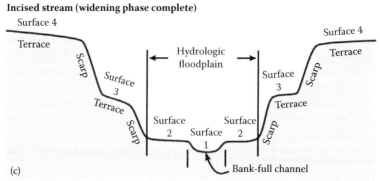

FIGURE 2.24 Terraces in (a) nonincised and (b,c) incised streams. (From FISRWG, *Stream Corridor Restoration: Principles, Processes, and Practices*, Federal Interagency Stream Restoration Working Group, 1998.)

2.2.3 ROSGEN CLASSIFICATION

A variety of classification systems has been designed to aid in assessing the relationships between channels, basin geomorphology, and watershed conditions, several of which are summarized in the NRCS *National Engineering Handbook* (Part 654, NRCS 2007). One of the more commonly used classification systems to identify the current status of channel reaches (rather than the entire system) for rivers is that developed by Rosgen (1994, 1996; Rosgen and Silvey 1996; NRCS 2007). The Rosgen scheme includes four levels of detail, ranging from qualitative descriptions to detailed qualitative comparisons, each with a successively more detailed or finer definition of the dimension, pattern, and profile of the stream reach being classified (NRCS 2007).

The first level of the scheme uses information on channel slopes and channel patterns to subdivide the rivers or streams into a series of broad stream types ("A" through "H"; Figures 2.25 and 2.26). Each of these types for North Carolina streams is illustrated in Figure 2.27 from Harman and Jennings (1999). The distinction between the stream types is made based on whether the streams are single-thread or multithread channels and on four morphological parameters: entrenchment ratio, width to depth ratio, sinuosity, and slope. The entrenchment ratio is a metric used to define the extent of incision, and it is estimated from the ratio of the flood-prone width (generally twice the elevation of the bank-full depth) to the bank-full width. The width to depth ratio is the bank-full width divided by the mean bank-full depth. The break between single-channel classifications is 12, meaning that the bank-full width is 12 times greater than the mean bank-full depth (Figure 2.26). The sinuosity is estimated by the channel length divided by a straight-line valley length. The water-surface slope is estimated from the top of a riffle to the top of another riffle at least 20 bank-full widths downstream (Harman and Jennings 1999). With the appropriate information, this level of classification can be determined using aerial photography and topographic information. The information needed is as listed by the U.S. Environmental Protection Agency (EPA) in the Watershed Academy training module on the Rosgen system. The Level I stream classification serves four primary functions:

1. Provides for the initial integration of basin characteristics, valley types, and landforms with stream system morphology.
2. Provides for a consistent initial framework for organizing river information and communicating the aspects of river morphology. The mapping of physiographic attributes at Level I can quickly determine the location and the approximate percentage of river types within a watershed and valley type.
3. Assists in the setting of priorities for conducting more detailed assessments and companion inventories.
4. Correlates similar general level inventories such as fisheries habitats, river boating categories, and riparian habitats with companion river inventories.

The second level of classification (Level II stream-type delineation) is based on the characteristics of the channel cross section, including the D50 or median particle size (see Figure 2.26). This classification also includes longitudinal profile measurements such as the slope and bed features (configuration of riffles/pools, rapids, steps/pools, etc.) and planform (pattern) measurements (such as sinuosity and meander width ratios).

The third level of Rosgen stream classification (Level III) is the stream state or condition assessment. This is an assessment and prediction of the stream's condition and its stability and requires an assessment of the channel erosion, riparian condition, channel modification, and other characteristics. The fourth level is a verification of the predictions made in Level III (Rosgen and Silvey 1996). While Levels I and II are used to describe the present condition of a channel, Levels III and IV are used to evaluate and validate an assessment of the stream's condition and its departure from the optimum or potential condition (NRCS 2007).

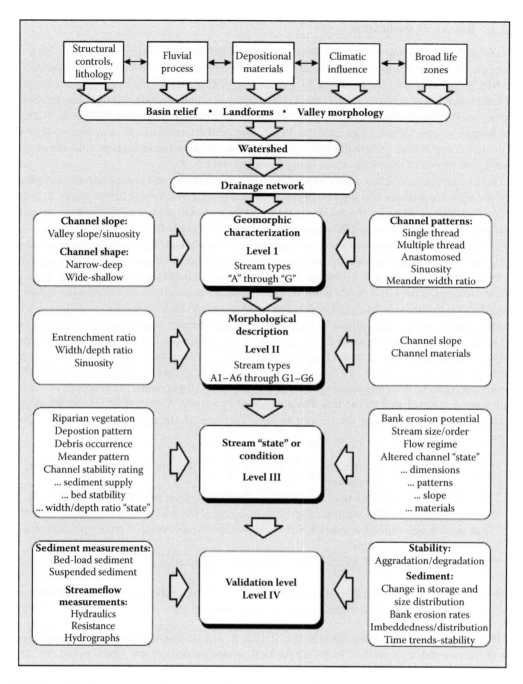

FIGURE 2.25 Levels of the Rosgen classification system. (From Harman, W.A. and Jennings, G.D., Application of the Rosgen stream classification system to North Carolina, North Carolina State University, North Carolina Cooperative Extension Service, Fact Sheet Number 2, 1999. With permission.)

2.2.4 VARIATIONS WITH TIME

Rivers are, by their nature, dynamic systems or systems that change over time. Streams are constantly changing their physical character (morphology), which, in turn, affects their ability to perform important ecological functions (Fischenich et al. 2000). This is natural, even in streams that are "stable,"

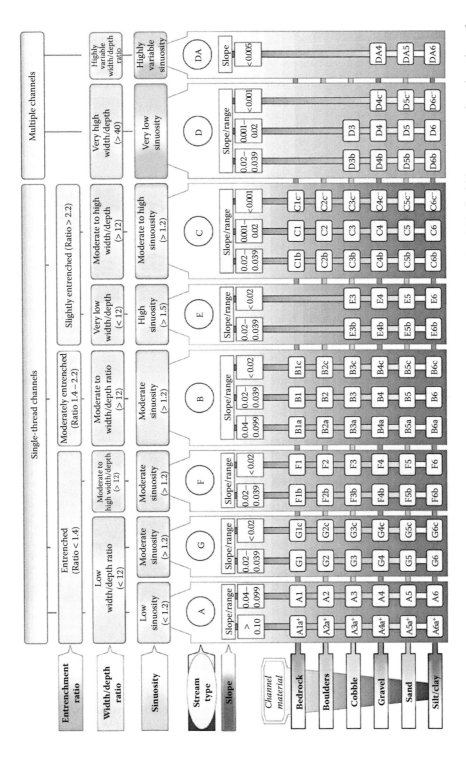

FIGURE 2.26 Key to the Rosgen classification of natural rivers. As a function of the "continuum of physical variables" within stream reaches, the values of entrenchment and sinuosity ratios can vary by ±0.2 units, while the values for width/depth ratios can vary by ±2.0 units. (From Rosgen, D.L., *Applied River Morphology*, Wildland Hydrology Books, Pagosa Springs, CO, 1996. With permission.)

Stream Type "A"

Type "A" streams are single-thread channels with a width/depth ratio less than 12, meaning that they are narrow and moderately deep. They are entrenched, high-gradient streams with step/pool bed features. "A" streams with a channel slope greater than 10% are classified as "Aa+." "A" streams flow through steep V-shaped valleys, do not have a well-developed floodplain, and are fairly straight.

Basin: Yadkin
Stream Type: A1

Basin: Yadkin
Stream Type: A1a+

Stream Type "B"

Type "B" streams are wider than "A" streams and have a broader valley but not a well-developed floodplain. These single-thread streams are moderately entrenched with moderate to steep slopes. Type "B" streams are often rapid dominated streams with step/pool sequences. Bank heights are typically low. The high width/depth ratios and moderate entrenchment ratios make this stream type quite resilient to moderate watershed changes.

Basin: Little Tennessee
Stream Type: B3

Basin: Catawba
Stream Type: B4c

FIGURE 2.27 Description of Rosgen's stream types illustrated using streams from North Carolina. (From Harman, W.A. and Jennings, G.D., Application of the Rosgen stream classification system to North Carolina, North Carolina State University, North Carolina Cooperative Extension Service, 1999. With permission.)

Stream Type "C"

Type "C" streams are riffle/pool streams with a well-developed floodplain, meanders, and point bars. These streams are wide with a width/depth ratio greater than 12. Type "C" streams are moderately entrenched, and therefore use their floodplain during large storms.

Basin: French Broad
Stream Type: C5

Basin: French Broad
Stream Type: C4

Stream Types "D" and "DA"

Type "D" streams are multichannel (three or more) streams. These braided streams are found in well-defined alluvial valleys. Braided channels are characterized by moderate to high bank erosion rates, depositional features such as transverse bars, and frequent shifts in bed forms. The channels are typically on the same gradient as their valley. There are few "D" streams in North Carolina.

The "DA" stream type is a stable braided stream with a low but highly variable width/depth ratio (for braided channels) and low slope (less than 0.5%). The DA stream types are found in wide alluvial valleys or deltas exhibiting interconnected channels and an abundance of wetlands. This stream type is often found in the coastal plain of North Carolina.

Basin: Chowan
Stream Type: DA6

Basin: Neuse
Stream Type: DA6

FIGURE 2.27 (Continued)

such as evidenced by changes in meander patterns. The process of change may also be anthropogenically accelerated. However, in many cases, river systems are managed as if they were static.

The changes in river systems can occur over a variety of timescales. Some of the changes occur over geological timescales, such as the evolution of river systems, and these timescales are often too large for practical consideration. Rivers are, of course, strongly impacted by and respond to changes

Stream Type "E"

For the single-thread channels, the "E" stream types are the evolutionary end point for stream morphology and equilibrium. The "E" stream type is slightly entrenched with low width/depth ratios, and moderate to high sinuosities. The bedform features are consistent riffle/pool sequences. Analyses of North Carolina streams determined that many "E" stream types in wide floodplains have been relocated to the edge of the floodplain and straightened. This has resulted in moderate entrenchment ratios and lower sinuosities. Dense vegetation has helped these streams remain as "E" stream types, but they do not function at their biological potential because of disruptions in the riffle/pool sequence. "E" stream types are generally found in wide alluvial valleys, ranging from mountain meadows to the coastal plain.

Basin: Holston (Virginia)
Stream Type: E4

Basin: Neuse
Stream Type: E4

Stream Type "F"

The "F" stream types are deeply entrenched, often meandering streams with a high width/depth ratio (greater than 12). These stream types are typically working to create a new floodplain at a lower elevation and will often evolve into "C" and then "E" stream types. This evolutionary process leads to very high levels of bank erosion, bar development, and sediment transport. The "F" stream types are found in low-relief valleys and gorges.

Basin: Watauga
Stream Type: F4

Basin: Watauga
Stream Type: F4

FIGURE 2.27   (Continued)

in climate, both local and global, which also commonly occur over long timescales. However, local and global climate changes may be of practical importance in river management due to the anthropogenic acceleration of climate change.

Patterns in precipitation, evaporation, and groundwater recharge obviously change with time and impact the flow in rivers. Not all rivers continuously flow. Based on their flows, rivers may

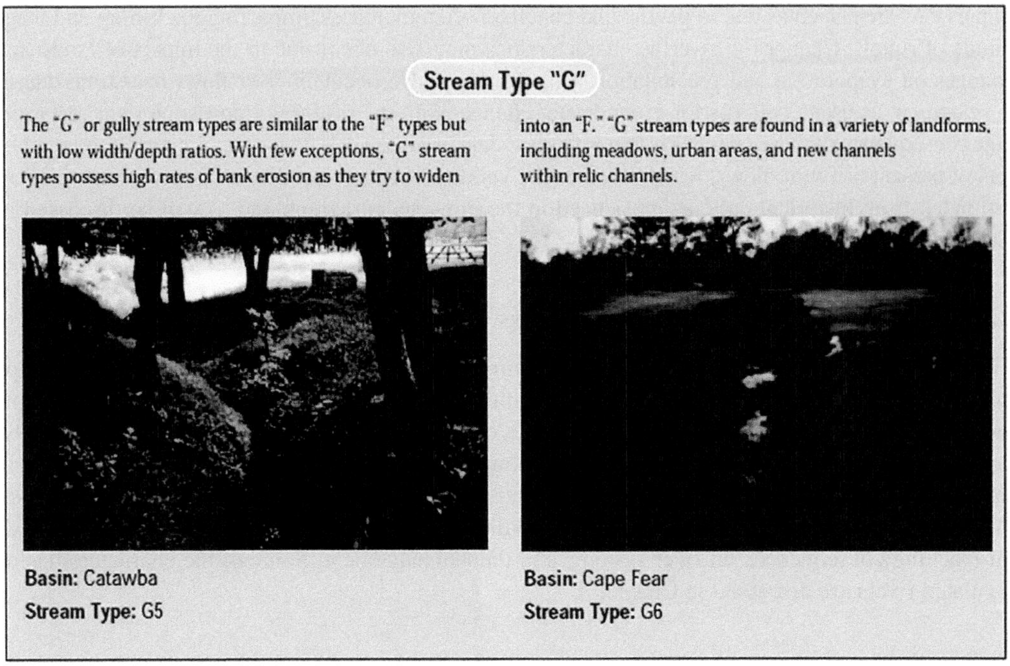

Stream Type "G"

The "G" or gully stream types are similar to the "F" types but with low width/depth ratios. With few exceptions, "G" stream types possess high rates of bank erosion as they try to widen into an "F." "G" stream types are found in a variety of landforms, including meadows, urban areas, and new channels within relic channels.

Basin: Catawba
Stream Type: G5

Basin: Cape Fear
Stream Type: G6

FIGURE 2.27 (Continued)

be classified as *perennial* or permanent, *periodic*, or *episodic*. Even in perennial rivers, flows vary over time, such as within and between seasons. Peak flows and flooding may occur, for example, in the springtime due to runoff and snowmelt. Flooding is a major concern in river management, as are low flow conditions. As will be discussed in Chapter 3, management design flows have historically been based on relatively rare high flows, for flooding. For example, the 100-year flow, or the flow with a probability of occurrence during any given year of 1%, has commonly been used to delineate the floodplain, with resulting regulatory implications. The design of restoration projects is often also based on being able to withstand the 100-year flow, or some other relatively rare flow event. The magnitude of the high-flow event used for design is usually based on the consequences of failure. For example, the design of dams in populated areas is usually based on very extreme and rare events. Similarly, low flows for the management of water quality or instream habitat have commonly been established based on relatively rare low flows, such as the 7-day average flow expected to occur once every 10 years (the 7Q10 flow). The determination of the magnitude of these relatively rare events, low or high flow, is typically based on a statistical analysis of historical data, as will be discussed in Chapter 3. However, as will also be discussed in Chapter 3, it is the maintenance of the seasonal variations in the flow hydrograph that is becoming the more common goal of maintenance or river restoration efforts. The maintenance of these seasonal changes may be necessary to maintain the characteristics of a river system or to satisfy the seasonal variations in the needs and life stages of biota within or contiguous to river systems.

Changes in the landscape and land use impact riverine characteristics over a variety of timescales. Changes in agricultural uses may vary over and between years, due to seasonal crop or pasture demands, and crop or pasture rotations or forestry practices. Changes may also occur within and between years as watersheds are developed, impacting infiltration and runoff. Large changes may occur during and postconstruction. Stormwater regulations typically require that construction projects reduce the peak runoff to that prior to construction. However, the timing and magnitude of runoff may vary during and postconstruction. For example, as watersheds are developed, the percentage of impervious area (such as due to paving and concrete) often increases, impacting the timing and magnitude of runoff. Changes in riverine characteristics may also occur due to the impact of landscape changes on evaporation and precipitation. Changes may also occur in river flows over time due to development as usage patterns for groundwater change, with the resulting impacts on river recharge. One consequence of these changes is that historical data, such as river flows, are often not good predictors of present or future flows, which impacts the validity of design conditions for river management estimated from historical data, as introduced in the previous paragraph and as will be discussed in Chapter 3.

2.2.5 REGULATED VERSUS UNREGULATED RIVERS

The goal of many projects is to restore the natural function of rivers. However, undisturbed or unregulated rivers or streams are rare. If the stream channel itself is not altered (such as by "realignment" or straightening), then hydraulic structures or other hydraulic modifications may impact the flows. Changes in the watershed that impact flows and sediment loads may also cause systems to be out of equilibrium with their sediment loads, resulting in bank sloughing and other factors impacting channel evolution. The degree of modification and the impacts of that regulation have far-reaching consequences on river systems and their management. Some of the characteristics of regulated rivers are discussed in Chapter 3.

REFERENCES

Anderson, E.W. 1987. Riparian area definition a viewpoint. *Rangelands* 9, 70.

Andrews, E.D. 1980. Effective and bankfull discharges of streams in the Yampa River basin, Colorado and Wyoming. *Journal of Hydrology* 46, 301–310.

Boulton, A.J., S. Findlay, P. Marmonier, E.H. Stanley, and H.M. Valett. 1998. The functional significance of the hyporheic zone in streams and rivers. *Annual Review of Ecology, Evolution and Systematics* 29, 59–81.

Brosofske, K.D. 1996. Effects of harvesting on microclimate from small streams to uplands in western Washington. MS thesis, Michigan Technological University.

Cushing, C.E. and J.D. Allan. 2001. *Streams: Their Ecology and Life*. Academic Press, New York.

Dick-Peddie, W.A. and J.P. Hubbard. 1977. Classification of riparian habitat in the Southwest. In R.R. Johnson, and D.A. Jones (tech. coords). *Importance, Preservation and Management of Riparian Habitat: A Symposium*. U.S. Forest Service General Technical Report RM-43, Tucson, AZ, pp. 85–90.

Dunne, T. and L.B. Leopold. 1978. *Water in Environmental Planning*. W.H. Freeman, San Francisco, CA.

ERDC. 2002. Technical and scientific considerations for upland and riparian buffer strips in the Section 404 permit process. ERDC TN-WRAP-01-06. U.S. Army Engineer Research and Development Center, Vicksburg, MS.

Federal Register. 2002. Issuance of nationwide permits; Notice. *Federal Register* 67 (10), 2020. Available at: http:\\www.gpo.gov.

Fischenich, C. and J.V. Morrow, Jr. 2000. Reconnection of floodplains with incised channels, ERDC TN-EMRRP-SR-09. U.S. Army Engineer Research and Development Center, Vicksburg, MS.

Fischer, R.A. 2002. The corps' role in riparian restoration and management. Presented at the U.S. Army Corps of Engineers Economic and Environmental Analysis Conference 2002. Balancing Economy and Environment, New Orleans, LA.

Fischer, R.A., C.O. Martin, J.T. Ratti, and J. Guidice. 2001. Riparian terminology: Confusion and clarification, ERDC TN-EMRRP-SR-25. U.S. Army Engineer Research and Development Center, Vicksburg, MS. Available at: http:\\www.wes.army.mil/el/emrrp.

FISRWG. 1998. *Stream Corridor Restoration: Principles, Processes, and Practices*, Federal Interagency Stream Restoration Working Group. GPO Item No. 0120-A; SuDocs No. A 57.6/2:EN 3/PT.653.

Gregory, S.V., F.J. Swanson, W.A. McKee, and K.W. Cummins. 1991. An ecosystem perspective of riparian zones. *BioScience* 41, 540–551.

Harman, W.A. and G.D. Jennings. 1999. Application of the rosgen stream classification system to North Carolina. North Carolina State University, North Carolina Cooperative Extension Service, Fact Sheet Number 2.

Horton, R.E. 1945. Erosional development of streams and their drainage basins: Hydrophysical approach to quantitative morphology. *Bulletin of the Geological Society of America* 56, 275–370.

Howard, A.D. 1967. Drainage analysis in geologic interpretation: A summation. *AAPG Bulletin Series* 51 (11), 2246–2259.

Ilhardt, B.L., E.S. Verry, and B.J. Palik. 2000. Defining riparian areas. In E.S. Verry, J.W. Hornbeck, and C.A. Doloff (eds), *Riparian Management in Forests of the Continental Eastern United States*. CRC Press, Boca Raton, FL, pp. 23–42.

Knighton, D. 1984. *Fluvial Forms and Processes*. Edward Arnold, Baltimore, MD.

Lane, E.W. 1955. The importance of fluvial morphology in hydraulic engineering. *Proceedings from the American Society of Civil Engineers* 81 (745), 1–17.

Leopold, L.B. and M.G. Wolman. 1957. *River Channel Patterns: Braided, Meandering and Straight*. U.S. Geological Survey Professional Paper 282-B. U.S. Government Printing Offices, Washington, DC, pp. 39–85.

Leopold, L.B., M.G. Wolman, and J.P. Miller. 1964. *Fluvial Processes in Geomorphology*. W.H. Freeman, San Francisco, CA.

Lunetta, R.S., B.L. Cosentino, D.R. Montgomery, E.M. Beamer, and T.J. Beechie. 1997. GIS-based evaluation of salmon habitat in the Pacific Northwest. *Photogrammetric Engineering and Remote Sensing* 63 (10), 1219–1229.

MacArthur, R.C., C.R. Neill, B.R. Hall, V.J. Galay, and A.B. Svidchenko. 2007. Overview of sedimentation engineering. In M. Garcia (ed.), *Sedimentation Engineering: Processes, Measurements, and Practice*. American Society of Civil Engineers, Reston, VA, pp. 1–20.

McManus, P.F. 1981. *A Fine and Pleasant Misery*. Henry Holt & Co., New York.

Miller, G.T. 1990. *Living in the Environment: An Introduction to Environmental Science*, 6th ed. Wadsworth Publishing Company, Belmont, CA.

National Research Council. 2002. Riparian areas: Functions and strategies for management. Committee on Riparian Zone Functioning and Strategies for Management. Water Science and Technology Board, National Research Council, Washington, DC.

North Carolina. 2003. Draft internal policy determination of the origin of perennial streams. Version 1.0. NC Division of Water Quality, Raleigh, NC.

NRCS. 1996. Riparian areas environmental uniqueness, functions, and values. RCA Issue Brief #11. U.S. Department of Agriculture, Natural Resources Conservation Service, Washington, DC.

NRCS. 2007. *National Engineering Handbook, Stream Restoration Design*. U.S. Department of Agriculture, Natural Resources Conservation Service, Washington, DC.

O'Keefe, T.C., S.R. Elliott, and R.J. Naiman. 2007. Introduction to watershed ecology. U.S. Environmental Protection Agency, Watershed Academy Web. Available at: http://cfpub.epa.gov/watertrain/pdf/modules/watershedecology.pdf.

Rosgen, D.L. 1994. A classification of natural rivers. *Catena* 22, 169–199.

Rosgen, D.L. 1996. *Applied River Morphology*. Wildland Hydrology Books, Pagosa Springs, CO.

Rosgen, D.L. and H.L. Silvey. 1996. *Applied River Morphology*. Wildland Hydrology Books, Fort Collins, CO.

Secretary of the Interior. 1994. *The Impact of Federal Programs on Wetlands*. Vol. 2. Report to Congress. Available at: http://www.doi.gov/oepc/wetlands2/.

Strahler, A.N. 1957. Quantitative analysis of watershed geomorphology. *Transactions American Geophysical Union* 38, 913–920.

Vannote, R.L., G.W. Minshall, K.W. Cummins, J.R. Sedell, and C.E. Cushing. 1980. The river continuum concept. *Canadian Journal of Fisheries and Aquatic Sciences* 37, 130–137.

Ward, J.V. 1989. The four-dimensional nature of lotic ecosystems. *Journal of the North American Benthological Society* 8 (1), 2–8.

Warner, R.E. and K.M. Hendrix (eds). 1981. *California Riparian Systems: Ecology, Conservation, and Productive Management*. University of California Press, Berkeley, CA.

Wetzel, R.G. 2001. *Limnology: Lake and River Ecosystems*, 3rd ed. Academic Press, San Diego, CA.

Winter, T.C., J.W. Harvey, O.L. Franke, and W.M. Alley. 1998. *Ground Water and Surface Water: A Single Resource*. U.S. Geological Service, Denver, CO.

3 Regulated Rivers

3.1 INTRODUCTION

Much of the previous discussion on the characteristics of rivers was related to rivers in their "natural" state. It is important to know the characteristics of natural or free-flowing rivers so that we can better manage and protect them. Also, as will be discussed in later chapters, much of the work done today is in the restoration of river systems. The National Research Council in its 1992 report, "Restoration of aquatic ecosystems" (NRC 2002), defined restoration as the "return of an ecosystem to a close approximation of its condition prior to disturbance." In order to do so, some knowledge of the undisturbed characteristics of river systems is required so we know what to restore the system to. However, by far the majority of rivers today are regulated or controlled to some degree. Free-flowing, or undisturbed, rivers are relatively rare. According to the World Wildlife Federation, in 1986 (WWF 1986) only 18% of the rivers in North America longer than 1000 km were free-flowing, including (length in parentheses) the Mackenzie (5472 km), Athabasca (1231 km), Liard (1115 km), Yellowstone (1080 km), Fraser (1370 km), and Kuskokwim (1050 km).

Since the majority of our river systems are controlled or regulated, it is important to have a fundamental understanding of those regulatory or control structures in order to determine their impact on the river's physical, chemical, and biological characteristics. The two primary sources of control or regulation are:

- Flow modification structures, such as dams and weirs
- Channel modification such as the dredging and straightening of rivers

The practice of river regulation using these sources of control has been around for thousands of years. The concept of "harnessing" and "controlling" rivers was the early goal of river management and river engineering, as expressed by Benjamin Franklin:

Rivers are ungovernable things, especially in hilly countries. Canals are quiet and very manageable …

Benjamin Franklin 1772

Channelization and control were the early common goals of engineering projects in Europe and elsewhere. For example, in 1817, the German engineer Colonel Johann Gottfried Tulla initiated channelization of the braided Alsatian section of the Rhine, and he is often quoted as stating, "as a rule, no stream or river needs more than one bed." By the 1900s, virtually all of the major rivers in Europe had been channelized.

Most of the early efforts were directed toward small rivers or sections of rivers. It was not until the last two centuries, and notably the last century, that complete control of rivers was commonly achieved, due in large part to advances in dam-building technology. The 1940s to the 1980s was the construction era of water resource management in the United States (Wurbs and James 2002). The period from 1950 to 1980 included the peak of dam building worldwide and during that period in North America over 200 large dams (over 15 m high) were constructed per year. While in more recent years the rate of construction of large dams has decreased dramatically, large dams have and are being constructed, such as the Three Gorges Dam in China, which became fully operational in 2012. Plans for even larger dams, such as in the Congo, are being considered.

Rivers are controlled or disturbed not only by direct flow and channel modification, such as was previously noted, but also indirectly by changes in their watershed and underlying groundwater. At issue, for example, is the natural flow hydrograph. Changes in the watershed impact the timing and magnitude of runoff. Changes in the groundwater impact recharge. Since the goal of restoration is to return the ecosystem to its undisturbed condition, the undisturbed characteristics should extend not only to the river's bed but also to the natural flow hydrograph. Unfortunately, because of the extent and long history of the modification of rivers and watersheds, there is often little information available to define and design the undisturbed or natural flow hydrograph.

3.2 FLOW MODIFICATION STRUCTURES

3.2.1 Dams

A dam is a barrier that is constructed to contain or constrain the flow of water or divide water, such as between freshwater and saltwater. Dams vary in size from those in small farm ponds or detention ponds in urban development projects to some of the largest engineered structures in the world. The impact of a dam, such as that on a river, is to increase the storage of water upstream and consequently increase the water-surface elevation. The extent to which the increased elevation extends upstream is called the backwater and typically varies with time and with the reservoir storage (e.g., as it impacts lake levels). The water-surface elevation within the zone is flatter horizontally (known in hydraulics as the M1 backwater curve; Figure 3.1) and in the reservoir pool the water is deeper and the velocities lower than in the upstream river.

The release from the dam may be controlled (such as for hydropower) or not (such as over a spillway) or both (such as hydropower with an emergency spillway). The impact of uncontrolled storage is to attenuate the peak flows in releases from that in the inflows. The dam impacts the physical, chemical, and biological characteristics of the impounded waters and the downstream receiving waters, either or both being impacted positively or negatively as will be discussed later.

3.2.1.1 So How Many Are There?

Because of the importance of dams, and the consequences associated with their failure, a National Inventory of Dams (NIDs; Figure 3.2) computer database is maintained by the U.S. Army Corps of Engineers, authorized by the National Dam Inspection Act (P.L. 92-367). The web-accessible database tracks information about a dam's location, size, use, type, proximity to the nearest town, hazard classification, age, and height. There are approximately 76,000 dams over 2 m and 2.5 million smaller dams on U.S. rivers. According to the U.S. Fish and Wildlife Service (www.rivers.gov), large dams across

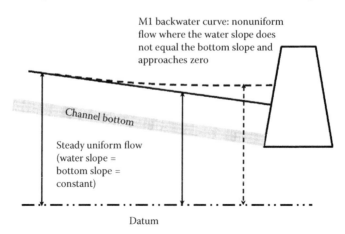

FIGURE 3.1   M1 backwater curve.

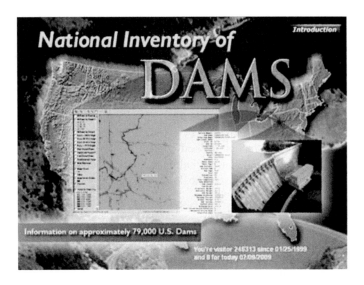

FIGURE 3.2 National Inventory of Dams database, Available at http://crunch.tec.army.mil/nidpublic/webpages/nid.cfm.

the country have modified at least 600,000 miles, or about 17% of American rivers. Virtually all large rivers (and most small rivers) in the United States are modified and fragmented by dams.

3.2.1.2 What Are They for (and How Are They Operated)?

Dams are typically designed and constructed for designated uses such as (the number of dams with their use is given in parentheses taken from NIDs) flood control and stormwater management (15,769), irrigation (9,405), water supply (7,430), navigation (693), recreation (33,934), and hydropower (2,551). By far the majority of dams are privately owned, as illustrated in Figure 3.3, with only 4% of U.S. dams being federally owned.

The purpose, function, and manner of a dam's operation will vary with its designated use. For example, some of the typical purposes and functions along with design considerations are:

- Municipal and industrial (M&I) water supply: Provide water of a satisfactory quality to meet the demand with adequate pressure at all times in all parts of the system.
  - Estimate the demand and supply based on historical data and expected trends.
  - Locate water sources as close to the demand as possible.
- Design surface storage to retain wet season flows to bridge dry periods; keep reservoirs as full as possible.
  - Design a distribution system to supply time-varying demand with adequate pressure.
  - Maintain source quality and treat as needed.
  - Treat waste flows.

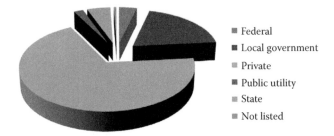

Federal
Local government
Private
Public utility
State
Not listed

FIGURE 3.3 Dam ownership. (Based on data from the National Inventory of Dams.)

- Irrigation: Provide water of a satisfactory quality to meet the allocated demand at specified times in all parts of the system.
  - Estimate the demand and supply based on historical data and expected trends.
  - Locate water sources as close to the demand as possible.
  - Design surface storage, if any, to retain wet season flows to bridge dry periods; keep reservoirs as full as possible.
  - Design a distribution system to supply time-varying allocations on time.
  - Manage runoff quantity and quality.
- Flood and storm damage reduction ("flood control"): Prevent loss of life and minimize property damage for events up to and including the design flood/storm.
  - Identify the design event(s) by statistical analyses, usually some rare event such as the maximum probable storm (MPS).
  - Map the affected areas.
  - Design damage reduction features to maximize benefits (prevent losses).
    - Source management.
    - Storage structures/reservoirs.
    - Levees/floodwalls.
    - Diversions and channel modifications.
    - Floodplain and emergency management.
  - Keep storage reservoirs as empty as possible to absorb flood flows.
- Navigation: Provide waterways that enable safe navigation for the design vessel, depth or traffic load.
  - Identify design vessel(s) and traffic loads from historical data and expected trends.
  - Analyze waterway flows and depths in light of navigation needs.
  - Design the required facilities to provide safe throughput for design vessels and traffic under given waterway conditions.
    - Channels, anchorages, and turning basins.
    - Structures (locks, dams, and training structures).
    - Ports.
  - Design reservoirs, if any, to retain water in wet seasons and release in dry seasons to maintain channel depth. Design run-of-river dams to maintain the required water level.
- Hydropower: Provide design levels of power on an allocated time schedule.
  - Identify base and peak demands from historical data and expected trends.
  - Design dams to provide needed head and storage capacity for effective power generation.
  - Design pumpback systems, if used, to increase the peak generation capacity.
  - Design dams to retain flows until needed for power generation (keep them full).
- Recreation: Provide water levels and flows that enable safe and desirable recreational use of the waterbody.
  - Identify expected recreational uses from historical data and expected population trends.
  - Design structures and watercourses that will support intended recreational uses; examples:
    - Lakes and rivers for fishing, boating, and skiing.
    - Rivers for white-water rafting.
  - Design for the predictability of water levels and flows—mimicking natural cycles for fishing or nearly constant cycles for other recreational uses.
- Environmental quality: Ensure a water supply of sufficient quantity and quality to meet the designated uses.
  - Identify designated uses, threatened and endangered (T&E) species, essential fish habitats (EFHs), and impairments (EPA 303D list), if any.

- Define the present conditions, including point and nonpoint pollution sources, expected flows and levels, and concentrations of pollutants.
- Define the total maximum daily loads (TMDLs) of pollutants and compare them with existing conditions.
- Develop corrective action implementation plans for impaired uses, excess loadings, and the protection of species or habitats.

A typical reservoir and reservoir storage zones are illustrated in Figure 3.4. Typically, there is a dead zone that cannot be managed, and the surcharge zone is to accommodate extreme events to prevent failure. The flood control and conservation pools are the managed pools. Conservation purposes include municipal, industrial, and agricultural water supplies, hydroelectric power, recreational uses, and instream flow maintenance. A reservoir may be operated only for conservation purposes or only for flood control or to designate a certain reservoir volume, or pool, for conservation purposes and a separate pool for flood control. In many cases, however, the construction of a reservoir cannot be justified by a single project purpose, and many are operated for multiple purposes. The conservation and flood control pools in a multipurpose reservoir are fixed by a designated top of conservation (bottom of flood control) pool elevation (Tibbets et al. 1985).

A reservoir's storage capacity and operating policies are generally established prior to its construction and tend to remain constant thereafter. For federal projects that are authorized by Congress, changing the operation usually requires an additional act by Congress. The operating procedures are usually implemented based on a rule or guide curve derived from historical data on tributary inflows and water demands. The rule curve shows the minimum water level elevation (and consequently storage) requirement in the reservoir over the year at a specific time to meet the particular needs for which the reservoir is designed and it controls the reservoir management. The goal of the management is typically to follow the rule curve, with exceptions allowed only under extreme conditions, such as during periods of extreme drought. The rule curve is usually bounded by an upper rule curve for handling surcharge and a lower rule curve based on the minimum level at which the design uses are satisfied (Figure 3.5).

Reservoir volumes typically increase in the spring and late fall or winter due to increased runoff during those periods. If a reservoir is operated to supply water, then the rule curve can be similar to that shown in Figure 3.6 where the goal would be to maintain as much water in the reservoir as possible. Similarly, for a peaking hydropower facility, the goal would be to have the maximum elevation possible, particularly during those periods of the year with the highest demand. For a flood control reservoir, as illustrated in Figure 3.7, the goal may be to have the reservoir as empty as possible during periods of high flows, to accommodate those flows and attenuate the downstream peak flow. For a facility operated for navigation purposes, the goal may be to maintain a constant water-surface elevation (Figure 3.8).

The designated uses of reservoirs are often in conflict. For example, the ideal rule curves are in conflict for reservoirs operated for both flood control and water supply, so the resulting rule curve

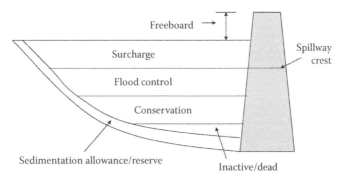

FIGURE 3.4 Typical reservoir zones.

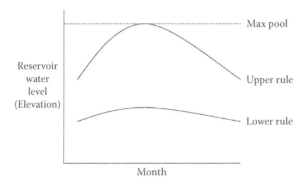

FIGURE 3.5 Reservoir rule curve.

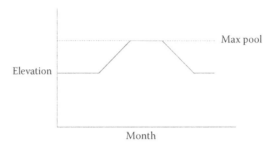

FIGURE 3.6 Water supply or hydropower (part of the rule curve).

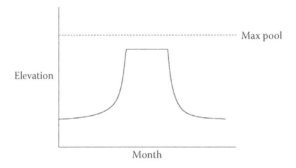

FIGURE 3.7 Flood control (part of the rule curve).

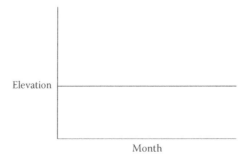

FIGURE 3.8 Navigation (part of the rule curve).

is a compromise. Operations implementing spring and fall drawdowns for flood control are also in conflict with recreation and fisheries management, where high water levels in the spring may be preferred. The operations of a reservoir also directly impact the immediate downstream river, the reservoir tailwater. Dams and their operations will be discussed in greater detail in a later chapter on the physical characteristics of reservoirs, while tailwaters are the subject of a separate chapter.

3.2.1.3 What Are Dam Impacts?

Dams impact both the characteristics of the impounded system and the dam tailwaters, and the impact varies with the design characteristics and the operation of the dam. Dams typically convert systems from flowing or lotic environments to standing or lentic environments, with consequent changes in the physical, chemical, and biological structures of the impounded system and its releases. With increases in the hydraulic retention time, the magnitude and the timing of release flows are altered. The thermal structure of the reservoir and its releases are altered. For example, during the summer months, in many southern reservoirs the bottom waters are colder than the surface, anoxic waters, and they have high concentrations of reduced materials, such as sulfides, metals, and nutrients. The release of bottom waters, such as from peaking hydropower facilities, has commonly changed tailwaters from warm to very productive cold-water fisheries. Dams also isolate communities, such as the tailwater and reservoir communities. Dams impact the ability of the system to store and transport sediments. The longer hydraulic retention time impacts seasonal temperature variation. Other impacts of dams are discussed in greater detail in later chapters.

3.2.2 Weirs and Dikes

Weirs are similar to dams, but are essentially low walls built across a river for the purpose of raising the upstream water level. Weirs are usually low-flow structures and may be completely submerged during high flows. The flow over a weir is uncontrolled (Figure 3.9).

Weirs are commonly used as an alternative flow measurement structure by measuring the height of the water flowing over the weir. These weir flow equations are used in estimating the flow over gated structures or spillways that act hydraulically like a weir. An example of a portable weir for measuring flow rates is provided in Figure 3.10.

Dikes are structures that protrude into a river channel, generally transverse to the flow, and are also called "groins," "jetties," "spurs," "wing dams" (Biedenharn et al. 1997), or "wing dikes."

FIGURE 3.9 Weir on the Big Sunflower River, Mississippi.

FIGURE 3.10 Measuring streamflow through a V-notch weir. (Photograph by Alan Cressler, USGS Multimedia Gallery.)

Wing dikes, which are common on many of our river systems, are rock structures extending from the riverbank out into the river channel for distances up to 1 mile or more. Dikes help to reduce sedimentation in the main channel by constricting flows and increasing velocities. Dikes can protect banks from erosion. They can also form protected areas, often with plunge pools, that provide fish habitats. Dikes are typically placed in straight reaches and long radius bends. While they are commonly used in wide and shallow streams, dikes are also used in large rivers to increase their depth for navigation and to stabilize banks and channel alignment.

Bendway weirs are similar to a dike but are directed at an angle to the flow. Bendway weirs were developed by the U.S. Army Corps of Engineers to increase the channel width in bends on the Mississippi River to improve navigation conditions and to reduce maintenance dredging requirements (Derrick et al. 1994). Bendway weirs are also commonly constructed on smaller streams and rivers (Figure 3.11) as a streambank protection measure, also providing a substrate for benthic organisms and cover for fish (Biedenharn et al. 1997).

FIGURE 3.11 Bendway weirs on Harland Creek. (From Biedenharn, D.S., Elliott, C.M., and Watson, C.C., *The Stream Investigation and Streambank Stabilization Handbook*, U.S. Army Corps of Engineers, Waterways Experiment Station, Vicksburg, MS, 1997.)

3.2.3 CULVERTS

Culverts are commonly used, hydraulically short conduits that are designed to transport water from one side to the other of a roadway, railroad, levee, or other structure. Culverts can be made of a variety of materials, including corrugated metal, concrete (reinforced or not), and thermoplastic pipe and there may be one or multiple barrels to the culvert (see Figures 3.12 and 3.13).

Engineers design culverts to pass a specific design high-flow discharge, as well as to maintain flows under low-flow conditions. Part of the hydraulic design process is to determine conditions where the inlet, outlet, or both are submerged. A typical low-flow design is for inlet control, where the flow capacity of the culvert entrance is less than the flow capacity of the culvert barrel. Under

FIGURE 3.12 "Open-bottom," arched culvert over Thunder Brook, Massachusetts, that enhances fish passage. (From Massachusetts Division of Ecological Restoration.)

FIGURE 3.13 Precast concrete box culvert. (Courtesy of the American Concrete Industries.)

inlet control, the flow in the culvert is controlled by the water-surface elevation at the inlet (head-water depth) and water can flow through and out of the culvert faster than it can enter it. Under inlet control, the head of the culvert acts like a weir or a sluice gate where the inlet controls the flow, not the length, roughness of the barrel, or outlet. Under inlet control, the flow passes through a critical depth at or just downstream of this location. Therefore, under inlet control the flow in the barrel of the culvert will always be in a state of shallow-velocity and high-velocity flows, known as supercritical flows where the velocity is greater than the celerity of a gravity wave $(gY)^{0.5}$, where g is the gravitational acceleration and Y is the depth and where downstream disturbances cannot travel upstream. A culvert flowing full is often considered the most "hydraulically efficient" since it is conveying the maximum possible flow.

Culverts are most commonly designed with the goal of maximizing hydraulic efficiency in order to maximize conveyance while minimizing pipe sizes and costs. Commonly not included in the development of the design is the impact of the culvert on the aquatic life and the geomorphic conditions of stream channels.

Of course, culverts do have many benefits, such as reducing flooding. However, culverts can also have many detrimental effects. One prevailing cause of those detrimental effects is basing the design of culverts on maximizing hydraulic efficiency, thereby minimizing pipe sizes and costs, without considering the environmental impacts. One of those environmental impacts is the isolation of communities by acting as barriers to fish movement. Five common conditions at culverts that create migration barriers are (Washington 2003):

- Excess drop at the culvert outlet
- High velocity within the culvert barrel
- Inadequate depth within the culvert barrel
- Turbulence within the culvert
- Debris and sediment accumulation at the culvert

The first four of these conditions could result from the design for inlet control. Other impacts of culverts include direct habitat loss, water quality degradation, upstream and downstream channel impacts, and impacts during channel maintenance and construction (Washington 2003).

So, how many culverts are there? This is presently not known and it has stimulated the question in the U.S. Department of Agriculture (USDA Forest Service Roads Analysis, [1999, p. 67, AQ(10)]: "How and where does the road system restrict the migration and movement of aquatic organisms; what aquatic species are affected and to what extent?" One result is the development of a procedure to identify barriers to fish passage, the "National inventory and assessment procedure for identifying barriers to aquatic organism passage at road-stream crossings," by Clarkin et al. (2005). One example of the methods developed to screen culverts for fish passage is the system developed by Taylor and Love (2002) for adult and juvenile anadromous salmonids. In this system, culverts are categorized as "red," "gray," or "green" as defined by (Clarkin et al. 2005) the following:

- Green: Conditions are assumed adequate for the passage of all salmonids. Even the weakest-swimming life stage (juveniles) can pass the crossing during the entire period of migration.
- Gray: Conditions may not be adequate for all salmonid species or life stages presumed present. Additional analyses are required to determine the extent of the barrier for each species and life stage.
- Red: Conditions do not meet passage criteria over the entire range of migration flows for even the strongest-swimming species and life stage (adults) presumed present. Assume "passage condition inadequate."

Examples of red, gray, and green culverts are illustrated in Figure 3.14.

Red

Green

Gray

FIGURE 3.14   Illustration of the good (green), bad (gray), and ugly (red) culverts. (Courtesy of USDA Forest Service, Alaska Division.)

One of the themes of this text is also "one generation's solution is the next generation's problem," or for engineers perhaps the next generation's job. Today, engineers are commonly involved in designing or retrofitting culverts to make them more environmentally friendly. Guidance for the design includes:

- FHWA Culvert repair practices manual, Volume 1 (FHWA 1995), pp. 3-58–3-61 and 5-39–5-50, and Volume 2, Appendix B-23, for a discussion on fish passage and fish passage devices
- NOAA Fisheries Draft: Culvert criteria for fish passage and the guidelines for salmonid passage at stream crossings
- Washington (2003) "Design of road culverts for fish passage"

3.2.4   LEVEES

Levees (as illustrated in Figure 3.15) are embankments that are primarily designed for flood protection from seasonal high waters with flood durations most commonly only a few days or weeks per year. Even though levees are similar to small earth dams, they differ from earth dams in the following important respects (USACE 2000):

- A levee embankment may become saturated for only a short period of time beyond the limit of capillary saturation
- Levee alignment is dictated primarily by flood protection requirements, which often results in construction on poor foundations
- Borrow is generally obtained from shallow pits or from channels excavated adjacent to the levee, which produce fill material that is often heterogeneous and far from ideal.

Some levee types according to their use are shown in Table 3.1.

FIGURE 3.15 Levee on the Mississippi River. (From U.S. Army Corps of Engineers photograph by Lisa Coghlan, USACE HQ Photostream.)

TABLE 3.1
Classification of Levees According to Use

Type	Definition
Mainline and tributary levees	Levees that lie along a mainstream and its tributaries, respectively.
Ring levees	Levees that completely encircle or "ring" an area subject to inundation from all directions.
Setback levees	Levees that are built landward of existing levees, usually because the existing levees have suffered distress or are in some way being endangered, as by river migration.
Sublevees	Levees built for the purpose of underseepage control. Sublevees encircle areas behind the main levee, which are subject, during high-water stages, to high uplift pressures and possibly the development of sand boils. They normally tie into the main levee, thus providing a basin that can be flooded during high-water stages, thereby counterbalancing excess head beneath the top stratum within the basin. Sublevees are rarely employed as the use of relief wells or seepage berms makes them unnecessary except in emergencies.
Spur levees	Levees that project from the main levee and serve to protect the main levee from the erosive action of stream currents. Spur levees are not true levees but training dikes.

Source: USACE, Design and construction of levees, U.S. Army Corps of Engineers, Waterways Experiment Station, Vicksburg, MS, 2000.

3.2.4.1 So, How Many Are There?

Who knows? Historically, large numbers of levees have been created and maintained by individuals and private and governmental organizations and the total number of levees is largely unknown. However, following and, in part, motivated by the levee failures resulting from Hurricane Katrina in 2005, a levee safety program was established by the National Levee Safety Program Act of 2007 to:

- Inspect each levee in the United States constructed or maintained by the secretary of the army or identified by a state governor, excluding levees whose failure would not pose a significant threat to human life or property
- Notify the governor of the state in which a levee is located of inspection results, hazardous conditions found, and necessary remedial measures

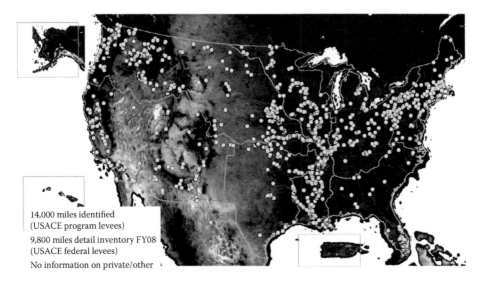

14,000 miles identified
(USACE program levees)

9,800 miles detail inventory FY08
(USACE federal levees)

No information on private/other

FIGURE 3.16 Locations from an initial survey to identify all levees in the USACE program. (From Shoffstall, G.D., USACE Levee Safety Program, New York State Emergency Management Association Winter Conference, 2009.)

- Submit, annually, a priority list of flood damage reduction studies and projects
- Periodically publish an inventory of U.S. levees, including inspection results

The act also established an interagency committee on levee safety to support and coordinate federal levee safety programs.

As of 2009, the U.S. Army Corps of Engineers had surveyed approximately 14,000 miles of primarily federally owned and maintained levees (Shoffstall 2009), as illustrated in Figure 3.16. That survey was expanded to greater than 14,700 miles of levee systems by 2012.

3.2.4.2 What Are the Impacts?

As with all such engineered structures, levees may have both positive and negative environmental effects. For example, levees have by design allowed for the isolation and drainage of wetlands for other purposes, such as agricultural uses. Levees have also been commonly used, such as in Louisiana, to isolate and protect wetlands from saltwater intrusion and to restore them. Some additional impacts include:

- Loss of riparian vegetation for levees immediately adjacent to the bank or stream edge
- Loss of sediments and nutrient replenishment by flooding
- Increase in velocities due to channel confinement, resulting in changes in erosional and sedimentation patterns
- Isolation of habitats and populations

3.3 CHANNEL MODIFICATIONS

3.3.1 OVERVIEW

The man-made alteration of rivers has been referred to as hydromodification or river engineering. Hydromodification, or river engineering, includes the construction of dams, levees, and other control structures, as discussed in the preceding section. One additional common goal particularly of early river engineering projects was either the construction of new channels or the alteration of existing natural channels to meet some design goal, such as for water supply, flood control or drainage, waterborne transportation, or to restrict flows to confined areas (Brookes 1981). Many of these engineered channels, new or altered, were, to the extent possible, straight with regular cross sections and with some near-constant design depth. Channels of this type are easier to design and build than natural channels. For example, one of the most common equations used in channel design, and one that all civil engineering students are exposed to, is Manning's equation:

$$Q = \frac{\delta}{n} R^{2/3} A\, S^{1/2} \qquad\qquad (3.1)$$

where
   δ is a unit conversion
   n is Manning's roughness coefficient
   R is the hydraulic radius of the channel
   A is the wetted area of the channel
   S is the channel slope

This equation assumes that the flow is constant (steady, or does not change with time) and the channel is prismatic (constant shape and slope) so that depths and velocities are constant throughout the channel (uniform); therefore, this equation can only be used to design channels with regular cross sections and a constant slope.

Examples of channels include those excavated for either a new or a reengineered river. Also, commonly, some portion of the existing channel or waterbody could be excavated to produce a navigation channel, as illustrated in Figure 3.17. In some cases, a confining structure is installed to raise the water level and minimize the excavation necessary, or to separate the navigable from the existing waterway as with a perched channel (Figure 3.18). As indicated in the preceding section, a goal of the operation of control structures in a navigation channel is nearly constant water-surface elevations. So, the control structures are designed to convert the natural river slope into a series of pools with a relatively constant depth, creating a "stair-stepped" water surface, with the structures connected by natural or excavated navigation channels.

One consequence of channelization is often the destabilization of the channel, as discussed in Section 3.3.2. One common consequence is aggradation, or the accumulation of sediments that tend to fill the channel. Many of the engineering practices involved in maintaining channelized rivers and streams are designed to protect embankments and channels from erosion. For example, layers of rock may be added to embankments (rock revetments) to prevent erosion. However, aggradation from upstream or watershed sources may fill the channel.

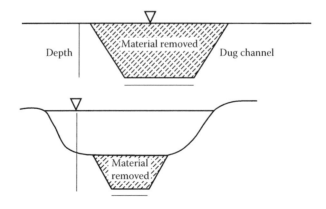

FIGURE 3.17   Excavated channels: a dug or excavated channel in an existing waterway.

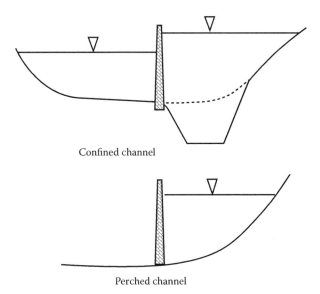

Confined channel

Perched channel

FIGURE 3.18 Confined and perched channels.

Channel design depths are often maintained by dredging. In the United States today, approximately 250,000 yd³ of sediments are removed from navigation channels annually. In many river systems, maintenance dredging is a nearly constant activity so that natural recovery is prevented and the benthos is continuously disturbed. Therefore, consideration of the dredging process and dredging impacts is an essential component of river management.

The process of maintenance dredging consists of the following stages:

- Excavation (loosening or dislodging) and removal of the loosened material to the dredge vessel
- Transportation of the material
- Placement or disposal of the material

The basic types of dredges include mechanical clamshell or backhoe dredges (Figure 3.19) that essentially have a large excavation bucket that is dropped or lowered to scoop the sediments.

FIGURE 3.19 Backhoe dredge. (Courtesy of the U.S. Army Engineer Research and Development Center, Vicksburg, MS.)

FIGURE 3.20   Self-propelled hopper (hydraulic) dredge. (Courtesy of the U.S. Army Engineer Research and Development Center, Vicksburg, MS.)

Alternatives include hopper dredges, as illustrated in Figure 3.20, cutter heads, and other dredge types. Disposal can occur in open water, confined disposal facilities (CDFs), or contained aquatic disposal (CAD), or increasingly the dredged material is used for beneficial use applications. The U.S. EPA and the U.S. Army Corps of Engineers have a cooperative program for the beneficial use of dredged materials, such as for beach nourishment, berm creation, capping, land creation, land improvement, replacement fill, and shore protection. Additional information is available through the Corps Dredging Operations Technical Support (DOTS) program, accessible online at: http://el.erdc. usace.army.mil/dots/.

3.3.2   WHAT ARE THE IMPACTS?

Channelization obviously has a number of positive economic and societal benefits. However, channelization can have a number of adverse environmental impacts as well. Many streams in agricultural landscapes have been straightened or otherwise modified to the point that they no longer provide ecological services such as habitat diversity, flood mitigation, or nutrient retention (Shields et al. 2008). Extreme channel incision often follows these modifications (Darby and Simon 1999).

One common consequence of channelization is the destabilization of the channel sediments. The impact of destabilization and the expected changes following destabilization may be illustrated using the channel evolution model of Simon and Hupp 1986 (Figure 3.21):

- Stage I: The waterway is a stable, undisturbed natural channel.
- Stage II: The channel is disturbed by some drastic change such as forest clearing, urbanization, dam construction, or channel dredging.
- Stage III: Instability sets in with scouring of the bed.
- Stage IV: Destructive bank erosion and channel widening occur due to the collapse of bank sections.
- Stage V: The banks continue to cave into the stream, widening the channel. The stream also begins to aggrade, or fill in, with sediment from eroding channel sections upstream.
- Stage VI: Aggradation continues to fill the channel, reequilibrium occurs, and bank erosion ceases. Riparian vegetation once again becomes established.

The channel evolution model is widely used to determine the successional stages of the evolution of rivers and streams, in order to identify impairments or to plan for river restoration or rehabilitation.

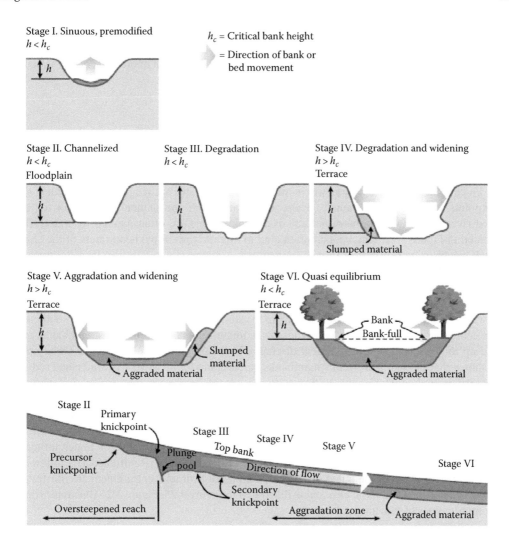

FIGURE 3.21 Stages of channel evolution. (From Simon, A. and Hupp, C.R., *Proceedings of the Forest Federal Interagency Sedimentation Conference*, Las Vegas, NV, U.S. Interagency Advisory Committee on Water Data, Washington, DC, 1986; from Langendoen, E.J., Concepts: Conservational channel evolution and pollutant transport system, Workshop presented at Mississippi State University, 2009.)

Destabilization may result from any change in the sediment load or flow regime. For example, increasing or decreasing the sediment loads to the system or the inflow hydrograph can result in destabilization, such as from watershed modifications, as discussed in the following section. Destabilization can also result from man-made channel modifications, or channelization. Channelization may introduce steeper gradients and the steepened channel can cause channel degradation, accelerated bank caving, and an increase in the downstream sediment regime (Schumm 1984; Brice 1982).

As illustrated by the channel evolution model, streams and rivers commonly respond to destabilization by changing vertically by scour or fill (degradation or aggradation) or horizontally by widening and migration. The Yalobusha River in north-central Mississippi illustrates the impacts of channelization. European agricultural development in the Yalobusha basin in the nineteenth century resulted in the restriction of the channel capacity due to the accumulation of silt and debris. In the early twentieth century, a 12-mile ditch was excavated through the river's valley, after which sand, silt, and debris closed the lower part of the channel. Clearing in the watershed continued in the 1940s. During the 1960s, additional channelization extended the ditch downstream, realigning and dredging the mouths of all major tributaries with the construction of erosion control structures. The channel response was:

- Headward-migrating knickpoints in all tributaries
- Extensive streambank erosion
- Recruitment of large woody debris, and the deposition of sands and debris at the terminus of channelization (Langendoen 2009)

A knickpoint is some location where there is a local, sharp change in the channel slope. Knickpoints may occur, for example, where a tributary enters the channelized river or the channelized river transitions to a more natural river. Knickpoints are often marked by waterfalls, as illustrated in Figure 3.22. Knickpoints often tend to migrate toward the headwaters of the channel through the drainage network.

Streambank erosion is a natural process. However, accelerated streambank erosion, which could occur due to channelization or other changes, is a major cause of nonpoint source pollution associated with an increased sediment supply (Rosgen et al. 2001), and often accounts for the majority of total watershed sediment yields. For example, in his studies of a selected stream in the southeastern plains ecoregion, Ramirez (2011) found that the sediment supply from incised streambanks represented up to 70% of the entire sediment discharge observed within the watershed. Other studies have shown that the sediment from streambanks can account for up to 90% of watershed sediment yields (Simon et al. 2002; Capello 2008).

Streambank erosion typically results from one of two processes: hydraulic action (also called hydraulic erosion, fluvial erosion, or tractive erosion) and gravitational mass failure (Ramirez 2011). These two processes are often linked where the hydraulic processes are often a precursor to gravitational failures (e.g., a hydraulic-induced mechanism, such as streambank undercutting, can cause a gravitationally induced collapse such as a cantilever failure [Ramirez 2011]) followed by the subsequent cleanout of the failed material. Examples of hydraulic entrainment are illustrated in Figures 3.23 and 3.24 and basal cleanout is illustrated in Figure 3.25. Figure 3.26 illustrates some of the common types of gravitational failure.

FIGURE 3.22 Headward-migrating knickpoints in a tributary. (From Langendoen, E.J., Concepts: Conservational channel evolution and pollutant transport system, Workshop presented at Mississippi State University, 2009.)

FIGURE 3.23 Examples of hydraulic or fluvial entrainment notch examples along the Town Creek, Mississippi. (From Ramirez, J.J., Assessment and prediction of streambank erosion rates in a southeastern plains ecoregion watershed in Mississippi, PhD dissertation in Civil and Environmental Engineering, Mississippi State University, 2011. With permission.)

FIGURE 3.24 Undercutting or scouring along tributaries within the Town Creek watershed. (From Ramirez, J.J., Assessment and prediction of streambank erosion rates in a southeastern plains ecoregion watershed in Mississippi, PhD dissertation in Civil and Environmental Engineering, Mississippi State University, 2011. With permission.)

FIGURE 3.25 Basal cleanout process after a streambank failure in a northern headwater of the Town Creek watershed. (From Ramirez, J.J., Assessment and prediction of streambank erosion rates in a southeastern plains ecoregion watershed in Mississippi, PhD dissertation in Civil and Environmental Engineering, Mississippi State University, 2011. With permission.)

FIGURE 3.26   Some selected examples of types of gravitational failure. (From Ramirez, J.J., Assessment and prediction of streambank erosion rates in a southeastern plains ecoregion watershed in Mississippi, PhD dissertation in Civil and Environmental Engineering, Mississippi State University, 2011. With permission.)

There are many beneficial aspects to large woody debris, and its reintroduction is a common goal of river restoration projects. However, the introduction of large woody debris, and the consequent loss of riparian vegetation, due to bank instability, as illustrated in Figure 3.27, often has negative impacts. For example, in the Yalobusha River, the recruitment of large woody debris resulted in the formation of a nearly 2-mile long plug of the river (Figure 3.28), which resulted in increases in the

FIGURE 3.27 Recruitment of large woody debris. (From Langendoen, E.J., Concepts: Conservational channel evolution and pollutant transport system, Workshop presented at Mississippi State University, 2009.)

FIGURE 3.28 Plug due to the recruitment of woody debris on the Yalobusha River in north-central Mississippi. (From Langendoen, E.J., Concepts: Conservational channel evolution and pollutant transport system, Workshop presented at Mississippi State University, 2009.)

frequency and magnitude of flooding in the area. Bennett and Rhoton (2003) discuss the formation and removal of the plug.

Channelization and channel maintenance largely impact the physical habitat of streams and rivers. For example, channelization tends to reduce the benthic habitat with the consequent impact of reducing species diversity. Sedimentation, resulting from alterations of rivers and their watersheds, is also a major cause of the impairment of rivers and streams (not meeting water quality criteria) in the United States today, as reflected by the number of impairments listed by states in their 303(d) reports (reported by the states to the U.S. EPA under Section 303(d) of the Clean Water Act [CWA]). In 2009, sediment/siltation was listed by the U.S. EPA in its national summary tables as the cause of nearly 6500 impairments, the fourth greatest cause of waterbodies not meeting water quality standards in the United States. Other impacts of channelization include (Schoof 1980):

- Draining of wetlands
- Cutting off oxbows and meanders
- Clearing of floodplain hardwoods
- Lowering of groundwater levels
- Reducing groundwater recharge from the streamflow
- Downstream flooding

Channelization not only impacts the physical characteristics of the stream or river, but also its chemical and biological characteristics.

Channelization has had many positive impacts, and has been and will continue to be a common engineering practice. However, in the past, the maintenance of aquatic habitats has not been a major design goal of channelization projects. In many cases today, however, ecological considerations are a component of many design studies, which is reflected by the number of academic and government institutions that are including *ecohydraulics* in their programs.

3.4 WATERSHEDS

The characteristics of rivers and streams are integrally linked with their watersheds. Therefore, changes in the watershed will have a direct impact on the physical, chemical, and biological characteristics of the receiving water.

The impact of watershed changes may be illustrated by considering the impacts of urbanization. In natural watersheds, the magnitude, duration, and timing of runoff are controlled not only by rainfall but also by the infiltration of water into surface soils, the storage of water in the watershed (such as depression storage), evaporation, and transpiration by plants. Infiltration and storage in forested areas, for example, can be quite high, thereby reducing runoff and increasing the time between peak rainfall and peak runoff, the lag time. With urbanization comes an increase in impervious areas, such as parking areas, sidewalks, roads, and other areas through which the water cannot infiltrate. Since more water runs off, the peak flow of the runoff hydrograph (plot of flow versus time) is greater. Also, since storage is less, the lag time is also less, as illustrated in Figure 3.29. Therefore, the natural flow hydrograph is altered.

Most cities, counties, and state agencies have stormwater regulations or ordinances that prohibit increases in the magnitude of runoff for some design rainfall event as a result of a construction project. For example, Ordinance Number 2006–7, an ordinance establishing stormwater control in the city of Starkville, Mississippi, states that "No development shall be undertaken that increases the rate of surface runoff to downstream property owners or drainage systems." For traditional stormwater engineering, postconstruction peak runoff is reduced using some type of storage facility, such as a detention or retention pond. However, it is the combined effects of all urbanization/construction activities that are reflected in the runoff hydrograph from the watershed to the receiving river or stream, so the cumulative effect of urbanization is an increase in peak flows and a change in the lag

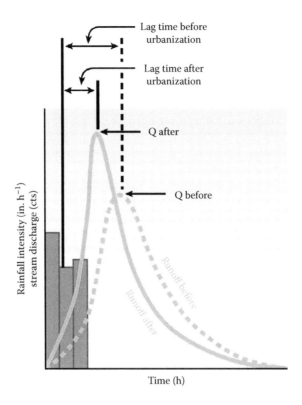

FIGURE 3.29 A comparison of hydrographs before and after urbanization (blue bars indicate rainfall rate and timing). The discharge curve is higher and steeper for urban streams than for nonurbanized streams due to faster and greater runoff. (From FISRWG, *Stream Corridor Restoration: Principles, Processes, and Practices*, Federal Interagency Stream Restoration Working Group, 1998.)

time, which from an ecological perspective may be just as important. Therefore, the natural flow hydrograph is altered.

The impact of urbanization is not only a change in the magnitude, timing, and duration of flows, but also a change in the runoff quality. For example, temperatures as well as sediment loads may be altered. As indicated in Section 3.3.2, a change in the sediment load may result in destabilization of the river or stream, such as changing it from a Stage I to a Stage II system, under the channel evolution model of Simon and Hupp (1986) (Figure 3.21), with the consequence of increasing sedimentation and erosion. Runoff of pesticides and herbicides from lawn maintenance, petroleum products and metals that accumulate on roadways and parking areas, and nutrients from fertilizers and other sources and materials has also increased due to urbanization.

3.5 ABSTRACTIONS AND AUGMENTATION

In addition to instream structures or channel modifications, the flow in rivers is often regulated based on abstractions or augmentation. Abstractions, or withdrawals, result in a reduction in discharge and also, commonly, a change in the timing or shape of the natural flow hydrograph. The common uses of water withdrawals include meeting domestic, agricultural, industrial, and other demands. As indicated in Chapter 1, thermoelectric power (48%), irrigation (34%), public water supply (11%), and industrial uses (5%) account for 98% of the U.S. water use (Hirsch 2006, based on data from Huston et al. 2000). Of these water uses, some may be consumptive (abstracted water not directly returning to the waterbody) and some not. For example, thermoelectric power is used in generating electricity with steam-driven turbine generators. In order to cool water using a condenser, water may be withdrawn from a source, circulated through heat exchangers, and then returned to a surface waterbody (once-through cooling). This practice results in an increase in the release temperature over the influent temperature, and this temperature increase may often be detrimental to some systems. In the 2003 temperature TMDL completed by the Georgia Environmental Protection Division (Georgia EPD 2003) for the Chattahoochee River, it was determined that once-through cooling operations resulted in violations of Georgia's temperature standard, with the resulting recommendation to remove the power plant heat loads from the river. In many areas, once-through cooling is being replaced by evaporating a portion of the cooling tower water and transferring the resultant heat into the air, which is a consumptive use.

The licensing or permitting of abstractions by state regulatory agencies is the primary method of control. However, the form and administration of these regulations differ widely. In part, this may be due to differences in the fundamental water rights resource doctrines adapted by states or regions, such as the doctrine of prior appropriation, also known as the "Colorado doctrine" of water law, common in western states, based on "first in time, first in right" as opposed to the riparian water rights in most eastern states. In the riparian rights system, based on English common law, water is allocated based on ownership of the land about its source.

One common difficulty in the regulation of flows, and the licensing or permitting of abstractions, is determining how much water there is to allocate. In order to manage water systems it is usually necessary to establish a water balance or water budget to determine all water sources and sinks. Instream flows are commonly measured in the United States by agencies such as the U.S. Geological Survey. However, information is often not available for sources and sinks due to tributary sources, precipitation and infiltration, nonpoint source runoff, point sources and abstractions, evapotranspiration, or exchanges of surface and groundwater, all of which are components of the total flow and may impact the effective management of a river or a stream. An additional complication is that the flows, along with the sources and sinks, vary with time at a variety of timescales. Healy et al. (2007) provide additional information and guidance on developing a water budget.

An additional complication is indirect abstractions due to groundwater uses. The pumping of groundwater, such as for municipal and agricultural uses, results in a depression in the original water table of the aquifer, as illustrated in Figure 3.30.

The area or zone of influence depends on the rate of pumping and the characteristics of the aquifer, but it can extend over many miles and the depth of the depression can be many feet. The

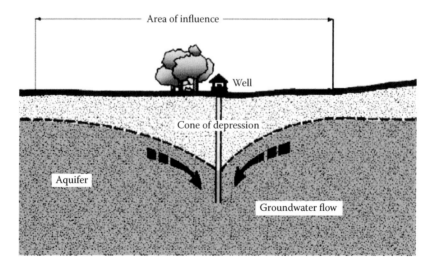

FIGURE 3.30 Cone of depression of groundwater. (From Cornell Cooperative Extension, Cornell University, 1988. With permission.)

result can be a reduction in the recharge of rivers and streams so that, as will be discussed elsewhere, the mean low flows are reduced over time. One mitigation practice in some areas, such as the Mississippi Delta, has been to supplement or augment river flows from other sources to maintain some minimum low flow.

3.6 INTRODUCTION TO U.S. WATER LAW

One of the factors largely affecting regulated rivers is the laws and regulations under which they are regulated. These laws may impact water and water availability and use either directly or indirectly. For example, the availability and use of water within a state is usually a state rights issue, while federal laws and regulations impacting the authorized purpose of reservoirs and navigable waterways within that state also impact the availability of water. As discussed in Chapter 1, the process impacting the availability of water can be described by the hydrologic cycle, a holistic framework, while the management of water is often based on typically fragmented and disconnected laws and regulations, such as those treating surface water and groundwater separately. The following sections will provide some background on the existing laws and regulations.

3.6.1 FEDERAL WATER LAWS

Typically federal laws do not deal directly with water supply. Instead, they may deal with specific issues such as water excess (e.g., protection from flooding), navigation, human health, water quality, and other issues that indirectly impact water supply. Exceptions include laws for water use to fulfill the designated uses of federal lands, including military reservations, national parks, and Indian reservations.

The typical process is that Congress passes laws that are subsequently assigned to some agency for implementation. That agency then has the responsibility for developing regulations to implement the law. The agency drafts regulations and posts them in the *Federal Register* for review. Following revision, the agency posts the final rules in the Code of Federal Regulations. The agency would then enforce the regulations. Following lawsuits, these regulations can be reviewed by the federal courts and, in some cases, overturned where the court determines that the agency did not properly interpret the intent of the law.

The agency responsible for developing regulations to implement federal laws is based on the authorized role and functions of that agency, its legal statutory authority. For example, the authorized civil functions of the Department of the Army Corps of Engineers pertain to (river and harbor) commercial navigation, flood and storm damage reduction, aquatic ecosystem restoration, and related efforts. Water supply is not an authorized function of the Corps. Several statutes also provide regulatory authority to the U.S. EPA, whose primary responsibilities have evolved to include the regulation of air quality, water quality, and chemicals in commerce; the development of regulatory criteria for the management and disposal of solid and hazardous wastes; and the cleanup of environmental contamination.

3.6.1.1 CWA

The CWA is a cornerstone law for the protection of surface water quality in the United States. The law deals primarily with the control of pollutant discharges to waterbodies, so although primarily targeted at the quality rather than the quantity of water, the flow in rivers is impacted by those discharges. For example, the relatively constant flows from wastewater treatment facilities have turned some ephemeral streams in the arid West into perennial streams.

The stated goal of the CWA is to restore and maintain the chemical, physical, and biological integrity of the nation's waters so that they can support "the protection and propagation of fish, shellfish, and wildlife and recreation in and on the water." The CWA does not deal directly with either groundwater or water quantity issues; generally, the agency concentrates on the "chemical" and "biological" components of the integrity of the nation's waters. However, the "physical" component of this integrity is critical, in that you cannot have good water quality or the "propagation of fish, shellfish and wildlife" if you do not have an adequate quantity of water. In the future, the agency may focus more on the physical component or water quantity. For example, the U.S. EPA's (2012) "Safe and Sustainable Water Resources, Strategic Research Action Plan 2012–2016" problem statement states:

> Increasing demands for sources of clean water, combined with changing land use practices, growth, aging infrastructure, and climate change and variability, pose significant threats to the Nation's water resources. Failure to manage the Nation's waters in an integrated, sustainable manner will limit economic prosperity and jeopardize human and aquatic ecosystem health.

In addition, the CWA has an indirect impact on water use if that use impacts water quality. For example, the federal courts have upheld that state administrators have a "public trust duty" to exercise authority to condition appropriations to accomplish water quality goals (e.g., in *California U.S. v. State Water Resources Control Board* [1986], commonly known as the *Racanelli* decision). In Mississippi, state law states "No use of water shall be authorized that will impair the effect of stream standards set under the pollution control laws of this state based upon a minimum stream flow."

3.6.2 STATE LAWS

Traditionally in the United States, water rights have been a state's right issue. That is, each state controls the water within that state. In general, water laws have been based on either statutory law (resulting from legislative action) or common law (resulting from court cases and judicial decisions). The states' rights issues result in differences in water law between states and often differences in water law between surface water and groundwater.

3.6.2.1 Surface Water Law

In the United States, four types of water law are used to control water and resolve conflicts between water users:

1. Riparian rights—which are most commonly used in eastern states, where water is more "plentiful," assign rights to water use to landowners and require equal sharing among users in times of shortage.
2. Appropriation system—which is dominant in the western states and allows the severance of water rights from landownership and allocates available water in order of priority. Priority of water use goes to those having senior water rights, those who used the water first (first in time, first in right).
3. Hybrid system—which contains elements of both systems.
4. Public Trust Doctrine—affirms public and state ownership of certain natural resources that benefit all.

3.6.2.1.1 Riparian Rights

Under riparian rights, the landowner adjacent to or abutting a stream (the riparian) has the rights to the natural flow of the stream at that location. That is, riparian water rights only entitle the owner of those lands to use the water. The "lands abutting" may be determined typically using one of two rules: (1) by lands that have been held as a single tract through the chain of title (so that any non-abutting lands sold or not part of the original title lose their riparian right forever, the source of title test); or (2) lands contiguous to the abutting tract are entitled to water use if they are all under single ownership, regardless of when they were acquired (unity of title rule). Since the rights go with the land, the rights transfer with the land when it is sold, subject to the aforementioned definitions of "lands abutting."

Riparian rights are based on the English rule of natural flow under which each landowner has the right to have the water flow past his lands undiminished in either quantity or quality. However, in the United States, rather than unrestricted rights, the use of water rights is limited to some "reasonable use," and that such use also must not significantly limit or damage the use by other riparians. So, for example, some downstream user could sue an upstream user if that use harms or prevents him or her from making use of the water, but the downstream user would have to prove not only that he or she was harmed, but also that the upstream water use was unreasonable. Riparian rights then view water as a public resource; however, in times of conflict, the problem may be to determine what a reasonable use is.

One problem associated with riparian rights is how to determine reasonable use and how to regulate and allocate water during times of shortage and water uses removed from their source. One of the tactics taken by a number of states is to implement a permit system, where a water user must obtain a permit from the state and where the permit system establishes by law what a reasonable use is. So, since riparian rights view water as a public resource, the waters then become the property of the state and are allocated by the state using permits, rather than the property of the riparian landowner.

3.6.2.1.2   Prior Appropriation

The water law of a number of Western states, where water is a scarce commodity, was based on water uses (gold mining, irrigation, etc.), which required large amounts of water for activities removed from the watercourse (not necessarily adjacent to it). So, the water rights went to those who first physically took the water (diverted it) from the watercourse for some beneficial use, and for however much water was needed for that use rather than some reasonable use. Thus, the ownership of water was not based on the ownership of the land but the use of the water. The first to use the water became the user with senior rights, and the use could be removed from the source (not necessarily riparian). Those with "senior" rights must then be satisfied before other users, who have "junior" rights. Since the ownership then goes with the user, rather than the land, water rights can be sold. An old saying in the Western United States is that "water flows uphill—towards money," since under this system water can be bought, sold, and leased for a profit and the person holding the water right gets to set the price and transfer the right to the highest bidder. Also, under this rule, removing water from one watershed to be used in another, or interbasin transfer, is generally permitted.

3.6.2.1.3   Hybrid Rights

Hybrid rights are somewhere in between the aforementioned water doctrines. For example, California is a hybrid state, which incorporates prior appropriation for allocating water rights among users on public land and riparian rights between landowners.

3.6.2.1.4   Public Trust Doctrine

The Public Trust Doctrine originates from Roman, Spanish, and English common law (Blumm 1995) based on the principle of public and state ownership of certain natural resources, such as water, that benefit all, with the state being the trustee to manage those resources. The state can then allocate those resources, commonly through permits. The Public Trust Doctrine is included in the water laws of a variety of states, stating for example that all waters:

- "belong to the public" (New Mexico)
- "are the property of the state for the use of its people" (Idaho)

- "property of the public" (Colorado)
- "property of the state for the use by its people" (Montana)
- "are reserved to the people for the common good" (Alaska)
- "held in trust for the public" (Texas)

3.6.2.2 Groundwater Law

3.6.2.2.1 Groundwater Doctrines

There are multiple standards or systems of water rights on which individual state groundwater laws are based. Some of the systems of water rights, as modified from Bowman (1991) and the Water Systems Council's (WSC) "Who Owns the Water" (www.watersystemscouncil.org) are:

- The Absolute Ownership Rule: This rule is based on English common law (Marvin and Little 2010) "specifically the doctrine of ad coelum, which says that a property owner is vested with property rights in all of the sky above his property up to the heavens and everything beneath his property to the center of the earth." This rule permits a landowner to intercept groundwater that would otherwise have been available to a neighboring water user and even to monopolize the yield of an aquifer without incurring liability. This is also, in some cases, somewhat modified and referred to as the "rule of capture" or the law of the biggest pump, which allows the legal pumping of whatever groundwater is available regardless of the impact that such pumping may have on neighboring users.
- The Reasonable Use Rule: This rule limits a landowner's use of water to those uses that have a reasonable relationship to the use of the overlying land. The rule is essentially the rule of absolute ownership with exceptions for wasteful and off-site use.
- Correlative Use Rule: This rule maintains that the authority to allocate water is held by the courts. The owners of overlying land and the nonowners or transporters have coequal or correlative rights in the reasonable and beneficial use of groundwater. A major feature of this doctrine is the concept that adjoining lands can be served by a single aquifer. Therefore, the judicial power to allocate water permits protects both the public's interest and the interests of private users.
- American Law Institute (ALI) Restatement of Torts Rule: The ALI redefined the rule of reasonable use, resulting in this doctrine, which holds that a landowner who uses groundwater for a beneficial purpose is not subject to liability for interference if certain conditions are met. The water withdrawal cannot cause unreasonable harm to a neighbor by lowering the water table or reducing the artesian pressure, it cannot exceed a reasonable share of the total store of groundwater, and it cannot create a direct and substantial effect on a watercourse or lake.
- The Prior Appropriation Rule: This rule maintains that the first landowner to beneficially use or divert water from a water source is granted priority of right. The amount of groundwater this priority, or senior, appropriator may withdraw can be limited based on reasonableness and beneficial purposes. Some states have replaced or supplemented the Prior Appropriation Doctrine with a permit system.

3.6.2.2.2 Tragedy of the Commons

One of the issues associated with groundwater can be described by the "tragedy of the commons":

> Freedom in the commons brings ruin to all, since every person is compelled to increase his/her individual benefit without limit. (Hardin 1968)

"Groundwater depletion is a logical consequence of a commons (a natural resource used jointly by many stakeholders) exploited in the absence of regulation or sustainable practices. As with any other natural resource held in common, an aquifer tends to be viewed by individuals pursuing their own self-interests as a resource to be exploited before others are able to get to it" (Ponce 2006). Some states, for example, California, have provisions to the water doctrines called "use it or lose it," such that if the owner does not use the water, someone else may acquire prescriptive rights. Classic examples resulting from the tragedy of the commons include:

- Borrego Valley, in Southern California's high desert, near Borrego Springs, in San Diego County. Water levels in Borrego Valley have declined 2 ft. per year (0.6 m yr^{-1}) over the past 20 years.
- South-central Arizona, where groundwater pumping to support the population growth, including the Tucson and Phoenix areas, has resulted in water-table drops of between 300 and 500 ft. (90 and 150 m) and land subsidence of as much as 12.5 ft. (3.8 m).

3.6.2.3 Sustainability, Conjunctive Use, and Water Law

One of the foundations for holistic water management is the hydrologic cycle, discussed in Chapter 1, and the connectivity between each of the components of that cycle. One of those components is the connectivity between surface water and groundwater as it impacts the flows in streams and rivers (see Figure 3.30). Another issue in water management and law is achieving the goal of sustainable use. Sustainability has been variously defined, but one definition applicable here is:

Sustainability is the ability to meet current needs without compromising the ability of future generations to meet their needs.

Generally, to attain sustainability a holistic approach to water law and management is required, or, at a minimum, recognizing the connectivity between surface water and groundwater, commonly referred to as conjunctive use or management (OWRB 2012).

States generally have water laws that impact both surface water and groundwater. However, states may or may not recognize the connectivity between the two, or have laws that allow for conjunctive use. Texas, for example, has a surface water permit based where the surface water is publicly owned and governed by the state of Texas and a permit from the state is required for any water use other than for domestic or livestock use. However, unlike surface water, in Texas the groundwater is the property of the landowner, under the rule of capture or the absolute ownership rule. Similarly, Louisiana also uses the rule of capture for regulating groundwater.

As an example, one of the more ecologically important rivers in southern Arizona is the San Pedro River, located just south of Sierra Vista. It is the last major, free-flowing undammed desert river in the American Southwest, and hosts two-thirds of the avian diversity in the United States. It also includes the 40-mile long San Pedro riparian natural conservation area created by Congress in 1988 (Figure 3.31). Groundwater depletion in the area has long been an issue, impacting the flows in the river. A plan for a 5900-home subdivision in the area was proposed, but a number of agencies and organizations opposed the plan because of its impact on groundwater depletion and on the river flows. However, in Arizona, water laws for surface water and groundwater are separate and do not recognize connectivity between surface water and groundwater and do not consider environmental needs (Megdal, et al. 2011). Under Arizona's Groundwater Management Act (GMA; Megdal et al. 2011), developers are required to demonstrate an assured water supply, which, according to this act, is to demonstrate that "a 100-year water supply to satisfy the subdivision's needs is physically, legally, and continuously available." A study by the developer, as reported in the *Arizona Daily Star* (August 30, 2012), indicated that the projected impact of the development would not lower the water table by more than 1200 ft. over the next 100 years. This met the requirements of the GMA, and since environmental impacts and connectivity could not be considered by law, as reported by the *Arizona*

FIGURE 3.31 Lower San Pedro River in Pinal County, Arizona, near the confluence with Aravaipa Creek. (From USFWS, June 2008 photograph by Carrie Marr, FWS.)

Star (July 24, 2012), the Arizona Department of Water Resources had no grounds on which to deny the permit and it was granted.

However, states laws can change, as illustrated by Bowman (1991), who followed trends in water law in eight midwestern states (Iowa, Illinois, Indiana, Michigan, Minnesota, Missouri, Ohio, and Wisconsin), between 1980 and 1987. While in 1980, several of these states had groundwater laws based on absolute ownership, none had such groundwater laws by 1987 and all were based on either statutory law or common law based on reasonable use and restatement of torts.

3.6.3 INTERSTATE WATER DISPUTES

One of the issues regarding water law and water rights is water flowing between states. State water laws, where statutes indicate that the waters are owned by the state, typically indicate that those waters are protected for use by the people of that state. But, in the use of that water, the state may impact the downstream states, and deprive them of water. Since states have no jurisdiction over water in their neighboring states, these conflicts are typically resolved either by congressional action, by interstate compact, or by the U.S. Supreme Court (Clemons 2004).

For example, the Georgia Water Control Act states that:

> O.C.G.A. §12-5-21 (a) 'The people of the State of Georgia are dependent upon the rivers, streams, lakes, and subsurface waters of the state for public and private water supply and for agricultural, industrial, and recreational uses. It is therefore declared to be the policy of the State of Georgia that the water resources of the state shall be utilized prudently for the maximum benefit of the people.'

The problem is that several of the major river basins in Georgia (the ACT and ACF basins, Figure 3.32) flow into Alabama and then Florida. The waters taken to support Georgia, such as the city of Atlanta, can then impact the environmental quality of estuaries in Florida. The result is what is generally known as the "tri-state water wars." The conflicts began in earnest in 1988 based on a proposed reallocation of the waters in Lake Lanier, on the Chattahoochee River in Georgia, by the U.S. Army Corps of Engineers, which initiated a series of lawsuits by Alabama and Florida. To try to resolve the issue, the three states and the Corps agreed to a comprehensive study of all of the water issues affecting the ACF and ACT basins in order to determine a fair allocation of water resources, which resulted in a halt to the legal battles. From these studies, two compacts were created for the

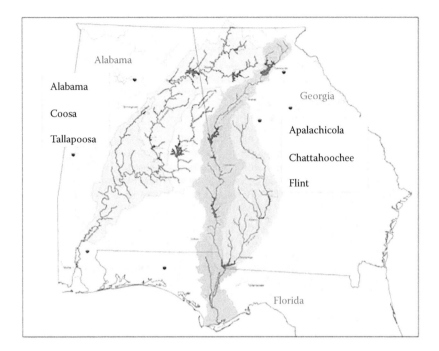

FIGURE 3.32   ACT and ACF basins. (From Atlanta Regional Commission, Tri-State Water Wars Resource Center.)

ACT and ACF basins, which were officially ratified by Congress in 1997. However, the states could not reach an agreement on these compacts and they have since expired without resolution, in 2003 for the ACF and in 2004 for the ACT, after which the legal battles and tri-state water wars resumed.

An example for groundwater is illustrated by a conflict between Mississippi and Memphis, Tennessee. Mississippi state water law (Code of 1972 as amended) states that:

§ 51-3-1 'All water, whether occurring on the surface of the ground or underneath the surface of the ground ... belong to the people of this state,'

A problem is that the city of Memphis, Tennessee, through its pumping of groundwater, creates a cone of depression that results in groundwaters flowing from Mississippi to Tennessee. In November 1998, the Memphis newspaper, *The Commercial Appeal*, reported that Memphis "is the largest user of Mississippi's ground water." In February 2005, Mississippi's attorney general filed an action against Memphis and Memphis Light, Gas & Water (MLGW) in the U.S. District Court for the Northern District of Mississippi, claiming that over a 40-year period MLGW diverted and unlawfully took over 363 billion gallons of groundwater owned by Mississippi to provide between 15% and 22% of Memphis' water supply and sales requirements and claiming 1.3 billion dollars in damage. In 2010, the suit went to the U.S. Supreme Court, which refused to hear the case without comment, but the water wars continue.

3.7   MANAGEMENT ALTERNATIVES

The majority of our streams and rivers are, and will remain, regulated. However, today, management efforts are directed toward restoring the remaining free-flowing rivers and streams or restoring regulated streams.

3.7.1 PRESERVATION

The preservation of existing free-flowing rivers is an alternative. A few of our free-flowing rivers have been protected and preserved as wild and scenic rivers under the National Wild and Scenic Rivers System:

> It is hereby declared to be the policy of the United States that certain selected rivers of the Nation which, with their immediate environments, possess outstandingly remarkable scenic, recreational, geologic, fish and wildlife, historic, cultural or other similar values, shall be preserved in free-flowing condition, and that they and their immediate environments shall be protected for the benefit and enjoyment of present and future generations. The Congress declares that the established national policy of dams and other construction at appropriate sections of the rivers of the United States needs to be complemented by a policy that would preserve other selected rivers or sections thereof in their free-flowing condition to protect the water quality of such rivers and to fulfill other vital national conservation purposes. (Wild & Scenic Rivers Act, October 2, 1968)

According to the U.S. Fish and Wildlife Service, Wild and Scenic Rivers program, as of 2008, the national system protects more than 11,000 miles of 166 rivers in 38 states and the Commonwealth of Puerto Rico. However, that is little more than one-quarter of 1% of the nation's rivers.

3.7.2 NATURALIZATION OR REHABILITATION

A management alternative for regulated or disturbed rivers may be to restore or rehabilitate them. While the terms are often used interchangeably, the goals of restoration or rehabilitation are often quite different. The goal of restoration is usually to reestablish the general structure and function of the river to that existing prior to the disturbance, the preexisting condition. Restoration typically requires a detailed understanding of the structure and function of not only the river but also its watershed prior to the disturbance. A problem is that regulation is so pervasive that many of the characteristics of natural rivers, such as the natural flow hydrograph, are poorly known or understood, thus making restoration problematic. An alternative strategy may be rehabilitation. To rehabilitate is to restore to useful purpose, or to make habitable again. Restoration and rehabilitation strategies are the subject of Chapter 8.

3.7.3 MANAGEMENT

As indicated at the beginning of this book, a common theme is often "one generation's solution is the next generation's problem." River regulation is a common solution that has existed for many generations and by far the majority of our rivers and streams are regulated to some degree. The regulation of rivers has provided a variety of benefits, including flood control, navigation, hydroelectric power, water supply, and other uses. However, the impact of regulated rivers is an ecological system that is often far removed in its characteristics from naturally occurring rivers. These systems are already altered from their natural condition, so preservation is not an option, nor is naturalization or restoration for the majority of regulated rivers. The only alternative is management since "like it or not we control the destiny of these streams" (Collier et al. 2000).

As indicated in "Dams and rivers: A primer on downstream effects of dams" (Collier et al. 2000), river managers have traditionally concentrated on efficiency, sometimes to the neglect of instream environmental requirements. For the majority of regulated rivers, engineering goals will continue to dictate the river engineering design and management procedures. However, these river systems, although regulated, do have an intrinsic environmental value. That environmental value can be enhanced if we choose to enhance and manage for those values.

River engineering design and management are often based on some optimization of the complex and diverging interests of water users, such as for navigation, hydropower, and water supply. Often, river ecology has not been a consideration in that design and optimization. To do so in some cases will require changes not only in attitudes, but also in the laws, regulations, codes, ordinances, and other rules that control and dictate river engineering design and management. For example, the rule curve by which a reservoir is operated may not have included fisheries management in the reservoir and its tailwaters during the rule curve development. Therefore, reservoir operators may then be precluded from operating that reservoir for fisheries management without a change in that rule curve, for some federal facilities this would require a reauthorization by Congress. Also, regulated rivers are impacted not only by instream modifications but also by modifications and developments in the watershed. In some cases, for example, local ordinances may preclude the use of some environmentally friendly management practices in watersheds, in favor of more traditional methods of handling stormwater runoff. In some cases, engineers would also need to be trained or retrained in methods and techniques to better consider environmental values in river design and management.

One recently emerging field of study that addresses "this generation's problem" with respect to the watershed is *low-impact development* (LID). In traditional development practices, the fraction of impervious area in the watershed is increased, with consequent changes in the magnitude and timing of runoff events. Traditionally, runoff is collected and conveyed through a series of impervious surfaces in storm drains and detention/retention ponds until the stormwater is treated, if necessary, and is finally discharged into the receiving stream. The goal of LID is to mimic a site's predevelopment hydrology to the extent possible by minimizing the impervious area and maximizing infiltration, using a series of best management practices (BMPs). The BMPs could be green roofs, infiltration trenches, porous pavements, or other practices designed to infiltrate, filter, store, evaporate, and detain runoff close to its source. LID can be a critical component of river management and/or rehabilitation.

A second emerging field of study that addresses "this generation's problem" is *ecohydraulics*. Ecohydraulics is becoming increasingly common in government programs and as a field of study in academic institutions. As an example of the objectives of ecohydraulics, the following is a list of the aims of the Eco Hydraulics Committee, Environmental and Water Resources Institute, American Society of Civil Engineers:

1. To promote the study of the fluid mechanics of ecosystems, with the aim of improving our understanding of such systems, and hence improving our ability to reliably predict and ameliorate the impact of human intervention in these ecosystems.
2. To encourage the transfer and application of understanding of fundamental transport processes and the consequent advancement of engineering practice and technology in the solution of problems with environmental consequences.

FIGURE 3.33   River. (From the American Society of Civil Engineers.)

3. To enhance communication and collaboration between practitioners in relevant technical fields, notably, hydraulicians, geologists, environmental chemists, and biologists, as well as the various professional societies, the broader community, including regulators, and society at large. Collective technical know-how is needed to best manage the water environment, including wetlands and small streams, such as that shown in Figure 3.33.

REFERENCES

ASDSO. 2009. Investment in infrastructure: Focus on dams, ASDSO Press Release. Association of State Dam Safety Officials, Lexington, KY.

Bennett, S.J. and F.E. Rhoton. 2003. Assessing sedimentation issues within the large woody debris plug along the Yalobusha River, Calhoun County, MS. USDA-ARS, National Sedimentation Laboratory Research Report. No. 34, p. 111.

Biedenharn, D.S., C.M. Elliott, and C.C. Watson. 1997. *The Stream Investigation and Streambank Stabilization Handbook*. U.S. Army Corps of Engineers, Waterways Experiment Station, Vicksburg, MS.

Blumm, M.C. 1995. Mono Lake and the evolving public trust in western water. *37 Arizona Law Review* 701, 709–714.

Brice, J.C. 1982. Stream channel stability assessment. Report No. FHWA-RD-82-021, Federal Highway Administration, U.S. Department of Transportation, Washington, DC, p. 45.

Brookes, A. 1981. Channelization in England and Wales, Discussion Paper. Geography Department, Southampton University.

Capello, S.V. 2008. Modeling channel erosion in cohesive streams of the Blackland Prairie, Texas at the watershed scale. Master's thesis, Baylor University.

Clarkin, K., A. Connor, M.J. Furniss, B. Gubernick, M. Love, K. Moynan, and S.W. Musser. 2005. National inventory and assessment procedure for identifying barriers to aquatic organism passage at road-stream crossings. U.S. Department of Agriculture Forest Service, National Technology and Development Program, San Dimas, CA.

Collier, M., R.H. Web, and J.C. Schmidt. 2000. Dams and rivers: A primer on downstream effects of dams. USGS Circular 1126, U.S. Geological Survey.

Darby, S.E. and A. Simon (eds). 1999. *Incised River Channels: Processes, Forms, Engineering, and Management*. Wiley, Chichester.

Derrick, D.L., T.J. Pokrefke Jr., M.B. Boyd, J.P. Crutchfield, and R.R. Henderson. 1994. Design and development of bendway weirs for the Dogtooth Bend reach, Mississippi River. Technical Report HL-94-10, U.S. Army Corps of Engineers, Waterways Experiment Station, Vicksburg, MS.

FHWA. 1995. *Culvert Repair Practices Manual*, Volume I, FHWA-RD-94-096. Federal Highway Administration, Washington, DC.

FISRWG. 1998. *Stream Corridor Restoration: Principles, Processes, and Practices*, Federal Interagency Stream Restoration Working Group. GPO Item No. 0120-A; SuDocs No. A 57.6/2:EN 3/PT.653.

Georgia EPD. 2003. Chattahoochee River temperature TMDL. Georgia Department of Natural Resources, Environmental Protection Division, Atlanta, GA.

Healy, R.W., T.C. Winter, J.W. LaBaugh, and O.L. Franke. 2007. Water budgets: Foundations for effective water-resources and environmental management. U.S. Geological Survey Circular 1308, p. 90.

Hirsch, R.M. 2006. USGS science in support of water resources management. Presented at Mississippi Water Resources Conference, Jackson, MS.

Hutson, S.S., N.L. Barber, J.F. Kenny, K.S. Linsey, D.S. Lumia, and M.A. Maupin. 2000. Estimated use of water in the United States in 2000. USGS Circular 1268, U.S. Geological Survey.

Langendoen, E.J. 2009. Concepts: Conservational channel evolution and pollutant transport system. Workshop presented at Mississippi State University.

NRC. 2002. Restoration of aquatic ecosystems: Science, technology, and public policy. Committee on Restoration of Aquatic Ecosystems: Science, Technology, and Public Policy. National Research Council, Washington, DC.

Ramirez, J.J. 2011. Assessment and prediction of streambank erosion rates in a southeastern plains ecoregion watershed in Mississippi. PhD dissertation in Civil and Environmental Engineering, Mississippi State University.

Raymond Jr., L. 1988. What is groundwater? New York State Water Resources Institute, Center for Environmental Research, Cornell University. Bulletin No. 1.

Rosgen, D.L. 2001. A practical method of computing streambank erosion rate. In *Proceeding of the 7th Federal Interagency Sedimentation Conference, Reno, Nevada, March 25–29 2001*, Vol. 2. Inter-agency Committee on Water Resources, pp. 9–15.

Schoof, R. 1980. Environmental impact of channel modification. *Journal of the American Water Resources Association* 16 (4), 697–701.

Schumm, S.A. 1984. River morphology and behavior: Problems of extrapolation. In C.M. Elliott (ed.), *River Meandering*. American Society of Civil Engineers, New York, pp. 16–29.

Shoffstall, G.D. 2009. USACE levee safety program. New York State Emergency Management Association Winter Conference.

Shields Jr., F.D., S.R. Pezeshki, G.V. Wilson, W. Wu, and S.M. Dabney. 2008. Rehabilitation of an incised stream with plant materials: The dominance of geomorphic processes. *Ecology and Society* 13 (2), 54.

Simon, A. and C.R. Hupp. 1986. Channel evolution in modified Tennessee channels. In *Proceedings of the Forest Federal Interagency Sedimentation Conference*, Las Vegas, NV. U.S. Interagency Advisory Committee on Water Data, Washington, DC, pp. 71–82.

Simon, A., R. Kuhnle, W. Dickerson, and M. Griffith. 2002. "Reference" and enhanced rates of suspended-sediment transport for use in developing clean-sediment TMDLs. In *Total Maximum Daily Load (TMDL) Environmental Regulations: Proceedings of the March 11–13, 2002 Conference,* Fort Worth, Texas. ASAE, St. Joseph, MI, pp. 151–162.

Taylor, R.N. and M. Love. August 2001. Fish passage evaluation at stream crossings. Prepared for California Department of Fish and Game.

Tibbets, M.N., C.L. Morris, and R.A. Wurbs. 1985. Optimum reservoir operation for flood control and conservation purposes. Texas Water Resources Institute TR-137, College Station, TX.

USACE. 2000. Design and construction of levees, EM 1110-2-1913. U.S. Army Corps of Engineers, Waterways Experiment Station, Vicksburg, MS.

USDA Forest Service. 1999. Roads analysis: Informing decisions about managing the National Forest transportation system. Misc. Rep. FS-643. U.S. Department of Agriculture Forest Service, Washington, DC.

Washington, F.W. 2003. Design of road culverts for fish passage. Washington Department of Fish and Wildlife, Olympia, WA.

Wurbs, R.A. and W.P. James. 2002. *Water Resources Engineering*. Prentice Hall, Upper Saddle River, NJ.

WWF. 1986. Free-flowing rivers: Economic luxury or ecological necessity? World Wildlife Fund Project No. 2085/February 2006.

4 Flows and Transport in Rivers
Measurement and Analysis

4.1 INTRODUCTION

Rivers are lotic systems (lotus, from *lavo*, to wash) and are generally dominated by flows, as opposed to lakes and reservoirs, which are lentic systems (*lenis*, to make calm). The flows in rivers have profound impacts on their physical, chemical, and biological characteristics. Therefore, the measurement and analysis of flows are a critical component of river management.

Flow simply refers to the flux of water passing though some defined area, such as a cross section of river, as illustrated in Figure 4.1, during some finite interval of time, in units of cubic length per time (e.g., cubic meter per second and cubic foot per second). The flow (Q), or flux, is the product of the water velocity (U, length/time) and the area (A, length^2); (Q = UA). To determine the flow, the velocity and area are typically measured and then used to compute it.

While theoretically simple, in practice the "devil is in the details." For example, velocities are not constant over the channel width or depth (cross section), so multiple area and velocity measurements are required to compute flows. One question then is: how many measurements are required and at what accuracy? Also, the variations in velocities over the cross section and length of the river have large impacts on the transport of materials carried by the flow.

Another part of the analysis of flows is determining which flows or which flow components are important. In many cases, continuous measurements are required and/or measurements corresponding to a water quality sampling event. Flows are required for water quantity/quality management in order to determine a flow budget or flow balance, to determine, for example, how much water is entering and leaving a river reach (section of river). Flows are required in water quality studies to determine both the flux of materials into a system and the transport of materials through it. The load or flux of materials (mass/time) into a river reach is computed from Q*C, where Q is the flow ($L^3 T^{-1}$) and C is the concentration of some material (mass L^{-3}) (suspended solids, dissolved oxygen, nutrients, etc.). So, if the flows are not known or not determined as part of a field study, the loads (fluxes) cannot be determined. For example, if the determination of monthly or yearly average loads into a river reach were of importance for addressing some management question, then both flows and concentration data would be required at a sufficient frequency to compute an average. Similar to a water budget, it is often necessary to construct a budget of materials of water quality interest, such as sediments, which requires both sediment and flow data at all points of inflow or outflow. Flows also impact the transport of materials and that transport is a fundamental process having large impacts of water quality variations (Martin and McCutcheon 1999).

The flows that are typically used for hydraulic design or the determination of regulated flows are some relatively rare flows as characterized by their flow magnitude and return interval. For example, in studies of floods, the 100-year flow is often used, so flow data are required to estimate the flow magnitude that would occur once every 100 years. For water quality studies, it is often the low flow that is important, such as the 7Q10 flow. The 7Q10 flow is the 7-day average flow that is expected to occur once every 10 years. For biota or for regulated flows, often some minimum flow is used. That minimum flow could be protective of biota or a flow that satisfies some downstream demand. More commonly today, for the protection of biota or to determine whether some hydrologic alteration has taken place, flow components include not only the magnitude and return interval, but also the duration, timing, and rate of change of flows (Poff et al. 1997; Richter et al. 1996,

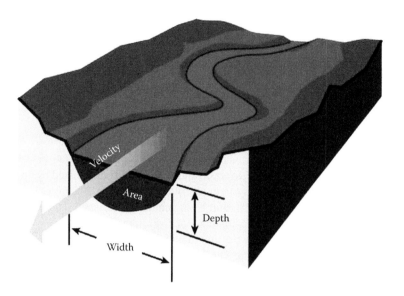

FIGURE 4.1 Characterization of flow. (From FISRWG, *Stream Corridor Restoration: Principles, Processes, and Practices*, Federal Interagency Stream Restoration Working Group, 1998.)

1997). The estimation of the flow components requires the availability of historical data and often some assumptions related to the underlying statistical distribution of the flows.

This chapter introduces the methods for the measurement and analysis of flows. First, the methods for the measurement of flows are presented, followed by the methods for analysis.

4.2 WATERSHED IMPACTS

The hydrology of a watershed is one of the primary factors influencing the physical and biological characteristics of rivers and streams. The impacts vary with the watershed's geology, including its topography, land use, and land cover. The magnitude and the duration of rainfall also have a profound impact, as illustrated by the typical rainfall runoff hydrograph illustrated in Figure 4.2. During periods

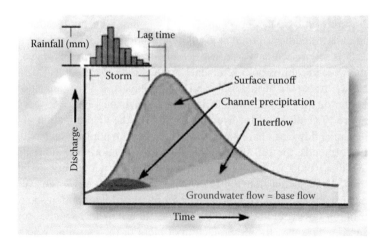

FIGURE 4.2 Stream hydrograph. (From Hebert, P.D. (ed.), Canada's Aquatic Environments, [Internet] CyberNatural Software, University of Guelph, 2002. Available at http://www.aquatic.uoguelph.ca/. With permission from the Biodiversity Institute of Ontario.)

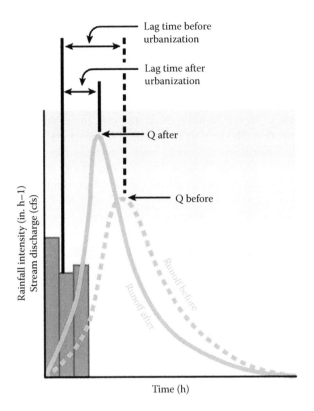

FIGURE 4.3 Comparison of hydrographs before and after urbanization. (From FISRWG, *Stream Corridor Restoration: Principles, Processes, and Practices*, Federal Interagency Stream Restoration Working Group, 1998.)

when rain does not occur, the flow from groundwater discharge or other inflow processes may make up the base flow. When rain does occur, rain falling on the portion of the watershed closest to the outlet runs off first, and rain falling on the most hydraulically distant part of the watershed will run off last. If rain continues to fall at the same magnitude, the runoff hydrograph will continue to increase (rising limb of the hydrograph). If rainfall continues to fall at the same rate, then the runoff would be constant as equilibrium is reached between the rainfall and runoff. Many of the simple methods for predicting runoff are based on the prediction of that peak equilibrium flow. When the rain stops, then the runoff will gradually return to the base flow, which is the falling limb of the hydrograph. However, in reality, for large watersheds, the shape of the hydrograph will depend on the shape of the basin, its land uses, and other factors impacting runoff, as well as the shape of the rainfall hyetograph. Also, watersheds are not static and changes in land uses, such as urbanization, can change the magnitude of the peak runoff and the shape of the hydrograph, as illustrated in Figure 4.3. A variety of techniques are available to estimate the runoff magnitude and duration as a function of watershed characteristics and these techniques are summarized in any good book on water resources engineering (e.g., Wurbs and James 2002) or hydrology (e.g., Bedient and Huber 2008).

4.3 STAGES OF MEASURING FLOW

4.3.1 LOCATION, LOCATION, LOCATION

The first step in measuring flows is to identify a suitable measurement location. The site should be accessible, for example, and, of course, it should represent a location where flow measurements are needed. The location could be different depending on the intended use of those measurements.

A typical practice is to obtain a series of direct measurements of flows and stages (water surface elevations) at a selected location. The direct measurement of flows is time consuming and expensive as compared to the measurement of water surface elevations or stages. A common practice is to use a series of direct measurements to obtain a relationship between flows and stages, which is known as a stage–discharge relationship or a rating curve (see Figure 4.4). Once established, the stage is measured and the rating curve is used to indirectly estimate the flows.

For a specific location to be usable for such an analysis, there must be a unique relationship between the stage and the flow. In many (if not most) areas of natural channels, such as in pools, behind dams (backwater areas), in tidal zones, and elsewhere, there is no unique relationship between water surface elevations and flows. Locations where there is such a unique relationship are known as control points, and the factors governing the relationship are known as controls. The control could be, for example, an artificial structure (artificial control), a change in channel shape such as downstream constriction (section control), or channel geometry and roughness, or other factors (channel control). The specific controls may also vary with the stage (see Rantz et al. [1982] and Kennedy [1984] for a more detailed explanation). Since in many (if not most) areas in rivers and streams there is no control, the selection of the location is critical.

According to Rantz et al. (1982), the ideal location is one that satisfies the following criteria:

- The general course of the stream is straight for about 300 ft. (approximately 100 m) upstream and downstream from the gage site.
- The total flow is confined to one channel at all stages, and no flow bypasses the site as subsurface flow.
- The streambed is not subject to scour and fills and it is free of aquatic growth.
- Banks are permanent, high enough to contain floods, and are free of brush.
- Unchanging natural controls are present in the form of a bedrock outcrop or other stable riffle for low flows and a channel constriction for high flows or a falls or cascade that is unsubmerged at all stages.
- A pool is present upstream from the control at extremely low stages to ensure a recording of the stage at extremely low flows, and to avoid high velocities at the streamward end of gauging-station intakes during periods of high flows.

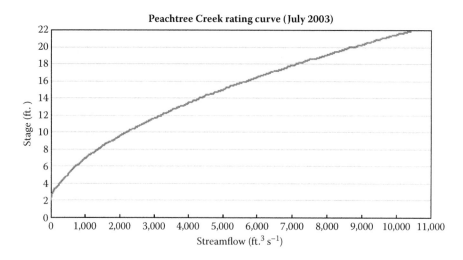

FIGURE 4.4 An example of a rating curve. (From USGS, Available at http://ga2.er.usgs.gov/peachtree/streamflow.cfm.)

- The gage site is far enough upstream from the confluence with another stream or from a tidal effect to avoid any variable influence from other streams or one that the tide may have on the stage at the gage site.
- A satisfactory reach for measuring the discharge at all stages is available within reasonable proximity of the gage site. (It is not necessary to measure low and high flows at the same stream cross section.)
- The site is readily accessible for easy installation and operation of the gaging station.

An ideal site is rare. Also, the controls and cross-section geometry may change in response to high-flow events or other factors, affecting the rating curve, as illustrated in Figure 4.5. Therefore, the rating curve must be updated periodically.

The accuracy of indirect estimates depends on the accuracy of the flow measurement, the accuracy of the rating curve, the completeness and accuracy of the gage-height record, and the degree to which changes occur, as depicted in Figure 4.5. The accuracies of the U.S. Geological Survey (USGS) discharge records for individual days are typically considered to be on the order of 90%–95% (errors of 5%–10%; Hirsch and Costa 2004), but can be much less in some locations.

4.3.2 Measurement of Morphometry

Once the location is identified, a field survey is typically done to measure the elevation of the channel bottom from some vertical datum at a series of horizontal locations across the channel, also georeferenced. General survey details can be found in Benson and Dalrymple (1984) and Kennedy (1990). Since the site may need to be rechecked periodically, a permanent benchmark should also be established to tie the cross-section and longitudinal profiles to an elevation control for future comparison. Channel surveys are also a critical component of river management, and the measurements taken will vary with the intended use of these data. For example, for flow measurements, only data at the flow measurement location may be of importance. However, field surveys for river morphology or hydraulic studies require multiple cross sections to establish the river profile, over some reach of hydraulic interest (such as in a flood zone determination) or management interest (such as multiple meander wavelengths).

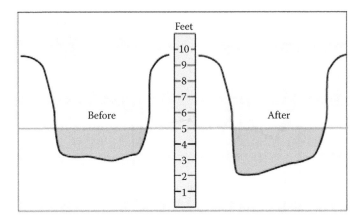

FIGURE 4.5 An example of change cross-section geometry. (From USGS, Available at http://ga2.er.usgs.gov/peachtree/streamflow.cfm.)

A river cross section is typically measured perpendicular to the flow direction, as illustrated in Figure 4.6. The cross-sectional geometry of a river is represented by a series of points, each point specified by a pair of X and Z values. The value X denotes the distance of the point along the cross section from some starting point, by convention on the left bank facing downstream. The ground elevation, with respect to a datum, is denoted by Z. A cross-section profile is typically identified by the name of the river, the station, and sometimes by geographic coordinates (Figure 4.7).

FIGURE 4.6   River cross sections. (From NRCS. Introduction to HecRAS. Available from http://www.nrcs. usda.gov/wps/portal/nrcs/detailfull/national/water/?cid=stelprdb1042484, 2004.)

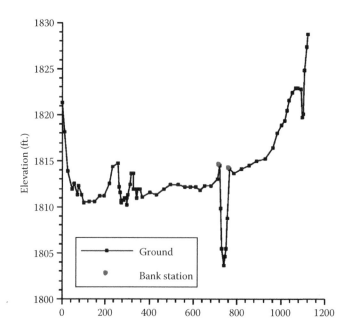

FIGURE 4.7   An example of a river cross section.

4.3.3 Measurement of Flow

Flows are typically measured by essentially computing an average velocity, which, when multiplied by the cross-sectional area, is the flow at that particular time and location. The average velocity for the cross section can be computed using an area-weighted average. That is, the average velocity can be computed by subdividing the cross section into many areas, each small enough that the velocity in that piece could be assumed constant (see Figure 4.8). The velocity for each piece would be determined, which, when multiplied by the area, yields a flow. The total flow would then be the sum of the individual flows ($U*A$, where U is the velocity and A is the area), or the average velocity (U) determined by the total flow divided by the total area.

$$U = \frac{1}{A} \int_A u(y,z)\, dApp \tag{4.1}$$

One practical question then is: how many lateral and vertical subdivisions are required to accurately determine the flow?

Stream gauging has a long history in the United States (see USGS 1995, 2000). The USGS is the agency primarily responsible for gauging flow in the United States and it maintains over 7000 gauging stations, constituting over 90% of the nation's gages (Hirsch and Costa 2004). Other agencies responsible for flow and/or water surface elevations measurements include the U.S. Army Corps of Engineers (the Corps), the National Weather Service, the Bureau of Reclamation, and other federal and state agencies. It is also not uncommon for industries to gage receiving waters as they impact permit compliance.

The traditional method used by the USGS for measuring the flow at each station is to subdivide the cross section into a minimum of 20 lateral sections and then determine the average velocity for each section; the total flow is the sum of the sectional flows. The USGS guideline for the number of points required to determine the vertically averaged velocity is based on an assumed velocity distribution. The assumptions include that the river flow is predominantly one-dimensional, the flow is constant with time (steady flow over the measurement period), and the river is wide, so that the velocity profiles in the river cross section are not affected by the presence of banks. The velocity

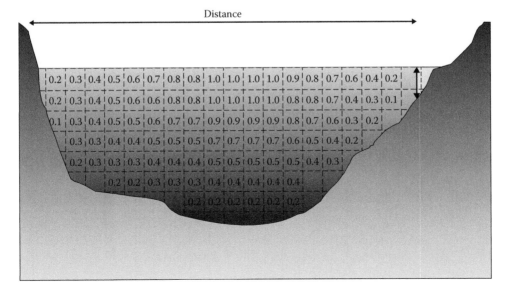

FIGURE 4.8 Flow characterization in a river. (From Winkler, M.F., Acoustic Doppler current profilers broadband band and narrowband technology explained, USACE Engineering Research and Development Center, Vicksburg, MS, 2006.)

distribution is then further assumed to be a fully developed turbulent boundary layer whose velocity profile can be approximated by the log-law-of-the-wall (Chow 1959). For these conditions, the average velocity occurs at 60% (0.6) of the depth (as measured from the surface).

So, for shallow rivers, the USGS guideline is that a single velocity measurement be taken at 60% of the depth of each section and used to represent the average velocity. For deeper sections, the average velocity is based on the average of two measurements taken at 20% and 80% of the depth. Where there are irregularities in the channel impacting the flow, a three-point method is recommended (Table 4.1).

Traditionally, a flow measurement would involve the USGS visiting the gauging site and taking direct measurements of the river depth, width, and velocity, using mechanical instruments such as a sounding rod or cable, a tagline, and a current meter. From these data, flow rates are computed (Figure 4.9).

The most common velocity meter traditionally used is the small Price (AA) current meter, which has a cup that, when placed in a current, rotates and the speed of the rotation is related to the velocity of the flow (Figure 4.10). The number of rotations in a measurement period may be manually or electronically measured and then converted to a velocity measurement. The meter could be used with a sounding rod for wadeable rivers. For deep rivers, the meter would traditionally (as well as presently) be suspended by a cable, with a sounding weight attached to help keep the cable as vertical as possible, as illustrated in Figure 4.11.

TABLE 4.1
Procedures for Determining Mean Vertical [$U(y)$] Velocities in a Lateral Section

No. of Points in Vertical	Vertical Depth of Measure	Application	Mean Velocity
1	0.6 D	Depth <0.5 m or quick measurement required	$U(y) = u(y, 0.6\,\text{D})$
2	0.2 and 0.8 D	Preferable where size of meter allows (D > 0.5 m)	$U(y) = \dfrac{u(y, 0.2\,\text{D}) + u(y, 0.8\,\text{D})}{2}$
3	0.2, 0.6, and 0.8 D	Where irregularities distort the velocity profile and depth is sufficient	$U(y) = \dfrac{u(y, 0.2\,\text{D}) + 2u(y, 0.6\,\text{D}) + u(y, 0.8\,\text{D})}{4}$

Source: Gordon, N.D., McMahon, T.A., Finlayson, B.L., Gippel, G.J., and Nathan, R.J., *Stream Hydrology—An Introduction for Ecologists*, 2004. Copyright Wiley-VCH Verlag GmbH & Co. KGaA.

Note: y = lateral section; D = vertical distance between the water surface and the streambed, measured downward.

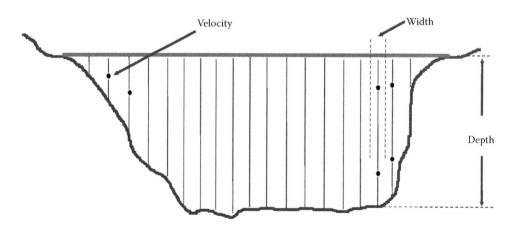

FIGURE 4.9   Cross-section profiling for flow measurements.

Gurley precision instruments 625 Pygmy
current meter

FIGURE 4.10 Price current meters. (From USGS. U.S. Geological Survey Streamgaging. Fact Sheet 2005–3131, March 2007. U.S. Department of the Interior, 2007; photo from Gurley Precision Instruments (GPI), available at http://www.gpi-hydro.com/. With permission.)

FIGURE 4.11 Stream gager from the 1940s making a streamflow measurement using a bridge crane and a current meter from a bridge near The Ohio State University. (From USGS, The history of stream gaging in Ohio, U.S. Geological Survey, Reston, VA, 2000.)

Measuring a stream discharge using the traditional method, such as with a Price current meter, is time consuming and expensive, with a single flow measurement often taking several hours.

A method that is replacing the traditional method is based on the use of acoustic Doppler current profilers (ADCPs). The development of ADCPs has been the most important development in streamflow measurement in the last 10 years (Simpson 2001; Hirsch and Costa 2004). Using acoustic energy (typically in the range of 300–3000 kHz) an ADCP measures the water velocity through the water column using the Doppler shift, or the shift in frequency, of the acoustic signals reflected from materials suspended in and moving with the water. The water depth is determined by the time-of-travel of the signals reflected from the channel bottom; boat velocities are determined using the Doppler shift of separate acoustic pulses reflected from the riverbed; and the channel width is computed using the instantaneous boat velocities and the time between each measurement (Figure 4.12).

The ADCP has made many direct contributions to flow measurements, including (Hirsch and Costa 2004):

- A conventional discharge measurement using a mechanical current meter can take several hours to complete, while measurements with an ADCP can take on the order of minutes and be equally accurate (Morlock 1996; Mueller 2003).
- The ADCP allows measurements of flows in areas where the traditional velocity area method does not work (no controls), such as the backwater areas of rivers or estuaries.
- The ADCP can be deployed to obtain continuous profiles of water velocities, rather than discrete measurements.

Also, measurements with ADCPs do not require any assumption regarding the vertical velocity profile, such as was used in the one-, two-, or three-point velocity measurement (Table 4.1). That is, the assumption of a vertical velocity distribution is no longer required where ADCPs are used.

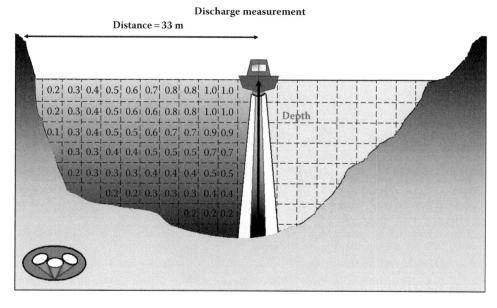

FIGURE 4.12   River flow profiling using a boat-mounted ADCP. (From Winkler, M.F., Acoustic Doppler current profilers broadband band and narrowband technology explained, USACE Engineering Research and Development Center, Vicksburg, MS, 2006.)

4.3.4 MEASUREMENT OF STAGE

The measurement of stages is a useful hydraulic characteristic for a number of purposes. Stage measurement methods include a variety of manually read or automatic stage recorders, and may be as simple as a graduated plate or rod that is set vertically in a streambed or attached to a solid structure. These manual gages do not produce a continuous record, but a single point in time measurement; they are of particular use in times of flooding. Another point-measurement gage is the crest gage, which records the maximum stage in rivers.

One of the most common uses of stage measurements is in the indirect estimation of flows using a rating curve (see Figure 4.4). Stage measurements for flow estimation typically include some type of water level measurement actuator, a recorder, and a stilling basin (Figure 4.13). A water level recorder is an instrument that records water levels in analog or digital form and it may be actuated by a float or by any one of several other sensor types. Historically, the recording device was a pen and graph, while more commonly today the stage is recorded electronically. For many USGS gauging stations, the electronic measurements, typically recorded at 15–60 minute intervals, are transmitted to USGS offices via satellite, telephone, and/or radio and these real-time data are available at their National Water Information System (NWIS) via web interface (Figure 4.13).

The general practice of the USGS for stage data collection is described in USGS Water-Supply Paper (WSP) 2175 (Rantz et al. 1982) and USGS *Techniques of Water-Resources Investigations*, Book 3, Chapter A-7 (Buchanan and Somers 1968).

4.3.5 MEASUREMENT OF FLOW USING TRACER TECHNIQUES

Tracer studies are used to provide flow estimates such as by using the dilution method as described by Kilpatrick and Cobb (1985). Dye studies are also commonly used to estimate the time of travel.

Tracer studies typically involve injecting a dye tracer such as Rhodamine WT at one location in a stream and later measuring the dye concentration at a downstream location. The principal property of commercial dyes such as Rhodamine WT is that they are highly fluorescent. For example, Rhodamine dyes (such as Rhodamine B and WT) have greatest excitation at a wavelength in the

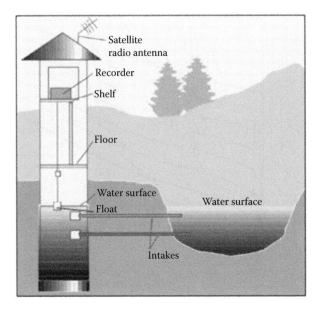

FIGURE 4.13 Stage recorder and stilling well. (From USGS, Stream-gaging program of the U.S. Geological Survey, U.S. Geological Survey Circular 1123, Reston, VA, p. 4, 1995.)

green band near 560 nm (nanometers or millimicrons) and emit light in the yellow–orange band near 580 nm. The relative intensity of the emitted light is also a function of the amount of fluorescent material present. Therefore, the combination of wavelengths and the intensity of the light absorbed and emitted can be used to measure the amount of material present using a fluorometer (Martin and McCutcheon 1999).

Two common types of dye studies are based on the manner in which the dye is injected: slug or continuous releases. The slug injection is more commonly used to estimate the time of travel and dispersion, while continuous injections offer advantages for estimating dilutions. Kilpatrick and Wilson (1989) provide guidance on conducting time of travel and dispersion studies. Story et al. (1994) discussed some of the considerations for planning and conducting dye studies in estuaries. The dye can be monitored either by measuring the dye concentrations as the water parcel passes a point in the river, or by periodically running through the dye cloud to determine the centroid and spread of the dye cloud (Figure 4.14).

The resulting time–concentration curves are generally Gaussian or bell-shaped but are skewed toward the leading edge. The curve shown in Figure 4.15 illustrates a time–concentration curve for a Mississippi river as reported by the Mississippi Department of Environmental Quality as part of "Time-of-travel of water tracing dyes in MS streams" (MDEQ 2007).

The time–concentration curve at a particular location downstream of the injection can be used to determine the time of travel, simply by measuring the distance downstream to that location and dividing by the elapsed time (described in more detail by Kilpatrick and Wilson [1989] and Jobson [1996]). The location of the centroid of the dye cloud is used to compute the time of travel, but the time of the leading and trailing edges of the dye cloud is also useful in dispersion and spill studies. The flow can also be estimated by first integrating under the time–concentration curve

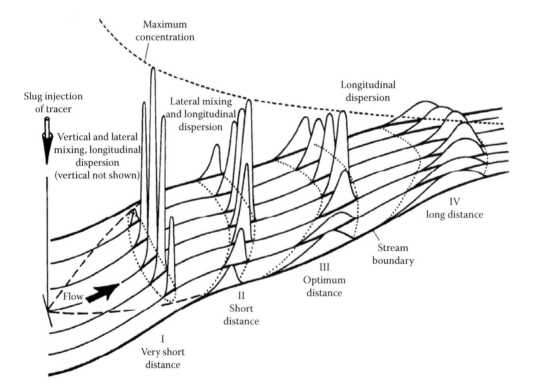

FIGURE 4.14 Lateral mixing and longitudinal dispersion patterns from a center slug injection of a tracer. (From Kilpatrick, F.A. and Wilson Jr. J.F., *Techniques of Water-Resources Investigations Reports*, Book 3 Applications of Hydraulics, United States Geological Survey, Washington, DC, p. 34, 1989.)

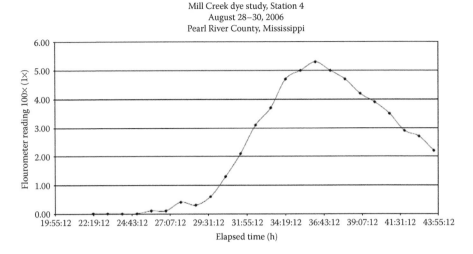

FIGURE 4.15 An example of a dye time–concentration curve. (From MDEQ, Time-of-travel of water tracing dyes in MS streams, Mississippi Department of Environmental Quality, Jackson, MS, 2007.)

and then dividing that into the product of the mass of dye injected and the dye's specific gravity, as described by Kilpatrick and Cobb (1985). Figure 4.16 illustrates some of the common results of dye studies.

Dye studies are often extremely useful in river studies. They allow measurements of flows where the standard method, as described previously, is not applicable. They also integrate the results of the river's features over their travel distance, thereby providing a good estimate of the average velocity, rather than a point velocity. Dye studies may also provide measures of mixing and dilution. A limitation is that they provide data only for the time and conditions under which the study is done, so they cannot be easily extrapolated to other conditions.

4.3.6 Measurement of Flows Based on Hydraulic Structures

The flow in reservoirs, many rivers and streams, and some estuaries is controlled by structures such as dams or weirs. In controlled releases, such as for hydropower, the releases are directly measured. In uncontrolled releases, such as over a spillway or an emergency spillway or through an orifice, the flow can be estimated from the type of structure and the water surface elevation. Uncontrolled structures, such as weirs, are often placed in channels for the purpose of measuring flows, and an example is the V-notch weir illustrated in Figure 4.17. A number of methods for measuring flows using culverts, weirs, flumes, and other hydraulics structures are well described in the USGS *Techniques of Water-Resources Investigations Reports. Book 3 Applications of Hydraulics*; Section A: Surface-water techniques (http://water.usgs.gov/pubs/twri/). The water measurement manual from the Bureau of Reclamation (2001) also provides a detailed discussion of measurement methods, and programs are available such as the Bureau of Reclamations WinFlume to aid in the design and analysis of flows from flumes and weirs as described by Clemmens et al. 2001.

4.3.7 Flow in Ungaged Rivers

There are many locations on gaged streams for which flows are not known and many ungaged locations, and, unfortunately, the number of such locations is increasing. For these locations, it is often necessary to estimate the flow or some flow characteristics. One common method is to use regression equations, which relate some streamflow statistic to basin characteristics, such as (Ries 2002)

Bear Creek

Location: near Canton
Date Collected: August 1981

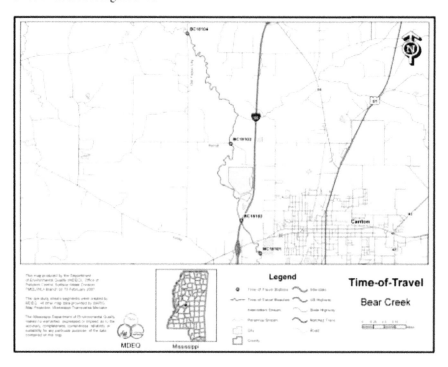

Reach	From Station	To Station	Reach Length (miles)	Solute Time of Travel (hrs)			Rate of Travel (fps)	
				Lead	Peak	Trail	Lead	Peak
BC181A	BC18101	BC18102	1.50	--	35.00	--	--	0.06
BC181B	BC18102	BC18103	3.00	--	65.00	--	--	0.07
BC181C	BC18103	BC18104	4.50	--	60.00	--	--	0.12

Station	Location	Latitude (DMS)	Longitude (DMS)	Stream Discharge (ft³/s)
BC18101	Lagoon	32° 36' 17"	-90° 03' 37"	Trace
BC18102	I-55	32° 36' 55"	-90° 04' 04"	Trace
BC18103	County Road	32° 38' 24"	-90° 04' 20"	4.50
BC18104	County Road	32° 40' 32"	-90° 05' 22"	5.00

FIGURE 4.16 Example time of travel results for a Mississippi stream. (From MDEQ, Time-of-travel of water tracing dyes in MS streams, Mississippi Department of Environmental Quality, Jackson, MS, 2007.)

$$Q_{100} = 0.471A^{0.715}E^{0.827}SH^{0.472} \tag{4.2}$$

where
    Q_{100} is the 100-year flow
    A is the drainage area
    E is the mean basin elevation
    SH is a shape factor

FIGURE 4.17   North Fork 120-degree compound V-notch weir and sampling bridge V-notch weir. (Courtesy of U.S. Forest Service Pacific Southwest Research Station.)

Some of the basin characteristics commonly used are listed in Table 4.2 (Ries 2002). Examples of methods used by specific states for estimating low-flow characteristics are given by Telis (1992) and Arihood and Glatfelter (1986). Pyrce (2004) reviewed the empirical equations used for determining the 7Q10 flows. Hirsch (1982) reviewed the methods for extending existing records and USGS gauging stations. These statistical methods have also been incorporated into a geographic information system (GIS)-based web application that is used to measure basin characteristics and estimate streamflow statistics for a site (gaged or ungaged), which is called StreamStats.

4.4   CHARACTERIZATION AND ANALYSIS OF FLOW

Given the availability of flow data, the determination of the characteristics of a flow is usually required for flow and/or water quality management, design, or analysis. Flows can be characterized according to their (Poff et al. 1997; Richter et al. 1996, 1997):

- Magnitude
- Return interval (or frequency)
- Duration
- Timing
- Rate of change

The magnitude of a flow is simply the amount of water passing a fixed point in the river at a specific point in time (e.g., 1000 ft^3 s^{-1} flow). The frequency describes how often a particular condition, such as a large flood, has occurred or is expected to occur (e.g., the 100-year flood). The duration is simply how long the event occurred (e.g., 7-day minimum flow, duration of zero-flow days). The timing could be when a particular event occurs (e.g., Julian date of annual 1-day minimum or maximum flow) or how the events occur in relation to each other. The rate of change, for example, could be the rise or fall rates of a particular event. Each of these components could have a different or combined hydrological or ecological significance.

4.4.1   ESTIMATION OF FLOW MAGNITUDE AND RETURN INTERVAL

Two of the most common characterizations of flows are the magnitude and the return interval (or frequency). Of the flow magnitudes, the one commonly used in hydraulic design and analysis is the

TABLE 4.2
Basin Characteristics Used for Peak Flows

Basin Characteristic	No. of States Using This (including PR[a])
Drainage area or contributing drainage area (mi.^{2})	51
Main-channel slope (ft. mi.$^{-1}$)	27
Mean annual precipitation (in.)	19
Surface water storage (lakes, ponds, swamps)	16
Rainfall amount for a given duration (in.)	14
Elevation of watershed	13
Forest cover (%)	8
Channel length (mi.)	6
Minimum mean January temperature (°F)	4
Basin shape ([length]^2 per drainage area)	4
Soils characteristics	3
Mean basin slope (ft. foot^{-1} or ft. mi.$^{-1}$)	2
Mean annual snowfall (in.)	2
Area of stratified drift (%)	1
Runoff coefficient	1
Drainage frequency (number of first-order streams per square mile)	1
Mean annual runoff (in.)	1
Normal daily May–March temperature (°F)	1
Impervious cover (%)	1
Annual PET (in.)	1

Source: Ries, K., StreamStats: A web site for stream information, *Symposium on Terrain Analysis for Water Resources Applications, December 16–18, 2002,* Center for Research in Water Resources, University of Texas, TX, 2002.

[a]   Commonwealth of Puerto Rico.

peak flow. For example, a common analysis would be the determination of the 100-year peak flow. The 100-year flow (return period $T_R = 100$ years) is that flow having a probability of occurring in any one year of 1%, or

$$P = \frac{1}{T_R} \tag{4.3}$$

To determine the magnitude of the peak flow requires historical data. For each year of record, the peak instantaneous is determined. That is, the peak flow in any one year provides one data point for the analysis. The probability of the peak flows is determined and the flow with a probability of occurrence of 0.01 would be the 100-year flow.

To illustrate, the 30-year peak flow record obtained from the USGS website for a station on the Mud Creek near Fairview, Mississippi, for the period of record (1954–1983) is shown in Table 4.3.

To determine the probability of each of the flows in Table 4.3, we could rank them (in descending order for a high-flow analysis) and then assign (Weibull distribution) a probability to each rank as computed from

$$P = \frac{R}{N+1} \tag{4.4}$$

TABLE 4.3
Peak Flow Record for Mud Creek near
Fairview, Mississippi

Year	Peak Flow (cfs)	Year	Peak Flow (cfs)
1976	613	1991	1390
1977	683	1992	1010
1978	631	1993	1510
1979	490	1994	843
1980	731	1995	1080
1981	196	1996	875
1982	497	1997	1080
1983	683	1998	NA
1984	760	1999	1040
1985	323	2001	907
1986	516	2002	815
1987	778	2003	489
1988	619	2004	800
1989	813	2005	830
1990	677	2006	737

where N is the number of peak flows and R is the rank (1 to N), and then tabulate the probability for each peak flow as illustrated in Table 4.4.

First, it is apparent that the highest flow in this record (1510 cfs) has a return period of only 30 years. Or, with 29 years of record, we can, using this specific method, only directly compute the flow with a return interval of 30 years ($N + 1$). So, to determine the 100-year flood, we would have to extrapolate.

An alternative is to assume some probability distribution for the data. Flow data are typically log-normally distributed and skewed and the statistical distribution typically used for peak-flow analyses is the Log-Pearson Type III distribution. To characterize that distribution, the statistics

TABLE 4.4
Peak Flow Probabilities for Mud Creek near Fairview, Mississippi

Rank	Prob.	T_R	Peak Flow (cfs)	Rank	Prob.	T_R	Peak Flow (cfs)
1	0.033	30	1510	16	0.533	1.9	737
2	0.067	15	1390	17	0.567	1.8	731
3	0.1	10	1080	18	0.6	1.7	683
4	0.133	7.5	1080	19	0.633	1.6	683
5	0.167	6	1040	20	0.667	1.5	677
6	0.2	5	1010	21	0.7	1.4	631
7	0.233	4.3	907	22	0.733	1.4	619
8	0.267	3.8	875	23	0.767	1.3	613
9	0.3	3.3	843	24	0.8	1.3	516
10	0.333	3	830	25	0.833	1.2	497
11	0.367	2.7	815	26	0.867	1.2	490
12	0.4	2.5	813	27	0.9	1.1	489
13	0.433	2.3	800	28	0.933	1.1	323
14	0.467	2.1	778	29	0.967	1	196
15	0.5	2	760				

needed are the mean, standard deviation, and coefficient of skewness of the common logarithms of the annual peak flows.

However, complications can result where data are missing in any given year or between years, due to outliers, annual peaks falling below the lower limit of measurements, and other factors that may impact the probability distribution. To account for these complications, a methodology was established by an Interagency Advisory Committee on Water Data (1982) and published in Bulletin 17B (Figure 4.18). This methodology has been implemented in a computer program available from the USGS, called PeakFQ (Flynn et al. 2006). So, peak flow data can be downloaded from the USGS NWIS database, loaded into the PeakFQ program and the output of that program includes tabular or graphical flow magnitudes, exceedance probability estimates, and 95% confidence for those estimates limits, as illustrated in Figures 4.19 and 4.20 for the Mud Creek gage.

In addition to peak flows, statistical data are available for selected gage locations from the USGS NWIS, including mean daily, monthly, and yearly flows, which are analyzed using the same methods to determine the return probabilities. It should be noted that the "year" for the USGS record is a water year and not a calendar year. The USGS defines a "water year" as the 12-month period from October 1 through September 30 of the following year. Other agencies may use different definitions; for example, for low-flow calculations, the U.S. Environmental Protection Agency (U.S. EPA) typically uses April 1 through March 31.

The flow conditions of interest for water quality, such as in setting discharge permits, are typically low flows rather than peak flows. Since the instantaneous minimum flow is not a realistic measure, typically a minimum flow over some specified duration is used, such as a XQT_R flow, where X is some averaging period and T_R is the return period. For example, a 7Q10 is the 7-day average low flow with a recurrence interval of 10 years (expected to occur once every 10 years). To determine the 7-day average flow for a particular year, a 7-day running average is used and then the low flow for the year determined, and this becomes a single data point. When compiled for consecutive years, the 7-day minimum flow can be used to estimate the magnitude of the flow with a 10-year recurrence interval. For example, a 7Q10 flow of 12 cfs (cubic feet per second) indicates that there is a 10% chance (return period = 1/0.1 = 10 years) that the flow for 7 consecutive days will be less than

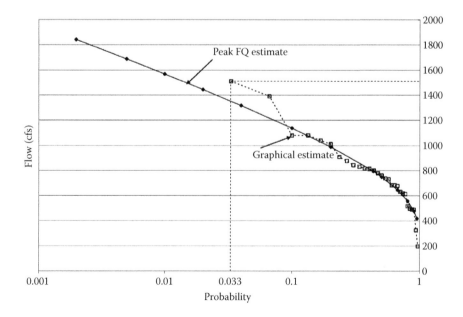

FIGURE 4.18 Peak flow probability curve for Mud Creek using graphical and Bulletin 17B estimate from PEAKFQ.

MUD CREEK.PRT - Notepad
File Edit Format View Help

ANNUAL FREQUENCY CURVE -- DISCHARGES AT SELECTED EXCEEDANCE PROBABILITIES

ANNUAL EXCEEDANCE PROBABILITY	BULL.17B ESTIMATE	SYSTEMATIC RECORD	'EXPECTED PROBABILITY' ESTIMATE	95-PCT CONFIDENCE LIMITS FOR BULL. 17B ESTIMATES LOWER	UPPER
0.9950	--	166.2	--	--	--
0.9900	--	203.0	--	--	--
0.9500	418.2	331.0	405.1	342.8	481.2
0.9000	477.5	415.2	468.3	403.1	540.8
0.8000	558.8	529.9	553.4	486.4	624.0
0.6667	645.3	646.6	642.7	573.5	715.9
0.5000	747.8	774.3	747.8	672.8	832.0
0.4292	794.2	827.4	795.3	715.8	887.3
0.2000	988.2	1018.0	997.0	884.7	1137.0
0.1000	1138.0	1133.0	1157.0	1005.0	1344.0
0.0400	1317.0	1238.0	1357.0	1144.0	1607.0
0.0200	1444.0	1296.0	1506.0	1239.0	1802.0
0.0100	1568.0	1341.0	1655.0	1329.0	1995.0
0.0050	1688.0	1376.0	1808.0	1415.0	2188.0
0.0020	1843.0	1411.0	2015.0	1525.0	2443.0

1

Program PeakFq
Ver. 5.2
11/01/2007

U. S. GEOLOGICAL SURVEY
Annual peak flow frequency analysis
following Bulletin 17-B Guidelines

Seq.001.003
Run Date / Time
09/23/2008 15:55

Station - 02430615 MUD CREEK NR FAIRVIEW, MS

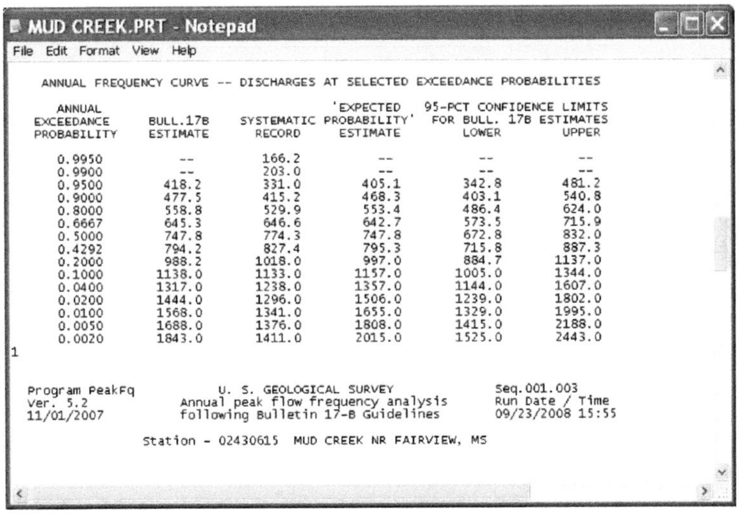

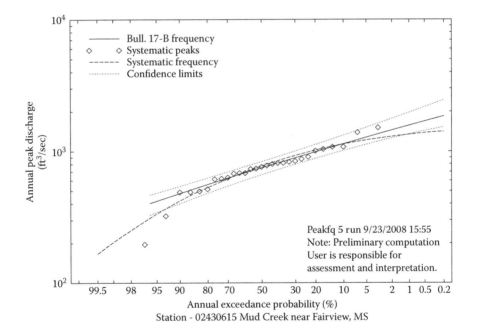

FIGURE 4.19 Notepad view of PeakFQ output table for Mud Creek.

FIGURE 4.20 PEAKFQ graphic plot for Mud Creek near Fairview, Mississippi.

or equal to 12 cfs. This extreme flow analysis can be computed similarly to the peak flows, assuming a Log-Pearson Type III distribution, but using the left (low probability) side of the frequency distribution. A program supported by the U.S. EPA for the computation of *XQT* flows from historical records is DFLOW. DFLOW directly incorporates the USGS SWSTAT Program (Program A193) that statistically analyzes time-series data and uses the Log-Pearson Type III estimation method. DFLOW also allows for the calculation of biological as well as hydrology-based design flows as described in Technical Guidance Manual for Performing Wasteload Allocations: Book VI: Design Conditions: Chapter 1: Stream Design Flow for Steady-state Modeling (USEPA 1986).

4.4.2 ESTIMATION OF FLOW DURATION

The duration of flows is typically analyzed as described in the previous section, but either daily, monthly, or annually to create a flow duration curve. A flow duration curve shows the percentage of time that a given flow rate is equaled or exceeded. The construction of a flow duration curve can be illustrated using the mean annual flows from the USGS Mud Creek Station (02430615 Mud Creek near Fairview, Mississippi), which are tabulated in Table 4.5 and sorted in descending order, ranked, and a probability assigned using Equation 4.4 (as graphically illustrated by Figure 4.21).

The flow duration is usually written as QX, where X represents the percentage that the flow is equaled or exceeded. For example, the Q50 for Mud Creek means that the annual flow is about 20 cfs (based on the preceding analysis), so the 20 cfs annual flow is equaled or exceeded 50% of the time. Flow duration curves are more commonly based on daily flows. Smakhtin (2001) indicates that the "design" low-flow range of a flow duration curve is in the 70%–99% range, or the Q70–Q99 range. The ratio of the discharge that is equaled or exceeded 90% of the time, to that of 50% of the time (Q90/Q50) is commonly used to indicate the proportion of streamflow contributed from groundwater storage (Nathan and McMahon 1990).

4.4.3 ESTIMATION OF TIMING AND RATE OF CHANGE

Timing describes the time of year at which particular flow events occur, such as the timing of floods or low-flow extremes, while the rate of change indicates how quickly the flow changes, as flows rise and fall from day to day.

The timing of an event is usually determined by two attributes: a magnitude and a time of occurrence. First, some flow condition is defined, such as the annual minimum or maximum flow, and then the date or Julian day for that event is recorded. The rate of change requires some "trigger" such as a high or low flow and then an algorithm to determine whether the flow is the rising or the descending limb of the hydrograph. For example, the following is the algorithm used for high flows in the Nature Conservancy's (2007) Indicators of Hydrologic Alteration (IHA) Program:

TABLE 4.5
Year Averaged Flow and Flow Probabilities for USGS Station 02430615 Mud Creek near Fairview, Mississippi

Rank	P (%)	Year	Flow (cfs)	Rank	P (%)	Year	Flow (cfs)
1	5.88	1991	45.20	17	100.00	1996	19.70
2	11.76	1997	30.10	18	105.88	1976	19.10
3	17.65	1989	24.50	19	111.76	1979	19.00
4	23.53	1983	24.30	20	117.65	1999	18.80
5	29.41	2005	23.80	21	123.53	1987	17.30
6	35.29	2003	22.50	22	129.41	2001	16.10
7	41.18	1994	22.20	23	135.29	1985	15.90
8	47.06	1993	22.00	24	141.18	1977	15.60
9	52.94	1980	21.40	25	147.06	1978	15.20
10	58.82	1992	21.40	26	152.94	2006	13.20
11	64.71	1984	20.80	27	158.82	1988	11.40
12	70.59	2004	20.70	28	164.71	2000	11.40
13	76.47	1990	20.30	29	170.59	1982	10.40
14	82.35	1995	20.10	30	176.47	1981	9.85
15	88.24	1998	20.00	31	182.35	1986	9.80
16	94.12	2002	19.90	32	188.24	2007	8.13

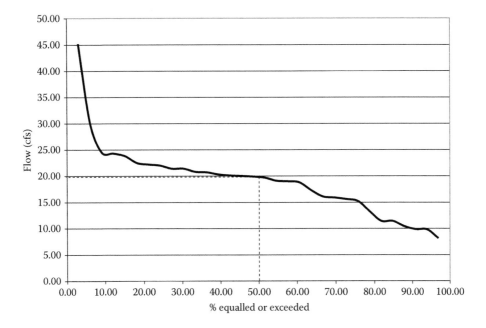

FIGURE 4.21   Flow duration curve for mean annual flows. (From USGS Station 02430615 Mud Creek near Fairview, MS.)

- Following a low-flow day, the next day is assigned to the ascending limb of a high-flow event if the daily flow is higher than the 75th percentile of all flows, or if the flow is higher than the 50th percentile of all flows and the daily increase is more than 25%. Otherwise, it continues as a low flow.
- The ascending limb of a high-flow event continues until the daily flow decreases by more than 10%, at which time the descending limb of the event is started.
- During the descending limb of a high-flow event, the ascending limb is restarted if the daily flow increases by more than 25%.
- During the descending limb of a high-flow event, the event is ended if the rate of decrease of the flow drops below 10% per day (meaning that the change in flow is between –10% and 25%), unless the flow is still above the 75th percentile of all flows, in which case the descending limb continues.

4.4.4   DATA REQUIREMENTS AND CONSIDERATIONS

An analysis of the aforementioned components (magnitude, return interval, frequency, duration, timing, and rate of change) shows that they all have at least one thing in common. They all require a lot of flow data. What is usually required is a long-term historical record of flows from a particular gauging location. The number of gauging stations with an adequate historical record is limited and is becoming increasingly limited. Hirsch and Costa (2004) indicated "the most crucial policy question remains how the stream gauging network should be supported to foster continued modernization and improved efficiency, and to assure the continuation of valuable long-term stream flow records." This is an important issue since, due largely to the lack of support, about 640 USGS stream gages with more than 30 years of records (Hirsch and Costa 2004) have been discontinued during the last decade. Making management decisions today based on flow data, the most recent records of which are a decade or more old (therefore not reflecting current conditions), may have costly consequences.

A second consideration is: How useful are historical data for predicting existing or future conditions? As discussed previously, rivers and streams are largely influenced by their watersheds, and those watersheds are rarely static. For example, as watersheds are developed, the percentage of impervious areas increases (such as parking lots), so the cumulative impact of all the development over the watersheds impacts the timing (e.g., time of concentration) and the magnitude of runoff (see Figure 4.3). Under some conditions, a compelling reason not to use 7Q10 low flows, or peak flows, based on historical conditions, is that the historical flows are not representative of present conditions and are not good predictors of future conditions. Also, 7Q10 flows have little meaning in a highly regulated system, and proposed, planned, and ongoing projects will or can alter the flow regime.

As illustrated in Figure 4.22, depending on the relationship between the groundwater and surface water levels, the river can be an influent or "losing" reach, which loses water to the aquifer, or an effluent or "gaining" reach, which receives discharges from the aquifer (FISRWG 1998). Pumping from aquifers for water supply usually results in a decrease in the groundwater levels that can change a gaining river to a losing one, impacting riverine low flows. For example, in the Mississippi Delta, the water level of the Mississippi Alluvia Aquifer, the primary source of water for agricultural use in the delta, has decreased precipitously in some areas since the 1950s, with a large zone of depression in groundwater levels over the period 1994–2004, as illustrated in Figure 4.23, and an approximately 27 ft. decline between 1950 and 2008 (Byrd 2011). As a result, significant water level declines and the lack of readily available water for catfish production and irrigated agriculture have had a potentially devastating influence on the delta's economy (NRCS 1998) and ecology. The reduced river recharge has resulted in declining base low flows in the Big Sunflower River and other riverine systems (see Figure 4.24) and has converted some delta streams from perennial to ephemeral or intermittent streams. For these cases, and for many regulated systems, flow characterization estimates based on historical conditions may not be appropriate. Some of the methods discussed in Section 4.6 can be used to characterize the degree of hydrologic alternation and environmental impacts. Groundwater, and surface water–groundwater interactions, also impacts the time and space scales of interest for water quality management. For example, the extent of the groundwater impact, illustrated in Figure 4.23, indicates that a basinwide rather than a waterbody-specific management approach may be required and that such an approach must consider changes that may occur over decades rather than months or seasons.

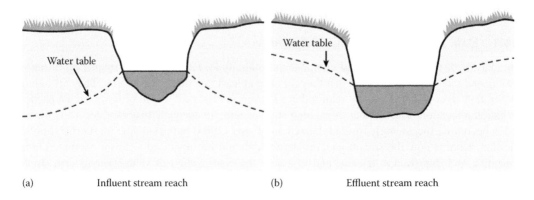

(a)   Influent stream reach   (b)   Effluent stream reach

FIGURE 4.22 Cross sections of (a) influent and (b) effluent stream reaches. (From FISRWG, *Stream Corridor Restoration: Principles, Processes, and Practices*, Federal Interagency Stream Restoration Working Group, 1998.)

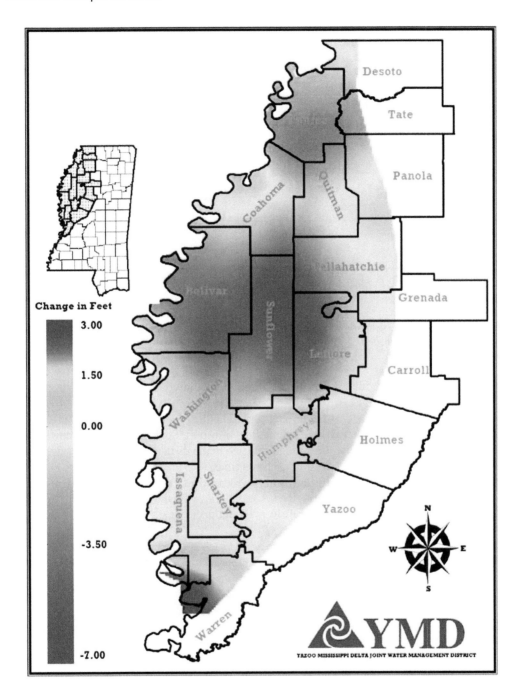

FIGURE 4.23 Fall water level changes in the Mississippi Alluvial Aquifer, Yazoo Basin, Mississippi from 1994 to 2004. (From YMD, YMD joint management district annual report 2004, YMD Joint Water Management District, Stoneville, MS, 2004.)

4.5 TRANSPORT PATTERNS

Generally, the transport patterns in streams and rivers are assumed to be dominated by advection (lotic system), as opposed to mixing processes, where the dominant transport is in the longitudinal direction. Streams and rivers are usually described using one-dimensional

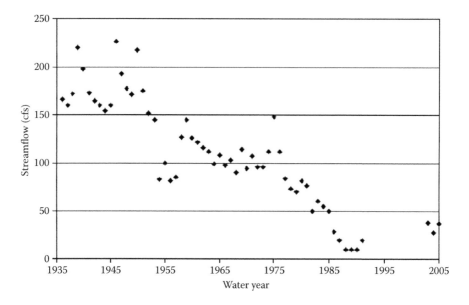

FIGURE 4.24 Minimum streamflows in the Big Sunflower River. (From Hirsch, USGS science in support of water resources management, Presented at Mississippi Water Resources Conference, Jackson, MS, 2006.)

(longitudinal) hydraulic models, with the assumption that lateral and vertical variations are relatively small.

Since flows are the total flux of water through a section (the product of velocity and area), flows only provide an estimation of that average longitudinal transport. So, for example, the flow divided by the area provides an estimate of the average velocity. Some distance down the river divided by the velocity provides an estimate of the average travel time.

But, in the estimation of a flow, it is necessary to measure velocities over the width and depth of a channel because they do vary. That is, the velocity varies with the depth and width. Variations may also occur longitudinally, since natural rivers do not occur in straight prismatic channels, as illustrated in Figure 4.25.

One consequence is that materials injected into a river, such as a dye, will move at different velocities depending on how and where they were injected. For example, if, initially, the dye were completely mixed over the cross section, then downstream, the dye located near the center of the river,

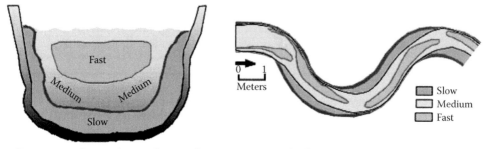

Cross section of a river showing the general pattern of current velocity

Example of current velocity patterns in a river

FIGURE 4.25 Variations in current velocities in natural rivers. (From Hebert, P.D. (ed.), *Canada's Aquatic Environments*, CyberNatural Software, University of Guelph, 2002. Available at http://www.aquatic.uoguelph.ca/. With permission from the Biodiversity Institute of Ontario.)

FIGURE 4.26 Discharge plume with lateral contraction. (From CORMIX image gallery, Available at http://www.cormix.info/picgal/rivers.php; original source unknown.)

and below the surface, would move faster than the dye on the banks, at the water surface, or at the bottom. The consequence would be that while the centroid of the dye would move with the average velocity, some dye would move faster and slower; therefore, as the dye cloud moves downstream, the concentration cloud would spread out and the peak concentration would be reduced (see Figure 4.14).

If, instead of being completely mixed over the cross section, the dye were injected at the bank, or at a specific location within the cross section, then it may be advected downstream at a faster rate than if it mixed laterally, so it may take a considerable distance downstream for the dye to mix over the width of the cross section (see Figure 4.26).

Similarly, when two tributaries join, their waters often do not mix completely at the junction, and they may be maintained separately for considerable distances downstream, as illustrated in Figure 4.27.

The velocity variations also produce changes in physical habitats, sedimentation rates, and other hydraulic and biotic river characteristics. For example, mussels and macrophytes may not occur in some sections of the river where shear stresses due to flows are greatest, but they may survive in other areas of the river. The variations in velocities and other hydraulic properties are also responsible for the formation of pools, riffles, bars, meanders, and other important physical characteristics of rivers and streams. Therefore, assuming that rivers are one-dimensional in terms of hydraulics, water quality, and ecology is misleading at best.

4.6 METHODS FOR DETERMINING INSTREAM FLOW REQUIREMENTS: ENVIRONMENTAL FLOWS

The streamflow requirements will vary with what we are trying to protect. The streamflow rates for hydraulic design are usually some rare high-flow event, such as the 100-year flow. If the consequences of structural failure are significant, and we are trying to prevent loss of life, then even more stringent conditions may be used in the design, such as a maximum probable storm. The streamflow requirements for habitats will vary with the location and the organisms, and the method used will vary from individual flow conditions to the stream hydrograph.

The instream flow can be defined as the amount of water flowing past a given point within a stream channel during 1 second (Estes 1994), while the instream flow requirements have most commonly been based on the habitat requirements for some species or groups of fish. Historically, these requirements have been expressed in terms of some minimum flow, such as a minimum flow that permitted water users or dams are required to release or pass, the nondepletable flow, or other appropriate instream flow limits. For example, minimum instream flows have commonly been established

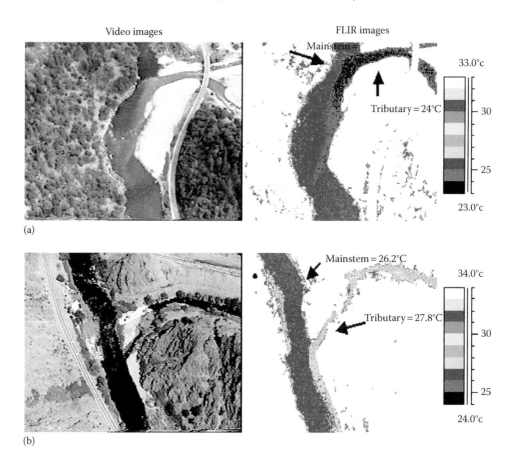

FIGURE 4.27 Examples of remote sensing of tributary mixing zones (TMZs). Typical FLIR output consists of GPS-tagged digital images covering approximately 100 × 150 m with less than 1 m of spatial resolution and ±0.5°C accuracy. (a) and (b) show cool and warm TMZ, respectively. Cool TMZs have been documented as critical salmonid refugia habitat in disturbed systems. (From CORMIX image gallery, Available at http://www.cormix.info/picgal/rivers.php, source of photographs Oregon DEQ.)

to estimate the impacts of diversions or to mitigate water withdrawals in western states. The most commonly used flow has been the 7Q10 flow previously discussed.

While minimum flows are common flows for design and regulatory purposes (many states specify minimum flows in their state statutes), there has long been a recognition that while minimum low flows may allow aquatic communities to subsist, they are not adequate to sustain fish communities and aquatic health (Baron et al. 2002; Poff et al. 2009; Richter et al. 2006 and others).

4.6.1 BASE FLOW AND LOW-FLOW INDICES

One method that has been commonly used to characterize flows, based solely on the flow magnitude, is the "low flow" or "minimum flow." Waddle (2001) indicated a typical description from when the concept of "low flow" was common:

> Water is taken out of the stream for a variety of uses, such as irrigated agriculture, municipal and industrial. Low flow means that amount of water that must be left in the stream for the fish. With anything less than the low flow, the fish will die.

The emphasis of this type of management was often more focused on how much water could be taken from a river rather than how much water is needed to be left in. Also, one major flaw of the "low-flow" approach was that there were only two relative conditions with respect to fish habitats for consideration by resource managers: "the level below which disaster would occur, and everything else" (Waddle 2001). In addition, for fisheries management, low flows were often established for other uses, such as water supply, navigation, etc. Low flows protective of aquatic life were sometimes referred to as environmental flows, which is not consistent with what today would be considered to be an environmental condition.

Today, in practice, while the term *minimum flows* is often used, particularly for regulatory purposes, today's minimum flow requirements are more commonly based on "instream flow uses," or how much water do we need to leave as opposed to how much water can we take out. For example, Waddle (2001) stated that with instream flows driving low-flow criteria, "biologists no longer were trying to find that magical flow level below which a stream should not be dewatered. Instead, they were in a position to assert instream flow needs for fish habitats and other environmental values. Furthermore, they could do so in terms of the seasonal life cycle needs of the fish (or other aquatic organisms) over the annual hydrograph."

The minimum condition may refer to instream flows (flows within the stream), and is also referred to as minimum instream flow requirements. The minimum condition may also refer to water levels. In Florida, for example, state statutes require that the state's water management districts establish minimum flows and levels (MFLs). Minimum flows are for rivers and streams, while minimum levels are for lakes and aquifers. Each management district develops MFLs for its waterbodies and reports their status annually to the state. The objective of the MFLs is to protect the waters of the state from significant harm due to permitted withdrawals. While the MFLs may be a magnitude, for some management districts, such as the St. Johns River Water Management District, they also include a frequency component (Figure 4.28).

As discussed previously, the 7Q10 flow is probably the most commonly used flow index in the United States. The 7Q10 flow is most often used as a "reasonable worst case" low-flow condition for allocating waste loads, but it has also been used in a number of other regulatory and management contexts, such as in an analysis of chronic toxicity, as an indicator of drought conditions, in the regulation of withdrawals, and for numerous other purposes as described by Pyrce (2004). In addition to the 7Q10 flow, other 7QX flows have been used for other purposes, such as the 7Q1, 7Q2, 7Q5, 7Q20, and 7Q25 flows as tabulated by Pyrce (2004) in Table 4.6. These 7QX flows are also computed by the DFLOW program discussed previously, using the methods described by USEPA (1986).

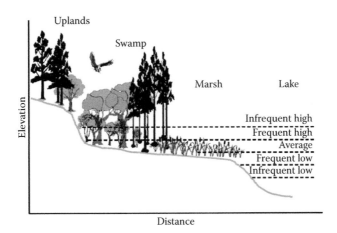

FIGURE 4.28 MFL frequency. (From SJRWMD, Minimum flows and levels: Fact Sheet, St. Johns River Water Management District, Palatka, FL, 2001.)

TABLE 4.6
7Q Flow Uses Source

7Q Flow	Uses	Source
7Q1	Known as the "dry weather flow"	Smakhtin (2001)
	Used for abstraction licensing	Smakhtin (2001), Smakhtin and Toulouse (1998)
	Used to remove the effect of minor river regulation	Matalas (1963)
7Q2	One of the most widely used design low-flow indices	Smakhtin (2001), Smakhtin and Toulouse (1998)
	Habitat maintenance flow (represents a period of stress on the system that causes some reduction in populations)	Ontario Ministry of Natural Resources (1994)
	Criteria for developing permits for wasteload allocations	Tortorelli (2002)
	Used as an instream flow	Caissie and El-Jabi (2003)
	Some use as a specific design application for stormwater holding facilities based on stormwater modeling	Odom (2004, personal communication)
	Not defined	Beran and Gustard (1977), Hayes (1991), Ries and Friesz (2000)
7Q5	Critical low flow for low-quality fishery waters (a stream classified for the beneficial use of warmwater semipermanent fish life propagation or warmwater marginal fish life propagation)	South Dakota Department of Environment and Natural Resources (1998)
7Q20	Used as a system's extinction flow (causes significant stress on the system)	Ontario Ministry of Natural Resources (1994)
	Used as an indicator of the minimum flow needed to maintain the ecosystem	Ontario Ministry of Natural Resources et al. (2003)
	Limiting condition for sewage treatment and wastewater disposal for a receiving waterbody	Ontario Ministry of the Environment (2000)
	Indicator of potential mortality of aquatic life for larger streams	Imhof and Brown (2003)
	Summer design low flow for effluent wastewater discharge and drought flow periods and volumes	Cusimano (1992)
	Flow for sustainable yield/carrying capacity for ecotourism	Shrivastava (2003)
7Q25	Critical low flow for high-quality fishery waters (surface waters designated for the beneficial use of cold-water permanent fish life propagation, cold-water marginal fish life propagation, or warmwater permanent fish life propagation)	South Dakota Department of Environment and Natural Resources (1998)

Source: Pyrce, R.S., Hydrological low flow indices and their uses, Watershed Science Centre, Peterborough, Ontario, 2004.

4.6.2 ENVIRONMENTAL FLOWS

While minimum flows, such as the 7Q10 flow, have commonly been used, it has long been recognized that maintaining some constant minimum flow is not conducive to the long-term health of aquatic communities. The sustainable management of rivers toward the long-term health of both the ecological and economic systems of a given watershed or country is one of the most crucial issues for current and future generations (Wali et al. 2009). A critical component of the allocation of waters among various competing land uses/covers is in identifying a desirable state for a river ecosystem and the flow requirements for maintaining that state, typically referred to as the environmental flow or the environmental flow requirement (EFR; Mazvimavi et al. 2007). Environmental flows have been variously defined. One useful definition is that included in the 2007 Brisbane Declaration:

Environmental Flows describes the quantity, quality and timing of water flows required to sustain freshwater and estuarine ecosystems and the human livelihoods and well-being that depend on these ecosystems.

Rather than identifying a single, low-flow criterion, environmental flows consider all aspects of the flow hydrograph as they impact aquatic health. For example, the National Research Council (NRC 2005) identified four critical components of a flow hydrograph (Figure 4.29; Hersh and Maidment 2006):

- Subsistence flow is the minimum streamflow needed during critical drought periods to maintain tolerable water quality conditions and to provide minimal aquatic habitat space for the survival of aquatic organisms.
- Base flow is the "normal" flow conditions found in a river in between storms, and base flows provide adequate habitats for the support of diverse, native aquatic communities and maintain groundwater levels to support riparian vegetation.
- High-flow pulses are short-duration, high flows within the stream channel that occur during or immediately following a storm event; they flush fine sediment deposits and waste products, restore normal water quality following prolonged low flows, and provide longitudinal connectivity for species movement along the river.
- Overbank flow is an infrequent, high-flow event that breaches riverbanks. Overbank flows can drastically restructure the channel and floodplain, recharge groundwater tables, deliver nutrients to riparian vegetation, and connect the channel with floodplain habitats that provide additional food for aquatic organisms.

The Brisbane Declaration presented summary findings at the 10th International River Symposium and International Environmental Flows Conference, held in Brisbane, Australia, in September 2007. The declaration also developed a global action agenda that included:

- Estimate environmental flow needs everywhere immediately.
- Integrate environmental flow management into every aspect of land and water management.
- Establish institutional frameworks.

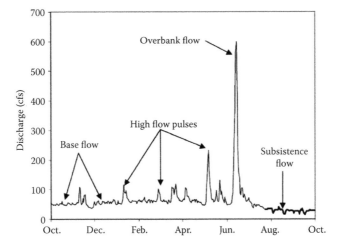

FIGURE 4.29 Streamflow hydrograph for the Guadalupe River at Victoria, Texas (USGS Gage No. 08176500) for water year 2000 with flow components identified. (From NRC, The science of instream flows: A review of the Texas instream flow program, Committee on Review of Methods for Establishing Instream Flows for Texas Rivers, National Research Council, The National Academies Press, Washington, DC, 2005.)

- Integrate water quality management.
- Actively engage all stakeholders.
- Identify and conserve a global network of free-flowing rivers.
- Learn by doing.

A wide variety of methods have been developed to establish EFRs. The development of an environmental flow is often referred to as an environmental flow assessment (EFA), defined by Tharme (2003) as

> an assessment of how much of the original flow regime of a river should continue to flow down it and onto its floodplains in order to maintain specified, valued features of the ecosystem hydrological regimes for the rivers, the environmental flow requirements, each linked to a predetermined objective in terms of the ecosystem's future condition.

A review by King et al. (2008) identified over 207 significantly different EFA methods implemented in 44 countries within 6 regions of the world. The following is a list of these methods, which can be loosely subdivided into four broad categories (from Acreman and Dunbar 2004; Dyson et al. 2008; King et al. 1999; Korsgaard 2006; Pyrce 2004; Tharme 2003; TRG 2008 and others), along with examples that will be discussed in greater detail in this section.

- Hydrologic index (desktop) models: these are simple and inexpensive and use flow as an indication of the biological condition. Examples include:
  - Tennant (Montana) method, U.S. Fish and Wildlife Service (1976)
  - Lyons method, Texas Parks and Wildlife Department (1979)
  - Biological flows
  - IHA, The Nature Conservancy (1997) and the range of variability approach (RVA) (Richter et al. 1997)
  - Hydrologic Assessment Tool (HAT) (Henriksen et al. 2006)
- Hydraulic models: these are used to compute and correlate available habitat areas based on river channel geometry. Examples include:
  - Wetted perimeter method, Montana Department of Fish, Wildlife, and Parks, 1970s
  - R2-Cross method, Colorado Division of Wildlife, 1980s
- Habitat models: these are generally complex and data intensive and use target species population data with hydraulic data to determine optimal habitats. An example is:
  - Instream flow incremental methodology (IFIM), U.S. Fish and Wildlife Service, 1970s, which includes the Physical Habitat Simulation Model (PHABSIM) (Stalnaker 1994; Stalnaker et al. 1995; Bovee et al. 1998)
- Holistic models: these are generally very complex, data intensive, and based on multidisciplinary scientific consensus. Examples include:
  - Building block methodology (BBM), South Africa Department of Water Affairs and Forestry and the University of Cape Town, 1990s (Tharme and King 1998)
  - Downstream response to imposed flow transformation (DRIFT), South Africa Department of Water Affairs and Forestry, the University of Cape Town, and Southern Waters Ecological Research and Consulting, 1990
  - Catchment abstraction management strategies (CAMS; Environment Agency 2010)
  - Expert panel or consensus, such as the Nature Conservancy's collaborative and adaptive process for developing environmental flow recommendations (Richter et al. 2006)

4.6.2.1 Hydrologic Index Models

Hydrologic index models are the simplest and perhaps the most commonly used models to establish EFRs and they rely largely on the availability of historical flow records. The models can be based on some fraction of an annual, seasonal, or monthly flow or they can include a large number of flow characteristics.

4.6.2.1.1 Montana (Tennant) Method

Perhaps the most widely used methods for establishing EFRs and instream flow requirements are the regression methods (Pyrce 2004), such as the Montana method developed by Tennant (1975, 1976). This method has the advantage that it can be implemented solely based on historical flow data, assuming that the stream morphology from Tennant's studies is similar to those streams that are targeted. The Montana method was developed based on the measurements of width, average depth, and average velocity in streams in Montana, Wyoming, and Nebraska.

Tennant found that the quality of instream habitats changed more rapidly from no flow to a flow of 10% of the average than in any higher range (Table 4.7). Tennant (1975, 1976) then concluded that 10% of the average annual flow is the minimum instantaneous flow needed to sustain short-term survival. The conditions at this low flow included:

- The depths and velocities were significantly reduced.
- The substrate was exposed.
- The gravel bars were dewatered.
- The streambank cover was diminished.
- The fish were crowded into pools.
- The riffles were too shallow for larger fish to pass.

Tennant determined that a flow of 30% of the average annual flow was required to maintain a good habitat for aquatic life and that an optimal habitat was provided by flows of 60%–100% of the average annual flow and flushing flows were 200% of the average annual flow. Therefore, the method could be implemented based on an analysis of annual flows to determine the minimal, good, optimum, or flushing flows.

4.6.2.1.2 Lyons Method

The Lyons method (Bounds and Lyons 1979) was derived from the Tennant method (1976), developed for the cold-water fisheries of Montana and other western states, but developed and validated

TABLE 4.7
Instream Flow Recommendations by Tennant Method

Description of Flows	Recommended Base Flow Regimes	
	October–March (%)	April–September (%)
Flushing or maximum	200	
Optimum range	60–100	
Outstanding	40	60
Excellent	30	50
Good	20	40
Fair or degrading	10	30
Poor or minimum	10	10
Severe degradation	0–10	

Note: All flows are defined as a percentage fraction of the average flow.

with data from two sites on the Guadalupe River, Texas, downstream from Canyon Dam on February 2, 1977, when flows were manipulated to provide a range of values (9–784 cfs). The environmental flow targets were as follows:

- During October–February: 40% of the median flows by month
- During March–September: 60% of the median flows by month

The 60% level was chosen to be more protective of the riverine ecosystem during the spring and summer periods, which are considered most critical to the warmwater fisheries found in Texas (Bounds and Lyons 1979; NRC 2005).

4.6.2.1.3   Biology-Based Design Flows

The biology-based design flow is based on first identifying exceedances. Exceedances are when unfavorable conditions occur, such as when a flow is less than some specified value. The exceedance condition is also termed *an excursion*. The premise is that if these exceedances occur rarely, then the population will have time to recover, or, conversely, if they occur frequently, then no recovery may occur. The biology-based design flow is then the lowest flow that will not cause exceedances to occur more often than allowed by some specified average frequency. The allowable frequencies of exceedance are specified so that they are sufficiently small, and far enough apart, that they will result in relatively small stresses on the biological community (USEPA 1986).

The biology-based design flows are computed so that they may be compared to the national, site-specific, and effluent toxicity criteria: the criteria continuous concentration (CCC) and the criterion maximum concentration (CMC; USEPA 1986). For comparison, the flow is based on site-specific durations (i.e., averaging periods) and frequencies specified in the aquatic life criteria (e.g., 1 day and 3 years for the CMC and 4 days and 3 years for the CCC). So, for example, for the CMC, a 4B3 flow is computed, where the 4B3 is a biology-based 4-day average flow event, which occurs (on average) once every 3 years. The computer program DFLOW (most recent version at the time of this writing was version 3.1 released in 2006) is available from the U.S. EPA for the computation of biology-based flows and XQY flows (e.g., 7Q10) (http://www.epa.gov/waterscience/models/dflow/).

4.6.2.1.4   IHA and RVA Approaches

The IHA approach, which was developed by the Nature Conservancy along with Brian Richter, is to first analyze a suite of parameters associated with the natural flow regime, using daily flow data such as from the USGS, and then to analyze and interpret those parameters to determine whether some impact has occurred or is occurring (a hydrologic alteration). The IHA approach includes a suite of hydrologic parameters (Table 4.8) that are ecologically meaningful and serve to characterize the range of inter- and intra-annual flow variations as characterized by

- Flow magnitude
- Return interval (or frequency)
- Duration and timing
- Rate of change

The IHA also calculates the parameters for five different types of environmental flow components: low flows, extreme low flows, high-flow pulses, small floods, and large floods. A total of 34 additional parameters are used to characterize the environmental flows (Nature Conservancy 2007). The strategy is implemented in an IHA software package distributed by the Nature Conservancy (2007), which can use historical data downloaded from the USGS as well as other data. The software can be used to compare between specific sites for which there are historical flow data or to compare contiguous periods of the flow record to each other to determine if a hydrologic alteration has occurred. These parameters can be analyzed using nonparametric or parametric statistics.

TABLE 4.8
IHA Parameters and Their Ecosystem Influences

IHA Parameter Group	Hydrologic Parameters	Ecosystem Influences
1. Magnitude of monthly water conditions	Mean or median value for each calendar month Subtotal 12 parameters	• Habitat availability for aquatic organisms • Soil moisture availability for plants • Availability of water for terrestrial animals • Availability of food/cover for fur-bearing mammals • Reliability of water supplies for terrestrial animals • Access by predators to nesting sites • Influences water temperature, oxygen levels, and photosynthesis in water column
2. Magnitude and duration of annual extreme water conditions	Annual minima. 1-day mean Annual minima. 3-day means Annual minima. 7-day means Annual minima. 30-day means Annual minima, 90-day means Annual maxima. 1-day mean Annual maxima. 3-day means Annual maxima. 7-day means Annual maxima. 30-day means Annual maxima. 90-day means Number of zero-flow days Base flow index: 7-day minimum flow/mean flow for year Subtotal 12 parameters	• Balance of competitive, ruderal, and stress-tolerant organisms • Creation of sites for plant colonization • Structuring of aquatic ecosystems by abiotic vs. biotic factors • Structuring of river channel morphology and physical habitat conditions • Soil moisture stress ill plants • Dehydration in animals • Anaerobic stress in plants • Volume of nutrient exchanges between rivers and floodplains • Duration of stressful conditions such as low oxygen and concentrated chemicals in aquatic environments • Distribution of plant communities in lakes, ponds, and floodplains • Duration of high flows for waste disposal, aeration of spawning beds in channel sediments
3. Timing of annual extreme water conditions	Julian date of each annual 1-day maximum Julian date of each annual 1-day minimum Subtotal 2 parameters	• Compatibility with life cycles of organisms • Predictability/avoidability of stress for organisms • Access to special habitats during reproduction or to avoid predation • Spawning cues for migratory fish • Evolution of life history strategies, behavioral mechanisms
4. Frequency and duration of high and low pulses	Number of low pulses within each water year Mean or median duration of low pulses (days) Number of high pulses within each water year Mean or median duration of high pulses (days) Subtotal 4 parameters	• Frequency and magnitude of soil moisture stress for plants • Frequency and duration of anaerobic stress for plants • Availability of floodplain habitats for aquatic organisms • Nutrient and organic matter exchanges between river and floodplain • Soil mineral availability • Access for waterbirds to feeding, resting, and reproduction sites • Influences bedload transport, channel sediment textures, and duration of substrate disturbance (high pulses)
5. Rate and frequency of water condition changes	Rise rates: Mean or median of all positive differences between consecutive daily values Fall rates: Mean or median of all negative differences between consecutive daily values Number of hydrologic reversals Subtotal 3 parameters Grand total 33 parameters	• Drought stress on plants (falling levels) • Entrapment of organisms on islands, floodplains (rising levels) • Desiccation stress on low-mobility stream edge (varial zone) organisms

Source: Nature Conservancy, Indicators of hydrologic alteration, Version 7 user's manual, The Nature Conservancy, Arlington, VA, 2007. With permission.

Once the IHA program has assessed the statistics, these metrics can then be used to determine whether there has been some hydrologic alteration to the system or to determine the natural range of variability approach (RVA) for management. A management strategy for the RVA would ensure that the annual values of each IHA parameter fall within or outside the selected range of natural variations for that parameter at the same frequency of occurrence as during the natural flow period (Richter et al. 1996, 1997, 1998). The three methods available in the IHA program for the analysis of the flow parameters are briefly described as follows (taken from Swanson 2002):

- IHA Analysis: This analysis provides a preimpact versus a postimpact comparison. A "scorecard" table and graphs of each IHA parameter are produced to quantify the changes in each parameter between the preimpact and postimpact flow regimes.
- Range of Variability Analysis: In this analysis, the user defines three categories that divide the range of preimpact data values. Delineation into three equal categories, for example, would result in parameters less than or equal to the 33rd percentile, parameters between the 33rd and 67th percentile, and parameters greater than or equal to the 67th percentile. The program then calculates the expected frequency at which the postimpact values should occur in each category. A comparison of these two data sets provides a preimpact and a postimpact measure of the hydrologic alteration between flow regimes.
- Trend Analysis: This analysis produces a complete graphical historical analysis of the IHA parameters, together with a linear regression analysis of the data.

4.6.2.1.5   HAT

The HAT is a primary component of the USGS National Hydrologic Assessment Tool (NATHAT). Based on daily and peak (optional) streamflow data for a period of record, it is designed to

- Establish a hydrologic baseline (reference time period)
- Establish environmental flow standards
- Evaluate past and proposed hydrologic modifications

The suite of software is composed of four components:

1. Hydrologic Index Tool (HIT; Henriksen et al. 2006) is a generic tool (i.e., not developed for any particular geographic region) to calculate the 171 statistical hydrologic indices presented in Olden and Poff (2003) based solely on the input of USGS streamflow data for any gaged site.
2. NATHAT, version 3.0, is a nationwide customization of the HIT based on the six stream classifications of Olden and Poff (2003). The 6 stream classes were pared down from 10 original classes developed by Poff (1996) based on a study of 420 gages across the contiguous United States. Although the NATHAT performs the full complement of 171 statistical routines, the default graphical presentation of the results is limited to the 10 nonredundant, critical indices identified by Olden and Poff (2003) for each of their 6 national stream classes.
3. New Jersey Hydrologic Assessment Tool (NJHAT) is a New Jersey-specific regionalization of the NATHAT incorporating the results of the New Jersey Stream Classification Tool (NJSCT) and the identification of 10 primary flow indices for each of the 4 stream classes.
4. NJSCT, version 1.0, is a New Jersey-specific tool to partition that state's gaged streams into four stream classes, termed A, B, C, and D, by their relative degree of skewness of daily flows (high versus low) and by the relative frequency of low-flow events (high versus low). Currently, there is no comparable national stream classification tool.

Six stream classes are available in the NATHAT, but no national stream classification tool is available. The typical procedure would be to review the national classification system (Poff 1996), based on six stream classes and ten primary indices and then assign the study stream to a stream class using the NATHAT (Poff 1996; Olden and Poff 2003). The NATHAT would then be used for the analyses. Hersh and Maidment (2006) compared the IHA and NATHAT for use in developing environmental flows for Texas streams and rivers.

4.6.2.2 Hydraulic Models

4.6.2.2.1 Wetted Perimeter Method

The wetted perimeter method assumes that there is a direct relation between the wetted perimeter in a riffle and fish habitats in streams (Annear and Conder 1984; Lohr 1993). The wetted perimeter is the width of the streambed and streambanks in contact with water for an individual cross section. The concept may be illustrated considering a rectangular cross section of width B and depth Y. The wetted area of the cross section would be B*Y, while the width of the wetted perimeter would be B + 2Y. Another common hydraulic characteristic is the hydraulic radius, which is the wetted area divided by the wetted perimeter.

The wetted perimeter method is used as a measure of the availability of aquatic habitats over a range of discharges, and the relationship can be computed using field data or hydraulic models (Annear and Conder 1984; Nelson 1984). A plot of the wetted perimeter and discharge is used to determine the streamflow required for habitat protection, often identified by the point of maximum curvature.

4.6.2.2.2 R2-Cross Method

The R2-Cross method is also known as the "Colorado method" or the "critical area method" (Nehring 1979; Espegren 1996). It has long been used by the Colorado Water Conservation Board to establish instream flow requirements to "preserve the natural environment to a reasonable degree." The method is based on one or more cross sections and the hydraulic modeling and selection of established width, depth, and velocity criteria intended to provide for the "reasonable" protection of aquatic resources.

The R2-Cross method requires the selection of a critical riffle along a stream and assumes that a discharge chosen to maintain a habitat in the riffle is sufficient to maintain habitats for fish in nearby pools and runs for most life stages of fish and aquatic invertebrates (Nehring 1979). The streamflow requirements for habitat protection in riffles are determined from flows that meet the criteria for three hydraulic parameters: mean depth, percent of bankfull wetted perimeter, and average water velocity. The R2-Cross method has been found to produce flow recommendations that are similar to those determined by more data-intensive techniques such as the instream-flow incremental methodology (Nehring 1979; Colorado Water Conservation Board 2001).

4.6.2.3 Habitat Models

4.6.2.3.1 IFIM

The IFIM is a commonly applied software system that is used to integrate microhabitat suitability and macrohabitat suitability into habitat units that are then related to flow over time (Schroeter et al. 2005). The habitat time series output displays the availability of a suitable habitat for the period of interest. For example, the IFIM uses five main steps in assessing the impacts of dams or abstractions (Acreman and King 2003):

1. "The problems are identified and broad issues and objectives are related to legal entitlement identification."
2. "The technical part of the project is planned in terms of characterizing the broad-scale catchment processes, species present and their life history strategies, identifying likely limiting factors, collecting baseline hydrological, physical and biological data."

3. "Models of the river are constructed and calibrated. IFIM distinguishes between micro-habitat, commonly modeled using an approach such as PHABSIM, and macro-habitat, which includes water chemistry/quality and physio-chemical elements such as water temperature. There is a structure for specifying channel and floodplain maintenance flows but there is little guidance on specific methods. Hydrological models of alternative scenarios, including a baseline of either naturalized or historical conditions drive the habitat models. The models are integrated, using habitat as a common currency."

4. "Alternative scenarios of dam releases or abstraction restrictions are formulated and tested using the models to determine the impact of different levels of flow alteration on individual species, communities or whole ecosystems."

5. "The technical outputs are used in negotiations between different parties to resolve the issues set out in step one."

The strength of the IFIM is in its prediction of environmental impacts and its assessment of tradeoffs. For example, the IFIM might result in monthly or weekly flow envelopes within which the flow might vary depending on other uses (Schroeter et al. 2005).

The IFIM incorporates the PHABSIM as a problem-solving outline for decision making. The PHABSIM was originally developed in the late 1970s by the U.S. Fish and Wildlife Service (FWS), through the U.S. EPA-funded Cooperative Instream Flow Service Group, organized to develop methods for quantifying the impacts of altered streamflows. The Instream Flow group developed the IFIM (Stalnaker et al. 1995; Bovee et al. 1998), of which the PHABSIM was a major component (Waddle 2001).

The PHABSIM is based on the assumption that aquatic organisms respond to changes in the hydraulic environment. The methodology first requires some relationship between the hydraulic conditions and habitat suitability. The relationship is usually expressed as univariate suitability curves for particular species and/or life stages of interest, with the suitability ranging from 0 to unity (Figure 4.30). Models are then used to determine the hydraulic (e.g., depth, velocity, and cover as a function of depth) and water quality characteristics (e.g., temperature) as a function of flow and other environmental conditions. The product of the surface area for a section of stream and the univariate suitability curve values result in a habitat index called the weighted usable area (WUA).

The PHABSIM method does not explicitly include the specification of some flow characteristic, such as magnitude, frequency, duration, timing, or rate of change. Instead, it allows for an assessment of the impact of those flow characteristics on the quality of the available habitat. For example, the WUA under one flow magnitude could be compared to another (such as perhaps due to a

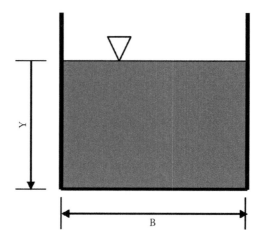

FIGURE 4.30   An example of wetted perimeter computation for a rectangular channel ($P_w = B + 2Y$).

withdrawal) to determine what impact the change may have on habitat suitability, providing a basis for mitigation or some other management action. The PHABSIM also provides a methodology that allows for the consideration of spatial variations in a habitat over the width and depth of a channel as well as along the channel length. The methodology also allows for a comparison of habitat changes (as reflected by the WUA) over time.

4.6.2.4   Holistic Models

4.6.2.4.1   BBM

The BBM is a functional analysis built on the links between the hydrology and ecology of a river system. The BBM was developed in South Africa and was designed to construct a flow regime to maintain a river in a predetermined condition (King et al. 2008). The basic premise of the BBM is that riverine species are reliant on basic elements of the flow regime that maintain the sediment dynamics and geomorphologic structure of the river. These building blocks allow for an acceptable flow regime to be established.

The BBM was developed by Jacqueline King at the University of Cape Town, South Africa (King and Louw 1998; Tharme and King 1998) and has been under development and revision since that time (King et al. 2008). Basically, the method is used to build a recommended instream flow hydrograph, or a set of hydrographs, using key pieces of information developed by a team of experts during technical studies (NRC 2005). The team of experts (physical scientists and biological scientists) follow a series of steps, assess the available data, use model outputs, and apply their combined knowledge and experience to a consensus on the building blocks of the flow regime. An example of an integrated hydrograph based on the building blocks of subsistence flows, base flows, high-flow pulses, and overbank flows is illustrated in Figure 4.31. Many other methods similar to the BBM have been developed and each

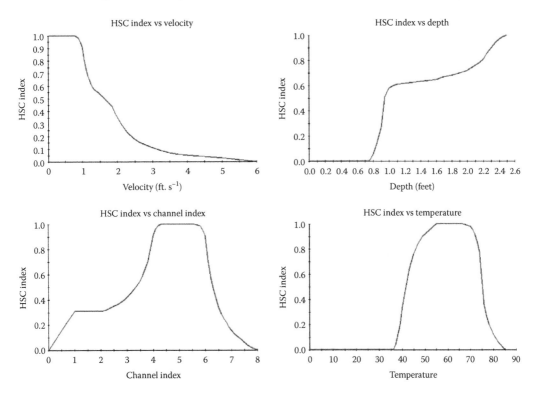

FIGURE 4.31   An example of habitat suitability criteria for adult brown trout (one life stage of one species). (From Waddle, T.J. (ed.), *PHABSIM for Windows: User's Manual and Exercises*, Midcontinent Ecological Science Center, CO, U.S. Geological Survey, Fort Collins, CO, p. 288, 2001.)

has involved the collaboration of experts in various hydrological and ecological fields. Judgments are made about the ecological consequences of various quantities and timings of flow in a river by using a mixture of existing and newly acquired data (Acreman et al. 2003).

4.6.2.4.2 European Water Framework Directive and CAMS

The CAMS was developed by the Environmental Agency (Environment Agency 2010) of England and Wales following the government's decision to apply more control on how much water is taken from rivers, reservoirs, lakes, and other waterbodies. The CAMS was developed based on the European Water Framework Directive (WFD), which was implemented into U.K. legislation in 2003. The WFD created new requirements for the protection of water resources and requires that water resources management is carried out in an integrated way.

For surface waters, the WFD requires that the impact of pressures is measured against natural flow conditions, where natural flow is defined as the flow that would occur if all artificial influences (abstractions, discharges, flow regulation) were not taking place. Surface waters are assessed as having high, good, moderate, poor, or bad ecological status. High ecological status (HES) waterbodies are those essentially undisturbed and must be maintained at that condition (not be allowed to deteriorate). The WFD requires that all surface waterbodies be at good ecological status (GES) by 2015, unless an alternative objective can be justified, where a GES basically means that flows must remain above the flows required by the biology. Figure 4.32 illustrates how the ecological status is determined in relation to the natural flow condition.

Groundwaters may be classed as either good or poor based on their chemical status and groundwater abstraction pressures and the WFD requires that all groundwater bodies achieve good status by 2015 unless alternative objectives are justified.

The CAMS process is a legal framework for the implementation of the WFD, which includes measures to control abstraction pressures and promote efficient and sustainable water use. The process is illustrated in Figure 4.33.

The CAMS process includes the participation of interested parties through a catchment stakeholder group and a resource assessment and management (RAM) framework (Acreman et al. 2003). The CAMS process (Figure 4.34) includes:

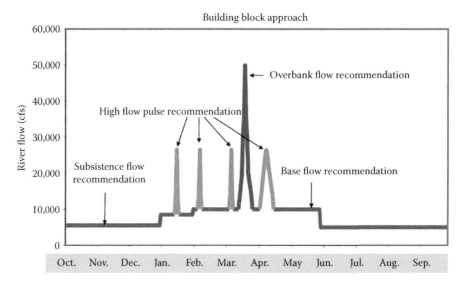

FIGURE 4.32 An example of an integrated hydrograph based on the building blocks of subsistence, base, high-flow pulses, and overbank flows. (From NRC, The science of instream flows: A review of the Texas instream flow program, Committee on Review of Methods for Establishing Instream Flows for Texas Rivers, National Research Council, The National Academies Press, Washington, DC, 2005. With permission.)

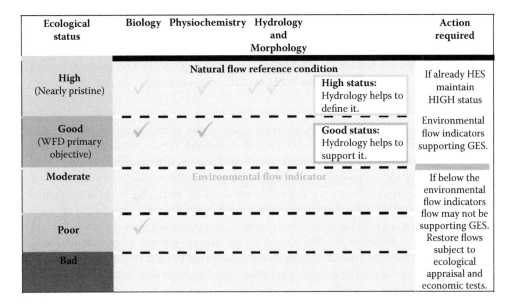

Ecological status	Biology	Physiochemistry	Hydrology and Morphology		Action required
High (Nearly pristine)	✓	✓	✓✓	**Natural flow reference condition** **High status:** Hydrology helps to define it.	If already HES maintain HIGH status
Good (WFD primary objective)	✓	✓		**Good status:** Hydrology helps to support it.	Environmental flow indicators supporting GES.
Moderate	✓			Environmental flow indicator	If below the environmental flow indicators flow may not be supporting GES.
Poor	✓				Restore flows subject to ecological appraisal and economic tests.
Bad					

FIGURE 4.33 Ecological status in relation to natural flow condition. (From Environment Agency, Managing water abstraction, Environment Agency, Bristol, UK, 2010.)

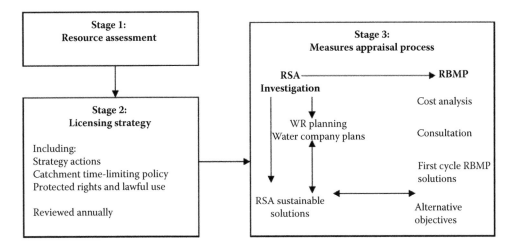

FIGURE 4.34 CAMS process. (From Environment Agency, Managing water abstraction, Environment Agency, Bristol, UK, 2010.)

- Stage 1. Resource Assessment: this step involves calculating a water balance for the study area (river flows, groundwater recharge, abstractions, etc.); a resource allocation for the environment and any other water uses or features that require protection; and a review of abstraction licenses to determine if there is potential for environmental damage and, if so, to identify options for a solution, to include the consideration of revoking, mitigating, or varying these abstraction licenses.
- Stage 2. Licensing Strategy: this states what resources are available; what conditions might apply to new licenses; whether licenses will be replaced with the same conditions; and an examination of the local impacts of the proposed abstraction or impoundment, ensuring that the rights of existing license holders and lawful water users are protected, as well as protecting the environment.

- Stage 3. Measures Appraisal Process: for waterbodies where flows are not supporting the WFD objective of GES (Figure 4.33), this process includes the implementation of a Restoring Sustainable Abstraction (RSA) program study and ultimately the development of a river basin management plan (RBMP) to set out the actions, or a program of measures, to ensure that these waters achieve WFD by 2015 and that there is no deterioration from their current status.

One of the components of the RSA program study is the participation of interested parties through a catchment stakeholder group and a RAM framework (Dyson et al. 2008), based on the assessment of four elements of the ecosystem: physical character, fisheries, macrophytes, and macroinvertebrates. Since the sensitivity of the ecosystem to reduced flows is related to the environmental flow needs of any river, rivers are given a flow status in the form of a RAM score from 1 to 5 (1 being the least sensitive to the reduction in flow and 5 being the most sensitive to the reduction in flow). Physical character is determined by comparing the river with photographs of typical river reaches in each class. Fisheries are determined by the expert opinions of environmental agency fisheries staff to classify the river according to the description of each of the RAM score classes. Macroinvertebrates and macrophytes scores are determined by using flow-sensitive metrics, such as LIFE scores. For example, in terms of physical character, rivers with steep slopes and/or wide shallow cross sections have a RAM score of 5, since very little flow reduction is needed to change the wetted perimeter of the cross section. Lowland river reaches that are narrow and deep have a RAM score of 1 (Acreman et al. 2003).

Once the scores have been determined, rivers are then categorized into one of five environmental weighting bands: Band A (most sensitive), Band B, Band C, Band D, and Band E (least sensitive). In a separate part of the RAM framework, a flow duration curve is produced where different allowable abstraction points can be found from this curve. The percentages of natural Q95 flows can be abstracted from the different bands, but more detailed methods, such as habitat modeling, are recommended where environmental flows need to be defined in more detail. The flow duration curve produced by the RAM framework focuses on ecologically acceptable flows and contains many characteristics of the flow regime. While the RAM framework contains the basic magnitudes of droughts, low flows, and floods, it does not retain characteristics such as temporal sequencing and the duration or timing of flows (Acreman et al. 2003). According to Acreman et al. (2003), an ecologically acceptable flow duration curve is most appropriate where the river ecosystem is controlled by the seasonal characteristics of dry season/wet season or winter/summer flows.

4.6.2.4.3 DRIFT

A notable example of a holistic environmental flow methodology is DRIFT, a scenario-based interactive approach to EFAs, in which a database is created that can be queried to describe the biophysical consequences of any number of potential future flow regimes (scenarios).

The DRIFT methodology focuses on identifying river water levels associated with a particular set of biophysical functions and specific hydrological and hydraulic characteristics. Similar to BBM, DRIFT uses multidisciplinary specialists to describe the consequences of reducing discharges through identified flow regimes in terms of a deterioration in biotic and abiotic conditions (King et al. 2008). DRIFT details and quantifies the links between changing river conditions and the social and economic impacts to those that rely on the river for subsistence. Acreman et al. (2003) state, "Probably it's most important and innovative feature is a strong socio-economic module, which describes the predicted impacts of each scenario on subsistence users of the resources of a river."

According to Brown and King (2000), the central idea behind DRIFT is that the various aspects of the river flow regimes result in different responses from the river ecosystem. The removal of all or part of an element in the river flow regime will have an effect on the river ecosystem differently than the removal of some other element.

Acreman et al. (2003) write that DRIFT has four modules: (1) biophysical, (2) socioeconomic, (3) scenario building, and (4) economics. Modules may be omitted according to the available common properties of scenarios. The biophysical component consists of first analyzing flow data to develop hydrologic statistics, such as seasonal flows and the number of days in seasons; seasonal flow duration curves; the magnitude, volume, and duration of high flows; and other statistics. The flow characteristics are then linked to cross-sectional characteristics. Scenarios are developed and incorporated into the database in the form of consequences due to additional possible flow regimes or subtle variations in the EFRs (King et al. 2008). For scenarios, the biotic and socioeconomic responses are then compiled and incorporated so that each scenario includes a modified flow regime, the resulting river condition, likely social and economic impacts, as well as the costs of mitigation and/or compensation for negative impacts. The DRIFT database can then be used to set flows for achieving specific objectives by incorporating a DRIFT solver. The DRIFT solver creates scenarios and an ultimate severity and integrity rating that describes the distribution of flows that would result in the best condition of the river ecosystem (Acreman et al. 2003).

4.6.2.4.4   *Nature Conservancy's Collaborative and Adaptive Process for Developing Environmental Flow Recommendations*

Another example of a method based on expert panel consensus is the Nature Conservancy's collaborative and adaptive process for developing environmental flow recommendations (Richter et al. 2006). The process includes five steps:

- An orientation meeting to inform and engage stakeholders.
- The selection of an institution or agency and the development of a literature review and a summary report on existing knowledge about flow-dependent biota and ecological processes of concern in order to identify key aspects of flow regimes important to the sustainable health of the system. Richter et al. (2006) define key questions to be addressed in the summary report.
- A workshop to develop ecological objectives and initial flow recommendations, and to identify key information gaps.
- The implementation of the flow recommendations on a trial basis to test hypotheses and reduce uncertainties.
- The monitoring of the system's response and conducting further research as warranted.

The last three steps are intended to be iterative and part of an adaptive management process.

REFERENCES

Acreman, M.C. and J. King. 2003. Defining water requirements. In M. Dyson, G. Bergkamp, and J. Scanlon (eds), *Flow: The Essentials of Environmental Flows*, Chapter 2. IUCN, Gland, Switzerland, pp. 11–30.

Acreman, M.C. and M.J. Dunbar. 2004. Defining environmental flow requirements—A review. *Hydrology and Earth Systems Sciences* 8 (5), 861–876.

Annear, T.C. and A.L. Conder. 1984. Relative bias of several fisheries instream flow methods. *North American Journal of Fisheries Management* 4, 531–539.

Arihood, L.D. and D.R. Glatfelter. 1986. Method for estimating low-flow characteristics of ungaged streams in Indiana. USGS Open File Report 86-323, p. 32.

Bedient, P.B. and W.C. Huber. 2008. *Hydrology and Floodplain Analysis*, 4th ed. Prentice Hall, Upper Saddle River, NJ.

Benson, M.A. and T. Dalrymple. 1984. Chapter A1: General field and office procedures for indirect discharge measurements. In *Techniques of Water-Resources Investigations. Book 3 Applications of Hydraulics*. United States Geological Survey, Washington, DC, p. 35.

Beran, M.A. and A. Gustard. 1977. A study into the low-flow characteristics of British rivers. *Journal of Hydrology* 35, 147–157.

Bounds, R. and B. Lyons. 1979. Existing reservoir and stream management: Statewide minimum streamflow recommendations. Texas Parks and Wildlife Department, Austin, TX.

Bovee, K.D., B.L. Lamb, J.M. Bartholow, C.B. Stalnaker, J. Taylor, and J. Henriksen. 1998. Stream habitat analysis using the instream flow incremental methodology. U.S. Geological Survey, Biological Resources Discipline Information and Technology Report USGS/BRD-1998-0004. USGS, Fort Collins, CO, p. 131.

Brown, C. and J. King. 2000. A summary of the drift process for environmental flow assessments for rivers. Southern Waters Information Report No. 01/00, Mowbray, South Africa.

Buchanan, T.J. and W.P. Somers. 1968. Chapter A7: Stage measurement at gauging stations. In *Techniques of Water-Resources Investigations. Book 3 Applications of Hydraulics.* United States Geological Survey, Washington, DC, p. 33.

Bureau of Reclamation. 2001 (Revised). *Water Measurement Manual.* U.S. Department of the Interior, Bureau of Reclamation, Water Resources Technical Publication in Cooperation with the U.S. Department of Agriculture, U.S. Government Printing Offices, Washington, DC.

Byrd, C. 2011. Water resources in the Mississippi delta. Presented at the Water for Fish and Farmers Water Research Scientists Meeting held in Stoneville, MS.

Caissie, D. and N. El-Jabi. 2003. Instream flow assessment: From holistic approaches to habitat modeling. *Canadian Water Resources Journal* 28, 173–184.

Chow, V. 1959. *Open-Channel Hydraulics.* McGraw-Hill, New York.

Clemmens, A.J., T.L. Wahl, M.G. Bos, and J.A. Replogle. 2001. *Water Measurement with Flumes and Weirs.* Water Resources Publications, LLC, Highlands Ranch, CO.

Colorado Water Conservation Board. 2001. Memorandum to Colorado Water Conservation Board Members, Available at: http://cwcb.state.co.us/isf/programs/docs/isfquantpolicy.pdf (accessed November 30, 2001).

Cusimano, B. 1992. Sumas river receiving water study, p. 19. Available at: http://www.dep.state.va.us/ (accessed January 2013).

Dyson, M., G. Bergkamp, and J. Scanlon (eds). 2008. *Flow—The Essentials of Environmental Flows*, 2nd ed. International Union for Conservation of Nature (IUCN), Gland, Switzerland.

Environment Agency. 2010. Managing water abstraction. Environment Agency, Bristol, UK.

Espegren, G.D. 1996. Development of instream flow recommendations in Colorado using R2CROSS. Colorado Water Conservation Board, Denver, CO.

Estes, C.C. 1994. Annual summary of Alaska Department of Fish and Game instream flow reservation applications. Alaska Department of Fish and Game, Fishery Data Series No. 94-37. Anchorage, Alaska.

FISRWG. 1998. *Stream Corridor Restoration: Principles, Processes, and Practices*, Federal Interagency Stream Restoration Working Group. GPO Item No. 0120-A; SuDocs No. A 57.6/2:EN 3/PT.653.

Flynn, K.M., W.H. Kirby, and P.R. Hummel. 2006. User's manual for program PeakFQ: Annual flood-frequency analysis using bulletin 17B guidelines, U.S. Geological Survey, *Techniques and Methods Book 4*, Chapter B4, p. 42.

Gordon, N.D., T.A. McMahon, B.L. Finlayson, G.J. Gippel, and R.J. Nathan. 2004. *Stream Hydrology—An Introduction for Ecologists*, 2nd ed., Wiley, Chichester, UK.

Hayes, D.C. 1991. Low-flow characteristics of streams in Virginia. U.S. Geological Survey Water-Supply Paper 2374, p. 55.

Hebert, P.D. (ed.) 2002. *Canada's Aquatic Environments*, CyberNatural Software, University of Guelph. Available at: http://www.aquatic.uoguelph.ca/.

Henriksen, J.A., J. Heasley, J.G. Kennen, and S. Niewsand. 2006. Users' manual for the hydroecological integrity assessment process software (including the New Jersey assessment tools). U.S. Geological Survey, Biological Resources Discipline, Open File Report 2006–1093, 71 p.

Hersh, E.S. and D.R. Maidment. 2006. Assessment of hydrologic alteration software. Center for Research in Water Resources, The University of Texas at Austin.

Hirsch, R.M. 1982. A comparison of four streamflow record extension techniques. *Water Resources Research* 18 (4), 1081–1088.

Hirsch. 2006. USGS science in support of water resources management. Presented at Mississippi Water Resources Conference, Jackson, MS.

Hirsch, R.M. and J.E. Costa. 2004. U.S. stream flow measurement and data dissemination improve. *Eos, Transactions, American Geophysical Union* 85 (20), 3.

Imhof, J.G. and D. Brown. 2003. Guaranteeing environmental flows in Ontario's rivers and streams. A Position Statement, prepared by Trout Unlimited Canada & Ontario Federation of Anglers and Hunters, p. 7.

Interagency Advisory Committee on Water Data. 1982. Guidelines for determining flood-flow frequency: Bulletin 17B of the hydrology subcommittee, Office of Water Data Coordination, U.S. Geological Survey, Reston, VA, p. 183. Available at: http://water.usgs.gov/osw/bulletin17b/bulletin_17B.html.

Jobson, H.E. 1996. Prediction of travel time and longitudinal dispersion in rivers and streams. USGS Water Resources Investigations Report 96-4013. U.S. Geological Survey, Washington, DC.

Kennedy, E.J. 1984. Chapter A10: Discharge ratings at gaging stations. In *Techniques of Water-Resources Investigations. Book 3 Applications of Hydraulics.* United States Geological Survey, Washington, DC, p. 69.

Kennedy, E.J. 1990. Chapter A19: Levels at streamflow gaging station. In *Techniques of Water-Resources Investigations. Book 3 Applications of Hydraulics.* United States Geological Survey, Washington, DC, p. 39.

Kilpatrick, F.A. and E.D. Cobb. 1985. Chapter A16: Measurement of discharge using tracers. In *Techniques of Water-Resources Investigations. Book 3 Applications of Hydraulics.* United States Geological Survey, Washington, DC, p. 52.

Kilpatrick, F.A. and J.F. Wilson Jr. 1989. Chapter A9: Measurement of time of travel in streams by dye tracing. In *Techniques of Water-Resources Investigations. Book 3 Applications of Hydraulics.* United States Geological Survey, Washington, DC, p. 34.

King, J.M. and D. Louw. 1998. Instream flow assessments for regulated rivers in South Africa using the building block methodology. *Aquatic Ecosystem Health and Management* 1, 109–124.

King, J.M., R.E. Tharme, and C. Brown. 1999. Definition and implementation of instream flows. World Commission on Dams Thematic Report, Vlaeberg, Cape Town, South Africa.

King, J.M., R.E. Tharme, and M.S. de Villiers. 2008. *Environmental Flow Assessments for Rivers: Manual for the Building Block Methodology* (Updated Edition), WRC Report No TT 354/08.

Korsgaard, L. 2006. Environmental flows in integrated water resources management: Linking flows, services and values. PhD. thesis. Institute of Environment and Resources, Technical University of Denmark.

Lohr, S.C. 1993. Wetted stream channel, fish food organisms and trout relative to the wetted perimeter inflection method. PhD dissertation. Montana State University.

Martin, J.L. and S.C. McCutcheon. 1999. *Hydrodynamics and Transport for Water Quality Modeling.* Lewis, Boca Raton, FL.

Matalas, N.C. 1963. Probability distribution of low flows. U.S. Geological Survey Professional Paper 434-A. United States Government Printing Office, Washington, DC.

Mazvimavi, D., E. Madamombe, and H. Makurira. 2007. Assessment of environmental flow requirements for river basin planning in Zimbabwe. *Physics and Chemistry of the Earth* 32, 995.

MDEQ. 2007. Time-of-travel of water tracing dyes in MS streams. Mississippi Department of Environmental Quality, Jackson, MS.

Morlock, S.E. 1996. Evaluation of acoustic Doppler current profiler measurements of river discharge. USGS Water Resources Investigations Report 95-4218. U.S. Geological Survey, Washington, DC.

Mueller, D.S. 2003. Field evaluation of boat-mounted acoustic Doppler instruments used to measure streamflow. In *Proceeding of the IEEE Seventh Working Conference on Current Measurement Technology 2003.* IEEE, New York, pp. 30–34.

Nathan, R.J. and T.A. McMahon. 1990. Evaluation of automated techniques for baseflow and recession analysis. *Water Resources Research* 26 (7), 1465–1473.

Nature Conservancy. 2007. Indicators of hydrologic alteration. Version 7 user's manual. The Nature Conservancy, Arlington, VA.

Nehring, R.B. 1979. *Evaluation of Instream Flow Methods and Determination of Water Quantity Needs for Streams in the State of Colorado.* Division of Wildlife, Fort Collins, CO.

Nelson, F.A. 1984. *Guidelines for Using the Wetted Perimeter (WETP) Computer Program of the Montana Department of Fish, Wildlife, and Parks.* Montana Department of Fish, Wildlife, and Park, Bozeman, MT.

NRC. 2005. The science of instream flows: A review of the Texas instream flow program. Committee on Review of Methods for Establishing Instream Flows for Texas Rivers, National Research Council. The National Academies Press, Washington, DC.

NRCS. 1998. Mississippi Delta comprehensive, multipurpose, water resource plan, study phase. USDA Natural Resources Conservation Service, Washington, DC.

NRCS. 2004. Introduction to HecRAS. Available from http://www.nrcs.usda.gov/wps/portal/nrcs/detailfull/national/water/?cid=stelprdb1042484.

Olden, J.D. and N.L. Poff. 2003. Redundancy and the choice of hydrologic indices for characterizing streamflow regimes. *River Research and Applications* 19, 101–121.

Ontario Ministry of Natural Resources. 1994. *Natural Channel Systems, An Approach to Management and Design.* Queen's Printer for Ontario, Ontario Ministry of Natural Resources, Peterborough, Ontario.

Ontario Ministry of Natural Resources, Ontario Ministry of the Environment, Ontario Ministry of Agriculture, Food and Rural Affairs, Ontario Ministry of Municipal Affairs and Housing, Ontario Ministry of Research and Innovation, Association of Municipalities of Ontario and Conservation Ontario. 2003 (Updated in 2010). Ontario low water response. Ontario Ministry of the Environment, Toronto, Ontario.

Ontario Ministry of the Environment. 2000. Guide for applying for approval of municipal and private water and sewage works, Sections 52 and 53, Ontario Water Resources Act, R.S.O.1990 (as amended by Services Improvement Act, s.o.1997). Environmental Assessment and Approvals Branch, Ontario Ministry of the Environment, Toronto, Ontario.

Poff, N.L. 1996. A hydrogeography of unregulated streams in the United States and an examination of scale-dependence in some hydrologic descriptors. *Freshwater Biology* 36, 71–91.

Poff, N., J. Allan, M. Bain, J. Karr, K. Prestegaard, B. Richter, R. Sparks, and J. Stromberg. 1997. The natural flow regime: A paradigm of river conservation and restoration. *BioScience* 47 (11), 769–784.

Pyrce, R.S. 2004. Hydrological low flow indices and their uses, WSC Report No. 04-2004. Watershed Science Centre, Peterborough, Ontario.

Rantz, S.E. et al. 1982. *Measurement of Stage and Discharge, Geological Survey Water.* Supply Paper 2175. U.S. Government Printing Office, Washington, DC.

Richter, B.D., J.V. Baumgartner, J. Powell, and D.P. Braun. 1996. A method for assessing hydrologic alteration within ecosystems. *Conservation Biology* 10, 1163–1174.

Richter, B.D., J.V. Baumgartner, R. Wigington, and D.P. Braun. 1997. How much water does a river need? *Freshwater Biology* 37, 231–249.

Richter, B.D., J.V. Baumgartner, D.P. Braun, and J. Powell. 1998. A spatial assessment of hydrologic alteration within a river network. *Regulated Rivers: Research & Management* 14, 329–340.

Richter, B.D., A.T. Warner, J.L. Meyer, and K. Lutz. 2006. A collaborative and adaptive process for developing environmental flow recommendations. *River Research and Applications* 22, 297–318.

Ries, K. 2002. StreamStats: A web site for stream information. *Symposium on Terrain Analysis for Water Resources Applications, December 16–18, 2002.* Center for Research in Water Resources, University of Texas, TX.

Ries, K.G. and P.J. Friesz. 2000. Methods for estimating low-flow statistics for Massachusetts streams. USGS Water Resources Investigations Report 00-4135. U.S. Geological Survey, Washington, DC.

Schroeter, H.O., A.D. Arthur, and W.S. Baskerville. 2005. Establishing environmental flow requirements for Big Creek. Long Point Region Conservation Authority, Tillsonburg, Ontario.

Shrivastava, G.S. 2003. Estimation of sustainable yield of some rivers in Trinidad. *Journal of Hydrologic Engineering* 8, 35–40.

Simpson, M. 2001. Discharge measurements using a broad-band acoustic Doppler current profiler, Open-File Rep. 01-01. U.S. Geological Survey, Sacramento, CA.

SJRWMD. 2001. Minimum flows and levels: Fact sheet. St. Johns River Water Management District, Palatka, FL.

Smakhtin, V.Y. 2001. Low flow hydrology: A review. *Journal of Hydrology* 240, 147–186.

Smakhtin, V.Y. and M. Toulouse. 1998. Relationships between low-flow characteristics of South African streams. *Water SA* 24, 107–112.

South Dakota Department of Environment and Natural Resources. 1998. Mixing zone and dilution implementation procedures. Available at: http://www.state.sd.us/denr/denr.html (accessed March 2004).

Stalnaker, C.B. 1994. Evaluation of instream flow habitat modeling. In P. Calow and G.E. Petts (eds), *The Rivers Handbook. Vol. 2. Hydrological and Ecological Principles.* Wiley, pp. 276–286.

Stalnaker, C., B.L. Lamb, J. Henriksen, K. Bovee, and J. Bartholow. 1995. The instream flow incremental methodology: A primer for IFIM. U.S. Geological Survey Biological Report 29. USGS, Washington, DC.

Story, A.H., R.L. McPhearson, and J.L. Gaines. 1974. Use of fluorescent dye tracers in Mobile Bay. *Journal of Water Pollution Control Federation* 46 (4), 657–665.

Swanson, S. 2002. Indicators of hydrologic alteration, Resource Note No. 58. Bureau of Land Management, Denver, CO.

Telis, P.A. 1992. Techniques for estimating 7-day, 10-year low flow characteristics for ungaged sites on streams in Mississippi. USGS Water Resources Investigations Report 96-4130. U.S. Geological Survey, Jackson, MS.

Tennant, D.L. 1975. Instream flow regimes for fish, wildlife, recreation, and related environmental issues. U.S. Fish and Wildlife Service, Billings, MT.

Tennant, D.L. 1976. Instream flow regimens for fish, wildlife, recreation and related environmental resource. *Fisheries* 1, 6–10.

Tharme, R.E. 2003. A global perspective on environmental flow assessment: Emerging trends in the development and application of environmental flow methodologies for rivers. *River Research and Applications* 19, 397–441.

Tharme, R.E. and J.M. King. 1998. Development of the building block methodology for instream flow assessments, and supporting research on the effects of different magnitude flows on river in ecosystems. Water Research Commission Report No. 576/1/98. Water Research Commission, Pretoria.

Tortorelli, R.L. 2002. Statistical summaries of streamflow in Oklahoma through 1999. USGS Water Resources Investigations Report 02-4025. U.S. Geological Survey, Washington, DC.

TRG. 2008. Review of desk-top methods for establishing environmental flows in Texas rivers and streams. Final report to the Texas Commission on Environmental Quality, Technical Review Group (TRG), Texas Commission on Environmental Quality, Austin, Texas.

USGS. 1995. Stream-gaging program of the U.S. Geological Survey, U.S. Geological Survey Circular 1123. USGS, Reston, VA.

USGS. 2000. The history of stream gaging in Ohio. USGS Fact Sheet 050–00. U.S. Geological Survey, Reston, VA.

USGS. 2007. U.S. Geological Survey streamgaging. Fact Sheet 2005–3131, March 2007. U.S. Department of the Interior, U.S. Geological Survey

USEPA. 1986. Chapter 1: Stream design flow for steady-state modeling, *Technical Guidance Manual for Performing Wasteload Allocations: Book VI: Design Conditions.* EPA document EPA440/4/86–014. U.S. EPA, Washington, DC.

Waddle, T.J. (ed.) 2001. *PHABSIM for Windows: User's Manual and Exercises.* Midcontinent Ecological Science Center, Open file Report 01-340. U.S. Geological Survey, Fort Collins, CO.

Wali, M.K., F. Evrendilek, and S. Fennessy. 2009. *The Environment: Science, Issues and Solutions.* CRC Press, New York.

Winkler, M.F. 2006. Acoustic Doppler current profilers broadband band and narrowband technology explained. USACE Engineering Research and Development Center, Vicksburg, MS.

Wurbs, R.A. and W.P. James. 2002. *Water Resources Engineering.* Prentice-Hall, Upper Saddle River, NJ.

YMD. 2004. YMD joint management district annual report 2004. YMD Joint Water Management District, Stoneville, MS.

5 Selected Water Quality Processes in Rivers and Streams

5.1 INTRODUCTION

A major issue for many rivers and streams is water quantity. However, it is not just the quantity of water that is important, for both human use and aquatic organisms, but also the water quality. Water quality has long been an issue in our streams and rivers. Some of the earlier concerns regarding water quality had more to do with navigation rather than aquatic health. For example, the Rivers and Harbors Act of 1899 was established to protect navigation by banning the dumping of refuse matter into waterways. This law led to a federal permitting process that is still used today by the U.S. Army Corps of Engineers (the Corps) to protect navigable waterways. Of course, implementation also requires a definition of "navigable waters." For example, one definition from the Corps (the legal definition of "traditional navigable waters") is

> A water body qualifies as a 'navigable water of the United States' if it is (a) subject to the ebb and flow of the tide, and/or (b) the water body is presently used, or has been used in the past, or may be susceptible for use (with or without reasonable improvements) to transport interstate or foreign commerce.

An alternative definition of a navigable waterbody is any waterbody that the courts have determined to be navigable. The definition of what is a navigable waterway is of particular recent importance with regard to the protection of wetlands, subject to them being directly connected to a navigable waterway, as will be discussed in a later chapter.

In addition to the impacts on navigation, other water quality problems (e.g., hypoxia and contamination by toxicants) have also long been an issue in the United States. However, only in relatively recent decades has water quality been enough of a concern that effective laws and regulations have been promulgated to protect not only human uses but also aquatic health.

During the early development of the United States and during the Industrial Revolution beginning in the mid-1800s, rivers were often used as a waste conduit, and raw wastewater from cities and towns and untreated industrial waste were often directly discharged into rivers and streams. An example is the Nashua River, Massachusetts. During the 1800s, mill cities such as Gardner, Fitchburg, Leominster, and Nashua rose up around the centers of industrial production. Paper, shoe, and textile factories along the river discharged untreated waste into the river with the result that the river changed color almost daily, due to the dyes released from paper production. By 1965, the Nashua was declared one of the most grossly polluted rivers in the nation. In other areas, river fires were becoming increasingly common. Some of the most famous were the fires on the Cuyahoga River. In 1936, 1952, and again in 1969, floating oil and debris caught fire on the Cuyahoga River near Cleveland, Ohio (Figure 5.1).

It was not until the 1940s and 1950s that water quality became of such national concern that federal legislation was passed to protect water quality. Some of the early acts included the Federal Water Pollution Control Act of 1948, the 1965 Water Quality Act, and the 1966 Clean Water Act (CWA). However, these early laws often lacked standards or methods to assess the extent of the pollution, methods or allocations of funds to reduce the pollution, and methods or means to enforce that pollution reduction.

FIGURE 5.1   1952 floating oil and debris fire on the Cuyahoga River near Cleveland, Ohio. (From NOAA, Available at http://oceanservice.noaa.gov/education/kits/pollution/02history.html.)

During the 1960s and 1970s, national concern with regard to the quality of the nation's waters increased. During this period, Rachel Carson's (1962) *Silent Spring* was published and helped launch a national environmental movement. Scientific reports and articles were common, detailing the pollution problems. Organizations such as the National Wildlife Federation, the Izaak Walton League, and the National Audubon Society campaigned for strong federal water quality bills. On April 22, 1970, approximately 20 million people participated in the first Earth Day. Earth Day was spearheaded by Wisconsin Senator Gaylord Nelson to draw attention to the environmental issues plaguing the planet and human health and it continues to be celebrated each April 22.

The results of the national environmental movement included a series of strong federal environmental laws, one of the most powerful of which was the 1972 Amendments to the Federal Water Pollution Control Act, now popularly known as the CWA, by the 92nd U.S. Congress. The framework of that act included (USEPA 2000c):

- The establishment of the National Pollutant Discharge Elimination System (NPDES), a program that requires that every point source discharger of pollutants obtains a permit and meets all the applicable requirements specified in the regulations issued under Sections 301 and 304 of the act. These permits are enforceable in both federal and state courts, with substantial penalties for noncompliance.
- The development of technology-based effluent limits, which serve as the minimum treatment standards to be met by dischargers.
- An ability to impose more stringent water quality-based effluent limits (WQBELS) where technology-based limits are inadequate to meet state water quality standards or objectives.
- The creation of a financial assistance program to build and upgrade publicly owned treatment works (POTWs).

Major water quality concerns at the time included the bacterial contamination of waters and low dissolved oxygen (DO) concentrations resulting from high loads of organic materials, such as from untreated wastewater. As bacteria degrade organic materials in waters that contain oxygen (oxic waters), DO is consumed.

In 1968, approximately 39% of the people served by POTWs in the United States received less than secondary treatment (raw and primary; USEPA 2000c). Primary treatment only removes floating and suspended solids, thereby removing only a fraction of the organic material impacting DO.

From 1970 to 1995, the U.S. Environmental Protection Agency (U.S. EPA) provided \$61.1 billion in Federal Construction Grants Program funds to help fund new or upgrade existing POTWs (Figure 5.2) to secondary or higher treatment. Secondary treatment includes a biological treatment process for the removal of dissolved organic matter from wastewater. The treatment plants also typically include a final treatment to disinfect the wastewater stream prior to its discharge, thereby treating two of the primary water quality concerns: bacterial contamination and DO impacts due to excess organic materials. As a result, the number of people served by POTWs with secondary or greater levels of wastewater treatment almost doubled from 1968 to 1996 (USEPA 2000a; Stoddard et al. 2002). Case studies for nine urban waterways with historically documented water pollution problems (the Connecticut River, Hudson-Raritan Estuary, Delaware Estuary, Potomac Estuary, James Estuary, Chattahoochee River, Ohio River, Upper Mississippi River, and Willamette River) indicated that following the implementation of pollution control policies as required by the CWA, worst-case DO levels increased from 1–4 mg L^{-1} for most of the case study sites during the period 1961–1970 to 5–8 mg L^{-1} during 1986–1995 (USEPA 2000a; Stoddard et al. 2002).

In addition to construction grants, the NPDES was a major component of the CWA, requiring all point source dischargers to obtain a discharge permit. In general, the formulation of regulations for the implementation of the CWA provisions and the enforcement of those regulations were the responsibilities of the U.S. EPA, but they were delegated to the states.

Also fundamental to the CWA and the establishment of permit limits is the feature requiring states to develop water quality standards for waterbodies, to monitor compliance with these standards, and to report impairments (Section 305). Water quality standards are the foundation of the water quality-based control program and they define the water quality goals for a waterbody. The standards also established water quality-based treatment practices for dischargers (WQBELS). The U.S. EPA publishes standards (under Section 304(a) of the CWA) that states can use or modify, or states can adopt criteria based on other scientifically defensible methods. A water quality standard consists of three basic elements:

- Beneficial designated uses (e.g., activities such as swimming, drinking water supply, and oyster propagation and harvest)
- Numeric criteria protective of designated uses
- Antidegradation policies

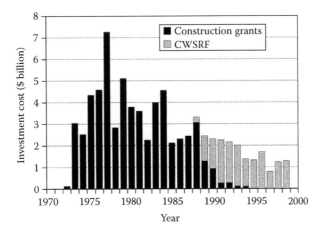

FIGURE 5.2 Annual funding provided by the U.S. EPA's construction grants and the Clean Water State Revolving Fund (CWSRF) Program to local municipalities for improvements in water pollution control infrastructure from 1970 to 1999 (costs reported in current year dollars). (From USEPA, Progress in water quality: An evaluation of the national investment in municipal wastewater treatment, Environmental Protection Agency, Office of Water, Washington, DC, 2000a.)

The standards may be narrative or specific (numeric) to a certain chemical. An example of a narrative criterion could be

> The waters of the State may not be polluted by ... substances ... that are unsightly, ... odorous, ... create a nuisance, or interfere directly or indirectly with water uses,

while an example of a numeric criterion could be

> the water quality standard for dissolved oxygen may be 5.0 milligrams per liter oxygen for some designated uses and 6.0 for others.

The standards, particularly the numeric standards, were and are used to determine allowable, or permitted, loads to waterbodies (the wasteload allocation process) or to determine if a waterbody is meeting water quality objectives. If waterbodies are not meeting standards, then states must report this to Congress (in the National Water Quality Inventory Report to Congress or the 305(b) report).

There is also a provision in the CWA for the case where, following the implementation of the best available technology and/or effluent limitations (for point sources), water standards were not achieved. Under those conditions, Section 303(d) required that states report impaired waterbodies to Congress and establish the total maximum daily loads (TMDLs) that were required to meet the water quality standards. That TMDL included the cumulative impacts of all point sources and nonpoint sources, and allowed for a margin of safety. Section 303(d) was largely unenforced until the early 1990s when lawsuits forced environmental agencies to implement its provisions. Today, as a result of including nonpoint sources, water quality control and management have shifted from point source control to watershed management.

5.2   LIGHT

Light is not something for which there is usually a water quality standard, but light (heat energy) does directly affect temperature and it is, of course, necessary for photosynthesis. To describe the impact of light on rivers and streams, it is first necessary to describe the characteristics and fate of light in the environment.

First of all, light is not "a" thing. For example, is light a "wave," having a wavelength and a frequency? Or, is light a particle (or a stream of particles, photons, or quanta)? The idea of light as a wave is sufficient for our purposes, and the range of radiation over all wavelengths is called the electromagnetic spectrum (Figure 5.3). For water quality impacts, we are typically concerned with only a small portion of the electromagnetic spectrum.

Our primary source of light is from the sun, and the incoming light available near the earth is relatively constant (solar constant = 440 BTU ft.$^{-2}$ h^{-1} = 1390 W m^{-2}). The amount of light reaching

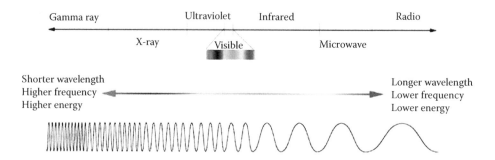

FIGURE 5.3   Electromagnetic spectrum. (From NASA, electromagnetic spectrum, Available at http://imagine.gsfc.nasa.gov/docs/science/know_l1/emspectrum.html.)

the upper atmosphere is impacted by the relative distance from the sun to the earth and declination, which vary over the course of a year. The amount of light reaching the upper atmosphere at a particular location depends on factors such as the latitude and the time of day. All of these factors are very predictable and are included in many heat budget models (Martin and McCutcheon 1999). Of the light reaching the upper atmosphere, some is reflected by the atmosphere and some is damped by scattering and absorption by dust and atmospheric moisture.

The light reaching the water surface of streams and rivers, in the form of heat energy, along with other processes, impacts the water temperature, as will be discussed in the following section. Some of the light reaching the water surface is direct solar radiation (shortwave radiation) and some of the light is reflected from clouds (typically longwave radiation) (Figure 5.4). Shortwave radiation is composed of ultraviolet, visible light and near-infrared radiation with wavelengths of about 200 and 4000 nm. Longwave radiation is composed of infrared radiation with wavelengths of about 4,000–100,000 nm.

Light also directly impacts photosynthetic organisms in rivers and streams. Chlorophylls a, b, and c are common pigments used by plants and chlorophyll concentrations are often used as an estimate of plant biomass since it is easier to measure concentrations of these pigments than to measure biomass. Chlorophylls absorb light most strongly in the red and violet parts of the spectrum, while green light is poorly absorbed. Typically, the conversion to energy involves absorbing a wavelength, extracting the energy, and then emitting light at a longer wavelength (less energy). The process is called fluorescence and it can be used to measure chlorophyll concentrations using devices such as a fluorometer. Essentially, light of a certain wavelength is emitted (an excitation beam) and the strength of the light fluoresced is an indication of the amount of chlorophyll present. The excitation beam is typically in the blue range (440–460 nm depending on the analysis) and the light emitted and measured is in the red range (685 nm). Only chlorophyll-a can use light to produce chemical energy, while chlorophyll-b and c are accessory pigments, which help capture and transfer the

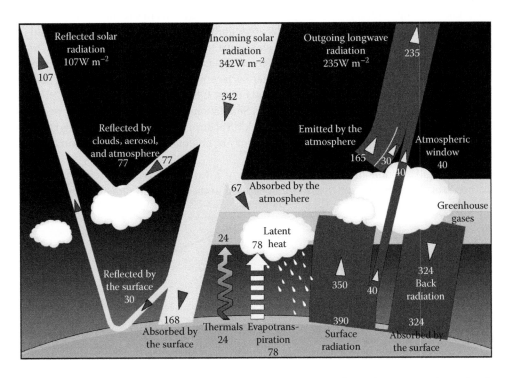

FIGURE 5.4 Solar radiation budget. (From NASA, Available at http://ceres.larc.nasa.gov/ceres_brochure.php?page = 2.)

energy. In cyanobacteria (blue-green algae), the phycobiliproteins are the predominant accessory pigment, giving the group their characteristic blue-green or red color.

Since the absorption of light in photosynthesis is dependent on the wavelength, only a portion of the electromagnetic spectrum is usable by plants and is referred to as photosynthetically available radiation (PAR). PAR is usually measured in the visible light spectrum, with wavelengths of about 400–700 nm.

The absorption of light can also cause the degradation of some compounds, such as some toxic organic chemicals, in a process called photodegradation or photolysis. Photolysis is a function of the quantity and wavelength distribution of incident light, the light adsorption characteristics of the compound, and the efficiency at which absorbed light produces a chemical reaction. Photolysis is classified into two types that are defined by the mechanism of energy absorption. Direct photolysis is the result of the direct absorption of photons by the chemical molecule. Indirect or sensitized photolysis is the result of energy transfer to the chemical from some other molecule that has absorbed the radiation (Wool et al. 2003). The sorptive properties are specific to the organic chemicals and are often available in chemical handbooks.

Processes such as heat transfer, photosynthesis, and photodegradation (photolysis) are functions of both the wavelengths and the intensity of light. For photosynthesis, productivity usually increases with light and reaches a maximum at some saturation light intensity, or the point where growth is at maximum (Figure 5.5). Beyond that point in some aquatic plants, photoinhibition can occur where growth (photosynthesis) decreases with increasing light intensity. The saturation light intensity varies with the aquatic plant. Since light varies over the course of a day, photoinhibition can vary with the season and over the course of a day. For example, during the summer months, optimal light may occur in the morning or afternoon and intensities may be great enough during the middle parts of the day to be photoinhibiting. Also, some plants are more shade or sun tolerant than others, so the saturation intensities vary. Diurnal cycles in production also affect variations in DO, pH, and nutrient cycling.

For studies of temperature, productivity, and photolytic chemicals in streams or rivers, estimates of the light intensity over specific wavelengths are required. This intensity can be estimated (predicted) or measured. Methods for predicting shortwave and longwave radiation for heat transfer in models of water temperature have been widely used and are described in detail by Martin and McCutcheon (1999). Often, the PAR at the water surface is assumed to be a fixed fraction (such 0.47) of the solar radiation (Szeicz 1974; Baker and Frouin 1987, as cited by Chapra 2007). For models of toxic chemicals, such as the Water Analysis Simulation Program (WASP) (Wool et al. 2003) and the Exposure Analysis Modeling System (EXAMS) (Burns and Cline 1985), the light intensity at each of a series of wavelengths (46 wavelengths) may be computed as a function of time, location, and atmospheric conditions.

Radiation is also commonly measured using one of a variety of available sensors. Depending on the needs of a particular study, sensors can be used to measure total radiation (pyranometers),

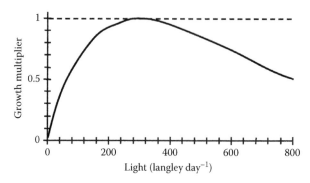

FIGURE 5.5 Light limitations to algal growth.

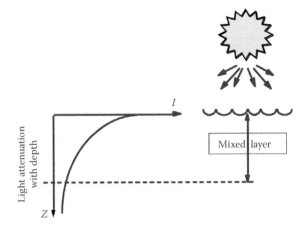

FIGURE 5.6 Light attenuation with depth.

longwave radiation (pyrgeometers), visible light (photometers), or PAR (Strangeways 2003). When planning a field study or using the results from measurements by others, attention should be given as to exactly what radiation measures will be (were) taken.

For aquatic studies, it is not only the light at the surface that is important, but also the light in the water column or that reaching the substrate. Light typically decreases exponentially from the surface (Figure 5.6, as described by Beer's law; Chapra 1997) as it is absorbed by the water and the materials suspended or dissolved in it. The sorption of light also varies with the wavelength. For example, red is absorbed most quickly, while blue light penetrates the farthest.

Light penetration is most commonly measured using an underwater photometer. A general procedure is to maintain a surface photometer for measuring surface light and measuring light as a function of depth in order to estimate the light extinction coefficient. Light attenuation in the water column is discussed in greater detail for lakes and reservoirs.

5.3 TEMPERATURE

Temperature is a master variable affecting many aspects of habitat and water quality. Temperature impacts the degradation of materials. For example, for materials degrading following the Q10 rule, the rate of degradation would double for each 10°C increase in temperature ($2 = 1.07^{T-20}$, where T is the temperature in degrees Celsius). Temperature impacts fisheries growth, reproduction, migration, susceptibility to diseases, and behavior. Temperature also impacts the timing of invertebrate reproductive cycles, such as for mayfly hatches that hatch at times of the year when temperatures are within certain ranges. Temperature also affects oxygen solubility and other processes related to the health of aquatic systems.

Heat variations occur largely as a result of exchanges of heat energy across a water surface. Models of heat exchange for predicting temperature are commonly used and methods are described in detail in Chapra (1997) and Martin and McCutcheon (1999). Shortwave and longwave radiation add heat energy, some of which is emitted back into the atmosphere (via blackbody radiation). Gradients between the water and the atmosphere in terms of the temperature and moisture content (as measured by relative humidity) result in exchanges due to conduction, convection, and evaporation (Figure 5.7).

The water temperature is a function of surface heat exchange and other factors. For example, shading by banks and riparian vegetation and processes such as fog formation in the early morning hours can reduce heat exchange, particularly in low-order streams and rivers. Also, as water movement slows, the impact of heat exchange is greater and the stream is warmer. In very large streams,

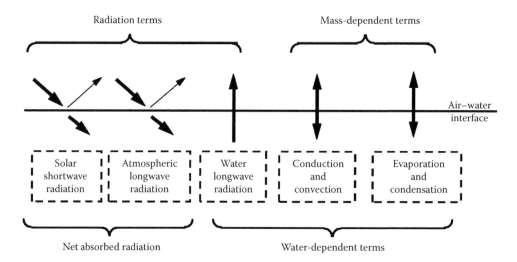

FIGURE 5.7 Heat exchange processes at the water surface. (After Chapra, S.C., *Surface Water-Quality Modeling*, WCB McGraw-Hill, New York, 1997.)

the water mass is usually great enough that much more energy is required to cause a change in temperature, so the impact is less. The result is that mid-order streams typically have a greater range of maximum daily temperatures than low-order or high-order streams (Allan and Castillo 2007) (see Figure 5.8).

Human impacts on the hydraulics of rivers and streams (hydromodifications) also have impacts on temperatures. For example, damming a river can change a cold-water habitat to a warmwater habitat. Conversely, releasing waters from the bottom of reservoirs in the south in hydropower operations has changed many warmwater fisheries to cold-water fisheries (much appreciated by southern trout fishermen). Thermal pollution, such as that due to temperature increases resulting from the cooling operations of power generation facilities, has negatively impacted aquatic systems. In the 2012 U.S. EPA's national summary of impaired waters and TMDL information, temperature or thermal pollution is the tenth largest cause of water quality impairments in the United States, as listed in the Causes of Impairment for 303(d) Listed Waters.

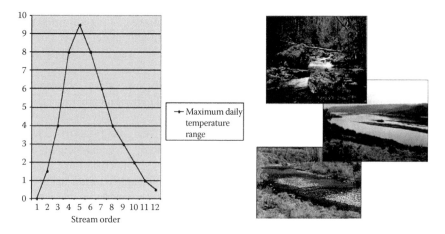

FIGURE 5.8 Relationship between water temperature and stream order. (From Water on the Web, Monitoring Minnesota Lakes on the Internet and training water science technicians for the future—A national on-line curriculum using advanced technologies and real-time data. University of Minnesota, Duluth, MN, 2004. Available at http://WaterOntheWeb.org. With permission.)

All states have standards for temperature, which vary according to the degree of cold-water or warmwater habitats. For example, for the Lower Columbia River in Oregon:

Temperature shall not exceed 20°C due to human activities. When natural conditions exceed 20°C no temperature increases will be allowed which will raise the receiving water temperature by greater than 0.3°C.

While for Mississippi, the regulation (in part) reads:

The maximum water temperature shall not exceed 90°F (32.2°C) in streams, lakes and reservoirs, except that in the Tennessee River the temperature shall not exceed 86°F (30°C). In addition, the discharge of any heated waters into a stream, lake or reservoir shall not raise temperatures more than 5°F (2.8°C) above natural background temperatures.

5.4 SEDIMENTATION

Sediments naturally occur in waters as a function of a variety of processes. However, excess sediments can result in degradation of the water quality. Under the CWA, state agencies are required to identify and report to the U.S. EPA any waters of their states that are impaired (not meeting water quality standards) under Sections 305(b) and 303(d) of the CWA. The U.S. EPA, in turn, prepares a summary of the quality of the nation's waters and submits it to Congress. In the 2002 report (Figure 5.9), sediment/siltation was listed as the number one cause of impairment in rivers and streams, with over 100,000 mi. of rivers and streams impaired. In the 2004 report (Figure 5.10), the

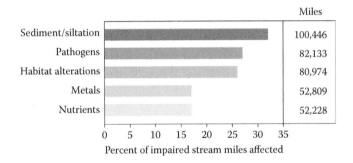

FIGURE 5.9 Listing of causes of impairments from the national water quality inventory: Report to Congress 2002.

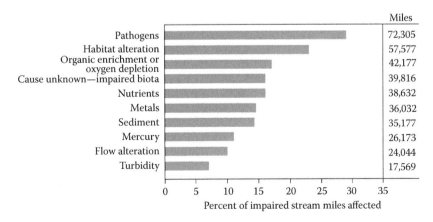

FIGURE 5.10 Listing of causes of impairments from the national water quality inventory: Report to Congress 2004.

number of impaired stream miles due to sediments decreased. However, this was primarily a result of differences in the listing methods and does not imply that those waters that were impaired in 2002 were somehow unimpaired by 2004. In the 2012 list, sediments were the fourth largest cause of waters not meeting water quality standards.

What is somewhat surprising, given the number of river miles identified as being impaired, is that among the state agencies there are few numeric criteria for sediments or siltation. Instead, in most cases, only narrative criteria are available, such as "The suspended sediment load and suspended sediment discharge rate of surface waters shall not be altered in such a manner as to cause nuisance or adversely affect beneficial uses." Therefore, most of the listings are based on qualitative rather than quantitative assessments. What is recognized though is the national scale of the problem, where excess sediments "smother stream beds, suffocate fish eggs and bottom-dwelling organisms, and interfere with drinking water treatment and recreational uses" (USEPA 2002). Some of the other impacts are illustrated in Figure 5.11.

The leading source of impairments as listed in the 2004 305(b) report was agricultural activities, such as crop production, grazing, and animal feeding operations (USEPA 2004; Figure 5.12). For example, some grazing practices with concentrated livestock distributions are known to adversely affect aquatic, riparian, and upland vegetation, resulting in bank destabilization and stream channel gullying. Diminished upland vegetative cover and soil compaction also result in increases in rill and sheet erosion.

The loss of soil from agricultural land has long been recognized as a problem. For example, the cartoons of Jay Norwood (Ding) Darling in the 1930s brought public attention to a number of environmental issues, including the erosion of the topsoil from his native Iowa (Figure 5.13). Ding stated that "the top soil which goes swirling by in our rivers at flood stage may look like mud to you but it is beefsteak and potatoes, ham and eggs and homemade bread with jam on it."

Sediments from agricultural runoff increase sedimentation and other related processes in streams, which, in turn, impact the stability of stream channels. As discussed in Chapter 2, undisturbed natural channels will reach an equilibrium condition, as described using the channel evolution model of

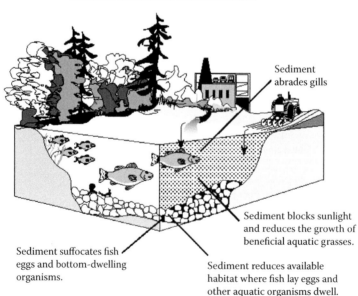

FIGURE 5.11   Impacts of siltation, from water quality assessments, part I. (From USEPA, National water quality inventory 2000 report, Environmental Protection Agency, Office of Water, Washington, DC, 2000b.)

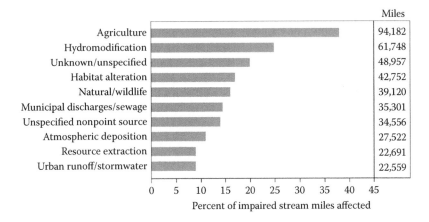

FIGURE 5.12 Listing of sources of impairments from the national water quality inventory: Report to Congress 2004.

FIGURE 5.13 "What that mud in our rivers adds up to each year" cartoon by J.N. (Ding) Darling from the 1930s. (Courtesy of the Jay N. "Ding" Darling Wildlife Society.)

Simon and Hupp (1986). If a channel is disturbed by some drastic change such as forest clearing, urbanization, dam construction, or channel dredging, then instability sets in with scouring of the bed, and destructive bank erosion and channel widening occur due to the collapse of sections of the bank. The banks will continue to cave into the stream, widening the channel, and the stream also begins to aggrade, or fill in, with sediment from eroding channel sections upstream. In the final stage, aggradation continues to fill the channel, reequilibrium occurs, bank erosion ceases, and riparian vegetation once again becomes established. See Chapter 3 for a more detailed discussion.

An important impact of destabilization is the consequent increase in destructive bank erosion, as illustrated in studies of the Goodwin Creek Watershed, Mississippi, by the National Sedimentation Laboratory (Simon 2009; Figure 5.14). For such a system, the contribution of sediments from bank erosion can often far exceed the contribution of sediments from channel sources, as illustrated in Table 5.1.

It is not only the magnitude of the sediments that is an issue, but also their size distribution. Some of the commonly used sediment size categories are listed in Table 5.2. The adverse effect of fine sediment deposits on benthic organisms is well documented in the literature. Suspended fine sediments reduce light transmission, which can limit the growth of aquatic plants, while deposited fine sediments can alter benthic habitats and entomb benthic organisms. In addition, contaminants are most closely associated with cohesive sediments, such as silts and clays.

The size distribution of sediments also impacts sediment transport. For example, the settling of particles in water is impacted by their size and density, so that in a high-energy stream channel, the bottom may be armored (coarser particles overlying finer) or consist primarily of coarse particles. Finer particles are more typically found in areas of lower energy, such as the inside of bends. As the characteristics of rivers change from their headwaters, changes also occur in sediment grain size, with bed material grain size decreasing with increasing distance from the headwaters (Figure 5.15).

FIGURE 5.14 Bank erosion in the Goodwin Creek watershed. (From Simon, A., Adjustment processes in fluvial systems implications for streambank instability and control, National Sedimentation Laboratory, Oxford, MS, from a workshop presented at the EWRI Water Resource Congress 2009, Kansas City, MO, 2009.)

TABLE 5.1
Comparison of Relative Contributions of Bank Erosion

Stream	Ecoregion	Dominant Bed Material	Contribution from Banks (%)
James Creek, MS	Southeastern Plains	Sand/clay	78
Shades Creek, AL	Ridge and Valley	Gravel	71–82
Goodwin Creek, MS	Mississippi Valley Loess Plains	Sand/gravel	64
Yalobusha River, MS	Southeastern Plains	Clay/sand	90[a]
Obion-Forked Deer River, TN	Mississippi Valley Loess Plains	Sand	81[a]

Source: Simon, A., Adjustment processes in fluvial systems implications for streambank instability and control, National Sedimentation Laboratory, Oxford, MS, from a workshop presented at the EWRI Water Resource Congress 2009, Kansas City, MO, 2009.

[a] Represents the contribution from banks relative to all channel sources.

TABLE 5.2
Sediment Size Categories

Category	Dia. (mm)	Wentworth Scale
Boulder	>256	<−8
Cobble		
Large	128–256	−7
Small	64–128	−6
Pebble		
Large	32–64	−5
Small	16–32	−4
Gravel		
Coarse	8–16	−3
Medium	4–8	−2
Small	2–4	−1
Sand		
Very coarse	1–2	0
Coarse	0.5–1	1
Medium	0.25–0.5	2
Fine	0.125–0.25	3
Very fine	0.063–0.125	4
Silt	<0.063	>5

Source: Water on the Web, Monitoring Minnesota Lakes on the Internet and training water science technicians for the future—A national on-line curriculum using advanced technologies and real-time data, University of Minnesota, Duluth, MN, 2004. Available at http://WaterOntheWeb.org.

The goal of many river management and restoration projects is to reduce sediment loadings and the adverse impacts of sediments. This is commonly done by the implementation of best management practices (BMPs). Guidance on sediment BMPs is available from a variety of agencies and organizations, including the Natural Resources Conservation Service (NRCS), the USDA Forest Service (USFS), the Bureau of Land Management (BLM), state transportation

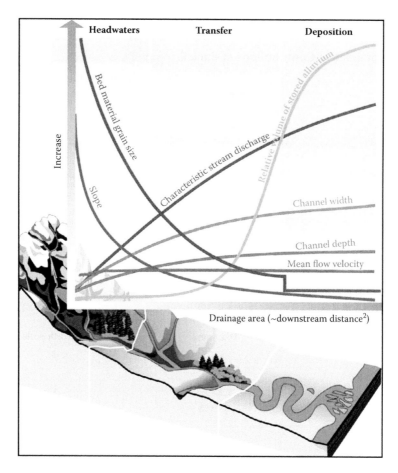

FIGURE 5.15 Longitudinal changes in river characteristics. (From FISRWG, *Stream Corridor Restoration: Principles, Processes, and Practices*, Federal Interagency Stream Restoration Working Group, 1998.)

departments, conservation districts, and many state water quality and forest management agencies.

Another devastating type of sediment flow is debris flow. Generally with sediment transport, the concentrations of the sediment are low enough that they do not dramatically alter the characteristics of the flow. In particular, sediments do not impact the Newtonian fluid properties. For Newtonian fluids, the relationship between stress and strain is linear, where the proportionality constant is the viscosity or "thickness" of the water. For a non-Newtonian fluid, the thickness varies with the applied stress. Another result is that water will continue to flow, while if a non-Newtonian material moves it may leave a hole behind. In hyperconcentrated flow, such as when large volumes of sand are transported in suspension, the amount of suspended sediment is sufficient to change the fluid properties and sediment transport mechanisms. Hyperconcentrated flow can be highly erosive. In debris flow, the sediments and water form a slurry capable of holding large particles and even boulders in suspension (Figure 5.16), and the impacts of these mud slides may be catastrophic (Costa 1988; Jakob and Hungr 2005), destroying bridges, culverts, roads, and aquatic habitats. Debris flows can occur, for example, from a rainfall event on a burned area, following liquefaction of hillsides, or volcanic eruptions such as Mount St. Helens (see Figure 5.17).

FIGURE 5.16 Aftermath of debris flow in Jones Gap State Park during the morning of Monday, June 26, 2006 (the large boulder in the middle of the image is several feet in diameter). (From Moore, P.D., Flash flooding along the Blue Ridge, National Weather Service Forecast Office, Greenville–Spartanburg, SC, 2006.)

FIGURE 5.17 Mount St. Helens debris flow. (From USGS, Available at volcanoes.usgs.gov/hazards/lahar/rain.php. Photograph by K. Scott on June 24, 1990.)

5.5 DISSOLVED OXYGEN

DO is a fundamental water quality parameter and one that is often used to assess the health of aquatic systems. While some rivers and streams are naturally low in DO, low DO is most commonly associated with pollution.

First, a little terminology and background related to oxygen in the environment. Some of the processes impacting DO are illustrated in Figure 5.18. Autotrophic (*auto* or self, *trophic* food or nutrition) organisms produce complex organic compounds from inorganic compounds in the presence of an energy source. Those that do so, such as plants, in the presence of light (photo) are referred to as phototrophs and the rate of biomass production is referred to as primary production (and the organisms as primary producers). Heterotrophs (*hetero* from different or other) utilize the organic matter to produce chemical energy, and produce inorganic compounds in the process. Both heterotrophs and autotrophs respire, during which carbon dioxide is produced. So, during daylight hours, phototrophs may both respire and produce, and only respire during dark hours, resulting in diel variations in oxygen and carbon dioxide concentrations.

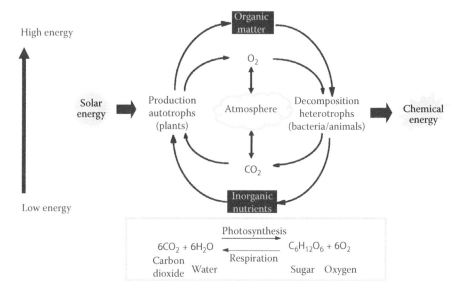

FIGURE 5.18 Production/decomposition cycle. (From Chapra, S.C., *Surface Water-Quality Modeling*, WCB McGraw-Hill, New York, 1997.)

During biomass production by heterotrophs (secondary production), energy is also required, generally produced by chemical oxidation. Oxygen is most efficient in terms of energy production, and organisms that use oxygen as an energy source (such as ourselves) are referred to as aerobic organisms, and an aquatic environment with oxygen is an aerobic environment. During aerobic processes, oxygen is consumed and carbon dioxide is released.

Anaerobic organisms, on the other hand, obtain chemical energy by reducing oxidized compounds such as NO_3, MnO_2, FOOH(s), SO_4, and CO_2, releasing soluble compounds such as ammonia (NH_4), hydrogen sulfide (H_2S), soluble Mn, Fe, and methane (CH_4).

Organisms have variously adapted to aerobic or anaerobic environments (Figure 5.19). Fish gills promote the diffusion of DO from water to the blood, similar to the lungs. However, there is considerably less oxygen in water than in air and oxygen diffuses more slowly in water than in air, making the process more difficult. Fish may increase the flow of oxygen by swimming or opening and closing their gill flaps (the *operculum*) to pump water across the gills. Some aquatic insects will surface and use atmospheric oxygen, using tubes or siphons to obtain atmospheric oxygen, or they will surface and bring air bubbles with them to the subsurface. Other aquatic insects have gills or hairs to facilitate oxygen transfer, or diffuse oxygen across the body wall. Chironomid midge larvae (bloodworms) have hemoglobin (rare in insects), allowing them to live in environments with low DO concentrations.

FIGURE 5.19 Oxygen-related adaptations. (From EFISH: The Virtual Aquarium. Available at http://cnre. vt.edu/efish/; and from State Hygienic Laboratory at the University of Iowa, Available at http://www.uhl. uiowa.edu/services/limnology. Photographs by Bob Jenkins and Noel Burkhead. With permission.)

Anaerobic environments are common in aquatic sediments, below the thin oxidized surface layer, but anaerobic conditions in the water column are very undesirable and often symptomatic of pollution. Anaerobic environments are also referred to as anoxic, and water with a diminished supply is referred to as hypoxic. Another phrase in common usage today to describe hypoxic waters is *dead zones*. Dead zones are becoming increasingly common in a variety of aquatic habitats.

So, how much oxygen is necessary for a healthy aquatic environment? The amount would vary with the type of environment and what could be considered healthy. One measure is the regulatory standards and criteria proposed by the U.S. EPA and adapted by state agencies. The ambient dissolved standards included in the U.S. EPA's quality criteria for water (USEPA 1986) are listed in Table 5.3. For cold-water environments and early life stages where intergravel (the hyporheic zone) concentrations are critical, the ambient criteria are those that would be expected to result in the intergravel concentrations given in parentheses. To reflect the diel variations that may occur, some states have adapted standards for both daily averages (such as 5 mg L^{-1}) and instantaneous minimums (such as 4 mg L^{-1}).

DO standards and criteria are also used to determine how much material may be discharged into a stream or river without violating those standards, used as a basis for establishing discharge permits (wasteload allocations). The typical result of a discharge of organic materials into a stream is a decrease in DO concentrations. However, the impact of the deoxygenation is not only a function of the amount of material discharged, but also the combined function of all of the instream processes impacting DO concentrations. The typical result is a longitudinal variation in DO concentrations, as illustrated by the idealized DO sag curve illustrated in Figure 5.20.

TABLE 5.3
Ambient Dissolved Oxygen Standards (mg L^{-1})

	Cold-Water Criteria		Warmwater Criteria	
	Early Life Stages 1,2	Other Life Stages	Early Life Stages 2	Other Stages
30-day mean	NA	6.5	NA	5.5
7-day mean	9.5 (6.5)	NA	6.0	NA
7-day minimum	NA	5.0	NA	4.0
1-day minimum	8.0 (5.0)	4.0	5.0	3.0

Source: USEPA, Quality criteria for water 1986, United States Environmental Protection Agency, Office of Water Regulations and Standards, Washington, DC, 1986.

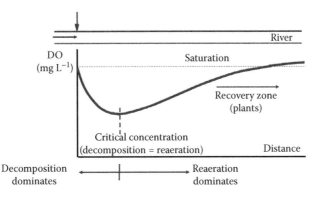

FIGURE 5.20 Dissolved oxygen sag. (From Chapra, S.C., *Surface Water-Quality Modeling*, WCB McGraw-Hill, New York, 1997.)

If there was no DO production or consumption in a stream or river, it would be expected that the DO concentration would be that in equilibrium with the atmosphere, the saturation concentration. If the water concentration was greater than the saturation concentration, then the DO net transfer would be out of the water, while if the water concentration was lower, the net transfer would be into the water until equilibrium was once again achieved, a process called reaeration. The rate of the transfer, or the reaeration rate, depends on the streamflow, wind, and other factors that will be discussed later.

When an organic material is introduced and heterotrophic organisms are available to consume it, oxygen is consumed (deoxygenation). In a stream or river, the material would also move downstream. Assuming uniform flow conditions, the distance downstream could be estimated as a product of the average velocity and travel time. If the rate of deoxygenation is greater than the rate of reaeration, the DO concentration would decline. The decline would continue as the oxygen-consuming materials move downstream with the flow until the rate of deoxygenation equals the rate of reaeration. This point would be the critical concentration point, where the DO concentration is lowest. Often, the goal of a wasteload allocation (or TMDL determination) would be to ensure that the critical concentration is not less than the water quality standard, often with some margin of safety included. The critical point is not static and could move further upstream or downstream as flows vary.

Beyond the point of the critical concentration, the DO concentration would increase until at some distance downstream equilibrium conditions with the atmosphere are achieved. However, other processes may occur that impact the DO profile, such as other oxygen demands (e.g., sediment demands) and plant productivity and respiration. Productivity and respiration may also result in diel variations, so that, for example, in some highly productive systems, oxygen concentrations may exceed saturation during the day and drop to their lowest levels, following nighttime respiration, in the early morning hours. Each of these processes will be discussed in greater detail in the following sections.

5.5.1 SATURATION

Air contains about 21% oxygen. However, the solubility of oxygen in water is much less. The water concentration will tend toward an equilibrium condition so that there is no gradient in the partial pressure of oxygen in the atmosphere and the partial pressure of air in water. The DO saturation concentration is that which is in equilibrium with atmospheric concentrations. The saturation concentration decreases with increasing temperature and dissolved solids concentrations, as illustrated in Figure 5.21. As a result, for example, waters in warm saline environments would be expected to have lower DO concentrations than cold freshwater mountain streams. Also, since the partial pressure of oxygen in the atmosphere decreases with altitude, there is a corresponding decrease in the saturation concentration, as illustrated in Figure 5.22.

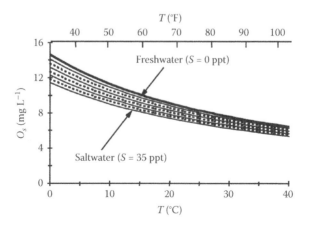

FIGURE 5.21   Variations in dissolved oxygen saturation concentrations as a function of temperature.

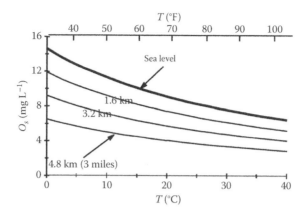

FIGURE 5.22 Variations in dissolved oxygen saturation concentrations (O_s) as a function of temperature.

There are a number of well-known methods to estimate saturation concentrations, and one commonly used formulation is published by the American Public Health Association (APHA 2005). Agencies such as the U.S. Geological Survey (USGS) also provide saturation calculation tables on their websites (DO saturation tables: http://water.usgs.gov/software/DOTABLES/).

5.5.2 REAERATION

Reaeration refers to the exchange of oxygen between water and the atmosphere. The net rate of reaeration depends on the factors influencing the exchange and the oxygen gradient between water and the atmosphere, as expressed by the saturation concentration discussed in Section 5.5.1. If the water concentration is at saturation, there is no gradient and the net exchange is zero. The greater the gradient is, the greater the next rate of exchange will be.

Since the exchange rate occurs at the water surface, any forcing, such as turbulence introduced by currents or wind that "stirs" the water, will increase the rate of exchange. For example, the rate of exchange in a stagnant reach of river would be slow, even if the gradient were large. However, wind and wave action would enhance the transfer. Similarly, the rate of reaeration in a stream or river is directly proportional to the current velocity, and inversely proportional to the depth.

Reaeration can also be induced by turbulence in riffle areas and waterfalls in a stream or river or by flows over structures such as weirs (Figures 5.23 and 5.24). In some cases, structures are designed for the specific purpose of increasing DO concentrations.

FIGURE 5.23 Reaeration over waterfall. (Photograph by J.L. Martin.)

FIGURE 5.24   Reaeration in flow over a weir. (Photograph by J.L. Martin.)

A variety of equations have been developed to predict reaeration from currents, winds, and over structures, and the reader is referred to Thomann and Mueller (1987), Chapra (1997), and Duan et al. (2009) for detailed explanations. Reaeration is frequently measured in the field, and field measurements have formed the basis for many of the commonly used reaeration equations.

Some of the methods previously used to measure reaeration are no longer an option. For example, Owens et al. (1964) and others deoxygenated water by adding sodium sulfite in order to estimate its reaeration. Tsivoglou (1967), Tsivoglou and Wallace (1972), and Wilhelms (1980) measured gas transfer at the water surface using injections of radioactive krypton-85 to mimic oxygen exchange while using tritium as a tracer to quantify the effects of mixing and dilution.

The method of using a continuous injection of a conservative tracer coupled with a nonconservative tracer to mimic oxygen is still used today for measuring reaeration; however, more commonly, rhodamine WT dye is used as the conservative tracer while nonradioactive krypton gas is used to mimic oxygen (Kilpatrick et al. 1989; USEPA SESD-EAB 2007). Since the noble gas krypton is inert and is already a component of air, it does not pose any ecological threat. This method also requires estimates of transport, thereby producing estimates of flow, time of travel, mean reach velocity, and depth. Other similar techniques have used chlorine (Cl⁻) or bromide (Br⁻) as a conservative tracer and propane or sulfur hexafluoride (SF_6) as a nonconservative tracer (Fellows et al. 2001; Grace and Walsh unpub., cited in Grace and Imberger 2006; LINX 2004). Other methods commonly used involve open water methods such as those based on diel variations in oxygen concentration. Grace and Imberger (2006) describe a variety of methods to estimate reaeration. Another method used for measuring reaeration is the floating diffusion dome method (Copeland and Duffer 1963; Hall 1970), illustrated in Figure 5.25, which Koenig and Murphy (2001) reported to be fast and inexpensive, but not as accurate as the tracer techniques.

5.5.3   BIOCHEMICAL OXYGEN DEMANDS

The decomposition of organic materials discharged into streams and rivers results in the subsequent decrease in DO concentrations, or deoxygenation, as a result of aerobic decomposition. The goal of wasteload allocations and permitting of discharges is to determine an allowable concentration of organic materials that could be discharged while still resulting in DO concentrations that meet standards and criteria. So, initially, a relationship between concentrations of organic material and consequent reductions in oxygen is needed. This could be accomplished with stoichiometry using a balanced chemical reaction and if the composition of the organic matter were known (e.g., as glucose, $C_6H_{12}O_6$). However, in the United States, a more common approach is to measure the oxygen

FIGURE 5.25   Diffusion dome. (From USEPA Region 4 Science and Ecosystem Support Division, Ecological Assessment Branch.)

consumption directly and the measure most commonly used by U.S. regulatory agencies is the biochemical oxygen demand (BOD).

Essentially, BOD is determined by measuring the decrease in oxygen concentrations over time in a water sample. First, an initial DO concentration is determined and a water sample is placed in a stoppered bottle, which is then incubated at a proscribed temperature and in the dark for a period of time, usually 5 days, after which the DO concentration is determined again. The difference between the two concentrations is the oxygen consumed (oxygen demand exerted) during the period. Since the DO can be consumed by both biological and chemical processes, the result is called biochemical oxygen demand and is expressed in oxygen units (milligrams per liter).

Most commonly in the United States, the incubation period is 5 days at 20°C, and the demand exerted is called BOD_5 (Figure 5.26). The demand could also result from the decomposition of carbonaceous materials or nitrogenous materials, and there are standard methods for determining either CBOD or NBOD, which is important since the timing and organisms impacting their degradation in aquatic systems are different. Most commonly today CBOD is determined by adding a chemical to the test that inhibits nitrogenous demands. Nitrogenous demands are estimated from ammonia concentrations and stoichiometric relationships between ammonia oxidation and oxygen consumption. The long-term impacts are also important, resulting in what is known as the ultimate BOD. The standard methods for 5-day test periods (standard method

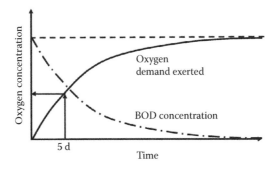

FIGURE 5.26   BOD curves.

5210B*), the oxygen consumed after 60–90 days of incubation (5210C), the continuous oxygen uptake (5210D), and the use of ammonia for estimating nitrogenous demands are provided in *Standard Methods for the Examination of Water and Wastewater* (APHA 2005).

While the amount of oxygen consumed is important, the rate at which it is consumed is also critical for estimating the impacts of BOD on DO in rivers and streams. The rate of decrease in the BOD concentration and the corresponding increase in the oxygen demand exerted (Figure 5.26) are typically exponential and can be described as a first-order process with a rate in units of days^{-1}. The typical rates for untreated sewage range from 0.2 to 0.5 day^{-1}, while the rates for sewage following primary treatment range from 0.1 to 0.3 day^{-1} (Chapra 1997). The more highly treated or resistant the materials are to decay (refractory), the lower the rate and the greater the ratio of BOD_U and BOD_5 will be.

Consider a first-order decay rate of 0.1 day, where after 1 day approximately 90% of the BOD would remain, and 10% of the BOD would be exerted. Similarly, on the second day, another 10% of the BOD remaining would be exerted. The time required for 50% of the BOD to decay (and/or be exerted) would be about 7 days [ln(0.5)/0.1], and the time required for 90% of the BOD to decay (or demand exerted) would be 23 days [ln(0.1)/0.1]. Assuming a constant discharge and a representative flow velocity of 1 ft. s^{-1}, the distance downstream where the BOD concentrations would be half and 10% of the initial concentrations would be 113 and 376 mi., respectively. So, depending on the rate of decomposition, the impact of a discharge can be far downstream of the source and varies with both the rates of flow of the system and the rates of deoxygenation.

5.5.4 PRODUCTIVITY AND RESPIRATION

The productivity of aquatic plants can add oxygen over the course of a day, while respiration consumes oxygen, resulting in sinusoid-like diel variations in DO concentrations. Depending on plant densities and other factors, the diel variations can be quite large, as illustrated in Figure 5.27 for the Cahaba River, Alabama. DO variations also occur as plant densities vary over seasons or years.

A variety of primary producers may occur in streams and rivers. In deeper and more quiescent waters, floating or planktonic plants that are suspended in the water column and carried by currents

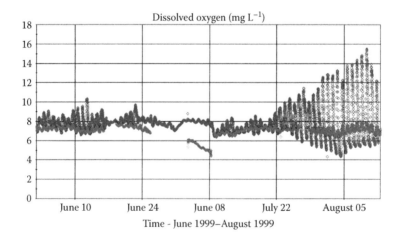

FIGURE 5.27 Diurnal fluctuation of dissolved oxygen concentrations at Piper Bridge (blue) and Shelby County Highway 52 (green) in the Cahaba River, Alabama. (From ADEM, Final nutrient total maximum daily loads (TMDLs) for the Cahaba River watershed, Alabama Department of Environmental Management, Water Quality Branch, Water Division, Montgomery, AL, 2006. With permission.)

---

* Numbers correspond to standard laboratory procedures from APHA.

may be dominated by phytoplankton. In shallower waters, plants may be dominated by aquatic macrophytes. Aquatic macrophytes may be emergent plants such as cattails, rooted plants with floating leaves such as lily pads, or submerged vegetation such as the invasive Eurasian milfoil. In shallower rivers and streams, plants that are attached to rocks or other substrates, the periphyton, may dominate. The periphyton can include algae growing on the surface of rocks and stones (epilithic forms), on submerged plants (epiphytic forms), or on the bottom sediments (epipelic forms, or the benthos) of rivers.

Primary productivity is the process by which chemical energy is stored as biomass, while producing oxygen as a by-product. The total rate is called gross primary production (GPP). Since some of the energy is used in respiration, the difference between the gross productivity and that remaining after respiration is called net primary production (NPP).

Various methods have been developed to estimate GPP and NPP, primarily by measuring variations in oxygen or carbon. One commonly used instream method is the total community oxygen metabolism. This method, originally described by Odum and Hoskins (1958), has been widely used in productivity studies. Generally, the method involves the monitoring of DO concentrations over a diel period (diurnally) and analyzing the variations graphically. The method involves estimating productivity, respiration, diffusion, and any accrual from other sources (a source of error). USEPA SESD-EAB (2007) described a variation in the method whereby diffusion is measured rather than estimated using methods originally described by Odum and Hoskins (1958). Grace and Imberger (2006) describe a variety of open water methods for measuring stream metabolism.

A second commonly used method to estimate water column productivity and respiration utilizes light and dark bottles. Essentially, bottles are suspended in a water column, where in the light bottle (clear bottles) productivity and respiration occur allowing an estimation of the NPP, while in the dark bottle (a covered bottle) only respiration occurs, allowing an estimation of the GPP. The light and dark bottle method is described in detail in USEPA SESD-EAB (2007) and *Standard Methods for the Examination of Water and Wastewater* (APHA 2005).

5.5.5 SEDIMENT DEMANDS

Sediment oxygen demand (SOD) is a process that has long been known to impact the water quality of surface waters. SOD, due to the mineralization (diagenesis) of organic materials in bottom sediments, can contribute to oxygen declines in waterbodies. While impacted by the flux, or deposition, of organic materials to sediments, the relationship is nonlinear (Hatcher 1986; Chapra 1997). This nonlinear relationship results from the complex physical, biological, and chemical cycling that occurs in anaerobic sediments and the interface between the sediments and the water column.

Methods are available to estimate the SOD based on the downward flux of carbon, nitrogen, and phosphorus from the water column. In a landmark paper, Di Toro et al. (1990) developed a model of the SOD that mechanistically arrives at the observed nonlinear relationship (Chapra 1997). Di Toro, in his book *Sediment Flux Modeling* (2001), provided a further description of the techniques. Martin (2004) described the implementation of the SOD algorithms into the WASP.

SOD is also commonly measured using *in situ* chambers or cores, or based on diel variations in oxygen concentrations. Figure 5.28 illustrates a SOD chamber developed by the U.S. EPA (Murphy and Hicks 1986), and the standard operating procedure for its deployment is described in USEPA SESD-EAB (2007). A typical procedure is to place and seal a chamber on the bottom sediments and measure the DO concentrations using a probe. The rate of DO utilization provides a measure of the SOD. Grace and Imberger (2006) describe similar SOD chambers and their applications.

An alternative to measuring the SOD is community substrate oxygen demand (CSOD), which is an assessment of the respiration of all substrates. Components of CSOD include the diel curve method (Odum and Hoskins 1958), the light and dark bottle method (water column productivity and respiration), and estimates of reaeration/diffusion. A community oxygen metabolism analysis

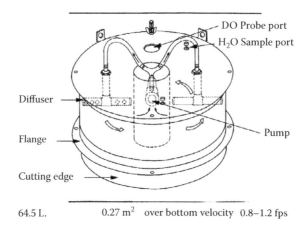

DO Probe port
H$_2$O Sample port
Diffuser
Pump
Flange
Cutting edge

64.5 L.   0.27 m^2 over bottom velocity 0.8–1.2 fps

FIGURE 5.28 *In situ* sediment oxygen demand chamber. (From USEPA SESD-EAB, Operating procedure: Sediment oxygen demand, U.S. Environmental Protection Agency, Region 4 Science and Ecosystem Support Division, Ecological Assessment Branch, Athens, GA, 2007.)

is a method for determining the oxygen production and respiration rates of a waterbody based on a graphical analysis of diurnal DO curves (Odum 1956). Essentially, the DO is monitored over a diel period and plotted, correcting for changes in saturation and diffusion (measured) and gross primary productivity and respiration. Grace and Imberger (2006) describe other open water community metabolism methods.

5.6   pH

The pH is a master variable impacting water quality. pH refers to the puissance or the power of hydrogen, or the negative log of the hydrogen ion concentration (H$^+$). So, a pH of 7 indicates that the hydrogen ion concentration is 10^{-7} mol.

In equilibrium, the product of the hydrogen ion and the hydroxide concentration (OH$^-$) would equal approximately 10^{-14} (the disassociation constant), so that if the pH is equal to 7, the pOH is 7. If the pH is less than 7, then the solution is acidic and if the pH is greater than 7, it is basic. Some typical pH values are provided in Figure 5.29.

Changes in the pH result from the addition of acidic or basic materials to aquatic systems. A change in the pH resulting from the addition of acids is not necessarily linear. Aquatic systems may have the ability to buffer or resist changes that may occur. The most common buffering system is due to the presence of carbonates and the buffering capacity is measured as carbonate alkalinity. Dissolved carbonates typically exist in one of three forms: carbonic acid, bicarbonate, and carbonate (Figure 5.30). As the pH changes, the dominant form of the dissolved carbonate also changes, in a nonlinear but easily predictable manner (Figure 5.31). So, at a pH below 4.5, all of the carbonates exist as carbonic acid (dissolved carbon dioxide). As the pH increases from 4.5 to 8.3, the balance shifts to the right and creates bicarbonate, and above 8.3, more carbonate is formed.

The impact of the carbonate species relationships can be illustrated using the alkalinity titration curve. For example, if we take a water sample that is alkaline (high pH) and add a strong acid to it (such as sulfuric or hydrochloric acid), the pH will go down, but only gradually, until we reach a pH of near 8.3, where there is a shift of the inorganic carbon from carbonates to bicarbonates. A color indicator, phenolphthalein, is commonly identified at this end point. The end point is the carbonate system and the shift from carbonates to bicarbonates will reduce the change in pH as an acid is added. This resistance to pH changes is called chemical buffering. Similarly, if we continue adding acid, we will reach another inflection point at a pH of 4.5, where the species shift is from bicarbonates to carbonic acid. A color indicator, methyl orange, is commonly identified at this end point. The total

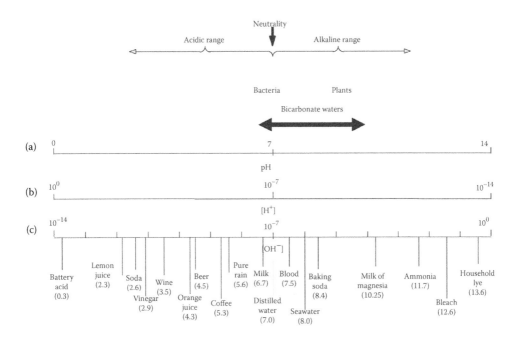

FIGURE 5.29 Typical pH values. (After Chapra, S.C., *Surface Water-Quality Modeling*, WCB McGraw-Hill, New York, 1997.)

FIGURE 5.30 Forms of inorganic carbon.

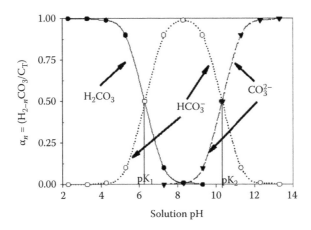

FIGURE 5.31 Variations in the fractions of inorganic forms of carbonate species as a function of pH.

amount of acid added (along with the normality of the acid and the size of the sample) is a measure of the alkalinity of the water (usually expressed as milligrams of $CaCO_3$ per liter) (Figure 5.32).

The alkalinity is then a measure of water's capacity to resist changes in pH that would make the water more acidic, commonly known as the "buffering capacity." For example, if we add acid to a system with an initial pH of 7 without any buffering capacity, the pH would drop immediately,

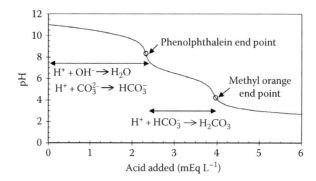

FIGURE 5.32 Alkalinity titration curve.

while if the system were highly buffered (high alkalinity), the pH would change very little or not at all until enough acid is added to overload the buffering capacity.

One prevalent problem due to emissions of sulfur dioxide (SO_2) and nitrogen oxide (NO_2) resulting from fossil fuel combustion into the atmosphere is the formation of acids, which are then deposited on surfaces by dry or wet deposition (acid rain) (Figure 5.33). Acid rain has resulted in the acidification of a large number of rivers, streams, and other waterbodies in the United States, Canada, and elsewhere, primarily those waterbodies with relatively little buffering capacity. Waterbodies in the southern United States, for example, where there is a greater prevalence of limestone ($CaCO_3$), have higher buffering capacities and a resistance to acidification. Acidification may have direct or indirect impacts on aquatic organisms. For example, in acid environments, toxic forms of metals such as aluminum may result in toxic impacts.

Acid mine drainage is another common problem affecting rivers and streams, resulting from the mining of coal or metals. Some have described the formation of acid drainage and the contaminants associated with it as the largest environmental problem facing the U.S. mining industry (USEPA 1994). In many areas, such as some western states, the very patchy distribution of tailings from abandoned mines and the resulting drainage are a pervasive problem.

Changes in pH also occur over seasons and over diel cycles in the water column and sediments due to productivity and respiration, during which CO_2 is taken up or released. In some cases, these

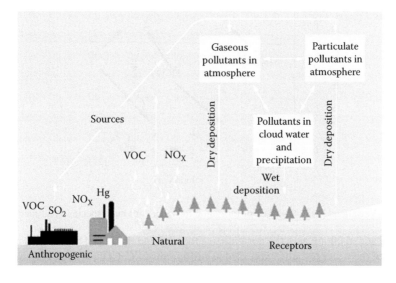

FIGURE 5.33 Acid rain. (From USEPA, Available at http://www.epa.gov/acidrain/what/index.html. With permission.)

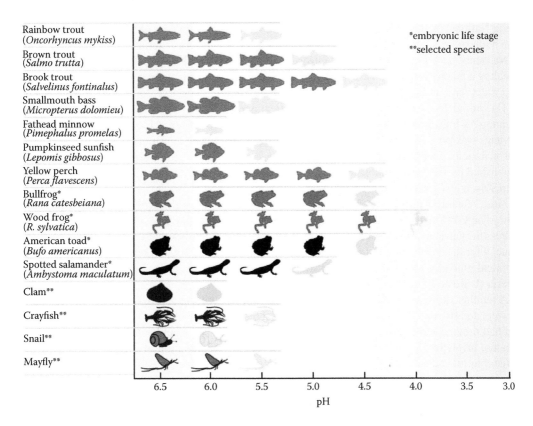

FIGURE 5.34 Effect of acidity on certain organisms and life stages. (From FISRWG, *Stream Corridor Restoration: Principles, Processes, and Practices*, Federal Interagency Stream Restoration Working Group, 1998.)

changes may be sufficient to result in impacts on aquatic organisms or toxicity. For example, a pH change of as little as 0.5 units may result in the complete sorption or desorption of some toxic metals.

The pH also directly impacts the biota of rivers and streams. Deviations toward acidity can cause direct or indirect mortality, and impacts can vary between specific species and life stages as illustrated in Figure 5.34.

As a result of the importance of pH and pH changes to aquatic health, water quality standards for pH have been promulgated by the U.S. EPA and states. For example, in Mississippi, the pH standard is given by:

> The normal pH of the waters shall be 6.0 to 9.0 and shall not be caused to vary more than 1.0 unit within this range. Variations may be allowed on a case-by-case basis if the Commission determines that there will be no detrimental effect on the stream's designated uses as a result of the greater pH change. In blackwater streams and in those watersheds with highly acidic soils, the pH may be lower than 6.0 due to natural conditions. (MDEQ 2007)

5.7 NUTRIENTS

As indicated by the National Water Quality Inventories (USEPA 2002, 2004), one of the leading causes of water quality impairment in rivers and streams is excess nutrients (see Figures 5.9 and 5.10). As with sediments, nutrients are recognized as a major cause of water quality impairment even in the absence of quantitative criteria to establish the degree or the extent of the impairment. The majority of the listings of impairments in the 305(b) reports to the U.S. EPA to date, as reflected in the National Water Quality Inventories, are based on qualitative or narrative standards, rather than numeric standards. All states are developing numeric nutrient criteria.

Nutrients, such as phosphorus and nitrogen, are required for plant growth and are necessary for a healthy aquatic ecosystem. However, excess nutrients can result in excessive plant growth that can clog waterways, reduce oxygen concentrations, and have other negative impacts on aquatic health.

The initial focus of regulation and water treatment was on DO concentrations in rivers and streams. As indicated earlier, considerable improvements have been achieved. The early focus on nutrients was primarily due to their impact on oxygen, by oxygen consumption due to nitrification, and on toxicity, such as due to unionized ammonia. Ammonia and ammonia toxicity standards were established for freshwater and saltwater as described in the U.S. EPA's ammonia aquatic life criteria (http://www.epa.gov/ost/standards/ammonia/99update.pdf).

More recently, nutrients as pollutants have become a nationwide regulatory and management issue. The CWA plan, a presidential initiative released in February 1998, included an initiative to address the nutrient enrichment problem. In 1998, the U.S. EPA developed a report entitled "National strategy for the development of regional nutrient criteria," and in 2000, it released "Nutrient criteria technical guidance manual: Rivers and streams." Today, all states are required to develop nutrient criteria (USEPA 2000).

A number of issues are involved in developing nutrient criteria and in managing nutrient loads to rivers and streams. One issue is that a large percentage of nutrient loads is from nonpoint sources that had previously been unregulated. Also, nutrient cycling is a complicated process in rivers and streams, involving the impacts of advective transport, the degradation of organic materials in the water column and sediments, and other processes impacting nutrient cycling. Also, it is not really the nutrients that are the issue, but the effect of excess nutrients. That is, since nutrients are essential to healthy ecosystems, the question becomes "how much is too much?" Most regulatory agencies and management organizations continually struggle with this issue.

5.8   TOXIC MATERIALS

The discharge of toxic materials into waters and streams without a clear understanding of the consequences of those discharges has also long been a problem, as discussed earlier in this chapter. In early U.S. history, streams and rivers were often treated as waste conduits, leading to the contamination of many rivers and streams that still persists today (Figure 5.35).

Many of the organic chemicals that are of major concern today are persistent organic pollutants (POPs), such as polychlorinated byphenyls (PCBs), dioxins, and dichlorodiphenyltrichloroethane (DDT), and other materials that have not been legally discharged for decades, in many cases not since the early 1970s. These materials often occur in very low concentrations in the water column (are hydrophobic) and are strongly associated with sediments, particularly organic sediments. Even though their concentrations in the water column are low, those concentrations are of environmental importance since the materials bioconcentrate and biomagnify. Also, many of the POPs are carcinogenic.

One example is DDT, an example of "one generation's solution is the next generation's problem" chemicals. DDT was initially heralded as an ideal insecticide. Paul Müller, who developed DDT, was awarded the Nobel Prize in Physiology or Medicine in 1948 "for his discovery of the high efficiency of DDT as a contact poison against several arthropods." The use of DDT greatly increased following World War II, and it was only later that its detrimental effects and extreme persistence in the environment were realized. Although DDT was banned in the United States in 1972, many areas in the United States remain contaminated. For example, the entire Mississippi Delta is presently under fish consumption advisories due to elevated concentrations of DDT in fish and the consequent human health risks associated with the consumption of contaminated fish.

Today, while many of the past "generation's solutions" are "this generation's problems," a new suite of toxic materials are emerging. These include a variety of pharmaceuticals and personal care products as pollutants (PPCPs). A 2002 report by the Toxic Substances Hydrology Program of the

*What Man Does To One Of The Most Beautiful Gifts
Of Nature — The River*

FIGURE 5.35 Cartoon from Jay Norwood (Ding) Darling from 1923. (Courtesy of the Jay N. "Ding" Darling Wildlife Society. With permission.)

U.S. Geological Survey (USGS 2002) indicated that testing of 139 streams in 30 states during 1999 and 2000 revealed a broad range of chemicals from human and veterinary drugs, among other household, industrial, and agricultural chemicals. The chemicals detected included insect repellant, caffeine, steroids, hormones, and other compounds (Buxton and Kolpin 2002). One of the associated problems is that wastewater treatment plants are not designed to remove these materials and may actually contribute to their concentrations in receiving streams. Figure 5.36 shows a framework of how PPCPs are introduced and cycle in the environment.

One class of PPCPs of particular concern is the endocrine disruptors, which are synthetic chemicals that mimic or block hormones, resulting in a disruption of the body's normal functions (Natural Resources Defense Council 1998). Human hormones, such as estrogen, were reported as early as 1965 and it was determined that the hormones were not completely removed by wastewater treatment (Stumm-Zollinger 1965). In more recent years, synthetic hormones from birth control have been detected (Tabak and Bunch 1970). Human and synthetic hormones have been linked to deformities (Routledge et al. 1998). Other known or suspected endocrine disruptors include some pesticides and plasticizers.

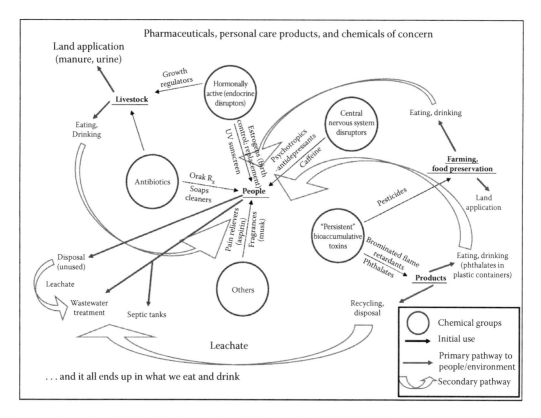

FIGURE 5.36 Framework of PPCPs. (From Florida Department of Environmental Protection, Pharmaceuticals, personal care products and chemicals of concern: A briefing on an emerging issue, retrieved November 9, 2008, from Florida Department of Environmental Protection, 2006, Available at http://www.dep. state.fl.us/waste/quick_topics/publications/shw/meds/PPCPBriefingForWeb112006.pdf. With permission.)

Another major problematic class of PPCPs is antibiotics, both from human and animal production usage. A major concern is that once discharged into the aquatic environment, bacterial strains will develop that are resistant to antibiotics (Hirsch et al. 1999). The Center for Disease Control (CDC) estimates that 70% of all hospital-contracted infections are resistant to at least one antibiotic (Chander et al. 2007). A major and controversial use of antibiotics in the environment is in the animal production industry (Phillips et al. 2004), and Denmark and the European Union have banned the use of antibiotics for growth production (Smith et al. 2005). The discharge of antibiotics to and from wastewater treatment facilities is also an issue since these facilities grow microorganisms to degrade waste. Recent research has suggested that "that wastewater treatment process contributes to the selective increase of antibiotic resistant bacteria and the occurrence of multi-drug resistant bacteria in aquatic environments" (Zhang et al. 2009) as reflected in a University of Michigan News Service Headline (May 8, 2009) "Bacteria create aquatic superbugs in waste treatment plants."

Other personal care products entering the environment, including cosmetics, fragrances, sunscreens, antibacterial products, and other materials, are essentially unregulated and their impacts are unknown. So, will PPCPs be our next generation's problem?

REFERENCES

ADEM. 2006. Final nutrient total maximum daily loads (TMDLs) for the Cahaba river watershed. Alabama Department of Environmental Management, Water Quality Branch, Water Division, Montgomery, AL.
Allan, J.D. and M.M. Castillo. 2007. *Stream Ecology*, 2nd ed. Springer, Dordrecht, Netherlands.

APHA. 2005. *Standard Methods for the Examination of Water and Wastewater*, 21st ed. American Public Health Association, American Water Works Association and Water Environment Federation, Washington, DC.

Baker, K.S. and R. Frouin. 1987. Relation between photosynthetically available radiation and total insolation at the ocean surface under clear skies. *Limnology Oceanography* 32, 1370–1377.

Burns, L.A. and D.M. Cline. 1985. *Exposure Analysis Modeling System: Reference Manual for EXAMS II.* EPA-600/3-85-038, U.S. Environmental Protection Agency, Athens, GA.

Buxton, H.T. and D.W. Kolpin. 2002. Pharmaceuticals, hormones and other organic wastewater contaminants in U.S. streams, 1999–2000: A national reconnaissance. *Environmental Science and Technology* 36 (6), 1202–1211.

Carson, R. 1962. *Silent Spring*. Houghton Mifflin, Boston, MA.

Chander, Y., S.C. Gupta, S.M. Goyal, and K. Kumar. 2007. Perspective. Antibiotics: Has the magic gone? *Journal of the Science of Food and Agriculture* 87 (5), 739–742.

Chapra, S.C. 1997. *Surface Water-Quality Modeling*. WCB McGraw-Hill, New York.

Chapra, S.C., G.J. Pelletier, and H. Tao. 2007. *QUAL2K: A Modeling Framework for Simulating River and Stream Water Quality, Version 2.07: Documentation and Users Manual*. Civil and Environmental Engineering Department, Tufts University, Medford, MA.

Copeland, B.J. and W.R. Duffer. 1963. Use of a clear plastic dome to measure gaseous diffusion rates in natural waters. *Limnology and Oceanography* 9, 494–499.

Costa, J.E. 1988. Rheologic, geomorphic, and sedimentologic differentiation of water floods, hyperconcentrated flows, and debris flows. In V.R. Baker, R.C. Kochel, and P.C. Patten (eds), *Flood Geomorphology*. Wiley-Intersciences, New York, pp. 113–122.

Di Toro, D.M. 2001. *Sediment Flux Modeling*. Wiley-Interscience, New York.

Di Toro, D.M., P.R. Paquin, K. Subburamu, and D.A. Gruber. 1990. Sediment oxygen demand model: Methane and ammonia oxidation. *Journal of Environmental Engineering ASCE* 116, 945–986.

Duan, Z., J.L. Martin, R.L. Stockstill, W.H. McAnally, and D.H. Bridges. 2009. Modeling streamflow-driven gas-liquid transfer rate. *Environmental Engineering Science* 26 (1), 155–162.

Fellows, C.S., H.M. Valett, and C.N. Dahm. 2001. Whole-stream metabolism in two montane streams. Contribution of the hyporheic zone. *Limnology and Oceanography* 46 (3), 523–531.

FISRWG. 1998. *Stream Corridor Restoration: Principles, Processes, and Practices*, Federal Interagency Stream Restoration Working Group. GPO Item No. 0120-A; SuDocs No. A 57.6/2:EN 3/PT.653.

Florida Department of Environmental Protection. 2006. Pharmaceuticals, personal care products and chemicals of concern: A briefing on an emerging issue. Available at: http://www.dep.state.fl.us/waste/quick_topics/publications/shw/meds/PPCPBriefingForWeb112006.pdf.

Grace, M.R. and S.J. Imberger. 2006. Stream metabolism: Performing and interpreting measurements. Water Studies Centre Monash University, Murray Darling Basin Commission, and New South Wales Department of Environment and Climate Change. Available at: http://www.sci.monash.edu.au/wsc/docs/tech-manual-v3.pdf.

Hall, C.A.S. 1970. Migration and metabolism in a stream. PhD dissertation. University of North Carolina.

Hatcher, K.J. 1986. Introduction to sediment oxygen demand modeling. In *Sediment Oxygen Demand Processes, Modeling, and Measurement*. Institute of Natural Resources, University of Georgia, Athens, GA, pp. 301–305.

Hirsch, R., T.A. Ternes, K. Haberer, and K.L. Kratz. 1999. Occurrence of antibiotics in the aquatic environment. *Science of the Total Environment* 225, 109–118.

Jakob, M. and O. Hungr. 2005. *Debris-Flow Hazards and Related Phenomena*, 1st ed. Springer-Praxis, Heidelberg, Germany.

Kilpatrick, F.A., R.E. Rathbun, N. Yotsukura, G.W. Parker, and L.L. Delong. 1989. Determination of stream reaeration coefficients by use of tracers. Chapter A18 in *Techniques of Water-Resources Investigations*. Book 3. U.S. Geological Survey, Reston, VA.

Koenig, M. and P. Murphy. 2001. Reducing model error through data collection. Presentation by USEPA Region 4 Science and Ecosystem Support Division, Ecological Assessment Branch, Athens, GA.

LINX. 2004. LINX II. Stream 15N experimental protocols. Revision 5. Accessed from the lotic intersite nitrogen experiment website, hosted at http://www.biol.vt.edu/faculty/webster/linx/.

Martin, J.L. 2004. WASP sediment diagenesis routines: Model theory and user's guide. Mississippi State University, prepared for Tetra Tech, Inc.

Martin, J.L. and S.C. McCutcheon. 1999. *Hydrodynamics and Transport for Water Quality Modeling*. Lewis Publishers, Boca Raton, FL.

MDEQ. 2007. State of Mississippi water quality criteria for intrastate, interstate, and coastal waters. Mississippi Department of Environmental Quality, Office of Pollution Control, Jackson, MS.

Moore, P.D. 2006. Flash flooding along the Blue Ridge, National Weather Service Forecast Office, Greenville–Spartanburg, SC.

Murphy, P.J. and D.B. Hicks. 1986. In-situ method for measuring sediment oxygen demand. In K.J. Hatcher (ed.), *Sediment Oxygen Demand—Processes, Modeling, and Measurement*. University of Georgia, Institute of Natural Resources, Athens, GA, pp. 307–322.

Natural Resources Defense Council. 1998. Endocrine disruptors. NRDC: Endocrine Disruptors FAQ. Available at: http://www.nrdc.org/health/effects/qendoc.asp#disruptor.

Odum, H.T. 1956. Primary production in flowing waters. *Limnology and Oceanography* 1, 102–117.

Odum, H.T. and C.M. Hoskins. 1958. Comparative studies on the metabolism of marine waters. *Publications of the Institute of Marine Science, University of Texas* 4, 115.

Owens, M., R. Edwards, and J.W. Gibbs. 1964. Some reaeration studies in streams. *International Journal of Air and Water Pollution* 8 (8-9), 469–486.

Phillips, I., M. Casewell, T. Cox, B.D. Groot, C. Friis, and R. Jones. 2004. Does the use of antibiotics in food animals pose a risk to human health? A critical review of published data. *The Journal of Antimicrobial Chemotherapy* 53 (1), 28–52.

Routledge, E.J., D. Sheahan, C. Desbrow, G.C. Brighty, M. Waldock, and J. Supmter. 1998. Identification of estrogenic chemicals in STW effluent. 2. In vivo responses in trout and roach. *Environmental Toxicology and Chemistry* 32, 1559–1565.

Simon, A. 2009. Adjustment processes in fluvial systems implications for streambank instability and control. National Sedimentation Laboratory, Oxford, MS, from a workshop presented at the EWRI Water Resource Congress 2009, Kansas City, MO.

Simon, A. and C.R. Hupp. 1986. Channel evolution in modified Tennessee channels. In *Proceedings of the Forest Federal Interagency Sedimentation Conference*, Las Vegas, NV. U.S. Interagency Advisory Committee on Water Data, Washington, DC, pp. 71–82.

Smith, D., J. Dushoff, and J. Morris. 2005. Agricultural antibiotics and human health. *PLoS Medicine* 2 (8), 731–735.

Stoddard, A., J.B. Harcum, J.T. Simpson, J.R. Pagenkopf, and R.K. Bastian. 2002. *Municipal Wastewater Treatment: Evaluating Improvements in National Water Quality*. John Wiley, New York.

Strangeways, I. 2003. *Measuring the Natural Environment*, 2nd ed. Cambridge University Press, Cambridge, UK.

Stumm-Zollinger, E. 1965. Biodegradation of steroid hormones. *Journal of the Water Pollution Control Federation* 37, 1506–1510.

Szeicz, G. 1974. Solar radiation for plant growth. *Journal of Applied Ecology* 11, 617–636.

Tabak, H.H. and R.L. Bunch. 1970. Steroid hormones as water pollutants. I. Metabolism of natural and synthetic ovulation-inhibiting hormones by microorganisms of activated sludge and primary settled sewage. *Developments in Industrial Microbiology* 11, 367–376.

Thomann, R.V. and J.A. Mueller. 1987. *Principles of Surface Water Quality Modeling and Control*. Harper and Row, New York.

Tsivoglou, E.C. 1967. Tracer measurement of stream reaeration. U.S. Department of the Interior, Federal Water Pollution Control Administration, Washington, DC.

Tsivoglou, E.C. and J.R. Wallace. 1972. Characterization of stream reaeration capacity, USEPA-R3-72-012. United States Environmental Protection Agency, Office of Research and Monitoring, Washington, DC.

USEPA. 1986. Quality criteria for water 1986. United States Environmental Protection Agency, Office of Water Regulations and Standards, Washington, DC.

USEPA. 1994. Acid mine drainage prediction, EPA 530-R-94-036. Environmental Protection Agency, Office of Solid Waste, Washington, DC.

USEPA. 2000a. Progress in water quality: An evaluation of the national investment in municipal wastewater treatment, EPA-832-R-00-008. Environmental Protection Agency, Office of Water, Washington, DC.

USEPA. 2000b. National water quality inventory 2000 report, EPA-832-R-00-008. Environmental Protection Agency, Office of Water, Washington, DC.

USEPA 2000c. Nutrient criteria technical guidance manual: Rivers and streams, EPA-822-B-00-002. U.S. Environmental Protection, Office of Science and Technology, Washington, DC.

USEPA. 2002. National water quality inventory: Report to Congress, 2002 reporting cycle: Findings, rivers and streams, and lakes, ponds and reservoirs. U.S. Environmental Protection Agency, Washington, DC.

USEPA. 2004. National water quality inventory: Report to Congress, 2002 reporting cycle: Findings, rivers and streams, and lakes, ponds and reservoirs. U.S. Environmental Protection Agency, Washington, DC.

USEPA SESD-EAB. 2007. Operating procedure: Sediment oxygen demand. U.S. Environmental Protection Agency, Region 4 Science and Ecosystem Support Division, Ecological Assessment Branch, Athens, GA.

USGS. 2002. Pharmaceuticals, hormones, and other organic wastewater contaminants in U.S. streams, USGS Fact Sheet FS-027-02. U.S. Geological Survey, Reston, VA.

Water on the Web. 2004. Monitoring Minnesota lakes on the Internet and training water science technicians for the future—A national on-line curriculum using advanced technologies and real-time data. University of Minnesota, Duluth, MN. Available at: http://WaterOntheWeb.org.

Wilhelms, S.C. 1980. Tracer measurement of reaeration: Application to hydraulic models, Technical Report E-80-5. U.S. Army Corps of Engineers, Waterways Experiment Station, Vicksburg, MS.

Wool, A.T., R.B. Ambrose, J.L. Martin, and E.A. Corner. 2003. Water quality analysis simulation program (WASP), version 6: Draft user's manual. Available at: http://www.epa.gov/athens/wwqtsc/html/wasp.html.

Zhang, Y., C.F. Marrs, C. Simon, and C. Xi. 2009. Wastewater treatment contributes to selective increase of antibiotic resistance among *Acinetobacter* spp. *The Science of the Total Environment* 407(12), 3702–3706.

6 Biota of Rivers and Streams
An Introduction

6.1 SPATIAL SCALE AND DISTRIBUTION

The physical and water quality characteristics of rivers and streams vary widely from their headwaters to their terminus, over their cross sections and over time, as described in Chapters 2 through 5. These physical and water quality variations also directly impact the presence and distribution of the biota of streams and rivers. Organisms may have specific ranges of depths, velocities, substrate types, temperatures, dissolved oxygen, and other physical and chemical factors that they can tolerate. The range of conditions from a preferred or optimal condition that can be tolerated may be broad or narrow depending on the organisms and, in many cases, depending on the particular life stage of the organisms. For example, organisms found in headwater or low-order streams may be less tolerant to high temperatures and temperature variations, than those found in high-order streams. In low-order, high-energy streams, the velocity may be sufficient to armor the bed, so the majority of the substrate may consist of gravel, rocks, and boulders. Some organisms may be more tolerant of higher velocities and resist being washed away by attaching to this coarse substrate, or by living in interstitial waters.

One of the goals of the management or restoration of rivers and streams is the maintenance or the establishment of a healthy biotic community, where a healthy biotic community it typically composed of a diverse assemblage of organisms, each with specific instream habitat requirements.

The instream habitat is impacted by changes that occur at a variety of spatial and temporal scales. For example, Frissell et al. (1986) described a hierarchical classification of instream habitats, ranging from the microhabitat where variations occur on the order of a foot or less, to the stream itself that may vary over scales of thousands of feet (Figure 6.1) (FISRWG 1998). The microhabitat could support fungi, small epiphytic plants, insects, and other biota and could include leaves and debris, rocks and cobbles, small patches of gravel, or other small-scale habitats as illustrated. Over a larger scale on the order of tens of feet are variations such as in steps and pools in steep systems, or pools and riffles in less steep systems (FISRWG 1998). Different organisms may inhabit the pool versus the riffle areas, and the frequency of the pools versus riffles, and their size, shape, and other factors affect the habitat and the ability of the organisms to move between habitats. On larger scales, changes may occur over scales of hundreds of feet (reaches) to thousands of feet (segments; Figure 6.2).

Changes occurring on an even larger scale resulting from physical and landscape changes are illustrated by the river continuum concept (RCC) developed by Vannote et al. (1980) (Figure 6.2). The RCC illustrates how watersheds and streams are connected and how variations in watershed and stream characteristics impact biological communities from the headwaters to the mouths of streams. For example, according to the RCC, productivity in low-order streams is more limited due to shading by riparian vegetation, so that most of the energy coming into these systems is from outside sources (allocthonous), such as leaf fall. As a result, the habitat favors organisms that can grow or feed on these coarse organic sources. Toward the mouth, rivers are wider, slower, deeper, and more of their productivity is derived from internal sources (autochthonous), such as phytoplankton.

The effect of scale on the distribution of organisms in rivers in streams is a critical component in evaluating aquatic health, and in maintaining or restoring that health. Healthy aquatic systems are typically diverse. For example, a habitat with a large number of few species is not considered

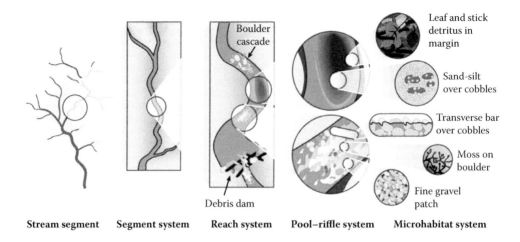

FIGURE 6.1 Hierarchical organization of a stream system and its habitat subsystems. (From FISRWG, *Stream Corridor Restoration: Principles, Processes, and Practices*, Federal Interagency Stream Restoration Working Group, 1998.)

as healthy as one with fewer numbers but with a greater diversity of organisms. Supporting a diverse assemblage of organisms typically requires a diverse habitat. Many of the metrics used to assess aquatic health as described in Chapter 7 and methods for stream restoration as described in Chapter 8 consider multiple spatial scales of land cover, geologic setting, hydrologic setting, aquatic habitat, and water chemistry. Some of the types of organisms that may occur in streams and rivers, and their typical habitats, are described in the following section.

6.2 AUTOTROPHS

6.2.1 Periphyton, or Benthic Autotrophs

Periphyton (from *peri* meaning around, about, or enclosing and *phyto* for plant; Figure 6.3) are algae or plants (autotrophs), a definition of which is "Algae attached to submerged substrate in aquatic environment" (USEPA SESD 2007).

As illustrated in Figure 6.2 and according to the RCC, periphyton are most common in low-order to mid-order rivers, where light reaches the bottom substrate. However, periphyton are found in virtually all streams and rivers on virtually all surfaces that receive light (Allan and Castillo 2007).

The periphyton are variously identified or classified based on their taxonomy or the substrate on which they reside. They are also distinguished in that they occur on substrates as opposed to other algae that flow with the current, for example, the phytoplankton or planktonic algae.

First, before continuing and to add to the confusion, these are not really plants, and not all of them are really algae. Organisms can be subdivided into two separate domains, the highest taxonomic level or rank, the Prokaryota and the Eukaryota. The general difference is that unlike the eukaryotic organisms, the prokaryotic organisms lack a cell nucleus. The Prokaryota includes bacteria. The Eukaryota may be further subdivided into kingdoms, such as Protista, Fungi, Plantae, and Animalia. True plants are in the Animalia kingdom, while most algae (such as the periphyton and phytoplankton) are protists. However, there are some planktonic and periphytic algae, known as blue-green algae which are not really algae. They fall within the domain Prokaryota and are also called Cyanobacteria. The most common forms of periphyton are diatoms and green algae (within the divisions and Chlorophyta, respectively, of the Protista kingdom), and Cyanobacteria. Each of these algal types may be composed of many thousands of subspecies. The level of taxonomic classification may differ, depending on the specific study goals; however, some level of classification is usually necessary to either assess the health or manage the quality of rivers and streams.

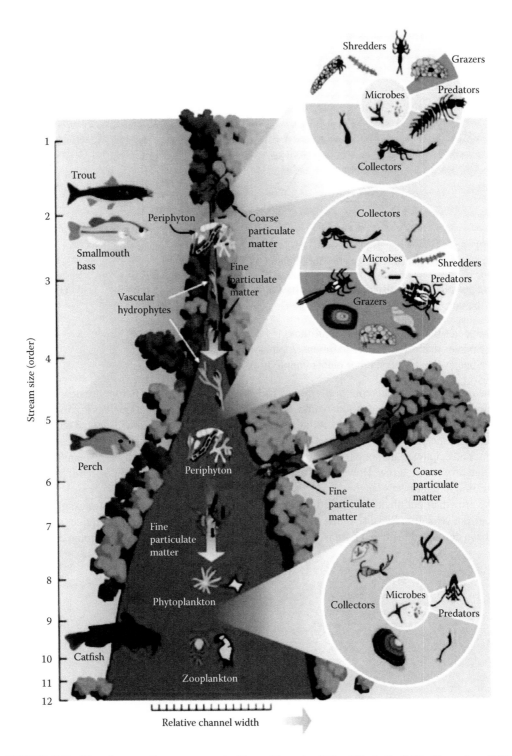

FIGURE 6.2 The river continuum concept. (From Vannote, R.L., Minshall, G.W., Cummins, K.W., Sedell, J.R., and Cushing, C.E., *Canadian Journal of Fisheries and Aquatic Sciences*, 37, 130–137, 1980. With permission from NRC Research Press.)

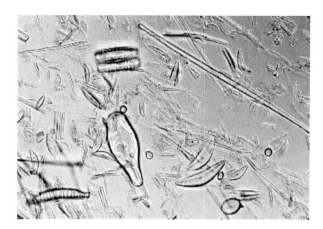

FIGURE 6.3 Periphyton assemblage—low magnification shot of periphyton in Lake Hovsgol. (From University of Michigan, micrograph by Mark Edlund. With permission.)

The periphyton are also distinguished based on the substrates on which they grow (Figure 6.4). They include (Allan and Castillo 2007):

- Species that grow on stones (epilithon)
- Species that grow on soft sediments (epipelon)
- Species that grow on other plants (epiphyton)

The epilithon would be commonly found in low-order, fairly high-energy streams with fairly armored substrates. The epipelon would be expected in more low-energy environments, since the sediment on which they reside could be swept away during high-flow events. The epiphyton are found in the presence of other aquatic plants, such as the macrophytes (macro meaning "big" plants). The epiphytes are not parasitic; they merely use the host plants, such as the plant leaves, as a substrate on which to grow. However, in excess, they can harm their host plant by reducing the light penetration to their host's leaves.

The growth of periphyton is limited by the availability of light, nutrients, substrates, grazing, and also current velocities. Because of the importance of periphyton, a number of techniques and models have been developed to predict periphyton variations. Velinsky et al. (2002) and Thuman et al. (2006)

FIGURE 6.4 Periphyton. (From U.S. Geological Survey, South Florida Information Access [SOFIA]. With permission.)

described an intensity study and a model application of periphyton growth on the Jackson River, Virginia, and also investigated the use of pulsed flows for periphyton management (Flinders and Hart 2009). Chapra et al. (2007) describe a water quality model that includes periphyton.

Periphyton surveys are used to determine periphyton composition or biomass. Methods for sampling vary from scraping or using other techniques to remove periphyton from substrates for analysis, to providing artificial substrates on which they would grow, and other techniques. USEPA SESD (2007) describe the methods for periphyton surveys in wadeable streams. Barry and Kilroy (1994) describe the sampling techniques used in New Zealand. Barbour et al. (1999) describe the periphyton sampling techniques for Rapid Bioassessment Protocols (RBPs), which will be described in more detail in Chapter 8 and are used as an indicator of aquatic health.

As with other aquatic plants, excess periphyton can be an environmental problem, and the presence of some forms is an indication of pollution, while the presence of a diverse and viable community is an indication of aquatic health, as will be discussed in Chapter 8. For example, diatoms are particularly useful indicators of biological conditions because they are ubiquitous and found in all lotic systems (Barbour et al. 1999). Conversely, Cyanobacteria are often an indication of pollution and are generally an undesirable form of algae. As such, the determination of the composition and biomass of the periphyton is often a critical component of stream or nutrient assessment studies.

6.2.2 Phytoplankton

Phytoplankton are floating plants, carried along with the currents. Like the periphyton, these are not true plants but consist of algae and Cyanobacteria. According to the RCC, phytoplankton dominate in higher-order streams, where velocities are less and the storage is greater. Phytoplankton types are in abundance and will be discussed in more detail in a subsequent chapter on the biota of lakes and reservoirs.

6.2.3 Macrophytes

Macrophytes (from *makros* meaning large and *phyton*, plant) are, as their name implies, large plants, either as an assemblage or individual, large enough to be seen without the aid of a microscope. Macrophytes may be:

- Macrophytic algae (such as some green algae, members of the Protista kingdom, such as *Chara*)
- Mosses (such as sphagnum, a member of the Bryophyte division of true plants or the Plantae kingdom)
- Fern allies (such as Salvinia and Iosetes, are also true plants of the Pteridophyta division)
- Angiosperms (true flowering plants of the Anthophyta division)

Of these, the angiosperms make up the majority of aquatic macrophytes. The aquatic angiosperms are a diverse group and include species native to the United States and many that are invasive and problematic.

Macrophytes are also commonly classified based on their growth form, and Westlak (1975, as given by WOW 2004) identified four primary forms (Figure 6.5):

1. Emergents (such as cattails) on riverbanks and shoals are typically rooted in soil that is near or below the waterline and have aerial leaves and reproductive structures. These are found in the shallowest waters.
2. Floating-leaved species, such as pond lilies, occupy slow-current areas, are rooted in submerged soils, and have aerial or floating leaves and reproductive structures.

FIGURE 6.5 (a–c) Aquatic macrophytes. (Photographs from Ohio Pond Management Plant Identification, Ohio Department of Natural Resources.)

3. Free-floating species, such as water hyacinth and duckweeds, are typically not attached to the substrate and often form mats that entangle other species in slow-flowing tropical rivers.
4. Submerged species, such as Isoetes and Potamogeton, are rooted to the substrate, have submerged leaves, and are located mid-channel to the point of insufficient light penetration. Submerged species that are not rooted include the bladderwort.

The aquatic macrophytes are typically confined to areas of streams and rivers with limited currents and suitable substrates, such as pools, side channels, embayments, backwater areas, and contiguous wetlands and, as a result, macrophyte distributions in rivers are often patchy (WOW 2004; Figure 6.6). The distribution and abundance of macrophytes often vary seasonally.

In addition to contributing to primary productivity, macrophytes also provide an aquatic habitat for a wide variety of organisms. However, as with other plants, excess nutrients can result in excessive macrophyte growth that can impede navigation, and cause dissolved oxygen declines and other water quality problems. Also, the growth and sloughing of macrophytes are not only impacted by, but also impact on the hydraulics of rivers and streams, such as by increasing drag and essentially reducing cross-sectional flow areas.

6.2.4 Interactions between Periphyton, Phytoplankton, and Macrophytes

Periphyton, phytoplankton, and macrophytes commonly coexist and interact in a number of ways. While all three consume and compete for water column nutrients, rooted macrophytes have the

FIGURE 6.6 Patchy distribution of macrophytes. (From Water on the Web, Monitoring Minnesota Lakes on the Internet and training water science technicians for the future—A national on-line curriculum using advanced technologies and real-time data, University of Minnesota, Duluth, MN, 2004. Available at http:// WaterOntheWeb.org. With permission.)

advantage of being able to obtain nutrients from their roots as well. Floating macrophytes limit light penetration, which also limits the growth of periphyton and phytoplankton. Conversely, excessive phytoplankton growth can cause shading that limits the growth of submerged macrophytes. Similarly, periphyton growth on the leaves of macrophytes may limit light penetration and limit macrophyte growth.

The interactions and dominance of particular forms may vary seasonally or as a function of other conditions. For example, on the Bow River below the city of Calgary, Alberta, macrophytes commonly dominate primary productivity. Since the river supports a world-class fisheries, a variety of agencies are involved in managing the nutrient loads from the expanding population of Calgary in order to protect those fisheries and the water quality of the river. A part of the target of a total loading management (TLM) program was focused on the macrophytes. Although periphyton were present, they were initially assumed to have a secondary impact. However, in 2005, a major flood event scoured the river, dramatically reducing the macrophyte densities. While following the flood the macrophyte density was greatly reduced, primary productivity as reflected in diel variations in dissolved oxygen remained relatively unchanged. The cause was attributed to a relatively rapid shift from a macrophyte-dominated system to a periphyton-dominated system, emphasizing the significance and previously unrecognized effects of periphyton (Robinson 2007).

6.3 HETEROTROPHS

6.3.1 Stream Invertebrates

There is an enormous variety of invertebrate (animals without a backbone), heterotrophic (consumers) organisms in streams and rivers, as shown in *Pennak's Freshwater Invertebrates of the United States* (Smith 2001). They vary widely in size and form, and some spend only a portion of their life cycle in water. Invertebrates include a variety of forms that live suspended in the water column, such as zooplankton. In contrast, in streams and rivers, particularly those of low order, organisms typically have to live in or cling to the bottom substrate (the benthos), and these are referred to as benthic invertebrates. Many of these benthic invertebrates are relatively large, such as mollusks, worms, insects, and crustaceans, and they are referred to as benthic macroinvertebrates. Some of the microscopic forms, such as the planktonic forms (zooplankton), are also important, particularly in higher-order streams, and these will be discussed in more detail in a subsequent chapter on the biota of lakes and reservoirs.

The benthic macroinvertebrates consume and transform organic matter and aid in the cycling of nutrients. These organisms are also of importance as indicator species for water quality, as will be discussed in Chapter 7.

6.3.1.1 Major Taxonomic Groups

All of the heterotrophic macroinvertebrates are included in the Animalia kingdom. This kingdom includes the phylum Chordata, which includes mammals and birds that have a backbone. Major macroinvertebrate phyla include the Arthropoda (e.g., insects and crustaceans), Mollusca (e.g., snails, clams, and mussels), Annelida (e.g., worms and leeches), and Platyhelminthes (e.g., flatworms).

6.3.1.1.1 *Arthropods (Insects and Crustaceans)*

Insects are typically the most numerous of the macroinvertebrates in unpolluted streams and rivers, comprising 70%–90% of the macroinvertebrates, andincluding 13 taxonomic orders (Cushing and Alan 2001), a few of the more common of which are illustrated in Figure 6.7.

Some of these insects, such as the mayfly, caddis fly, and stone fly, are very sensitive to pollution and, as such, are used as indicators of good watershed health, as will be discussed in Chapter 7. Conversely, some fly larvae, such as the midge larvae, are most commonly found in highly polluted waters. Also called bloodworms, they are reddish in color since they have hemoglobin in their blood, which allows them to live in waters with low dissolved oxygen concentrations.

Some of the insects spend only part of their lives in streams and rivers. For example, true flies (those with only one pair of wings) occur primarily in their larval stage in streams and rivers. The fly's life cycle development involves undergoing complete metamorphosis (referred to as holometabolous). It passes through four stages of development: egg, larvae, pupa, and adult; and the larval stage does not resemble the adult stage. This development can take only a week for some

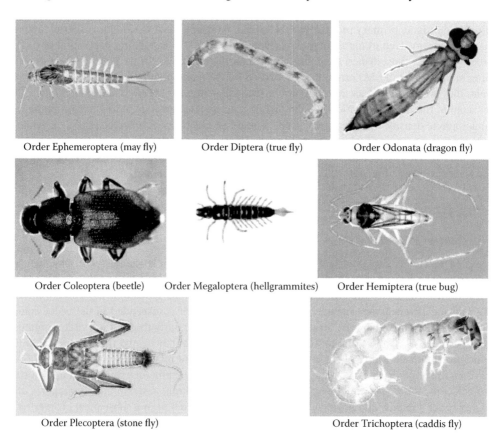

FIGURE 6.7   Common orders of insects. (Images from Australian Fresh Water Invertebrates. Available at http://www.mdfrc.org.au/bugguide/resources/rr_clearedhills.htm.)

mosquitoes, to a year or longer (Cushing and Allan 2001), while the adult may commonly live for a month or less.

In incomplete metamorphosis (hemimetabolous development), there are typically three main stages of development: the eggs, immature insects called nymphs that undergo a series of molts until they become adults. There is no intermediate pupae stage where transformation occurs. The nymphs typically resemble the adults except for their wing development. The damselfly, mayfly, and dragonfly are examples of insects that undergo incomplete metamorphosis. Mayflies undergo incomplete metamorphosis, where they may spend 11–24 months as a nymph, and undergo 20–30 molts. They then rise to the surface, take flight, and undergo a final molt to become an adult with no functioning mouthparts, so they cannot eat; subsequently, they reproduce and die.

Some insects can have several generations per year (referred to as multivoltine), one generation per year (univoltine), or one generation every 2–3 years (semivoltine). The cycle is based largely on temperature, where a certain temperature history (number of days with a certain temperature) will trigger the transformation of an adult. The temperature history is usually measured in degree-days, the cumulative product of the temperature and the time between some development threshold. Mayflies typically require 2000+ degree-days before emerging (Cushing and Allan 2001). Insects are also variously adapted to live in rivers and streams. Some of the common adaptations are listed in Table 6.1.

Crustaceans, also in the phylum Arthropoda, include the orders Isopoda (aquatic sow bug), Decapoda (crayfish), and Amphipoda (scud/sideswimmers) (Figure 6.8). Crustaceans are found in a

TABLE 6.1
Insect Adaptations to Living in Streams and Rivers

Adaptation	Significance	Representative Groups and Structures	Comments
Dorsoventrally flat	Allows crawling in slow-current boundary layer on substrate	Odonata—Gomphidae Trichoptera—Glossosoma	
Streamlining	Fusiform body minimizes resistance to current	Ephemeroptera—*Baetis* Diptera—Simulium	Relatively rare body form
Reduced projecting structures	Reduces resistance to current	Ephemeroptera—*Baetis*	Large lateral structures exist in some groups
Suckers	Attach to smooth surfaces	Diptera—Blephariceridae	Rare adaptation
Friction pads	Increased contact reduces chances of being dislodged	Coleoptera—*Psephinus*	
Small size	Allows use of slow-current boundary layer on top of substrate		Stream animals are smaller than stillwater relatives
Silk and sticky secretions	Attachment to stones in swift current	Diptera—*Simulium* Trichoptera—Hydropsychidae	
Ballast	Cases made of large stones	Trichoptera—*Goera*	
Attachment claws/ dorsal processes	Stout claws aid in attachment to plants	Ephemeroptera—*Ephemerella*	
Reduced power of flight	Prevents emigration from small habitats	Plecoptera—*Allocapnia*	Reduces dispersal ability
Hairy bodies	Keeps sand/soil particles away while burrowing	Ephemeroptera—*Hexagenia*	Allows water flow over body
Hooks or grapples	Attachment to rough areas of substrates	Coleoptera—Elmidae	

Source: Developed by Merrick, R.; Water on the Web, Monitoring Minnesota Lakes on the Internet and training water science technicians for the future—A national on-line curriculum using advanced technologies and real-time data, University of Minnesota, Duluth, MN, 2004. Available at http://WaterOntheWeb.org.

Scud (*Crangonyx* sp.)   Isopod (*Caecidotea* sp.)

FIGURE 6.8 Crustaceans. (Photographs from the Department of Environmental Protection, Bureau of Land and Water Quality, State of Maine, Available at http://www.maine.gov/dep/blwq/docmonitoring/ biomonitoring/sampling/bugs.htm. With permission.)

wide variety of habitats, including pools, riffles, and the hyporheic zone. They are typically scavengers, eating both dead and live plant and animal debris.

6.3.1.1.2 Phylum Mollusca

This phylum includes snails (class Gastropoda), and clams and mussels (class Bivalvia). The gastropods, or snails, are common and most easily recognized by their shells. Some have true gills and an operculum, a plate that can seal the shell after the soft body has been retracted (prosobranchs; Figure 6.9), while others take up oxygen through a vascularized mantle cavity that acts like a lung (pulmonates; Figure 6.9). The pulmonates can trap air, such as from the water surface, in their mantle cavity and obtain oxygen directly and as a result, they are not as sensitive to polluted conditions. The presence of the pulmonate pouch snail is used by the U.S. Environmental Protection Agency (U.S. EPA) as an indicator of nutrient-enriched conditions and poor water quality (U.S. EPA's Biological Indicators of Watershed Health). Most of the aquatic snails are herbivorous, scraping algae from substrates using a tonguelike structure covered by a ribbon of teeth, the radula (Figure 6.10; Clifford 1991).

Mussels and clams are bivalves, having a shell separated into two symmetrical valves. They have a soft body with enlarged gills and a muscular extendable foot that assists them in burrowing or moving. They typically burrow into bottom sediments, and feed by filtering fine particles from the water. Clams and mussels are found around the world, but native mussels of the family Unionidae are more diverse in North America than anywhere else, with nearly 300 species documented (USGS 1999).

Native mussels have a unique life history (Figure 6.11). During spawning, probably prompted by temperature, males release sperm into the water column, which is taken in by females and fertilizes their eggs. The fertilized eggs develop into a larval stage, the glochidia, which are stored in the female's gills for several weeks or months and then released into the water column. The glochidia first drift until they encounter and attach to the fins or gills of a suitable fish host. There may be

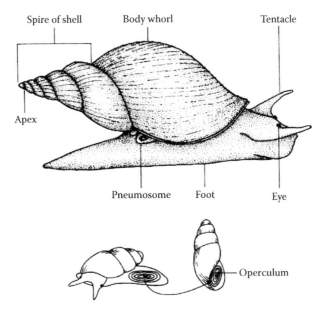

FIGURE 6.9 A pulmonate snail (*upper*) and a prosobranch snail with an operculum (*lower*). (From Clifford, H., *Aquatic Invertebrates of Alberta*, pp. 638, University of Alberta Press, Edmonton, Alberta, Canada, 1991. With permission.)

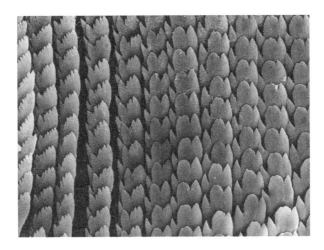

FIGURE 6.10 Radula of *Helisoma trivolvis* (0.03 mm). (From Clifford, H., *Aquatic Invertebrates of Alberta*, University of Alberta Press, Edmonton, Alberta, Canada, 1991.)

200,000–17,000,000 glochidia released per mussel per year (McMahon and Bogan 2001). The large number aids in survival since the glochidia can only survive for a few days without finding a host. Also, there are only certain fish that the glochidia can parasitize, and these and other factors result in high glochidia mortality. However, while their mortality rates are high, the glochidia life stage and the parasitism on fish provide a means of transporting the mussels to new habitats.

In addition to drifting, some species use structures or shapes to attract fish hosts to the female, such as a flap lure (Figure 6.12). Once the mantle lure is struck, the female expels the glochidia to infect the fish.

Once attached to the fish, the glochidia remain for a period of several weeks to a month and then drop off to the riverbed as juveniles. If dropped to an unfavorable habitat, they do have some

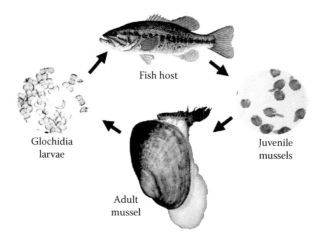

FIGURE 6.11   Native mussel life history. (From Barnhart, C. and Neves, D., Overview of North American freshwater mussels, USEPA Mussel Toxicity Testing Workshop, Chicago, IL, 2005.)

FIGURE 6.12   Mantle flap lure of the plain pocketbook (*Lampsilis cardium*). (Photographs by Jeff Grabarkiewicz and Todd Crail; From Grabarkiewicz, J.D. and Davis, S.W., An introduction to freshwater mussels as biological indicators, including accounts of interior basin, Cumberlandian, and Atlantic slope species, U.S. Environmental Protection Agency, Washington, DC, 2008.)

ability to move, but mortality rates are high. The surviving juveniles become adults in 5–12 years, and under favorable conditions they can live for 20 or more years, with some species surviving up to 140 years (McMahon and Bogan 2001; Morales-Chaves 2004). Mussels are commonly used as biological indicators, since they are sensitive to habitat changes and pollution. Some of the attributes that make them good biological indicators are (Grabarkiewicz and Davis 2008):

- Long lived.
- Sedentary: juvenile and adult mussels move little during their entire lifetime.
- Burrowers: some species and juveniles burrow deep into the streambed.
- Filter feeders: mussels obtain food and oxygen from the water column and via interstitial flow.
- Fairly large: mussels contain ample soft tissue for chemical analysis.
- Spent valves: dead mussels leave a historical record.

Freshwater mussels are also one of the most imperiled groups of organisms in the world, and about 70% of North American species is considered threatened, endangered, or extinct (Lydeard et al. 2004; Strayer et al. 1999; Figure 6.13). One of the causes of mussel decline in the United States was overharvesting of mussels for making buttons or for using shell fragments for seeds in the cultured pearl industry. Habitat alterations such as due to dams or channelization for navigation also contributed to the decline of the mussel populations in the United States, as did pollution and excess sedimentation.

Another factor that resulted in the decline of mussels in some areas was an invasive species, the zebra mussels. The zebra mussels, originally native to the Caspian Sea, were released into the Great Lakes in 1988. Presumably, the zebra mussels were transported and released into ballast water. According to the USGS Great Lakes Science Center, in less than 10 years following their original release, the zebra mussels had spread to all of the Great Lakes, as well as the Hudson and Ohio River basins, and the entire length of the Mississippi (Kirk et al. 2001). The zebra mussel distribution of May 2011 is indicated in Figure 6.14, from the U.S. Zebra Mussel Sightings Distribution web page (Benson 2011).

Zebra mussels affect native clams and other species in a variety of ways. The reproductive cycle of the zebra mussels differs from native clams in that there is no parasitic glochidia stage. The fertilized eggs develop into larvae (the veliger), which float in the water column (are planktonic) from 2 weeks to several months, they then settle, develop into juveniles and can become adults within a year. Their high reproductive rate and rapid growth give them a competitive advantage (Morales-Chaves 2004). Zebra mussels form thick encrustations on a variety of substrates, including clams

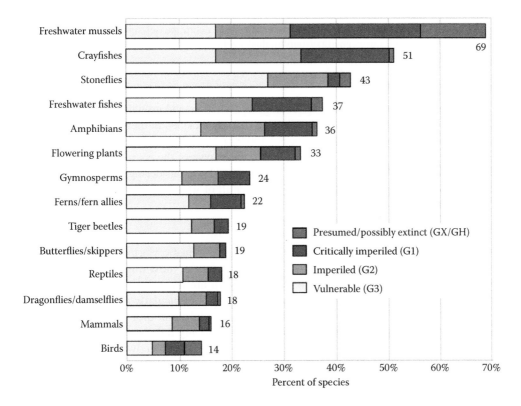

FIGURE 6.13 Proportion of species at risk by plant and animal group. (From Grabarkiewicz, J.D. and Davis, S.W., An introduction to freshwater mussels as biological indicators, including accounts of interior basin, Cumberlandian, and Atlantic slope species, U.S. Environmental Protection Agency, Washington, DC, 2008.)

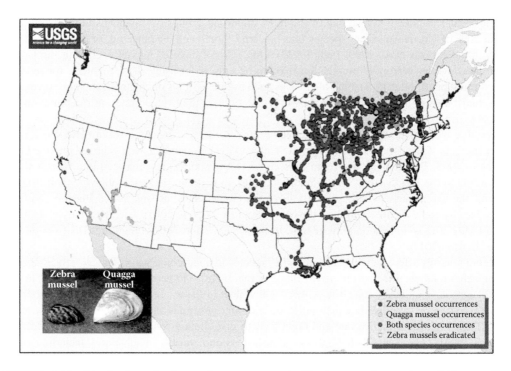

FIGURE 6.14 Distribution of zebra and quagga mussels (*Dreissena polymorpha* and *D. rostriformis bugensis*) sightings as of July, 2011. (From Benson, A.J., Zebra mussel sightings distribution, 2011. Retrieved from http://nas.er.usgs.gov/taxgroup/mollusks/zebramussel/zebramusseldistribution.aspx. With permission.)

and crustaceans (Figure 6.15), impacting feeding, reproduction, and growth, and some native mussels have been found with more than 10,000 zebra mussels attached to them. They also encrust on water pipes and other surfaces, and densities of 100,000–500,000 mussels m^{-2} have been reported in Lake Erie (Alexander et al. 1997). In addition to the impact of the encrustations, zebra mussels are filter feeders, each filtering up to a quart of water per day. Considering the enormous numbers often present, zebra mussels can then essentially filter all of what is the food for other organisms, such as native mussels.

6.3.1.1.3   *Phylum Annelida (Worms and Leeches)*

This group includes the Hirudinea (leeches), Oligochaeta (aquatic worms), and Nematomorpha (horsehair worms), of which the Oligochaetes are perhaps the most important in stream habitats (Figure 6.16). Of this group, perhaps the best known are the aquatic sludge worms (family Tubificidae). The sludge worms are very tolerant of pollution, particularly pollution resulting from sewage effluents with high concentrations of organic materials, and they are commonly used as environmental indicators. Like the midge larvae, also commonly found in highly polluted waters, the sludge worms have hemoglobin in their blood and are also called bloodworms. In polluted areas, the sludge worms may dominate the aquatic population (low numbers of species and large numbers of sludge worms).

6.3.1.2   Functional Groups

In addition to their taxonomic classification, benthic macroinvertebrates are also often classified by their method of feeding, as illustrated in the RCC (Figure 6.2). Basing identification on function rather than taxonomy is often done since identification may be more rapidly accomplished. Often,

(c) Zebra mussels on clam

(b) Zebra mussels on crayfish

(a) Zebra mussels cover a current
meter in Lake Michigan

FIGURE 6.15 (a–c) Zebra mussels. ((a) From NOAA, http://www.noaa.gov/features/earthobs_0508/zebra.
html; (b,c) From USGS, http://sfbay.wr.usgs.gov/benthic_eco/exotic_species/next.html.)

Ologichaetes (left is
Lumbriculidae, right
Specaria josinae)

Leeches (left to right *Placobdella montifera*,
ventral view of a glossiphoniid leech carrying
young, *Erpobdella punctata* and *Nephelopis
obscura*)

Nematomorpha
(*Gordius*)

FIGURE 6.16 Examples of worms and leeches. (From Clifford, H., *Aquatic Invertebrates of Alberta*,
University of Alberta Press, Edmonton, Alberta, Canada, 1991. With permission.)

a detailed taxonomic identification is expensive and time consuming, requiring on the order of
10 hours or more in the laboratory for each hour in the field collecting samples.

Some methods for measuring aquatic health, as will be discussed in Chapter 7, require a
certain level of taxonomic classification. For example, the U.S. EPA's Rapid Bioassessment
Protocols (RBP; Barbour et al. 1999) indicates that most organisms are identified at the "lowest
practical level (generally genus or species) by a qualified taxonomist using a dissecting microscope." The RBP also include, rather than a taxonomic assessment, a functional characterization

TABLE 6.2

Aquatic Insect Functional Feeding Group Categorization and Food Resources

Functional Groups	Feeding Mechanisms	Dominant Food Resources	Particle Size Range of Food (mm)
Shredders	Chew conditioned or live vascular plant tissue, or gouge wood	Decomposing (or living hydrophyte) vascular plant tissue-coarse particulate organic matter	>1.0
Filtering-Collectors	Suspension feeders—filter particles from the water column with nets or adapted body parts	Decomposing fine particulate organic matter; detrital particles, algae, and bacteria	0.01–1.0
Gathering-Collectors	Deposit feeders—ingest by gathering sediments or brush loose surface deposits	Decomposing fine particulate organic matter; detrital particles, associated detritus, and microflora and fauna	0.05–1.0
Scrapers	Graze mineral and organic surfaces	Periphyton-attached algae and associated detritus and microflora and fauna	0.01–1.0
Plant piercers	Herbivores—pierce tissues or cells and suck fluids	Macroalgae	0.01–1.0
Predators	Capture and engulf prey or ingest body fluids	Prey-living animal tissue	>0.5

Source: Modified from Merritt, R.W., Cummins, K.W., and Berg, M.B. (eds), *An Introduction to the Aquatic Insects of North America*, 4th ed., Kendall/Hunt Publishing Company, Dubuque, IA, 2008.

based on methods of feeding (e.g., percentages of filterers, grazers, and scrapers) and habitat (e.g., percentage of clingers).

Functional feeding-type mechanisms are listed in Table 6.2. As illustrated by the RCC (Figure 6.2), all feeding types would be expected in low-order streams and rivers, where most of the organic matter is coarse, such as leaves and debris. In higher-order streams, where fine particulate matter dominates, the collectors and predators dominate the benthic macroinvertebrates. Typically, a decrease in the percentage of the aquatic macroinvertebrate population represented by each of these functional feeding groups, or the number of taxa within each group, is an indicator of a decline in aquatic health. Conversely, an increase in the percentage population of filterers is evidence of a decline in aquatic health (Barbour et al. 1999).

The characteristics of functional habitat designations are also used to differentiate macroinvertebrates. The U.S. EPA's Biological Indicators of Watershed Health includes the following habitat groups:

- Clinger—able to remain stationery on bottom substrates in flowing waters
- Climber—feeds on submerged aquatic vegetation (SAV) by climbing
- Sprawler—can be found on both the surface of SAV and substrates
- Burrower—feeds on fine organic matter while buried in sediments of lakes and streams
- Swimmer—can control the direction and velocity of its movements
- Diver—able to swim from the surface to the bottom of the water column

These habitat types reflect macroinvertebrate adaptations for maintaining their position and moving about in the aquatic environment (Merritt et al. 2008). Generally, a decline in any of these feeding types in terms of the number of taxa or the percentage of the population is an indicator of a decline in aquatic health (Barbour et al. 1999).

Since habitat types are a component of the indicators of aquatic health, such as reflected in the RBPs and indices of biotic integrity, as discussed in Chapter 7, organizations and state agencies have identified native macroinvertebrates and their feeding and habitat classification to aid in those assessments. For example, the Georgia Environmental Protection Division (Georgia EPD 2007) provides a common taxa list that includes feeding and habitat types as part of a guide to conducting macroinvertebrate biological assessments in Georgia. The guide, *An Introduction to the Aquatic Insects of North America* by Merritt and Cummings (1998), is also a commonly used source of habitat types associated with specific macroinvertebrate species.

6.3.2 Stream Vertebrates

There are a variety of vertebrates (kingdom Animalia, phylum Chordata) that live in or depend on rivers and streams for food, water, and cover. These include a variety of mammals (class Mammalia), such as beavers; a variety of birds (class Aves) such as ducks, geese, egrets, and others; reptiles (class Reptilia) such as snakes, lizards, and alligators; amphibians (class Amphibia) such as frogs and salamanders; and fish (class Osteichthyes, for bony fish).

All vertebrates may be adversely impacted by processes such as eutrophication, habitat modification, and other impacts on their aquatic environment. Many of the vertebrate populations are in decline and their population densities and distributions can serve as indicators of aquatic health.

As an example, amphibian populations are declining around the world. Amphibians are particularly sensitive to pollution and changes in their environment since they have an aquatic larval stage, and lack a protective epidermis. Chemicals, for example, can often be readily transferred through their semipermeable skin, making them susceptible to pollution. Population declines prompted the president and Congress in 2000 to undertake a national amphibian research and monitoring initiative (ARMI; http://armi.usgs.gov/index.asp) to measure, develop, and understand the response to the effects of environmental change on amphibians.

In addition, many avian species depend on the riparian zone of streams and rivers. While riparian zones comprise only about 1% of avian habitats in the western United States, these areas are used by more avian species than any other habitat in the United States (Knopf 1985).

While other vertebrates are important components of the aquatic environment as indicators of aquatic health, in this section we will primarily concentrate on fish. Fish are perhaps most commonly used in metrics for aquatic health, as discussed in Chapter 7. Also, similar to other aquatic vertebrates, the diversity of fish species is declining. As of 2008, there are 139 fish species listed as threatened or endangered.

The characteristics and habitat of fish are often reflected in their anatomy (see Figure 6.17). Most fish have fins that aid in their movement, which are located along the centerline on the top (dorsal side), or along the bottom (ventral side) in the pelvic or anal region. Exceptions are the lampreys (Petromyzontidae), which lack paired fins. The dorsal and ventral fins primarily add stability when swimming. The caudal or tail fin provides locomotion, while fins along the side and in the pectoral area add stability and aid in side-to-side movement. In some unusual fish such as the "walking catfish," which is an invasive species in the United States, the pectoral fin is used to aid in navigation on land. The number and location of the fins and fin rays, and the presence and structure of spines are commonly used in fish identification.

Most fish also have scales, which overlap and provide a protective barrier. Another important protective barrier is fish slime, produced by the secretions of slime cells.

Respiration is generally through gills, which, in some fish, are covered by an opercular flap that may open and close to aid in pumping water over the gills. Internally, most fish also have a hollow, gas-filled bladder (swim or air bladder) that is used to aid in swimming by maintaining a neutral buoyancy in water. In some fish, such as the African lungfish, the swim bladder is also used in respiration.

In addition to sight and smell, one of the fish's primary sense organs is the lateral line, which detects underwater vibrations and is capable of determining the direction of their source. The

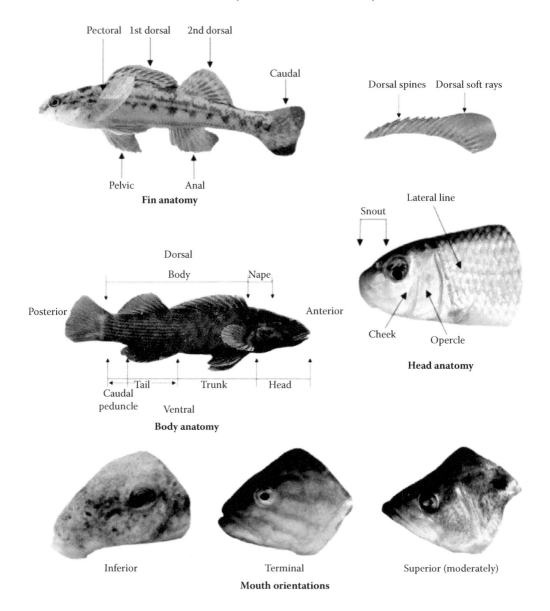

FIGURE 6.17 Fish anatomy. (From U.S. EPA, Biological indicators of watershed health, Available at http://www.epa.gov/bioiweb1/html/fish_id.html.)

barbels, whiskers, or mouth projections on some bottom-feeding fish also provide a sense of touch as well as taste.

The body shape of fish is a good indicator of their environment. For example, slender and torpedo-shaped fish are adapted to fast-moving waters. The location of the mouth in a fish is also an indication of its feeding habits. Surface-feeding fish usually have an upturned (superior) mouth, while bottom-feeding fish generally have a downturned or inferior mouth. Predatory fish usually have a wide mouth, while omnivorous fish have smaller mouths. Some bottom-dwelling fish that feed on algae, such as the Loricariidae or suckers, have a suction-cup-like mouth.

Some of the common taxonomic orders of fish are illustrated in Figure 6.18. Grabarkiewicz and Davis (2008) describe each order's characteristics, as well as methods of collection. Some fish are

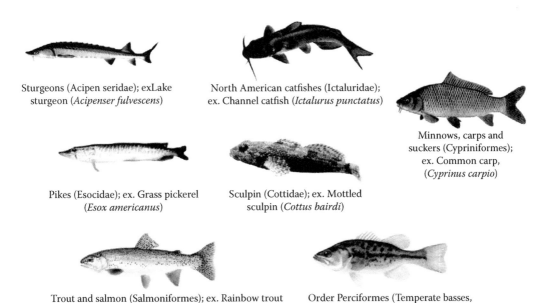

Sturgeons (Acipen seridae); exLake sturgeon (*Acipenser fulvescens*)

North American catfishes (Ictaluridae); ex. Channel catfish (*Ictalurus punctatus*)

Minnows, carps and suckers (Cypriniformes); ex. Common carp, (*Cyprinus carpio*)

Pikes (Esocidae); ex. Grass pickerel (*Esox americanus*)

Sculpin (Cottidae); ex. Mottled sculpin (*Cottus bairdi*)

Trout and salmon (Salmoniformes); ex. Rainbow trout (*Oncorhynchus mykiss*)

Order Perciformes (Temperate basses, sunfish, perches, drums); ex. largemouth bass (*Micropterus salmoides*)

FIGURE 6.18 Freshwater fish by order. (Photographs from U.S. EPA, Biological indicators of watershed health, Available at http://www.epa.gov/bioiweb1/html/fish_id.html.)

more commonly associated with low-order, colder streams and rivers, such as smallmouth bass and trout. Fish such as catfish and other species are commonly associated with higher-order streams, as illustrated by the RCC (Figure 6.2).

Fish populations are adversely impacted by pollution, hydromodification, and a variety of other habitat modifications. Hydromodification, such as channelization and dam construction, has impacted fish populations in a number of ways. One impact of dams is changing the thermal regime, both within and downstream of the dam, depending on the size of the dam and the dam operations. For example, peaking hydropower facilities with the release of colder bottom waters have converted many tailwater systems from warmwater to cold-water fisheries.

Dams also have an impact as barriers to fish movement. Most commonly known are the impacts of dams on the upstream movement of anadromous fish, those that live in the ocean and then reproduce in streams. Perhaps most widely known is the plight of Pacific salmon species. However, dams impede the migration of other anadromous fish as well, such as the American shad, herring, striped bass, and Atlantic sturgeon. There are also a number of freshwater fish that migrate up rivers and tributaries to spawn, such as the white bass, and are adversely impacted by dams.

One of the stated goals of the Clean Water Act was to make the nation's waters "fishable and swimmable." One difference between fish and other indicator organisms is that often fisheries management is targeted toward specific "fishable" or sports fisheries, rather than native species, such as artificial stocking and fish management programs targeted toward fish species of recreational importance. Examples include stocking programs for Florida bass, striped bass, and the introduction of hybrids, such as white bass–striped bass hybrids, which sometimes occur at the expense of species native to those stocked waters. The introduction of rainbow and brown trout has resulted in dramatic declines of native brook trout and other species in some rivers and streams.

Fish are commonly used as indicators of aquatic health. Grabarkiewicz and Davis (2008) describe the importance and use of fish as biological indicators.

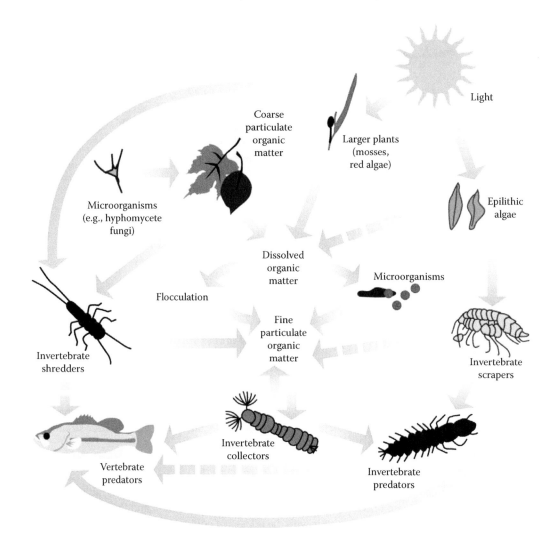

FIGURE 6.19 General organic matter pathway. (From FISRWG, *Stream Corridor Restoration: Principles, Processes, and Practices*, Federal Interagency Stream Restoration Working Group, 1998.)

6.4   SOURCES AND DISTRIBUTION OF ORGANIC MATTER

The general pathway of organic matter in aquatic systems is illustrated in Figure 6.19. The sources of organic matter are autochthonous (instream sources) or allochthonous (out of stream) and the relative importance of each varies, for example, with the stream order, as indicated by the RCC (Figure 6.2).

In low-order streams, most of the organic matter is allochthonous. The organic materials are produced outside the streams and then carried to them, such as coarse-sized leaves and litter falling into the river from riparian vegetations, dead organisms, and other particulate materials transported to the stream. Coarse particulate organic matter (CPOM) is considered to have a particle size greater than about 1 mm. The organic material can also be dissolved organic compounds (dissolved organic matter; DOM), such as materials leached from decaying materials. DOM can comprise the largest pool of organic carbon in streams. In some Southern streams, highly colored tannic acid and DOM derived from wetlands and often complexed with iron give the streams a characteristic stained color and they are often called blackwater streams.

Of the POM entering streams and rivers, about one-quarter is stored in the bank or channel, one-quarter is metabolized, and one-half is exported downstream (WOW 2004). In mid-order streams, shading and the contribution of coarse organic matter decrease and primary productivity is greater, due to periphyton (a greater percentage of the autotrophs are composed of periphyton). The export of processed coarse organic materials provides an input of fine particulate organic materials (FPOMs). In higher-order streams, macrophytes and phytoplankton become dominant and the majority of the carbon production is autochthonous, as described by the RCC (Figure 6.2).

POM varies seasonally in streams and rivers. For example, POM is typically highest during low flows during the summer months and lowest during high-flow conditions. The contribution of allocthonous POM also varies with processes in the watershed, such as seasonal growth cycles, and with flows. For example, the flood pulse theory of Junk et al. (1989) describes the process of material exchange between a floodplain and a river during floods, impacting the productivity of both.

POM is both a measure and a source of stream and river productivity. It impacts, and is impacted by, stream biota. POM and the RCC emphasize that the management, or restoration, of rivers and streams cannot be accomplished by attempting to manage the biotic population alone. While the biotic population, vertebrate and invertebrate, can provide an indication of aquatic health, that health depends on the cumulative impact of all processes, physical, chemical, and biological, impacting the stream or river.

REFERENCES

Alexander, J.E., J.H. Thorp, and J.C. Smith. 1997. *Biology and Potential Impacts of Zebra Mussels in Large Rivers*. AWWA Research Foundation and American Water Works Association, Denver, CO.

Allan, J.D. and M.M. Castillo. 2007. *Stream Ecology*, 2nd ed. Springer, New York.

Barbour, M.T., J. Gerritsen, B.D. Snyder, and J.B. Stribling. 1999. *Rapid Bioassessment Protocols for Use in Streams and Wadeable Rivers: Periphyton, Benthic Macroinvertebrates and Fish*, 2nd ed., EPA 841-B-99-002. U.S. Environmental Protection Agency, Office of Water, Washington, DC.

Barnhart, C. and D. Neves. 2005. Overview of North American freshwater mussels. USEPA Mussel Toxicity Testing Workshop, Chicago, IL.

Barry, J., F. Biggs, and C. Kilroy. 1994. Stream periphyton monitoring manual. Prepared for The New Zealand Ministry for the Environment by NIWA, Christchurch NIWA, Christchurch, New Zealand.

Benson, A.J. 2011. Zebra mussel sightings distribution. http://nas.er.usgs.gov/taxgroup/mollusks/zebramussel/zebramusseldistribution.aspx.

Chapra, S.C., G.J. Pelletier, and H. Tao. 2007. *QUAL2K: A Modeling Framework for Simulating River and Stream Water Quality*, Version 2.07: Documentation and Users Manual. Civil and Environmental Engineering Department, Tufts University, Medford, MA.

Clifford, H. 1991. *Aquatic Invertebrates of Alberta*, p. 638. University of Alberta Press, Edmonton, Alberta, Canada.

Cushing, C.E. and J.D. Allan. 2001. *Streams: Their Ecology and Life*. Academic Press, San Diego, CA.

FISRWG. 1998. *Stream Corridor Restoration: Principles, Processes, and Practices*, Federal Interagency Stream Restoration Working Group.

Flinders, C.A. and D.D. Hart. 2009. Effects of pulsed flows on nuisance periphyton growths in rivers: A mesocosm study. *River Research and Applications* 25 (10), 1320–1330.

Frissell, C.A., W.J. Liss, C.E. Warren, and M.D. Hurley. 1986. A hierarchical framework for stream habitat classification: Viewing streams in a watershed context. *Environmental Management* 10 (2), 199–214.

Georgia EPD. 2007. Macroinvertebrate biological assessment of wadeable streams in Georgia, standard operating procedures, Version 1.0. Georgia Department of Natural Resources, Environmental Protection Division, Watershed Protection Branch.

Grabarkiewicz, J.D. and S.W. Davis. 2008. An introduction to freshwater mussels as biological indicators, including accounts of interior basin, Cumberlandian, and Atlantic slope species, EPA-260-R-08-015. U.S. Environmental Protection Agency, Washington, DC.

Junk, W.J., P.B. Bayley, and R.E. Sparks. 1989. The flood pulse concept in river-floodplain systems. In D.P. Dodge (ed.), *Proceedings of the International Large River Symposium*. Canadian Special Publication of Fisheries and Aquatic Sciences, 106, Ottawa, pp. 110–127.

Kirk, J.O., K.J. Kilgore, and L.G. Sanders. 2001. Potential of North American molluscivorous fish to control *Dreissenid* mussels. Zebra Mussel Research Program. U.S. Amery Engineer Waterways Experiment Station, Vol. 1, No. 1, June 2001.

Knopf, F.L. 1985. Significance of riparian vegetation to breeding birds across and altitudinal cline. River ecosystems and their management, Reconciling conflicting uses. U.S. Forests Service Technical Report RM-120, Fort Collins, CO.

Lydeard, C., R.H. Cowie, W.F. Ponder, A.E. Bogan, P. Bouchet, S. Clark, K.S. Cummings, et al. 2004. The global decline of nonmarine mollusks. *BioScience* 54 (4), 321–330.

McMahon, R.F. and A.E. Bogan. 2001. Mollusca: Bivalvia. In J.H. Thorp and A.P. Covich (eds), *Ecology and Classification of North American Freshwater Invertebrates*, 2nd ed. Academic Press, New York, pp. 331–429.

Merritt, R.W. and K.W. Cummins. 1998. *An Introduction to the Aquatic Insects of North America*, 4th ed. Kendall/Hunt Publishing Company, Dubuque, IA.

Merritt, R.W., K.W. Cummins, and M.B. Berg (eds). 2008. Introduction. Chapter 1 in *An Introduction to the Aquatic Insects of North America*, 4th ed. Kendall/Hunt Publishing Company, Dubuque, IA.

Morales-Chaves, Y. 2004. Analysis of mussel population dynamics in the Mississippi River. Doctoral dissertation, University of Iowa.

Robinson, K.L. 2007. Modeling aquatic vegetation and dissolved oxygen in the Bow River, Alberta. Master's thesis, Schulich School of Engineering, Calgary.

Smith, D.G. 2001. *Pennak's Freshwater Invertebrates of the United States: Porifera to Crustacea*, 4th ed. Wiley, New York.

Strayer, D.L., N.F. Caraco, J.J. Cole, S. Findlay, and M.L. Pace. 1999. Transformation of freshwater ecosystems by bivalves: A case study of zebra mussels in the Hudson River. *Bioscience* 49 (1), 19–27.

Thuman, A.J., T.W. Gallagher, and M.M. Timothy. 2006. Jackson River modeling: 50-year perspective. *Journal of Environmental Engineering—ASCE* 132, 1051–1060.

U.S. EPA SESD. 2007. Periphyton sampling and algae surveys in wadeable streams, operating procedures. SESDPROC-510-R1m USEPA Region 4 Science and Ecosystem Support Division, Athens, GA.

USGS. 1999. Ecological status and trends of the upper Mississippi River system. Long Term Research and Monitoring Program, Report LTRMP 99-T001, Upper Midwest Environmental Sciences Center, La Crosse, WI.

Vannote, R.L., G.W. Minshall, K.W. Cummins, J.R. Sedell, and C.E. Cushing. 1980. The river continuum concept. *Canadian Journal of Fisheries and Aquatic Sciences* 37, 130–137.

Velinsky, D.J., C.A. Flinders, D.F. Charles, T.L. Bott, T.W. Gallagher, D.D. Hart, and R.L. Thomas. 2002. Periphyton dynamics in the Jackson River (VA): A multi-disciplinary study. Presented at the NABS Annual Meeting, Pittsburgh, PA.

Water on the Web. 2004. Monitoring Minnesota lakes on the internet and training water science technicians for the future—A national on-line curriculum using advanced technologies and real-time data. University of Minnesota, Duluth, MN. Available at: http://WaterOntheWeb.org.

7 Measures of the Health of Rivers and Streams

7.1 INTRODUCTION

The stated objective of the Clean Water Act (CWA) is to "restore and maintain the physical, chemical and biological integrity of the Nation's waters" (United States Code title 33, sections 1251–1387). In order to promulgate regulations to meet the objectives of the CWA, one challenge has been to develop well defined and enforceable metrics and operational conditions that can be used to both assess the condition of the nation's waters and determine when the objective of the CWA is met.

One difficulty is the scale of the problem and the diversity in waterbody types, habitats, and aquatic organisms across the United States. For example, the major rivers and streams of the conterminous United States, as illustrated in Figure 7.1, comprise only 10% of the length of U.S. flowing waters.

Changes occurring on an even larger scale resulting from physical and landscape changes are illustrated by the river continuum concept (RCC) developed by Vannote et al. (1980) (Figure 7.2). The RCC illustrates how watersheds and streams are connected and how variations in watershed and stream characteristics impact biological communities from the headwaters to the mouth of streams. For example, according to the RCC, productivity in low-order streams is more limited due to shading by riparian vegetation, so that most of the energy coming into these systems is from outside (allochthonous) sources, such as leaf fall. As a result, the habitat favors organisms that can grow or feed on these coarse organic sources. Toward the mouth, rivers are wider, slower, and deeper, and more of the productivity is derived from internal sources (autochthonous), such as phytoplankton.

In addition to the complexity of the spatial distribution of organisms, variations in time are also important. For example, often a particular life stage may be impacted by conditions at specific times of the year, such as the conditions for flow and temperature.

A wide variety of methods have been developed and implemented to assess the "physical, chemical and biological integrity" of our rivers and streams, a number of which are discussed in the following text. Some of the most commonly used methods (chemical integrity) have been based on (water quality criteria) establishing a designated use for a waterbody that is used to develop goals for that waterbody, usually indicated by the concentrations of materials dissolved or suspended in the water column and specified in water standards. In addition to the concentrations of materials in water, flow and habitat metrics have been developed (physical integrity). Flow criteria have ranged from setting minimum flows to establishing goals based on the natural flow regime (see Section 4.6.2, Chapter 4). The presence or absence and the numbers and distributions of specific organisms or groups of aquatic organisms, the indicator species, have also been used to assess the biological integrity. Metrics have also included the distribution or diversity of organisms within the biotic community as an indicator of biotic health or integrity. Finally, multimetric methods (e.g., the biological condition gradient [BCG], the index of biotic integrity [IBI], and the index of biological condition) have been developed (and will be introduced in this chapter) to assess the "physical, chemical, and biological" conditions and the integrity of our rivers and streams.

FIGURE 7.1   Major rivers and streams of the conterminous United States. (From USEPA, *Wadeable Streams Assessment: A Collaborative Survey of the Nation's Streams*, U.S. Environmental Protection Agency, Office of Research and Development, 2006.)

7.2   AMBIENT WATER QUALITY CRITERIA

As introduced in Chapter 5 on water quality processes, one metric used to evaluate whether waterbodies of the United States are "fishable and swimmable" is the ambient concentrations of specific chemicals. The premise is that target concentrations of certain materials could be established, which, if met, would protect both human and aquatic health. Therefore, the concentrations themselves, rather than the aquatic biota or biotic community, could provide an indication of aquatic health.

A fundamental provision of the CWA (Section 304(a)(1)), and amendments, was the requirement that states develop water quality standards for waterbodies. The U.S. Environmental Protection Agency (U.S. EPA) publishes standards (under 304(a) of the CWA) that states can use or modify, or states can adopt criteria based on other scientifically defensible methods (subject to approval by the U.S. EPA). States are also required to monitor for compliance with these standards, and report impairments (Section 305 of the CWA) to Congress.

Water quality standards are the foundation of a water quality-based control program and define the water quality goals for a waterbody, again with the presumption that water quality goals provide a metric for aquatic and human health. Standards have also established water quality-based treatment practices for dischargers (WQBELS, or water quality-based effluent limits). A water quality standard consists of three basic elements:

- Beneficial designated uses (e.g., designated uses include activities such as swimming, drinking water supply, oyster propagation and harvest, and other uses)
- Numeric criteria protective of designated uses
- Antidegradation policies

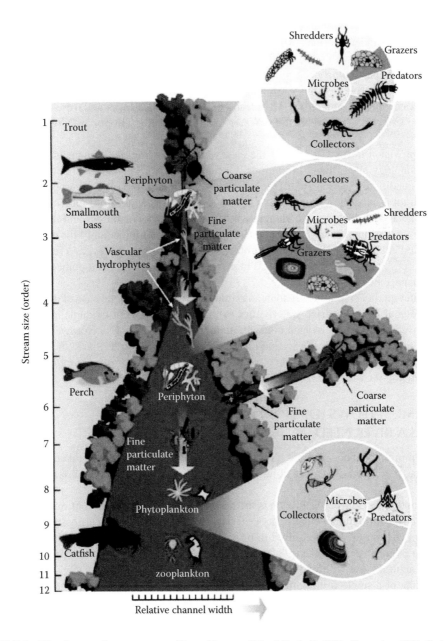

FIGURE 7.2 The river continuum concept. (From Vannote, R.L., Minshall, G.W., Cummins, K.W., Sedell, J.R., and Cushing, C.E., *Canadian Journal of Fisheries and Aquatic Sciences*, 37, 130–137, 1980. With permission.)

The standards may be narrative or specific (numeric) to a certain chemical. An example of a narrative criterion could be

> The waters of the State may not be polluted by … substances … that are unsightly, … odorous, … create a nuisance, or interfere directly or indirectly with water uses,
> while an example of a numeric criterion could be
> the water quality standard for dissolved oxygen may be 5.0 milligrams per liter oxygen for some designated uses and 6.0 for others.

To ensure that the standards are met, a permitting system was established, the National Pollutant Discharge Elimination System (NPDES), which is a program that requires that every point source discharger of pollutants obtains a permit and meets all the applicable requirements specified in the regulations issued under Sections 301 and 304 of the act. These permits are enforceable in both federal and state courts, with substantial penalties for noncompliance.

The standards, particularly the numeric standards, are used to determine allowable, or permitted, loads to waterbodies (the wasteload allocation process) or to determine if a waterbody is meeting water quality objectives. If waterbodies are not meeting standards, then states must report this to Congress (in the National Water Quality Inventory Report to Congress or 305(b) report).

Another provision in the CWA is for the case where, following the implementation of the best available technology and effluent limitations (for point sources), water standards are not achieved. Under these conditions, Section 303(d) requires that states report impaired waterbodies to Congress and establish total maximum daily loads (TMDLs), which are required to meet the water quality standards. The TMDL includes the cumulative impacts of all point sources and nonpoint sources, and allows for a margin of safety. Section 303(d) was largely unenforced until the early 1990s when lawsuits forced environmental agencies to implement its provisions. Today, as a result of including nonpoint sources, water quality control and management have shifted from point source control to watershed management.

While water quality standards set some maximum (or minimum in the case of dissolved oxygen) standard, the goal is not to drive waterbodies to those standards. For example, a common minimum daily average dissolved oxygen concentration criterion is 5 mg L^{-1}. However, the goal of the CWA is not to permit waste loads so that all waterbodies are driven to that standard, as reflected in antidegradation policies (as described in the U.S. EPA's *Water Quality Handbook*, Chapter 4: Antidegradation [40 CFR 131.12]).

7.3 MINIMUM FLOWS (OR HOW MUCH WATER DOES A RIVER NEED, AND WHEN?)

Similarly to minimum concentrations, an analogous approach has often been taken for flows. Flows are, of course, a critical component of any aquatic environment. The question is, how much flow, and when?

The first and most common approach is the determination and regulation of minimum flow criteria, analogous to minimum water quality criteria. Minimum flow criteria are assumed to represent the minimum amount of water required to protect defined criteria that often address the needs of aquatic biota (Annear and Conder 1984). Minimum flows also represent the maximum allowable depletion of natural flows without impairing the ecological services of rivers (Silk et al. 2000).

The concept of a single minimum flow was developed from western U.S. water law to reserve an amount of water from future legal consumptive use appropriations, to provide an instream water right for fish. For example, minimum instream flows were established to determine the maximum allowable abstractions. The general paradigm of minimum flows was "how much water can we take out of a river or stream" rather than "how much water do we need to leave" (Hirsch 2006).

The concept of minimum instream flows is still in common use today. For example, minimum instream flow requirements are commonly established for dam releases to protect tailwaters and downstream segments, for both aquatic health and human uses. The minimum release requirements for dams involve a constant release of a small, specified volume of water from below each dam during nongeneration or nonflood releases in order to protect downstream uses. One example is to protect downstream fisheries below hydropower operations. Many of the peaking hydropower reservoirs in the south, by releasing bottom waters from the reservoirs, have created cold-water fisheries below the dams. In addition to the cold-water releases, during nongeneration the only flow in downstream rivers is often due to seepage from the dam. The cold-water releases, while often not cold enough to support the reproduction of cold-water fish, such as rainbow and brown trout, are cold enough to support viable put-and-take trout fisheries. However, the fisheries are dependent on releases from the dam. If the dam does not release, such as during drought periods, the tailwater cold-water fisheries are diminished or cease to exist due to lack of nutrients and increases in temperatures. Therefore, minimum flow requirements are often established for such structures protective of the downstream fisheries. Similarly, minimum flows are established to protect other downstream uses, such as water supply.

An informative example of studies resulting in the establishment of minimum release requirements is that for the White River system in Arkansas. The White River system consists of five U.S. Army Corps of Engineers (the Corps) project dams (Bull Shoals, Norfork, Beaver, Table Rock, and Greers Ferry lakes) built between 1941 and 1965 and authorized for flood control and hydropower, and, to a lesser degree, water supply. Each of these reservoirs has highly productive tailwater trout fisheries and a well-developed tailwater fishing and tourism industry. Since management of the tailwaters was not a component of the original authorized use, and implementation of minimum flows would reduce income from hydropower, the implementation of minimum flows required years of study and several acts of Congress to implement. After many years of efforts to establish and implement minimum tailwater flows, the Water Resource Development Acts (WRDA) of 1999 (Section 374) and 2000 (Section 304) first modified the basic authorization and operation for the five multipurpose reservoirs "to provide minimum flows necessary to sustain tail water trout fisheries, and authorized feasibility studies" (USACOE 2009). An environmental impact study was required as well as studies to determine the costs and mitigation measures for losses of income due to hydropower. Once these studies were completed, the Energy and Water Development Appropriations Act (H.R. 3183) allocated funding in 2006 (Section 132) for Bull Shoals and Norfork lakes, but did not authorize the implementation of the proposed minimum flows at Beaver, Table Rock, or Greers Ferry lakes. Construction was initiated with the implementation of minimum flow releases from Norfolk and Bull Shoals Dams beginning in 2012–2013.

One problem with minimum flow requirements is that they vary with the particular use of a waterbody, and the uses are often in conflict. For example (available at Watershed Academy Web: Protecting Instream Flows, http://www.epa.gov/watertrain):

- Pollutant Concentration: Higher flows are important for the dilution of pollutants; in fact, many rivers and streams that violate water quality standards for common pollutants do so when flows are abnormally low.
- Aquatic Habitat: Pools, runs, and secondary channels are deeper, more varied, and more abundant when the flow is higher, and this allows a river or a stream to support more abundant and diverse aquatic life.
- Water Temperature: The amount of water in a stream or a river affects its resistance to becoming warm, because more water takes longer to heat. Higher flows protect sensitive, cold-water species, such as trout and salmon, from harmful or even lethal water temperatures.
- Recreational Uses: Sports such as white-water rafting and canoeing depend on certain levels of flow for the number of days per year that outfitters can make a living. Flows also significantly affect other sports such as fishing.

While minimum flows are still used in water management, the current environmental flow paradigm asserts that all flows are important (King et al. 2003; Hirsch 2006) and that healthy aquatic and wetland populations and communities require variable flow regimes to protect habitats and life history processes (Poff et al. 1997; Richter et al. 1996, 1997; Neubauer et al. 2008; see Chapter 4). The biota of river ecosystems have evolved in response to natural variations (Bunn and Arthington 2002) and a variety of management strategies have been developed to evaluate "how much water a river needs" and to protect the flow regime. One adaptive management method described by the U.S. EPA is the range of variability approach (RVA; available at: http://www.epa.gov/watertrain) as discussed in the following section.

7.4 HABITAT REQUIREMENTS

7.4.1 INSTREAM FLOW INCREMENTAL METHODOLOGY

The instream flow incremental methodology (IFIM) (Stalnaker et al. 1995) is really a broad conceptual and analytical framework for addressing streamflow management issues, consisting of five phases:

- Phase I—problem identification
- Phase II—study planning
- Phase III—study implementation
- Phase IV—alternatives analysis
- Phase V—problem resolution

The overall incremental approach is illustrated in Figure 7.3.

One component of the IFIM that may be used if habitat is an issue is the physical habitat simulation (PHABSIM) model previously discussed in Chapter 4. While the IFIM provides a problem-solving outline, the PHABSIM is a modeling component of the IFIM, designed to calculate the amount of microhabitat available for different life stages at different flow levels. That is, the method is based on an assessment of available habitats, as an indicator of stream integrity. This could be considered a "if you build it they will come" method, in that it addresses habitats and is not a direct measurement of water quality or biological integrity. This differs from methods discussed in the following sections based on flow (such as the RVA), or methods discussed later based on biota.

The basic PHABSIM methodology is illustrated in Figure 7.4. First, using a combination of surveys and hydraulic models, variations in habitat indicators, such as velocity, depth, and cover (bottom type or substrate), are determined for lateral sections of a reach of river as a function of flow conditions. Each section is represented by an area. Habitat suitability indices or criteria are developed or obtained for each of these habitat indicators (e.g., depth, velocity, and cover) for the species

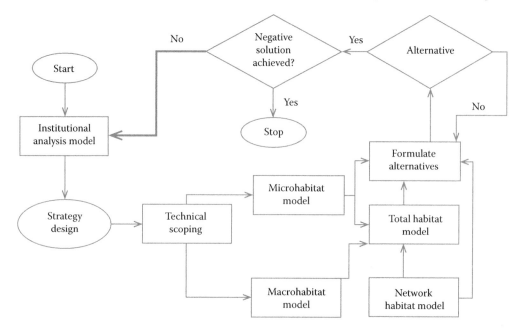

FIGURE 7.3 Overview of incremental methodology. (Redrawn from USGS, Available at http://www.fort.usgs.gov/Products/Software/ifim/5phases.asp.)

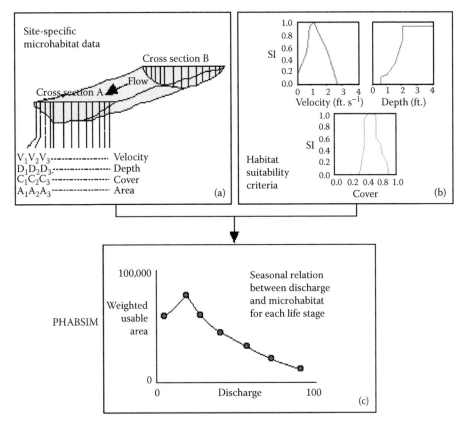

FIGURE 7.4 Overview of PHABSIM. (From USGS, Available at http://www.fort.usgs.gov/Products/ Software/ifim/5phases.asp.)

of interest, or the target species. For example, a particular target species may have an optimal range of stream velocities, and below or above that range, the species may not occur. The criteria are typically scaled from one to unity. Based on the measured or predicted habitat indicators, a specific value is determined for each of the habitat suitability criteria, which when multiplied by the cross-sectional area yields a wetted usable area for a specific flow condition. The wetted usable area can be recomputed for another flow condition, such as following a proposed flow abstraction for irrigation, and compared to the base condition to assess the change in habitat that would be expected. That change can then be used as the basis for management or mitigation decisions. The methodology and application of PHABSIM are discussed in Chapter 4 and in detail by Waddle (2001).

7.4.2 RVA

The RVA is based on the premise that the natural integrity and diversity of streams and rivers have developed in response to natural variations in flow, and the best way to maintain that integrity and diversity is to maintain the natural flow conditions (Armstrong et al. 2004). So, the method recommends that flows be managed to maintain the natural

- Flow magnitude
- Return interval (or frequency)
- Duration, timing
- Rate of change

In order to assess the natural condition, Richter et al. (1996, 1997) developed the indicators of hydrologic alteration (IHA) method and the corresponding program distributed by The Nature Conservancy (2007) and discussed in Chapter 4. The IHA method characterizes the range of variation of discharge at a site by using a suite of 33 hydrologic indices. The RVA typically involves (available at: http://www.epa.gov/watertrain):

- Characterizing the range of flows using a series of hydrologic parameters (such as the IHA method described in Chapter 4; The Nature Conservancy 2007).
- Selecting flow management targets based on these parameters.
- Designing a management system that will attain these targets.
- Implementing the management system and monitoring its effects.
- Repeating characterization yearly and comparing new values to the management targets.
- Incorporating new monitoring information and revising either the management system or the RVA targets as necessary.

7.4.3 HYDROECOLOGICAL INTEGRITY PROCESS

The hydroecological integrity assessment process (HIP) (Henriksen et al. 2006) developed by the U.S. Geological Survey (USGS) was also designed (as was the IHA method discussed in Section 7.4.2 and in Chapter 4) to determine hydrologic alteration, or to manage streamflow and establish flow standards. The HIP method involves an evaluation and classification of relatively unmodified streams in a geographic area, based on 171 ecologically relevant indices computed from the available data for daily mean discharges and peak annual flows.

Once the indices are computed, they are evaluated based on a review by Olden and Poff (2003) of the available hydrologic indices as tested using data from 420 sites around the United States. Following the analysis of the indices, a series of principal components analyses are used to identify the most significant indices associated with 10 subcomponents of the flow regime for each of the 6 stream types:

- Magnitude—low, average, high
- Frequency—low, high
- Duration—low, high
- Timing—low, high
- Rate of change—average

The indices can then be used for state or regional stream classification analyses.

7.5 INDICATOR ORGANISMS

As opposed to habitat or flows, an alternative approach commonly used is to identify certain species that, by their presence, are indicators of a healthy aquatic system. Typically, however, it is not an individual species that is used as an indicator, but groups or types of organisms, where within each group the individual species can be used as a metric, such as the percentage of occurrence. An example may be the percentage of a species such as *Achnanthes minutissima* (a diatom species) or groups of species (e.g., a number of Ephemeroptera/Plecoptera/Trichoptera [EPT] taxa). Some of the more commonly used indicators include fish, aquatic algae, plants (e.g., periphyton and macrophytes), and invertebrates. Mammals, reptiles and amphibians, birds, woody or wetland plants, and other organisms may also be used. While there are many rare or endangered species, it is usually the more common species that are preferable indicators, since they are easier to find and monitor. That is, rare and endangered species should be monitored, but probably not for this specific purpose. The considerations for the selection and sampling of indicators include their level of tolerance for specific conditions (such as low-dissolved oxygen, toxicity, habitat, and geographic area), the spatial and temporal (such as based on reproductive cycles, etc.) availability of reference conditions, and the degree of taxonomic classification. For example, depending on the required level of taxonomic classification, many hours may be required in the laboratory for each hour in the field.

7.5.1 FISH

Fish have commonly been used as biological indicators, and it is typically not the more well-known sports fish that are used, but often the less commonly known species. Certain fish species are sensitive to changes in habitat or water quality, and the presence or absence of fish can impact other aquatic organisms. For example, the parasitic life phase of freshwater mussels depends on fish hosts, often of specific species, and the fish contribute to the distribution of the mussels.

The following lists of the advantages and disadvantages of fish as indicators are provided from Grabarkiewicz and Davis (2008) based on Karr (1981) and Hocutt (1981):

Advantages

1. Long lived: some families possess long lifespans.
2. Ubiquitous: fishes occur in a wide variety of habitats.
3. Extensively studied: there is a large amount of published information regarding the occurrence and habitats of fishes.
4. Diversity: North American fishes exhibit a wide range of feeding habits, reproductive traits, and tolerances to environmental perturbations.
5. Easily identified: relative to other groups of aquatic biota, fishes are among the easier groups to identify to the species level.
6. Well known: many fish species are familiar to the general public and provide recreational opportunities.
7. Toxicity trends: presence/absence, growth, and recruitment data analysis may detect acute and sublethal effects.

Disadvantages

1. Manpower: with most sampling equipment, a three-person crew is required to effectively and safely sample fish communities.
2. Migratory: the movement of fishes may provide misleading data.
3. Sampling bias: each sampling method (electroshocking, seining, etc.) has associated biases.

7.5.2 PERIPHYTON AND MACROPHYTES

Periphyton and macrophytes have also been used as indicators of aquatic health. Both can be impacted by changes in currents or substrates, and both respond to excess nutrients, resulting in large increases in biomass and large diel variations in dissolved oxygen concentrations, as a function of production during the day and respiration during nighttime hours. The advantages of macrophytes as indicators are that they (available at: http://www.epa.gov/bioiweb1/html/indicator.html):

- Respond to nutrients, light, toxic contaminants, metals, herbicides, turbidity, water level change, and salt
- Are easily sampled through the use of transects or aerial photography
- Do not require laboratory analysis
- Are easily used for calculating simple abundance metrics
- Are integrators of environmental conditions

Similarly, the advantages of periphyton are that they

- Have a naturally high number of species
- Have a rapid response time to both exposure and recovery
- May be identified to a species level by experienced biologists
- Are easily sampled, requiring few people
- Have a tolerance for or a sensitivity to specific changes in environmental conditions that is, for many species, well known.

7.5.3 INVERTEBRATES

Perhaps one of the more common groups of indicators is the invertebrates, particularly the bottom-living or benthic macroinvertebrates, as discussed in the preceding chapter. For example, taxa such as the EPT are generally intolerant, so a high number of EPT taxa is usually an indication of an undisturbed condition.

Generally, invertebrates are good indicators of aquatic health since they (available at: http://www.epa.gov/bioiweb1/html/indicator.html):

- Live in the water for all or most of their life
- Stay in areas suitable for their survival
- Are easy to collect
- Differ in their tolerance to the amount and type of pollution
- Are easy to identify in a laboratory
- Often live for more than 1 year
- Have limited mobility
- Are integrators of environmental conditions

7.6 RAPID BIOASSESSMENTS

Rapid bioassessments were initially developed in the 1980s to reduce the time (field, laboratory, and analytical costs) associated with bioassessments. The advantage of the rapid bioassessment is, as the name implies, that it can be conducted rapidly with minimal time and cost. Basically, a rapid bioassessment compares the biota, habitat, and chemistry of selected sites to those of a similar but undisturbed (or relatively undisturbed) site, the reference condition. Indices are then developed to quantitate the characteristics of the study sites and compare them to the reference condition. The study sites are then classified based on the deviation from the reference or acceptable condition (such as nonimpaired, moderately impaired, or severely impaired). To keep the method "rapid," visual inspections, simple chemical analysis (such as using probes), and a relatively coarse taxonomic classification of biota (such as insect family) are largely relied upon. Detailed guidance on conducting rapid bioassessments is provided by Barbour et al. (1999) and Hughes et al. (2009).

One of the first steps in applying a rapid bioassessment is the selection of the reference condition. Ideally, the reference condition would be an undisturbed (unimpaired) stream or river. However, since completely undisturbed systems are rare, some acceptable condition may be defined. Examples of criteria that are useful in the selection of a reference condition are provided in Table 7.1.

An assessment of the habitat is a key component of a rapid bioassessment. For rapid bioassessment, a visually based approach is used (rather than a rigorous survey method), where specific parameters are defined that represent the habitat. Some of the parameters that may be used are indicated in a survey sheet provided as part of the U.S. EPA's rapid bioassessment protocols (Barbour et al. 1999), as summarized in Table 7.2. Each of the categories could be checked in terms of general habitat limitation or specifically as limiting to biota (such as periphyton, fish, or macroinvertebrates). For a more detailed assessment, examples of rating sheets for habitats are provided in Figures 7.5 and 7.6.

TABLE 7.1

Examples of Criteria for Selection of Reference Conditions for a Rapid Bioassessment

Example Criteria for Reference Sites (Must Meet all Criteria)

- pH $\geq$ 6; if blackwater stream, then pH $\leq$ 6 and DOC >8 mg L^{-1}
- Acid neutralizing capacity $\geq$50 μeq L^{-1}
- Dissolved oxygen $\geq$4 ppm
- Nitrate $\leq$300 μeq L^{-1}
- Urban land use $\leq$20% of catchment area
- Forest land use $\geq$25% of catchment area
- Instream habitat rating optimal or suboptimal
- Riparian buffer width $\geq$15 m
- No channelization
- No point source discharges

Source: USEPA. Watershed Academy Web: Rapid Bioassessment Protocols, USEPA Distance learning modules on Watershed Management at www.epa.gov/waetrtrain (last accessed 11 August 2013), n.d.

TABLE 7.2

Metrics from Survey Approach for Compilation of Historical Data

Limiting Factor	Probable Cause
Insufficient instream structure	Agriculture
Insufficient cover	Silviculture
Insufficient sinuosity	Mining
Loss of riparian vegetation	Grazing
Bank failure	Dam
Excessive siltation	Diversion
Insufficient organic detritus	Channelization
Insufficient woody debris for organic detritus	Urban encroachment
Frequent scouring flows	Snagging
Insufficient hard surfaces	Other channel modifications
Embeddedness	Urbanization/impervious surfaces
Insufficient light penetration	Land use changes
Toxicity	Bank failure
High water temperature	Point source discharges
Altered flow	Riparian disturbances
Overharvest	Clear cutting
Underharvest	Mining runoff
Fish stocking	Stormwater
Nonnative species	Fishermen
Migration barrier	Aquarists
Other (specify)	
Not limiting	

Source: Barbour, M.T., Gerritsen, J., Snyder, B.D., and Stribling, J.B., *Rapid Bioassessment Protocols for Use in Streams and Wadeable Rivers: Periphyton, Benthic Macroinvertebrates and Fish*, U.S. Environmental Protection Agency, Office of Water, Washington, DC, 1999.

The chemical parameters selected for rapid bioassessments are usually those that can be measured quickly in the field, such as by using probes. Examples include temperature, specific conductance, dissolved oxygen, pH, and turbidity. Other water quality indicators, such as odors or the presence of surface oils or deposits, are usually determined by visual inspection.

A variety of groups of biota may be used for rapid bioassessments, such as assemblages of fish, periphyton, and macroinvertebrates, with perhaps macroinvertebrates being most commonly used.

Stream name		Location	
Station # _____ Rivermile _____		Stream class	
Lat _____ Long _____		River basin	
Storet #		Agency	
Investigators			
Form completed by		Date _____ Time _____ AM PM	Reason for survey

	Habitat parameter	Condition category			
		Optimal	Suboptimal	Marginal	Poor
Parameters to be evaluated in sampling reach	1. Epifaunal substrate/ available cover	Greater than 50% of substrate favorable for epifaunal colonization and fish cover; mix of snags, submerged logs, undercut banks, cobble or other stable habitat and at stage to allow full colonization potential (i.e., logs/snags that are *not* new fall and *not* transient).	30%–50% mix of stable habitat; well-suited for full colonization potential; adequate habitat for maintenance of populations; presence of additional substrate in the form of newfall, but not yet prepared for colonization (may rate at high end of scale).	10%–30% mix of stable habitat; habitat availability less than desirable; substrate frequently disturbed or removed.	Less than 10% stable habitat; lack of habitat is obvious; substrate unstable or lacking.
	Score	20 19 18 17 16	15 14 13 12 11	10 9 8 7 6	5 4 3 2 1 0
	2. Pool substrate characterization	Mixture of substrate materials, with gravel and firm sand prevalent; root mats and submerged vegetation common.	Mixture of soft sand, mud, or clay; mud may be dominant; some root mats and submerged vegetation present.	All mud or clay or sand bottom; little or no root mat; no submerged vegetation.	Hard-pan clay or bedrock; no root mat or vegetation
	Score	20 19 18 17 16	15 14 13 12 11	10 9 8 7 6	5 4 3 2 1 0
	3. Pool variability	Even mix of large-shallow, large-deep, small-shallow, small-deep pools present.	Majority of pools large-deep; very few shallow.	Shallow pools much more prevalent than deep pools.	Majority of pools small-shallow or pools absent.
	Score	20 19 18 17 16	15 14 13 12 11	10 9 8 7 6	5 4 3 2 1 0
	4. Sediment deposition	Little or no enlargement of islands or point bars and less than <20% of the bottom affected by sediment deposition.	Some new increase in bar formation, mostly from gravel, sand or fine sediment; 20%–50% of the bottom affected; slight deposition in pools.	Moderate deposition of new gravel, sand or fine sediment on old and new bars; 50%–80% of the bottom affected; sediment deposits at obstructions, constrictions, and bends; moderate deposition of pools prevalent.	Heavy deposits of fine material, increased bar development; more than 80% of the bottom changing frequently; pools almost absent due to substantial sediment deposition.
	Score	20 19 18 17 16	15 14 13 12 11	10 9 8 7 6	5 4 3 2 1 0
	5. Channel flow status	Water reaches base of both lower banks, and minimal amount of channel substrate is exposed.	Water fills >75% of the available channel; or <25% of channel substrate is exposed.	Water fills 25%–75% of the available channel, and/or riffle substrates are mostly exposed.	Very little water in channel and mostly present as standing pools.
	Score	20 19 18 17 16	15 14 13 12 11	10 9 8 7 6	5 4 3 2 1 0

FIGURE 7.5 Example of a rapid bioassessment field data sheet for habitat (front page), for low gradient streams. (From Barbour, M.T., Gerritsen, J., Snyder, B.D., and Stribling, J.B., *Rapid Bioassessment Protocols for Use in Streams and Wadeable Rivers: Periphyton, Benthic Macroinvertebrates and Fish*, U.S. Environmental Protection Agency, Office of Water, Washington, DC, 1999.)

Habitat parameter	Condition category			
	Optimal	Suboptimal	Marginal	Poor
6. Channel alteration	Channelization or dredging absent or minimal; stream with normal pattern.	Some channelization present, usually in areas of bridge abutments; evidence of past channelization, i.e., dredging, (greater than past 20 years) may be present, but recent channelization is not present.	Channelization may be extensive; embankments or shoring structures present on both banks; and 40%–80% of stream reach channelized and disrupted.	Banks shored with gabion or cement; over 80% of the stream reach channelized and disrupted. Instream habitat greatly altered or removed entirely.
Score	20 19 18 17 16	15 14 13 12 11	10 9 8 7 6	5 4 3 2 1 0
7. Channel sinuosity	The bends in the stream increase the stream length 3 to 4 times longer than if it was in a straight line. (Note - channel braiding is considered normal in coastal plains and other low-lying areas. This parameter is not easily rated in theses areas.)	The bends in the stream increase the stream length 1 to 2 times longer than if it was in a straight line.	The bends in the stream increase the stream length 1 to 2 times longer than if it was in a straight line.	Channel straight; waterway has been channelized for a long distance.
Score	20 19 18 17 16	15 14 13 12 11	10 9 8 7 6	5 4 3 2 1 0
8. Bank stability (score each bank)	Banks stable; evidence of erosion or bank failure absent or minimal; little potential for future problems. <5% of bank affected.	Moderately stable; infrequent, small areas of erosion mostly healed over. 5%–30% of bank in reach has areas of erosion.	Moderately unstable; 30%–60% of bank in reach has areas of erosion; high erosion potential during floods.	Unstable; many eroded areas; "raw" areas frequent along straight sections and bends; obvious bank sloughing; 60%–100% of bank has erosional scars.
Score __ (LB)	Left bank 10 9	8 7 6	5 4 3	2 1 0
Score __ (RB)	Right bank 10 9	8 7 6	5 4 3	2 1 0
9. Vegetative protection (score each bank) Note: Determine left or right side by facing downstream.	More than 90% of the streambank surfaces and immediate riparian zone covered by native vegetation, including trees, understory shrubs, or nonwoody macrophytes; vegetative disruption through grazing or mowing minimal or not evident; almost all plants allowed to grow naturally.	70%–90% of the streambank surfaces covered by native vegetation, but one class of plants is not well-represented; disruption evident but not affecting full plant growth potential to any great extent; more than one-half of the potential plant stubble height remaining.	50%–70% of the streambank surfaces covered by vegetation; disruption obvious; patches of bare soil or closely cropped vegetation common; less than one-half of the potential plant stubble height remaining.	Less than 50% of the streambank surfaces covered by vegetation; disruption of streambank vegetation is very high; vegetation has been removed to 5 centimeters or less in average stubble height.
Score __ (LB)	Left bank 10 9	8 7 6	5 4 3	2 1 0
Score __ (RB)	Right bank 10 9	8 7 6	5 4 3	2 1 0
10. Riparian vegetative zone width (score each bank riparian zone)	Width or riparian zone >18 meters; human activities (i.e., parking lots, roadbeds, clear-cuts, lawns, or crops) have not impacted zone.	Width of riparian zone 12–18 meters; human activities have impacted zone only minimally.	Width of riparian zone 6–12 meters; human activities have impacted zone a great deal.	Width of riparian zone <6 meters; little or no riparian vegetation due to human activities.
Score __ (LB)	Left bank 10 9	8 7 6	5 4 3	2 1 0
Score __ (RB)	Right bank 10 9	8 7 6	5 4 3	2 1 0

Parameters to be evaluated broader than sampling reach

Total score _____

FIGURE 7.6 Example of a rapid bioassessment field data sheet for habitat (back page), for low gradient streams. (From Barbour, M.T., Gerritsen, J., Snyder, B.D., and Stribling, J.B., *Rapid Bioassessment Protocols for Use in Streams and Wadeable Rivers: Periphyton, Benthic Macroinvertebrates and Fish*, U.S. Environmental Protection Agency, Office of Water, Washington, DC, 1999.)

Similarly to habitats, the specific taxa are assigned a numerical rating. The rating may be based on measures of species richness (specific numbers of taxa) or specific groups (such as Ephemeroptera or Plecoptera), the composition or percentage of distributions, the presence/absence of specific tolerant species, or trophic/habit measures (such as percentages of clingers, filterers, or scrapers; Barbour et al. 1999).

Once completed, all of the ratings (habitat, water quality, and biota) are totaled and used to determine the degree of impairment, as compared to the reference condition. For example, the stream or river could then be classified as unimpaired, moderately impaired, or severely impaired, based on the rating, which could then be used to guide the management or listing of impaired waterbodies.

7.7 BIOLOGICAL DIVERSITY

Biodiversity is a metric of aquatic health that refers to the diversity of the biotic assemblage. The metric has been applied to diverse species, populations, and habitats.

The most biodiverse communities are those that support a variety of species, but none is dominant. A homogeneous assemblage, one that, for example, has large numbers of only a few or a single species, is not diverse (see Simpson 1949). A commonly proclaimed relationship in ecology is that between biodiversity and stability, where the more diverse an assemblage is, the more stable it will be, which may be thought of as Mother Nature's way of "not putting her eggs in one basket," so to speak. Some of the founders of modern ecology such as Eugene Odum in his *Fundamentals of Ecology* (Odum 1953) noted that simplified communities are characterized by more violent fluctuations in population density than diverse communities. Similarly, Charles (1958) argued that "simple communities were more easily upset than that of richer ones; that is, more subject to destructive oscillations in populations, and more vulnerable to invasions." Elton's (1958) phrase "richer communities" refers to species richness or the number of species in a community. For example, taxonomic richness is a commonly used component of a biological assessment and it is simply a count of the distinct number of taxa within selected taxonomic groups. Another component of diversity is the relative abundance or equitability, which is the evenness with which the individuals are spread out among the species in a community.

The relationship between diversity, richness, and relative abundance is shown in Table 7.3, where the species richness is indicated by the number of taxa. The relative abundance (H) may be computed using Shannon's index (H), also referred to as Shannon's diversity index (Shannon 1948), and it is computed from

$$H = -\sum_{i=1}^{S} P_i \ln(P_i)$$

where S is the number of species and P is the number of individuals in a species divided by the total number of individuals.

For this example (see Table 7.3), Community 1 is dominated by a single species and has only five species, resulting in a low diversity index (H). Community 2, with the same richness, has a greater diversity since the individuals are evenly spread among the species. Community 3, with an even greater species richness, also has a low diversity, since it is dominated by a single species. Community 4 has the greatest diversity, since like Community 3 it has a greater species richness (than Communities 1 or 2) and the individuals are evenly distributed among those species.

In habitat studies, low evenness or high percentage dominance by a few taxa is an indication that environmental conditions favor a limited type of organism, which suggests the presence of stressors.

According to Connell's (1978) intermediate disturbance hypothesis, it is not only the stressor or disturbance that impacts biodiversity, but also the frequency of the disturbance. Maximum diversity may be obtained when disturbances are neither too rare nor too frequent (Figure 7.7). This forms the basis of some water quality criteria based on the magnitude, frequency, and duration of exposure. For example, if a system is disturbed too frequently, it may never have time to recover from that disturbance.

TABLE 7.3

Example Computation of Species Richness and Diversity

Species	Community 1 No.	Community 1 P	Community 2 No.	Community 2 P	Community 3 No.	Community 3 P	Community 4 No.	Community 4 P
Species A	96	0.96	20	0.2	81	0.81	5	0.05
Species B	1	0.01	20	0.2	1	0.01	5	0.05
Species C	1	0.01	20	0.2	1	0.01	5	0.05
Species D	1	0.01	20	0.2	1	0.01	5	0.05
Species E	1	0.01	20	0.2	1	0.01	5	0.05
Species F	–	–	–	–	1	0.01	5	0.05
Species G	–	–	–	–	1	0.01	5	0.05
Species H	–	–	–	–	1	0.01	5	0.05
Species I	–	–	–	–	1	0.01	5	0.05
Species J	–	–	–	–	1	0.01	5	0.05
Species K	–	–	–	–	1	0.01	5	0.05
Species L	–	–	–	–	1	0.01	5	0.05
Species M	–	–	–	–	1	0.01	5	0.05
Species N	–	–	–	–	1	0.01	5	0.05
Species O	–	–	–	–	1	0.01	5	0.05
Species P	–	–	–	–	1	0.01	5	0.05
Species Q	–	–	–	–	1	0.01	5	0.05
Species R	–	–	–	–	1	0.01	5	0.05
Species S	–	–	–	–	1	0.01	5	0.05
Species T	–	–	–	–	1	0.01	5	0.05
Sum	100	1	100	1	100	1	100	1
Total Species	5	5	5	5	20	20	20	20
Relative Abundance, *H*		0.22		1.61		1.05		3.00

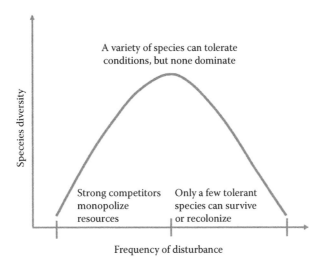

FIGURE 7.7 Intermediate disturbance hypothesis. (Redrawn from Connell, J.H., *Science*, 199, 1302–1310, 1978.)

While there is considerable evidence of a relationship between diversity and stability, the relationship depends not only on the number and distribution of species, but also on the types of organisms present. For example, it has been shown that the relationship between diversity and stability is not only a function of diversity but also the presence of function groups that can adapt to a changing environment. See McCann (2000) for a discussion of diversity and a discussion of the relationships between stability and diversity. Biodiversity and stability then depend not only on the number of entities and the evenness of the distributions (in space and time), but also on the differences in their functional traits (activities or behaviors that change ecosystem properties) and the differences in their interactions (Hooper et al. 2005).

One major environmental issue is the loss of biodiversity. It is not only the rapidly increasing rate of extinctions that is of concern, but also the decline in populations and ranges, and the homogenization of populations, in terms of species composition and genetics. Studies reported by Naeem et al. (1999) indicated that:

- Human impacts on global biodiversity have been dramatic, resulting in unprecedented losses in global biodiversity at all levels, from genes and species to entire ecosystems.
- Local declines in biodiversity are even more dramatic than global declines, and the beneficial effects of many organisms on local processes are lost long before the species become globally extinct.
- Many ecosystem processes are sensitive to declines in biodiversity.
- Changes in the identity and abundance of species in an ecosystem can be as important as changes in biodiversity in influencing ecosystem processes.

7.8 BIOLOGICAL INTEGRITY

As in the introduction to this chapter, in response to degrading conditions in aquatic systems, a national goal was established as reflected in the CWA to "restore and maintain the chemical, physical and biological integrity of the Nation's waters." While established as a national goal, the CWA did not include language to define "biological integrity" or to determine either how to accomplish that maintenance or restoration or to determine when the goals were achieved. To do so requires defining specific and scientifically supportable ecological end points and measures of how those end points can be used to quantify the condition of the nation's waters.

In order to design regulations to implement the CWA and establish a definition of "biological integrity," in 1975, the U.S. EPA hosted a national forum on the "Integrity of Water" and invited well-known experts in several disciplines (Ballentine and Guarraia 1977). Two definitions of biological integrity that were informally proposed at the forum were:

Biological integrity may be defined as the maintenance of community structure and function characteristic of a particular locale or deemed satisfactory to society. (Cairns 1977)

the capability of supporting and maintaining a balanced, integrated, adaptive community of organisms having a composition and diversity comparable to that of the natural habitats of the region. (Frey 1977)

Since then, the concept of biological integrity has evolved to mean a balanced, integrated, adaptive system having a full range of ecosystem elements (e.g., species and assemblages) and processes (e.g., biotic interactions and nutrient and energy dynamics) expected in areas with no or minimal human influence (Karr and Dudley 1981; Karr 1991; Karr and Chu 2000; Davies and Jackson 2006).

The methods for the evaluation of biological integrity that have evolved from this effort are based on the establishment of base or reference conditions, which can then be compared with the impacted condition in order to determine the state of well-being of the system. The methods developed for establishing, evaluating, and communicating biological integrity include the BCG and the IBI, which will be described in the following sections.

7.8.1 BCG

The BCG is a conceptual model that describes how the ecological integrity of aquatic systems changes in response to increasing levels of human disturbance. The BCG is consistent with ecological theory and has been verified by aquatic biologists throughout the United States. The BCG model is based on first identifying the critical attributes of an aquatic community that can be used to evaluate its biological integrity and then evaluate each attribute on a scale indicating the level of disturbance (USEPA 2005; Davies et al. 1993; Davies 2003; Davies and Jackson 2006).

Ten attributes are used to assess the changes in an aquatic ecosystem's response to increasing levels of stressors (USEPA 2005):

1. Historically documented, sensitive, long-lived, or regionally endemic taxa
2. Sensitive rare taxa
3. Sensitive ubiquitous taxa
4. Taxa of intermediate tolerance
5. Tolerant taxa
6. Nonnative or intentionally introduced taxa
7. Organism condition
8. Ecosystem function
9. Spatial and temporal extent of stressor effects
10. Ecosystem connectance

The 10 attributes are based on the spatial or temporal scales at which data are collected for an analysis. Attributes 1–6 are ecological attributes based on taxonomic composition and structure. Attributes 7 and 8 deal with organism condition and system performance. Attributes 9 and 10 are scale-dependent for physical–biotic interactions of importance in evaluating long-term impacts, restoration potential, and recoveries.

The BCG is further divided into six tiers of biological conditions along the stressor–response curve, ranging from observable biological conditions found at no or low levels of stress to those found at high levels of stress (Davies and Jackson 2006; Figure 7.8):

1. Native structural, functional, and taxonomic integrity is preserved; the ecosystem function is preserved within the range of natural variability.
2. Virtually all native taxa are maintained with some changes in biomass, abundance, or both; ecosystem functions are fully maintained within the range of natural variability.
3. Some changes in structure due to loss of some rare native taxa; shifts in the relative abundance of taxa but sensitive ubiquitous taxa are common and abundant; ecosystem functions are fully maintained through redundant attributes of the system.
4. Moderate changes in structure due to the replacement of some sensitive ubiquitous taxa by more tolerant taxa, but reproducing populations of some sensitive taxa are maintained; overall a balanced distribution of all the expected major groups; ecosystem functions are largely maintained through redundant attributes.
5. Sensitive taxa are markedly diminished; a conspicuously unbalanced distribution of major groups from that expected; organism condition shows signs of physiological stress; system function shows reduced complexity and redundancy; increased buildup or export of unused materials.
6. Extreme changes in structure; wholesale changes in taxonomic composition; extreme alterations from normal densities and distributions; organism conditioning is often poor; ecosystem functions are severely altered.

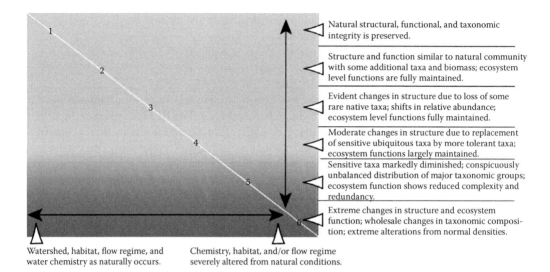

FIGURE 7.8   Condition gradients. (From USEPA, Use of biological information to better define designated aquatic life uses in state and tribal WQS: Tiered aquatic life uses, U.S. Environmental Protection Agency, Washington, DC, 2005.)

The model provides a common framework for interpreting biological information regardless of methodology or geography. The model is typically applied and calibrated to a particular state or region and is then used to evaluate the current and potential biological condition of its waters (USEPA 2005). Local taxa are assigned indicative of attributes and are calibrated to assign aquatic systems to tiers in the BCG. For example, some of the taxa associated with Tier I in Maine include mollusks such as the brook floater (*Alasmodonta varicosa*), triangle floater (*A. undulata*), yellow lampmussel (*Lampsilis cariosa*), fishes such as the brook stickleback (*Culaea inconstans*), and the swamp darter (*Etheostoma fusiforme*). In Washington, Tier I organisms may include fishes such as the steelhead (*Oncorhynchus mykiss*) and amphibians such as the spotted frog (*Rana pretiosa*; Davies and Jackson 2006). See Gerritsen and Leppo (2005) for an example application to streams in New Jersey.

7.8.2   INDEX OF BIOTIC INTEGRITY

A second method that evolved based on the definitions of biological integrity is the widely used IBI, which is used to assess the degree of human disturbance on streams and rivers. Based on his original definition of biological integrity, Karr (1981) developed the original multimetric index IBI for use in small warmwater streams in central Illinois and Indiana. The original idea was that there was a gradient between a severe disturbance (such as where everything dies) and a pristine condition, or a gradient between a very unhealthy and nonsustainable condition and a healthy, sustainable condition or one with biological integrity (Karr and Chu 2000). The biological condition could be assessed by evaluating, in the original version (Karr 1981), 12 metrics that reflected fish species richness and composition, the number and abundance of indicator species, the trophic organization and function, reproductive behavior, fish abundance, and the condition of individual fish:

- Species richness and composition metrics
  - Total number of fish species (total taxa)
  - Number of Catostomidae species (suckers)
  - Number of darter species
  - Number of sunfish species

- Indicator species metrics
  - Number of intolerant or sensitive species
  - Percentage of individuals that are *Lepomis cyanellus* (Centrarchidae)
- Trophic function metrics
  - Percentage of individuals that are omnivores
  - Percentage of individuals that are insectivorous Cyprinidae
  - Percentage of individuals that are top carnivores or piscivores
- Reproductive function metrics
  - Percentage of individuals that are hybrids
- Abundance and condition metrics
  - Abundance or catch per effort of fish
  - Percentage of individuals that are diseased, deformed, or have eroded fins, lesions, or tumors (DELTs).

In the original IBI, each metric received a score of:

- Five points: for a value similar to that expected for a fish community characteristic of a system with little human influence.
- One point: for a value similar to that expected for a fish community that departs significantly from the reference condition.
- Three points: for an intermediate value.

A cumulative score was determined, ranging from 0 (usually 12 is considered the "worst") to 60 (best), which was then used as an indicator of the degree of human disturbance.

More recent IBIs have generally retained the original metrics but they have been modified to reflect conditions in particular geographic regions or states. Kerans and Karr (1994) also demonstrated the use of a benthic IBI that includes metrics based on benthic macroinvertebrates, which was applied to rivers in the Tennessee Valley.

Generally, the applications of specific organizations today start with the available and documented metrics (see, for example, the U.S. EPA website on metrics and the IBI; Simon and Lyons [1995] and select the applicable metrics). The metrics are then standardized to a numeric scale, from worst to best. Once the metrics are scored, they can then be used as a measure of the condition of the waterbody. An example is provided in Table 7.4 for California streams and rivers (Harrington and Born 1999).

Another example is the M-BISQ methodology in Mississippi (MDEQ 2003), an IBI for Mississippi freshwater wadeable streams based on benthic macroinvertebrates. Formerly, the state of Mississippi used water quality criteria to determine water quality impairments as required under Section 303(d) of the CWA; that is, a list of waterbodies of the state that are not meeting water quality standards. The difficulty, as described previously, is having sufficient water quality data to support or determine the listing. An alternative that was implemented by the Mississippi Department of Environmental Quality was the determination of aquatic health or biotic integrity based on a biological rather than a chemical monitoring plan, under the assumption that biota integrates impacts, which, if less than some criteria, can be used to list waters that are impaired. Developing the M-BISQ (MDEQ 2003) involved the following steps:

1. Develop a database
2. Delineate preliminary site classes
3. Develop criteria for the designation of least and most disturbed sites
4. Calculate metrics
5. Delineate final site classes
6. Test metrics
7. Develop index

TABLE 7.4
IBI Metrics for the Russian River, California

Biological Metric	Score 5	Score 3	Score 1	How to Use the Russian River Index of Biological Integrity
Taxa richness	>35	35–26	<26	Obtain a sample of benthic macroinvertebrates following the state
% Dominant taxa	<15	15–39	>39	standard procedures in *California Stream Bioassessment*
EPT taxa	>18	18–12	<12	*Procedure. Protocol Brief for Biological and Physical/Habitat.*
Modified EPT Index	>53	53–17	<17	*Assessment in Wadeable Streams* by CA Department of Fish and
Shannon diversity	>2.9	2.9–2.3	<2.3	Game dated 2003. There must be at least three replicate samples
Tolerance value	<3.1	3.1–4.6	>4.6	collected at each monitoring location. The samples should be

processed by a professional bioassessment laboratory using the
Level 3 Taxonomic Effort. Determine the mean values for the six
listed biological metrics, compare them to the values in the
columns, and add the scores listed in the column headings. The
total score will be between a low of 6 and a high of 130.
Determine the biotic condition of the monitoring location from
the following categories:

Excellent	Good	Fair	Poor
30–24	23–18	17–12	11–6

Source: Harrington, J. and Born, M, *Measuring the Health of California Streams and Rivers: A Methods Manual for Water Resource Professionals, Citizen Monitors, and Natural Resources Students*, Sustainable Land Stewardship International Institute, Sacramento, CA, 1999.

Under Step 1, data were collected from over 450 stream locations. In Step 2, a series of 10 preliminary classes was developed based on the variability of the physical and chemical parameters among potential least and most disturbed sites. Five site classes or bioregions were established. The spatial distribution of the biological metric values was calculated in Step 4. The ability of the metrics to discriminate was statistically evaluated in Step 6 through a comparison of the least and most disturbed sites, and the best performing metrics within each site classes were standardized and incorporated into the final indices (Step 7), resulting in five indices (one for each bioregion), each with six or seven metrics, as shown in Table 7.5.

Index scores were established and then used to determine whether a site was biologically impaired. Using the established metrics, if the site was identified as being impaired, then additional monitoring was implemented to determine the cause of the impairment, and to develop strategies, such as TMDLs, to remove the impairment. A stressor identification process was established to identify the cases of impairment.

The IBI processes have been adapted by a wide variety of agencies to aid in determining whether aquatic systems are biologically impaired. As of 2002 (USEPA 2002), the majority of states in the United States either have or are developing multimetric biological indices (Figure 7.9).

7.8.3 INDEX OF BIOLOGICAL CONDITION

The biological condition could be considered the primary indicator of the ecological quality, where stressors indicating that quality include the physical habitat, water chemistry, land use, etc. The IBI approach described earlier is often used as an indicator of the biological condition.

A biological condition approach was used in the National Wadeable Streams Assessment (WSA). The WSA, as part of an assessment of the quality of the nation's waters, was designed to provide a consistent and statistically valid assessment of wadeable streams (USEPA 2006). The assessment, a collaborative effort with other private, state, and federal organizations, was designed to address the following questions (USEPA 2006):

- Is there a water quality problem?
- How extensive is the problem?
- Does the problem occur in "hotspots" or is it widespread?
- Which environmental stressors affect the quality of the nation's streams and rivers, and which are most likely to be detrimental?

TABLE 7.5

Mississippi Benthic Index of Stream Quality (M-BISQ) Indices for Macro-Invertebrate IBIs by Bioregion

Metrics for Specific Bioregions				
Black Belt	**East**	**Northwest**	**Northeast**	**West**
No. of collector taxa	Caenidae (%)	No. of Chironomidae taxa	Clingers (%)	Hydropsychidae/ Trichoptera
Beck's Biotic Index	No. of Tanytarsini taxa	Clingers (%)	Diptera (%)	Beck's Biotic Index
No. of Plecoptera taxa	Filterers (%)	Ephemeroptera (no Caenidae) (%)	Filterers (%)	No. sprawler taxa
Total taxa	Beck's Biotic Index	No. of Filterer taxa	Tanytarsini (%)	EPT (no Caenidae) (%)
No. of sprawler taxa	Hilsenhoff Biotic Index	Beck's Biotic Index	Hilsenhoff Biotic Index	No. Coleoptera. taxa
No. of Coleoptera taxa	EPT (no Caenidae) (%)	Hilsenhoff Biotic Index	No. of Trichoptera taxa	No. predator taxa
Caenidae (%)	Clingers (%)	Tanytarsini (%)		

Source: MDEQ, Development and application of the Mississippi Benthic Index of Stream Quality (M-BISQ), Prepared by Tetra Tech, Inc., Owings Mills, MD, for the Mississippi Department of Environmental Quality, Office of Pollution Control, Jackson, MS, 2003.

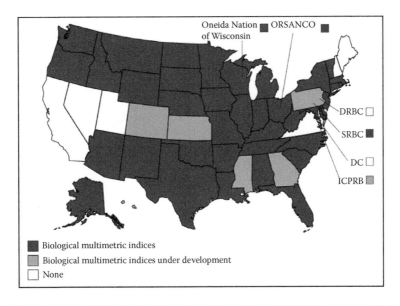

FIGURE 7.9 Development of biological multimetric indices. (From USEPA, Summary of biological assessment programs and biocriteria development for states, tribes, territories, and interstate commissions: Streams and wadeable rivers, U.S. Environmental Protection Agency, Office of Environmental Information and Office of Water, Washington, DC, 2002.)

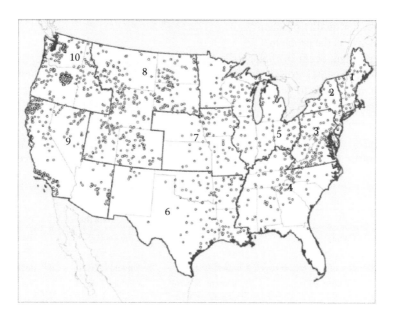

FIGURE 7.10 Sites surveyed as part of the wadeable streams assessment. (From USEPA, *Wadeable Streams Assessment: A Collaborative Survey of the Nation's Streams*, U.S. Environmental Protection Agency, Office of Research and Development, 2006.)

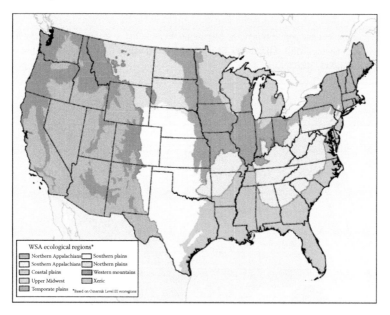

FIGURE 7.11 Ecoregions surveyed for the wadeable streams assessment. (From USEPA, *Wadeable Streams Assessment: A Collaborative Survey of the Nation's Streams*, U.S. Environmental Protection Agency, Office of Research and Development, 2006.)

The sites surveyed during the assessment are illustrated in Figure 7.10, where the sites were selected and analyzed representing nine ecoregions, as illustrated in Figure 7.11. "Least-disturbed" sites were identified, surveyed, and then used as a reference condition.

The WSA biological condition metrics concentrated on one biological assemblage, the macroinvertebrates, as an indicator of the biological condition, as with some of the macroinvertebrate IBIs. Specific macroinvertebrate metrics included (USEPA 2006):

- Taxonomic richness
- Taxonomic composition
- Taxonomic diversity
- Feeding groups
- Habits
- Pollution tolerance

The taxonomic richness refers to the number of distinct taxa within a sample, while the diversity compares all taxa and the distribution of organisms among taxa. Healthy communities should be both rich and diverse. Also, specific taxa (the composition and pollution tolerance) serve as indicators of aquatic health, both positive and negative. For example, the presence of pollution-tolerant taxa and/or the absence of nontolerant species would represent a degraded condition. Changes in the expected feeding groups or habitats from a reference condition would also represent an impact. As illustrated by the RCC (Figure 7.2), certain functional feeding groups may be representative of specific habitats or stream orders. A change, such as from a periphyton-dominated to a planktonic-dominated algae community, along with the associated changes in feeding groups, would represent a negatively impacted stream segment perhaps following the loss of riparian vegetation or some other cause. Also, a loss of organisms representative of a certain habitat could be an indication of degradation, such as could result from burial due to excess sedimentation.

In addition to biological conditions, the WSA also considered the chemical, physical, and biological indicators of aquatic stress. The chemical indicators included:

- Total phosphorus
- Total nitrogen
- Salinity
- Acidification

Excess nutrients can result in excessive algal growth, hypoxia, and a variety of other problems, as discussed in Chapter 5. Freshwater organisms are generally intolerant of salinity, so high salinity can cause stress, and acidification can reduce or eliminate aquatic populations. The physical indicators of aquatic stress included in the WSA (USEPA 2006) are:

- Relative stability of streambed sediments
- Instream habitat
- Riparian vegetative cover
- Riparian disturbance

The relative stability of streambed sediments is the ratio of the observed mean streambed particle diameter to a "critical diameter" representing the largest particle size that the stream can move as bedload during storm flows, where the critical diameter is a function of the physical stream characteristics (size, slope, etc.; Kaufmann et al. 1999). A low ratio would indicate that the streambed is unstable, such as could occur as a result of excess sedimentation, an increase in flows, or other causes. Similarly, changes in habitat such as burial by fine sediments, or changes in riparian habitat could indicate aquatic stress.

The WSA (USEPA) also included an evaluation of the biological causes of aquatic stress. One of the causes of stress results from invasive species, which will be discussed in the following section.

The result of the WSA (USEPA 2006) indicated that 45% of the wadeable streams (measured in stream lengths) in the western ecoregions were in good condition, while only 8% of stream lengths was in good condition in the eastern ecoregions. More than one-half of the stream lengths in eastern ecoregions was in poor condition (Figure 7.12). The plains and lowland ecoregions were intermediate, with nearly 30% in good condition and 40% in poor condition (USEPA 2006).

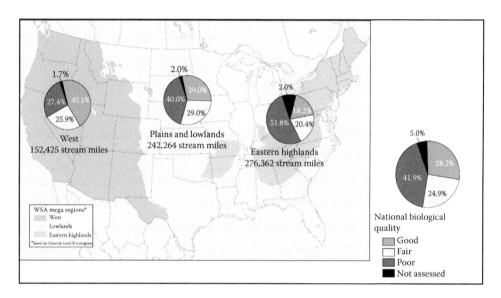

FIGURE 7.12 Condition of wadeable streams. (From USEPA, *Wadeable Streams Assessment: A Collaborative Survey of the Nation's Streams*, U.S. Environmental Protection Agency, Office of Research and Development, 2006.)

7.9 INVASIVE SPECIES

Eugene Odum, one of the great pioneers of modern ecology and the author of *Fundamentals of Ecology* (Odum 1953) introduced the concept of the interdependence of all the actors on the stage, biotic and abiotic. Those concepts are essentially the basis for some of the techniques discussed in this chapter, such as the concepts of aquatic condition and integrity.

One of the greatest causes of disturbance is the introduction of new actors onto the stage. Introduced or invasive species often cause great disruptions in native aquatic (or terrestrial) populations. Several of these new actors, such as the zebra mussel, were discussed in the preceding chapter.

The USGS maintains the Nonindigenous Aquatic Species (NAS) information system as a central repository for spatially referenced biogeographic accounts of nonindigenous aquatic species. A search of the NAS database (February 2013; http://nas.er.usgs.gov/) indicated that species that have been introduced into the United States include 668 species of fish, 49 species of amphibians

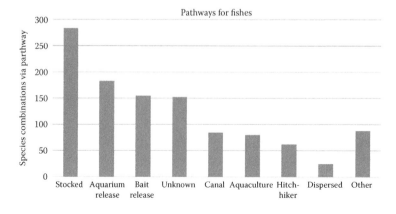

FIGURE 7.13 Major pathways for the freshwater introduction of exotic fish. (Based on data from the USGS Nonindigenous Aquatic Species Database, 2013.)

and frogs, 44 species of reptiles and turtles, 43 species of mollusks and gastropods, and lesser numbers of other species, with a total of 994 introduced species. Of the 668 fish recorded, 276 were exotic and 309 were native transplants. Of the 241 exotics, the majority were introduced by either aquaculture, stocking, or aquarium release (Figure 7.13).

REFERENCES

Annear, T.C. and A.L. Conder. 1984. Relative bias of several fisheries instream flow methods. *North American Journal of Fisheries Management* 4, 531–539.

Armstrong, D.S., G.W. Parker, and T.A. Richards. 2004. Evaluation of streamflow requirements for habitat protection by comparison to streamflow characteristics at index streamflow-gaging stations in southern New England. U.S. Geological Survey, Water-Resources Investigations Report 03-4332, p. 108.

Ballentine, R.K. and L.J. Guarraia (ed.). 1977. *The Integrity of Water. Proceedings of a Symposium, March 10–12, 1975.* U.S. Environmental Protection Agency, Washington, DC.

Barbour, M.T., J. Gerritsen, B.D. Snyder, and J.B. Stribling. 1999. *Rapid Bioassessment Protocols for Use in Streams and Wadeable Rivers: Periphyton, Benthic Macroinvertebrates and Fish*, 2nd ed. EPA 841-B-99-002. U.S. Environmental Protection Agency, Office of Water, Washington, DC.

Bunn, S.E and A.H. Arthington. 2002. Basic principles and ecological consequences of altered flows regimes for aquatic biodiversity. *Environmental Management* 30 (4), 492–507.

Cairns, J. Jr. 1977. Quantification of biological integrity. In R.K. Ballentine and L.J. Guarraia (eds), *The Integrity of Water. Proceedings of a Symposium, March 10–12, 1975.* U.S. Environmental Protection Agency, Washington, DC, pp. 171–187.

Charles, E. 1958. *The Ecology of Invasions by Animals and Plants.* Methuen, London (reprinted 2000, University of Chicago Press, Chicago).

Connell, J.H. 1978. Diversity in tropical rain forests and coral reefs. *Science* 199, 1302–1310.

Davies, S.P. 2003. The biological condition gradient. Available at: http://www.epa.gov/waterscience/ bio-criteria/modules/talu101-05-biological-conditiongradient.pdf.

Davies, S.P. and S.K. Jackson. 2006. The biological condition gradient: A conceptual model for interpreting detrimental change in aquatic ecosystems. *Ecological Applications* 16 (4), 1251–1266.

Davies, S.P., L. Tsomides, D.L. Courtemanch, and F. Drummond. 1993. Maine Biological Monitoring and Biocriteria Development Program. Maine Department of Environmental Protection, Bureau of Water Quality, Augusta, ME, p. 61.

Elton, C.S. 1958. *Ecology of Invasions by Animals and Plants.* Chapman & Hall, London.

Frey, D. 1977. Biological integrity of water: An historical approach. In R.K. Ballentine and L.J. Guarraia (eds), *The Integrity of Water. Proceedings of a Symposium, March 10–12, 1975.* U.S. Environmental Protection Agency, Washington, DC, pp. 127–140.

Gerritsen, J. and E.W. Leppo. 2005. Biological condition gradient for tiered aquatic life use in New Jersey. Prepared by Tetra Tech for New Jersey Department of Environmental Protection.

Grabarkiewicz, J. and W. Davis. 2008. An introduction to freshwater fishes as biological indicators, EPA-260-R-08-016. U.S. Environmental Protection Agency, Office of Environmental Information, Washington, DC.

Harrington, J. and M. Born. 1999. Measuring the health of California streams and rivers: A methods manual for water resource professionals, citizen monitors, and natural resources students. Sustainable Land Stewardship International Institute, Sacramento, CA.

Henriksen, J.A., J. Heasley, J.G. Kennen, and S. Nieswand. 2006. Users' manual for the Hydroecological Integrity Assessment Process software (including the New Jersey Assessment Tools). U.S. Geological Survey Open-File Report 2006-1093, p. 72.

Hirsch, R.M. 2006. USGS science in support of water resources. Presented at Mississippi Water Resources Conference, Jackson, MS.

Hocutt, C.H. 1981. Fish as indicators of biological integrity. *Fisheries* 6 (6), 28–30.

Hooper, D.U., F.S. Chapin III, J.J. Ewel, A. Hector, P. Inchausti, S. Lavorel, J.H. Lawton, et al. 2005. Effects of biodiversity on ecosystem functioning: A consensus of current knowledge. *Ecological Monographs* 75 (1), 3–35.

Hughes, D.L., J. Gore, M.P. Brossett, and J.R. Olson. 2009. *Rapid Bioassessment of Stream Health.* CRC Press, Boca Raton, FL.

Karr, J.R. 1981. Assessment of biotic integrity using fish communities. *Fisheries (Bethesda)* 6, 21–27.

Karr, J.R. 1991. Biological integrity: A long-neglected aspect of water resource management. *Ecological Applications* 1, 66–84.

Karr, J.R. and E.W. Chu. 2000. Sustaining living rivers. *Hydrobiologia* 422, 1–14.

Karr, J.R. and D.R. Dudley. 1981. Ecological perspective on water quality goals. *Environmental Management* 5, 55–58.

Kaufmann, P.R., P. Levine, E.G. Robison, C. Seeliger, and D.V. Peck. 1999. Quantifying physical habitat in wadable streams, EPA/620/R-99/003. U.S. Environmental Protection Agency, Washington, DC, p. 149.

Kerans, B.L. and J.R. Karr. 1994. A benthic index of biotic integrity (B-IBI) for rivers of the Tennessee Valley. *Ecological Applications* 4 (4), 768–785.

King, J.M., C. Brown, and H. Sabet. 2003. A scenario-based holistic approach to environmental flow assessments for rivers. *River Research and Applications* 19, 619–639.

McCann, K.S. 2000. The diversity–stability debate. *Nature* 405, 228–233.

MDEQ. 2003. Development and application of the Mississippi benthic index of stream quality (M-BISQ). Prepared by Tetra Tech, Inc., Owings Mills, MD, for the Mississippi Department of Environmental Quality, Office of Pollution Control, Jackson, MS.

Naeem, S., F.S. Chapin III, R. Costanza, P.R. Ehrlich, F.B. Golley, D.U. Hooper, J.H. Lawton, et al.. 1999. Biodiversity and ecosystem functioning: Maintaining natural life support processes. *Issues in Ecology* 4, 12.

Neubauer, C.P., G.B. Hall, E.F. Lowe, C.P. Robison, R.B. Hupalo, and L.W. Keenan. 2008. USA Minimum flows and levels method of the St. Johns River water management district, Florida, USA. *Journal Environmental Management* 42 (6), 1101–1114.

Odum, E.P. 1953. *Fundamentals of Ecology*, 1st ed. W.B. Saunders, Philadelphia, PA.

Olden, J.D. and N.L. Poff. 2003. Redundancy and the choice of hydrologic indices for characterizing streamflow regimes. *River Research and Applications* 19, 101–121.

Poff, N.L., D. Allan, M.B. Bain, J.R. Darr, K.L. Prestegaard, B.D. Richter, R.E. Sparks, and J.C. Stromberg. 1997. The natural flow regime: A paradigm for river conservation and restoration. *Bioscience* 47 (11), 769–784.

Richter, B.D., J.V. Baumgartner, J. Powell, and D.P. Braun. 1996. A method for assessing hydrologic alteration within ecosystems. *Conservation Biology* 10 (4), 163–1174.

Richter, B.D., J.V. Baumgartner, and D.P. Braun. 1997. How much water does a river need? *Freshwater Biology* 37, 231–249.

Shannon, C.E. 1948. A mathematical theory of communication. *Bell System Technical Journal* 27, 379–423; 623–656.

Silk, N., J. McDonald, and R. Wigington. 2000. Turning instream flow water rights upside down. *Rivers* 7 (4), 298–313.

Simon, T.P. and J. Lyons. 1995. Application of the index of biotic integrity to evaluate water resource integrity in freshwater ecosystems. In W.S. Davis and T.P. Simon (eds), *Biological Assessment and Criteria: Tools for Water Resource Planning and Decision Making.* CRC Press, Boca Raton, FL, pp. 245–262.

Simpson, E.H. 1949. Measurement of diversity. *Nature* 163, 688.

Stalnaker, C., B.L. Lamb, J. Henriksen, K. Bovee, and J. Bartholow. 1995. The instream flow incremental methodology—A primer for IFIM. Biological Report 29, March 1995. U.S. Department of the Interior, National Biological Service, Fort Collins, CO.

The Nature Conservancy. 2007. *Indicators of Hydrologic Alteration Version 7 User's Manual.* The Nature Conservancy, p. 75. Available at: http://www.nature.org/initiatives/freshwater/conservationtools/art17004.html.

USEPA. n.d. Watershed Academy web: Rapid bioassessment protocols. USEPA distance learning modules on watershed management. Available at: http://www.epa.gov/watertrain (last accessed 11 August 2013).

USEPA. 2002. Summary of biological assessment programs and biocriteria development for states, tribes, territories, and interstate commissions: Streams and wadeable rivers, EPA-822-R-02-048. U.S. Environmental Protection Agency, Office of Environmental Information and Office of Water, Washington, DC.

USEPA. 2005. Use of biological information to better define designated aquatic life uses in state and tribal WQS: Tiered aquatic life uses, EPA-822-R-05-001. U.S. Environmental Protection Agency, Washington, DC.

USEPA. 2006. *Wadeable Streams Assessment: A Collaborative Survey of the Nation's Streams*, EPA 841-B-06-002. U.S. Environmental Protection Agency, Office of Research and Development, p. 113.

Vannote, R.L., G.W. Minshall, K.W. Cummins, J.R. Sedell, and C.E. Cushing. 1980. The river continuum concept. *Canadian Journal of Fisheries and Aquatic Sciences* 37, 130–137.

Waddle, T.J. (ed.). 2001. *PHABSIM for Windows: User's Manual and Exercises.* U.S. Geological Survey, Fort Collins, CO.

8 Introduction to Stream Restoration

8.1 INTRODUCTION

The stated objective of the Clean Water Act (CWA) was to "restore and maintain the physical, chemical and biological integrity of the Nation's waters" (United States Code title 33, Sections 1251–1387). As discussed in Chapter 7, following the implementation of the CWA, a variety of methods were developed to evaluate the (physical, chemical, and biological) "integrity of the Nation's waters." The methods developed ranged from criteria as expressed in concentration standards, to physical characteristics, to the more complex indices of biotic integrity (IBI). In this chapter, we will concentrate on the first part of the CWA's stated objective to "restore and maintain" that integrity (e.g., Figure 8.1).

As with the variances in the methods used to define and determine "physical, chemical, and biological integrity," a number of definitions have been developed for restoration. One common difference in the definitions is in the stated goals of what condition, by restoration, a river should be returned "to." For example, the U.S. Environmental Protection Agency (U.S. EPA) (USEPA 2000) defined restoration as "the return of a degraded ecosystem to a *close approximation of its remaining natural potential.*" Cairns (1991) defined restoration as the "complete structural and functional return [of the river] to a pre-disturbance state." Similarly, the Society for Ecological Restoration Science and Policy Working Group (2002) indicates that restoration may be thought of as an "attempt to return an ecosystem to its historic (predegradation) trajectory." In addition, a variety of other terms have been associated with river restoration, such as (Shields et al. 2003a, 2008; National Research Council 1992; Brookes and Shields 1996; FISRWG 1998):

- Rehabilitation: Activities to facilitate the partial recovery of ecosystem processes and functions, making them useful again, such as restabilizing a riverbank
- Preservation (or protection, maintenance): Activities to maintain the current functions and characteristics of an ecosystem or to protect it from change or loss
- Mitigation: Activity to compensate for or alleviate environmental damage, such as creating wetlands to mitigate the loss of wetlands elsewhere
- Naturalization: Activity that results in creating a natural and sustainable environment, although not necessarily that which existed prior to the disturbance
- Creation: Activities to create a new system that does not currently exist, such as a wetland or a riparian zone
- Enhancement: Activities to improve some aspect of the existing system's quality, such as riparian vegetation
- Reclamation: Traditionally used to adapt systems to human needs (such as "reclaim" or drain wetlands), but now commonly used to refer to bringing about a stable, self-sustaining ecosystem (but not necessarily the original ecosystem)

Restoration projects do not necessarily return a river or a stream to its undisturbed or natural condition, because either that condition is not well described or a complete return is not feasible. As a result, Copeland et al. (2001) suggested that a realistic goal of a restoration project should be a partial recovery of the natural geomorphic, hydraulic, and ecological functions of the stream. In

FIGURE 8.1 Example of wetland mitigation project for a bottomland forest, a cooperative project between the Ohio Department of Transportation and Ohio State's Wilma H. Schiermier Olentangy River Wetland Research Park, which is one of 24 Ramsar Wetlands of International Importance in the United States.

this chapter, we will first review some of the common causes of the degradation of the "physical, chemical, and biological integrity" and then some of the common methods used in restoration.

Since about 1990, river restoration projects, and the literature related to those projects, have increased nearly exponentially in the United States, with annual expenditures of $1 billion or more (Bernhardt et al. 2005; Palmer et al. 2007; Figure 8.2), and the need for restoration of rivers and streams will increase for the foreseeable future (Palmer et al. 2007). Restoration has become an integral component of environmental policy in many areas of the world (e.g., White et al. 1999; Bernhardt et al. 2005; Palmer et al. 2005; Lake et al. 2007).

The National Biological Information Infrastructure has compiled the National River Restoration Science Synthesis (NRRSS) database, containing records of 37,000 stream restoration projects throughout the United States, with concentrations in seven regions of the country—California,

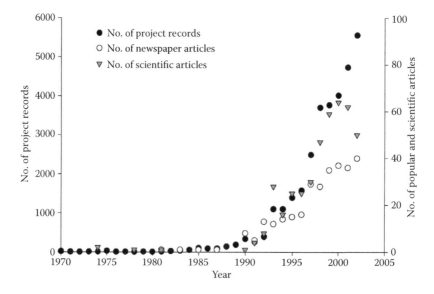

FIGURE 8.2 Rates of increase in restoration projects. (From Bernhardt, E.S. et al., *Science*, 308, 636–637, 2005. Reprinted with permission of AAAS.)

TABLE 8.1
Project Intent Information for All NRRSS Nodes

Project Intent	Number of Projects	Expenditure (millions of dollars)
Water quality management	4816	2374
Floodplain reconnection	323	1984
Flow modification	476	1645
Riparian management	5141	1468
Aesthetics/recreation/education	1131	1463
Channel reconfiguration	557	1358
Bank stabilization	2546	925
Instream habitat improvement	2292	922
Stormwater management	536	727
Dam removal/retrofit	468	549
Land acquisition	247	541
Instream species management	369	449
Other	759	292
Fish passage	859	122

Source: National River Restoration Science Synthesis Database, from March 28, 2006.

Central United States, Chesapeake Bay, Pacific Northwest, Southeast, Southwest, and Upper Midwest. The goals of the NRRSS were to summarize restoration trends and assess project effectiveness. The most common types of projects and costs (for those projects where the intent and costs were specified) are illustrated in Table 8.1. One result of this compilation and analysis was that only 10% of the projects that were reviewed had documented any form of project monitoring, and little if any of that information was either appropriate or available for assessing the ecological effectiveness of the restoration activities (Bernhardt et al. 2005). The lack of monitoring often precludes determining whether restoration projects actually achieve their design goals.

8.2 ANTHROPOGENIC IMPACTS

Historically, much of river management was based on the alteration and control of rivers and streams for human benefit, rather than maintaining their ecological integrity. For example, the traditional management paradigm for flows was based on how much water could be taken out of a river, rather than how much should be left in (Hirsch 2006). Rivers were controlled to reduce flooding and were altered for some specific purpose, such as channelization for navigation, agricultural abstractions, etc. As stated by Boon (1992): "rivers have been abstracted from, fished in, boated on, discharged into; their headwaters have been diverted; their middle reaches dammed; their floodplains developed."

Examples of common engineering modifications (Soar and Thorne 2001) include:

- Resectioning: Designed to alter the cross section to increase the in-bank discharge capacity. This is usually accomplished by enlarging the cross-sectional area and/or elevating the slope, usually by dredging and widening. Resectioned rivers are commonly trapezoidal and wider and deeper than natural rivers. Resectioned rivers also usually require maintenance, such as dredging, and stabilization of their banks, such as by using riprap.
- Realignment: Usually included in resectioning, realignment involves straightening rivers and increasing the slope of the bed, or bottom, in order to increase the flow velocity.

Realignment usually removes the natural sinuosity of rivers and streams and removes pool and riffle areas and natural shoals.

- Diversion Channels: Designed to, as the name implies, divert flows from some area. Diversion channels are commonly constructed to reduce flooding in urban areas and are commonly prismatic (constant shape, such as trapezoidal, and slope).
- Flood Banks (levees or embankments): Designed to increasing the flow capacity of streams and rivers and reduce overbank flows, thereby protecting floodplain development.
- Control Structures (e.g., dams and weirs): Designed to impound water for a variety of purposes such as water supply, irrigation, recreation, hydropower, or other uses.

An example of a channelized river is the Kissimmee River, Florida (Figure 8.3). The Kissimmee River was originally about 100 mi. long between Lake Kissimmee and Lake Okeechobee. It is the largest tributary to Lake Okeechobee, supplying about one-half of its inflow. The distance between these two lakes is only approximately 50 mi., indicating the degree of winding of this river. In addition, the river had a large floodplain, which was inundated most of the time (Audubon 2005).

In response to deadly hurricanes in the 1920s, the Central and Southern Flood Control Project was constructed with the design intended to reduce flooding, resulting in the channelization of the Kissimmee River. The main canal (C-38) constructed was 30 ft. deep and 110 yd. wide; at least three times the width and depth of the natural river channel (USACE 1991). As a result of the project, more than 30,000 acres of wetlands were lost, resulting in the decline of a variety of wildlife, including waterfowl and bald eagles. In addition, channelization, such as its impacts on increased velocities, resulted in an increase in pollutant loads to Lake Okeechobee, causing eutrophication of that waterbody. A variety of restoration projects have been completed or are underway to restore the system, such as by removing structures and restoring its natural sinuosity (Audubon 2005). At the time of its authorization, the Kissimmee River restoration was the largest and most expensive restoration ever attempted.

Currently, there are approximately 235,000 mi. of channelized streams and rivers in the United States and only approximately 2% of U.S. rivers are *natural or relatively undisturbed* (American Rivers 2009). There are over 2.5 million dams in the United States (Johnston Associates 1989), with 600,000 stream miles under reservoirs (Echeverria et al. 1989). In addition to channelization, there has been an approximately 50% nationwide reduction in wetlands, a 98% loss in ripar-

Before channelization   After channelization

FIGURE 8.3 Kissimmee River before and after channelization. (From USACE, Kissimmee River Restoration, U.S. Army Corps of Engineers Jacksonville District, Jacksonville, Florida, 2008, Available at http://www.saj.usace.army.mil/Divisions/Everglades/Branches/ProjectExe/Sections/UECKLO/KRR.htm.)

ian zones in the southwest, and 40% of U.S. waterbodies currently do not meet water quality standards.

In addition to channel alterations, watershed impacts such as those due to urbanization, agriculture, or forestry practices also directly impact the functioning of rivers and streams. For example, urbanization usually results in increases in impervious areas (roads, parking lots, etc.) that may increase cumulative runoff flows and the rate of runoff. Individual changes in land uses or the cumulative impacts of land uses in a watershed directly impact receiving waters. Wesche (1985) describes the chain of events that lead to (FISRWG 1998) (Figure 8.4):

- Changes in land use that lead to changes in geomorphology and hydrology
- Changes in stream hydraulics, sediment transport and storage
- Changes in the functions of stream habitats

The potential effects of changes in land use activities as tabulated by the Federal Interagency Stream Restoration Working Group (FISRWG 1998) are listed in Figure 8.5a,b.

The lack of natural and unimpaired streams and rivers introduces an additional complexity to restoration in that since there are so few undisturbed streams, many of the characteristics of natural systems, such as the natural flow regime, are only poorly understood and quantified. In addition, determining the extent of impacts and establishing restoration goals are often based on a comparison of existing conditions in a stream or river with some reference stream or river considered to represent an undisturbed or acceptable condition. The paucity of undisturbed streams makes locating representative reference conditions difficult.

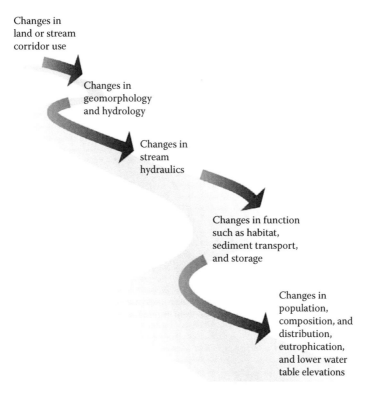

Changes in land or stream corridor use

Changes in geomorphology and hydrology

Changes in stream hydraulics

Changes in function such as habitat, sediment transport, and storage

Changes in population, composition, and distribution, eutrophication, and lower water table elevations

FIGURE 8.4   Chain of events due to disturbance. (From FISRWG, *Stream Corridor Restoration: Principles, Processes, and Practices*, Federal Interagency Stream Restoration Working Group, 1998.)

Potential effects, with column activities:

- Vegetative clearing
- Channelization
- Streambank armoring
- Streambed disturbance
- Withdrawal of water
- Dams
- Levees
- Soil exposure or compaction
- Irrigation and drainage
- Contaminants
- Hard surfacing
- Overgrazing
- Roads and railroads
- Trails
- Exotic species
- Utility crossings
- Reduction of floodplain
- Dredging for mineral extract.
- Land grading
- Bridges
- Woody debris removal
- Piped discharge/cont. outlets

Potential effects (rows):

- Homogenization of landscape elements
- Point source pollution
- Nonpoint source pollution
- Dense compacted soil
- Increased upland surface runoff
- Increased sheetflow w/surface erosion rill and gully flow
- Increased levels of fine sediment and contaminants in stream corridor
- Increased soil salinity
- Increased peak flood elevation
- Increased flood energy
- Decreased infiltration of surface runoff
- Decreased interflow and subsurface flow
- Reduced ground water recharge and aquifer volumes
- Increased depth to ground water
- Decreased ground water inflow to stream
- Increased flow velocities
- Reduced stream meander
- Increased or decreased stream stability
- Increased stream migration
- Channel widening and downcutting
- Increased stream gradient and reduced energy dissipation
- Increased or decreased flow frequency
- Reduced flow duration
- Decreased capacity of floodplain and upland to accumulate, store and filter materials and energy
- Increased levels of sediment and contaminants reaching stream
- Decreased capacity of stream to accumulate and store or filter materials and energy
- Reduced stream capacity to assimilate nutrients/pesticides
- Confined stream channel w/little opportunity for habitat development

(a) ■ Activity has potential for direct impact.   ▓ Activity has potential for indirect impact.

FIGURE 8.5 (a,b) Potential effects of major land use activities. (From FISRWG, *Stream Corridor Restoration: Principles, Processes, and Practices*, Federal Interagency Stream Restoration Working Group, 1998.)

	Disturbance activities																					
Potential effects	Vegetative clearing	Channelization	Streambank armoring	Streambed disturbance	Withdrawal of water	Dams	Levees	Soil exposure or compaction	Irrigation and drainage	Contaminants	Hard surfacing	Overgrazing	Roads and railroads	Trails	Exotic species	Utility crossings	Reduction of floodplain	Dredging for mineral extract.	Land grading	Bridges	Woody debris removal	Piped discharge/cont. outlets
Increased streambank erosion and channel scour	■	■	░	■	░	■	■	■	░	░	░	■	░	░	░	░	░	■	░	░	■	■
Increased bank failure	■	■	░	■	■	■	■	■	░	░	░	■	░	░	░	░	░	■	░	░	■	■
Loss of instream organic matter and related decomposition	■	■	■	■	░	■	■	░	░	░	░	■	░	░	░	░	■	■	░	░	■	░
Increased instream sediment, salinity, and turbidity	■	■	░	■	░	■	■	░	■	■	■	■	■	■	░	■	■	■	■	░	■	■
Increased instream nutrient enrichment, siltation, and contaminants leading to eutrophication	■	░	░	■	░	■	■	■	■	■	■	■	■	░	░	■	■	■	░	░	■	■
Highly fragmented stream corridor with reduced linear distribution of habitat and edge effect	■	■	■	░	■	■	░	░	░	░	■	■	░	░	░	░	■	■	■	■	■	░
Loss of edge and interior habitat	■	■	■	■	░	░	■	░	░	░	■	■	■	░	■	░	■	■	■	░	■	░
Decreased connectivity and width within the corridor and to associated ecosystems	■	■	■	■	░	■	■	░	░	░	■	■	■	■	■	■	■	■	■	■	░	░
Decreased movement of flora and fauna species for seasonal migration, dispersal, and population	■	■	■	■	■	■	■	░	░	░	■	■	■	■	■	■	■	■	■	■	░	░
Increase of opportunistic species, predators, and parasites	■	■	■	░	░	■	■	░	░	■	░	■	░	■	■	■	░	░	░	░	■	■
Increased exposure to solar radiation, weather, and temperature extremes	■	■	■	■	■	░	■	■	░	░	■	■	■	■	░	■	■	■	■	░	■	░
Magnified temperature and moisture extremes throughout the corridor	■	░	░	░	░	░	■	■	░	░	■	■	░	░	░	░	░	░	░	░	░	░
Loss of riparian vegetation	■	■	■	■	■	■	■	■	░	░	░	■	■	■	■	■	■	■	░	■	■	░
Decreased source of instream shade, detritus, food, and cover	■	■	■	■	■	░	■	░	░	░	░	░	░	░	░	░	░	■	░	░	■	░
Loss of vegetative composition, structure, and height diversity	■	■	■	░	■	░	■	░	■	░	■	■	■	░	■	░	■	■	░	░	░	░
Increased water temperature	■	■	■	■	■	■	■	░	░	░	■	■	░	░	░	░	■	■	░	░	■	░
Impaired aquatic habitat diversity	■	■	■	■	■	■	■	░	■	■	■	■	░	░	■	░	■	■	░	░	■	■
Reduced invertebrate population in stream	■	■	■	■	■	■	■	░	■	■	■	■	░	░	■	░	■	■	░	░	■	■
Loss of associated wetland function including water storage, sediment trapping, recharge, and habitat	░	■	░	■	■	■	■	░	■	░	■	■	■	░	■	░	■	■	■	░	░	░
Reduced instream oxygen concentration	■	■	■	■	■	■	■	░	░	░	■	■	░	░	░	░	■	■	░	░	■	■
Invasion of exotic species	■	■	■	■	■	░	■	░	░	░	░	■	░	░	■	░	■	░	░	░	░	■
Reduced gene pool of native species for dispersal and colonization	■	■	■	■	■	░	■	░	░	░	░	■	░	░	■	░	■	░	░	░	■	░
Reduced species divversity and biomass	■	■	■	■	■	░	■	■	■	░	■	■	░	░	■	░	■	░	░	░	■	░

(b) ■ Activity has potential for direct impact. ░ Activity has potential for indirect impact.

FIGURE 8.5 (Continued)

8.3   RESTORATION GOALS AND GUIDING PRINCIPLES

The first step in a restoration project is establishing the goals for that project, and then recognizing that those goals may evolve and change over the course of the project. Restoration projects are often complex and their impacts are uncertain, particularly since many of the conditions that we attempt to restore a system "to" are not completely known, and are time consuming and expensive. For example, the restoration projects for the Kissimmee River, discussed earlier, were many times more expensive and took much longer than the original channelization project. As a result, many restoration projects involve "learning while doing," also commonly referred to as adaptive management.

As indicated by Shields et al. (2003a,b, 2008; Figure 8.6), setting general goals includes establishing achievable end points that are measurable by project stakeholders (FISRWG 1998), and the process should be iterative. Setting goals typically requires first establishing the nature, cause, and extent of the problem to be restored, such as using the habitat metrics described in previous chapters and then defining the *desired future condition* (FISRWG 1998). Again, to be usable for design, the desired future condition must be quantifiable, such as the use of hydraulic techniques to restore channel stability (Shields et al. 2003b; Figure 8.6).

The FISRWG (1998) indicated that restoration goals should not only include the desired future condition but also consider the political, social, and economic values and constraints. The desired future condition may be represented by some undeveloped or reference condition that may or may not be achievable. For example, if the scale of the impact is large, such as in the watershed, it may be impossible to restore the watershed to predevelopment conditions. The political, social, and economic values and constraints as promulgated in local, state, or federal laws, or financial constraints, or constraints due to human values or welfare, would also impact restoration goals. For example, would an environmental impact assessment be required under the National Environmental Policy Act (NEPA; which may impact the time and costs of a restoration)? Also, if the restoration were to result in an impact such as increased flooding that may have occurred under the undeveloped condition, that plan may not be feasible. For example, the original goal of the Kissimmee River channelization project was flood protection, which was viewed at that time as necessary to promote and protect economic development. The 1992 Water Resources Development Act authorized the restoration of the Kissimmee River with the established goal to "restore ecological integrity to a portion of the ecosystem." The act further indicated that the restoration must also be accomplished "while retaining existing levels of flood protection to surrounding communities." The U.S. Environmental Protection Agency (U.S. EPA) (USEPA 2000) Office of Wetlands, Oceans, and Watersheds has assembled a list of principles, which are listed in Table 8.2 (and discussed in more detail in USEPA 2000), on which project goals and designs can be based and which have contributed to the success of a wide range of aquatic resource restoration projects. Some of the methods for restoration are discussed in the following sections.

8.4   RESTORATION INTENT AND TECHNIQUES

A variety of methods and techniques are available for restoration. Guidance documents such as "Stream corridor restoration: Principles, processes, and practices" (FISRWG 1998) and "Part 654, National engineering handbook, stream restoration design" (NRCS 2007a) review the principles and methodologies used in stream restoration design. This section is intended to review and provide an introduction to some of the most commonly used restoration practices in the United States. The selection of those practices to be included was based on the NRRSS. In that synthesis, based on a representative sample of restoration projects from various regions within the United States, the top ten activities for a series of project intents were identified, as listed in Table 8.3. Each of these activities will be briefly described in the following sections.

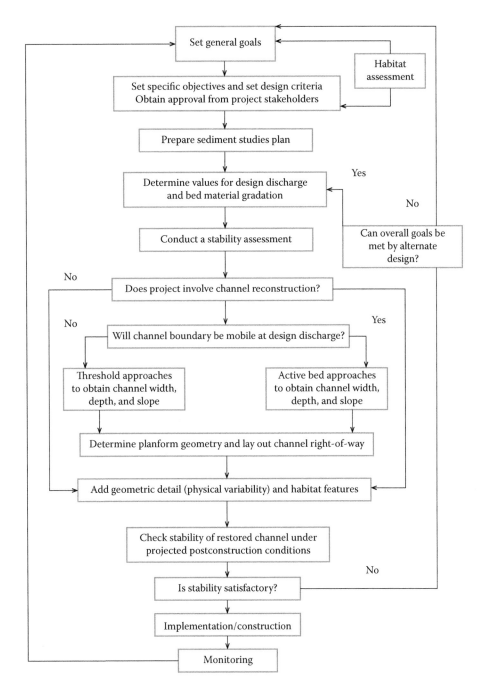

FIGURE 8.6 Flow chart for hydraulic engineer's aspects of restoration projects. (Redrawn from Shields, F.D., Cooper Jr., C.M., Knight, S.S., and Moore, M.T., *Ecological Engineering*, 20, 441–454, 2003a; Shields Jr., F.D., Copeland, R., Klingeman, P.C., Doyle, M.W., and Simon, A., *Sedimentation Engineering*, American Society of Civil Engineers, Reston, VA, 2008. With permission from the American Society of Civil Engineers.)

8.4.1 Agricultural Best Management Practices

Best management practices (BMPs) are generally understood as practices that are intended to reduce the pollutant content of nonpoint source discharges, such as pathogens or indicator species (e.g., coliform bacteria), pesticides, organic materials, sediments, and nutrients.

TABLE 8.2
Restoration Guiding Principles

Preserve and protect aquatic resources	Use reference sites
Restore ecological integrity	Anticipate future changes
Restore natural structure	Involve a multidisciplinary team
Restore natural function	Design for self-sustainability
Work within the watershed/landscape context	Use passive restoration, when appropriate
Understand the potential of the watershed	Restore native species, avoid nonnative species
Address ongoing causes of degradation	Use natural fixes and bioengineering
Develop clear, achievable, and measurable goals	Monitor and adapt where changes are necessary
Focus on feasibility	–

Source: USEPA, Principles for the ecological restoration of aquatic resources, Office of Water (4501F), United States Environmental Protection Agency, Washington, DC, 2000.

However, BMPs may also be used to reduce air pollution (or odor) or for water conservation management.

BMPs encompass a wide variety of practices, such as structural modifications or nonstructural methods such as education and public involvement (Table 8.3). Many of the methods that are used for stream restoration discussed elsewhere in this chapter may also serve to reduce pollutant loads, such as the restoration of riparian zones. Livestock exclusion, discussed separately, is also an agricultural BMP. In addition, while BMPs focus on pollutant load reductions or water management, their impact is often a critical component of stream restoration projects. Thus, while the focus of the projects differs, the practices for pollutant load reduction and stream restoration produce synergistic benefits.

This section will focus on agricultural BMPs, while urban BMPs will be discussed separately, although many of the BMPs are applicable to both environments.

The USEPA (2010) maintains a website for agricultural management practices for water quality protection. The National Resources Conservation Service (NRCS) maintains a national handbook of conservation practices (maintained as a table), updated continuously, which details nationally accepted management practices (160 at the time of this writing; see http://www.nrcs.usda.gov/technical/standards/nhcp.html). The NRCS, for each conservation practice, provides information on the conservation practice standards, information sheets, conservation practice physical effects (CPPE) work sheets, job sheets, templates for statements of work associated with each conservation practice, effects diagrams, and conservation practice standards. The conservation practice standard contains information on why and where the practice is applied, and it sets forth the minimum quality criteria that must be met during the application of that practice in order for it to achieve its intended purpose(s). The NRCS conservation practice code-numbering scheme is commonly used to identify particular practices. For example, some of the general farm practices (and conservation practice numbers) are: access road (560), forage harvest management (511), pasture and hay planting (512), pond (378), and roof runoff structure (558) (Figure 8.7).

With the increased interest in sediment and nutrient management and water conservation, a large body of guidance has been developed for agricultural BMPs. Many of the available guidance documents contain information on design and implementation, estimates of removal efficiencies, cost effectiveness considerations, and references to assist end users in implementation. Guidance on agricultural BMPs is available from a number of federal agencies. The U.S. EPA (USEPA 2003b) provided the guidance manual "National management measures to control nonpoint source pollution from agriculture." The U.S. EPA also provides a training module in their online Watershed Academy on "Agricultural management practices for water quality protection," and "Forestry BMPs

TABLE 8.3

Most Commonly Used Restoration Practices

	Aesthetics/ Education	Bank Stabilization	Channel Reconfiguration	Dam Removal/ Retrofit	Fish Passage	Floodplain Reconnection	Flow Modification	Instream Habitat Improvement	Instream Species Management	Land Acquisition	Riparian Management	Stormwater Management	Water Quality Management
Agricultural BMPs	X	–	–	–	–	–	–	–	–	–	–	–	X
Backwater sediment dredging	–	–	–	–	–	–	X	–	–	–	–	–	–
Bank or channel reshaping	X	X	X	X	–	X	X	–	X	X	X	X	X
Boulders added	–	X	–	–	X	X	X	X	–	–	X	–	X
Cleaning (e.g., trash removal)	X	–	–	–	–	–	–	–	–	–	–	–	–
Dam removal	–	–	–	X	X	–	–	–	X	–	–	–	–
Deflectors	–	X	–	–	–	–	–	X	–	–	–	–	–
Education	X	–	–	–	–	–	–	–	–	–	–	–	X
Eradication of weeds/nonnative plants	X	–	–	–	–	–	–	–	–	–	X	–	–
Fencing	–	–	–	–	–	–	–	–	X	X	X	–	X
Fish ladders installed	–	–	–	X	X	–	–	–	–	–	–	–	–
Fish passage installed	–	–	–	X	X	–	–	–	–	–	–	–	–
Fish screens installed	–	–	–	–	X	–	–	–	–	–	–	–	–
Flow regime enhancement	–	–	–	X	–	–	X	–	X	–	–	–	–
Grading-banks	–	X	X	–	–	X	–	–	–	–	–	X	–
Irrigation (increased efficiency)	–	–	–	–	–	–	–	–	–	X	–	–	–
Land acquisition or purchase	–	–	–	–	–	–	–	–	–	X	–	–	–
Livestock exclusions	–	–	–	–	–	–	–	–	–	X	X	–	X
LWD added	–	X	–	–	X	–	–	X	–	–	X	–	X
LWD removed	–	–	–	–	X	–	–	–	–	–	–	–	–
Meander creation	–	–	X	X	–	X	–	–	–	–	–	X	–
Monitoring biota	–	–	–	–	–	–	–	–	X	X	–	–	–
Monitoring flow	–	–	–	–	–	–	X	–	X	–	–	–	–
Native species protection	–	–	–	–	–	–	–	–	X	–	–	–	–

(Continued)

TABLE 8.3 (Continued)
Most Commonly Used Restoration Practices

	Aesthetics/ Education	Bank Stabilization	Channel Reconfiguration	Dan Removal/ Retrofit	Fish Passage	Floodplain Reconnection	Flow Modification	Instream Habitat Improvement	Instream Species Management	Land Acquisition	Riparian Management	Stormwater Management	Water Quality Management
Native species reintroduction	–	–	–	–	–	–	–	–	X	X	–	–	–
Pools created	–	–	–	–	X	–	–	X	–	–	–	–	–
Reinstating/maintaining hydraulic connections	–	–	–	–	–	X	X	–	–	–	–	–	–
Research	X	–	–	–	–	–	–	–	–	–	–	–	–
Revegetation live stakes	X	X	X	–	–	X	–	X	–	–	X	–	X
Revegetation seedlings/saplings	X	X	X	X	X	X	X	X	X	–	X	X	X
Revegetation unspecified	X	X	X	X	–	X	X	X	X	X	X	X	X
Revegetation-seeds	–	X	–	X	–	–	–	–	–	–	–	X	–
Riprap	–	X	X	X	–	–	–	X	–	–	–	X	–
Riparian buffer creation/ maintenance	X	–	X	–	–	–	–	–	–	X	X	X	X
Rock vanes	–	–	X	–	–	–	–	X	–	–	–	–	–
Root wads	–	X	X	–	–	–	–	X	–	–	–	–	–
Sand traps	–	–	–	–	–	–	–	X	–	–	–	–	–
Urban BMPs	–	–	–	–	–	–	–	–	–	–	–	X	–
Water level control/maintenance	–	–	–	–	–	X	X	–	–	–	–	–	–
Weirs (rock vortex)	–	–	–	–	X	–	–	–	–	–	–	–	–
Wetland conservation	–	–	–	–	–	–	–	–	–	X	–	–	–
Wetland construction	–	–	–	–	–	X	X	–	–	–	–	X	–

Source: National River Restoration Science Synthesis.

Forage harvest management: Practice Code 511

Access road: Practice Code 560

Pasture and hay planting: Practice Code 512

Roof runoff structure: Practice Code 558

Pond: Practice Code 378

FIGURE 8.7 Examples of NRCS conservation practice codes. (From Individual CP network effects diagrams.)

in watersheds" (available at: http://cfpub.epa.gov/watertrain/index.cfm). Agencies such as the U.S. Department of Agriculture (USDA), the U.S. Geological Survey (USGS), and the U.S. Army Corps of Engineers (the Corps) also produce guidance documents or regulations. For example, the Corps prescribes specific BMPs for 404 permitting. In addition, many state agencies produce agricultural BMP guidance. Selected examples include:

- GASWCC (2007). *Best Management Practices for Georgia Agriculture*. The Georgia Soil and Water Conservation Commission, Athens, GA.
- King County (2010). "King County manual of BMPs for maintenance of agricultural waterways, first edition," Washington State University.
- MDAFRR (2007). "Manual of BMPs for Maine agriculture," Maine Department of Agriculture, Food and Rural Resources, Division of Animal Health and Industry.
- NHDAMF (2008). "Manual of BMPs for agriculture in New Hampshire," New Hampshire Department of Agriculture, Markets, and Food, Concord, NH.
- TWDB (2005). "Water conservation BMPs guide," Texas Water Development Board, Report 362, Austin, TX.

The regulatory climate with regard to agricultural BMPs is changing. Historically, some agricultural practices have been excluded from environmental regulations and permitting (such as for some stormwater pollution prevention plans) and the establishment of BMPs has been voluntary. However, with more stringent requirements for load reductions, in compliance with the CWA, such as in the execution of watershed implementation plans (for example in the Chesapeake Bay), the use of agricultural BMPs will become mandatory.

There are a wide variety of agricultural BMPs (e.g., 160 NRCS conservation practices), and a complete discussion would not be possible in the context of this text. However, a few representative methods are discussed in the following sections, including conservation tillage, crop nutrient management, pest management, and conservation buffers, which are the four core practices used in training by the Conservation Technology Information Center (CTIC). In addition, tailwater recovery and reuse will be discussed as a common example of a water conservation (and load reduction) practice.

8.4.1.1 Conservation Tillage

This practice includes leaving crop residue (plant materials from past harvests) on the soil surface in order to reduce runoff and soil erosion, conserve soil moisture, help keep nutrients and pesticides on the field, and improve soil, water, and air quality. Three primary conservation tillage practices include:

8.4.1.1.1 Residue and Tillage Management No Till/Strip Till/Direct Seed (Code 329)

This practice applies to all cropland and consists of managing the amount, orientation, and distribution of crop and other plant residue on the soil surface year-round, while limiting soil-disturbing activities to only those necessary to place nutrients, condition residue, and plant crops (NRCS Practice Code 329; Figure 8.8).

8.4.1.1.2 Residue Management, Mulch Till (Code 345)

This practice applies to all cropland and consists of managing the amount, orientation, and distribution of crop and other plant residue on the soil surface year-round, where the entire field surface is tilled prior to planting. The residue is partially incorporated using chisels, sweeps, field cultivators, or similar farming implements (NRCS Code 345, Figure 8.9).

8.4.1.1.3 Residue and Tillage Management: Ridge Till (Code 346)

This practice consists of managing the amount, orientation, and distribution of crop and other plant residue on the soil surface year-round, while growing crops on preformed ridges alternated with furrows protected by crop residue (NRCS Code 346; Figure 8.10).

FIGURE 8.8 Residue and tillage management no till/strip till/direct seed. (From NRCS Code 329 network effects diagram, revised 4/2010.)

FIGURE 8.9   Residue management, mulch till. (From NRCS Code 345.)

FIGURE 8.10   Residue and tillage management: ridge till. (From NRCS Code 346.)

FIGURE 8.11   Crop nutrient management. (From NRCS Code 590.)

8.4.1.2 Crop Nutrient Management (Code 590)

This practice (Figure 8.11) manages the amount, source, placement, form, and timing of the application of plant nutrients and soil amendments. The practice helps to ensure that nutrients are available to meet crop needs while reducing nutrient movements off fields. The purpose of this practice is to budget and supply nutrients for plant production; properly utilize manure or organic by-products as a plant nutrient source; minimize agricultural nonpoint source pollution of surface water and groundwater resources; protect air quality by reducing nitrogen emissions (ammonia and NOx compounds) and the formation of atmospheric particulates; and maintain or improve the physical, chemical, and biological condition of the soil (NRCS Code 590).

8.4.1.3 Integrated Pest Management (Code 595)

This practice consists of a variety of methods for keeping insects, weeds, disease, and other pests below economically harmful levels while protecting the soil, water, and air quality. The purposes of this practice include the prevention or mitigation of off-site pesticide risks to water quality from leaching, solution runoff, and adsorbed runoff losses; risks to soil, water, air, plants, animals, and humans from drift and volatilization losses; on-site pesticide risks to pollinators and other beneficial species through direct contact; and cultural, mechanical, and biological pest suppression risks to soil, water, air, plants, animals, and humans (NRCS Code 595).

8.4.1.4 Conservation Buffers

Conservation buffers, from simple grassed waterways to riparian areas, provide an additional protective barrier by capturing potential pollutants that might otherwise move into surface waters. Ten of the common methods are listed as follows (definitions and photographs are from NRCS standard for specific conservation practice code; Figures 8.12 and 8.13):

- Alley Cropping (Code 311): Trees or shrubs are planted in sets of single or multiple rows with agronomic, horticultural crops or forages produced in the alleys between the sets of woody plants that produce additional products.
- Contour Buffer Strips (Code 332): Narrow strips of permanent, herbaceous vegetative cover established around a hill slope, and alternated down the slope with wider cropped strips that are farmed on the contour.
- Crosswind Ridges (Code 588): Ridges formed by tillage, planting, or other operations and aligned across the direction of erosive winds.
- Field Border (Code 386): A strip of permanent vegetation established at the edge or around the perimeter of a field.
- Filter Strip (Code 393): A strip or area of herbaceous vegetation that removes contaminants from overland flow.
- Grassed Waterway (Code 412): A shaped or graded channel that is established with suitable vegetation to carry surface water at a nonerosive velocity to a stable outlet.
- Herbaceous Wind Barriers (Code 603): Herbaceous vegetation established in rows or narrow strips in a field across the prevailing wind direction.
- Riparian Forest Buffer (Code 391): An area of predominantly trees and/or shrubs located adjacent to and up-gradient from watercourses or waterbodies.
- Vegetative Barrier (Code 601): Permanent strips of stiff, dense vegetation established along the general contour of slopes or across concentrated flow areas.
- Windbreak/Shelterbelt Establishment or Renovation (Codes 380 and 650): Windbreaks or shelterbelts are single or multiple rows of trees or shrubs in linear configurations.

8.4.1.5 Tailwater Recovery and Reuse System (Code 447)

This is a planned irrigation system in which all the facilities that are utilized in the collection, storage, and transportation of irrigation tailwater and/or rainfall runoff for reuse have been installed in order to conserve irrigation water supplies and improve off-site water quality (NRCS Practice Standard).

Here, the tailwater refers to the irrigation water (or stormwater) that runs off the end of an irrigated field. Tailwater systems consist of ditches or pipelines that are used to collect this water and divert it to a storage location or other system for reuse (Figure 8.14). Tailwater recovery and reuse systems are applicable to any irrigated agricultural system (typically flood or furrow irrigation) in

Alley cropping   Contour buffer strips

Crosswind trap strips   Field border

Filter strip   Grassed waterway

FIGURE 8.12 Examples of conservation buffers. (From NRCS Conservation Practice Standards.)

which a significant quantity of irrigation water, as a result of the irrigation method, runs off the end of the irrigated field. Most tailwater systems also collect rainfall that may run off the irrigated field. Natural reservoirs, such as the playa lakes located in plains areas, may serve to capture both irrigation runoff and rainfall runoff and may be used as part of a tailwater system. Also, the capture and reuse of tailwater can improve the water quality of downstream reaches of rivers, streams, or waterways, but may also reduce agricultural drain flow and the amount of water in those downstream reaches.

8.4.2 BACKWATER SEDIMENT DREDGING

River development projects have commonly resulted in losses of secondary channels and backwater areas. In addition, backwater areas are generally depositional and are degraded by excessive

Herbaceous wind barriers   Riparian forest buffer

Vegetative barrier   Windbreak/shelter belt establishment or renovation

FIGURE 8.13   Additional examples of conservation buffers. (From NRCS Conservation Practice Standards.)

FIGURE 8.14   Example of a tailwater recovery/reuse system. (From NRCS Conservation Practice Standard 447.)

amounts of sediment. For example, Marlin (2001) indicated that during the past century, sediment has filled over 70% of the volume of the Illinois River backwaters and side channels.

To restore backwater habitats, dredging is used to remove the sediment and aid in maintaining water movement. Backwater dredging typically consists of dredging channels that extend out from the main cut. The depth and size of the dredge cut depend on several site-specific factors. Two common dredging methods are mechanical and hydraulic. Mechanical dredging involves using some form of device (e.g., backhoe, clamshell, or dragline) to scrape and remove sediment from the stream. It works well with hard compacted sediment but can leave fine soils behind. Hydraulic dredges (such as the cutterhead dredge) generally pump the sediment from the river and, unlike mechanical dredging, they can place the material directly into a placement site. Hydraulic dredging also allows for the ability to pump almost continuously, which results in higher production rates

FIGURE 8.15 Dry dredge operation in backwater restoration of the Illinois River. (From Marlin, J.C., *Potential Use of Innovative Dredge Technology and Beneficial Use of Sediment for River Restoration*, Waste Management and Research Center, Illinois Department of Natural Resources, Champaign, IL, 2001.)

than mechanical dredging. Dry dredges (high solids dredges; Figure 8.15) lift sediment from the bottom and use a displacement pump to move it without adding water.

8.4.3 BANK OR CHANNEL RESHAPING

Bank or channel reshaping is a component of channel reconfiguration or alignment and floodplain reconnection activities. The reshaping is intended to reintroduce or create a stable configuration, such as in channels that are incised and/or eroding, straightened (channelized), over wide, or have a loss of floodplain function or connection. Channel reshaping may also be necessary for the creation or maintenance of riparian buffers.

One case for channel reshaping is for incising channels. Deeply incised channels are often unstable, with accelerated bank erosion, land loss, aquatic habitat loss, lower water tables, reduced land productivity, and accelerated downstream sedimentation. In addition, in an incised condition, the river may have lost access to the floodplain and is therefore no longer receiving the benefits from overbank flooding. Riprap is commonly used to reduce the consequences of channel incision. However, riprap generally results in negative values for fish and wildlife. An alternative may be to excavate and construct a new floodplain at the incised channel elevation or higher (but not at the original level), which is a proactive acceleration of the natural progression of incised channels. Where reach alterations are the primary cause of incision, another alternative would be to restore the historic channel grade and elevation to reestablish a connection with the floodplain by raising the channel bed or moving the channel to a new or former location on the old floodplain surface (Saldi-Caromile et al. 2004).

Channel or bank reshaping is also an intervention method for over-widened rivers, which are common in western states in flow-depleted streams due to a reduction in the stream power and an increase in sedimentation and consequently bank erosion. Reshaping usually requires the use of heavy equipment and is usually a very publicly visible restoration methodology. Reshaping is also commonly completed in conjunction with other structural measures to restore and maintain stream stability, as described by Rosgen (1997).

8.4.4 BOULDERS ADDED

The addition of boulders is a commonly used method of enhancing instream habitats. The pools and eddies around the boulders create fish habitats and the boulder surface is important to clinging invertebrates, periphyton, and other organisms.

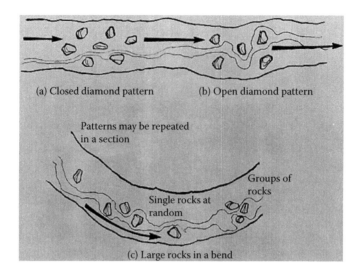

FIGURE 8.16 Placement of rocks. (Modified from Barton, J.R. and Cron, F.W., *Restoration of Fish Habitat in Relocated Streams*, U.S. Department of Transportation, Federal Highway Administration, Washington, DC, 1979.)

Boulders are typically placed randomly or in clusters and the recommended boulder size is a function of the stream hydraulic characteristics and the equipment available (Figure 8.16). Generally, in larger streams, a minimum diameter of 2 ft. is recommended, while smaller boulders may be appropriate in smaller streams (see Table 8.4).

8.4.5 CLEANING (TRASH REMOVAL)

One common problem for degraded streams is that they often serve as illegal dumping sites for trash, from depositing grass clippings or debris from yards to household garbage and appliances, impacting not only the stream's characteristics, but also its aesthetics and recreational value. The removal of trash from streams and rivers is a common restoration practice. It is also a restoration practice that can usually benefit from public involvement and education, such as in Adopt-A-Stream projects. The public may be involved in trash removal and/or monitoring. For example, the Friends of the Rappahannock hosts annual spring and fall cleanups, in which typically over 300 individuals

TABLE 8.4
Approximate Basis for Selecting Large Rocks

Channel Width at Normal Summer Flow	Depth of Water Where Rock is Placed	Dimension of Long Side of Rock
Up to 20 ft (6 m)	1.0–2.5 ft (0.3–0.75 m)	2–4 ft (0.6–1.2 m)
20–40 ft (6–12 m)	1.0–3.0 ft (0.3–0.9 m)	3–8 ft (0.9–2.4 m)
40–60 ft (12–18 m)	1.5–4.0 ft (0.45–1.2 m)	4–12 ft (0.6–3.6 m)
Above 60 ft (18 m)	1.5–5.0 ft (0.45–1.5 m)	5 ft (1.5 m) to as large as can be handled with available equipment

Source: Barton, J.R. and Cron, F.W., *Restoration of Fish Habitat in Relocated Streams*, U.S. Department of Transportation, Federal Highway Administration, Washington, DC, 1979.

FIGURE 8.17   Trash cleanup activities on the Rappahannock River. (From Friends of the Rappahannock, FOR.)

participate and collect over 6000 lb. of trash over a 15 mi. reach, making a significant, positive impact on the river (Figure 8.17).

8.4.6   Dam Removal

Dam removal is becoming a more commonly used restoration practice and is often crucial in reconnecting river habitats, restoring passage of fish and other aquatic organisms, and restoring the free flow of water and sediment (Woodworth et al. 2010) (Figure 8.18). In 2010, the organization American Rivers (2011) reported that 60 dams were dismantled in the United States, 30 of which were in Pennsylvania, and that 450 dams have been removed since 1999.

Dam removal is a complex issue. Due to physical and biotic alterations, dam removal cannot ensure the restoration of flooded river reaches to their preexisting natural system in rivers and dam removal could cause social, economic, and ecological disruptions that greatly complicate decisions regarding dam removal (AFS 2004). Some of these issues and concerns are addressed in the American Fisheries Society (AFS 2004; Bigford 2004) draft policy statement on dam removal:

FIGURE 8.18   Dam removal and river restoration. (From USDA Blog on Dam Removal, 2010.)

The American Fisheries Society (AFS) recognizes that dams and associated aquatic communities provide many important societal benefits but that river blockages may cause adverse environmental impacts and societal costs. The net costs and benefits of dams should be compared to traditional values that were affected by altered habitat and ecology. American Fisheries Society believes that dam removal can be a legitimate alternative to mitigate the adverse environmental effects of dams and their operation. Decisions about dam removal should rely on the best available scientific information give full, objective consideration to local costs and benefits and broader, regional considerations. The AFS supports dam removal when it is determined that both: (1) the benefits of dam removal outweigh the costs associated with societal, cultural, environmental, economic, engineering, and technical issues; and (2) dam removal is the best approach to restore fish habitat and the fish populations and fisheries they supported. Removal decisions should be selected with full stakeholder involvement.

When deemed to be the preferred alternative, dam removal should minimize impacts to aquatic and riparian resources. The AFS recognizes that adverse impacts to fisheries and impounded ecosystems are an unavoidable consequence of dam removal, but a well designed removal can minimize short-term impacts. Over the longer term, removal is often warranted where temporary impacts are outweighed by the long-term benefits of dam removal. When the decision to preserve or rebuild a dam is made, effective and efficient fish passage facilities should be included at the structure to mitigate dam induced fragmentation of the river ecosystem and resulting impacts to aquatic communities.

Since 1999, more than 185 dams have been removed across the country, while in 2005 alone, 56 dams in 11 states were removed or slated for removal in order to restore the natural river conditions (Clark 2007). The dams that are removed are often obsolete or past their design life, often requiring repair or rehabilitation, are of only marginal value (such as for hydropower and other uses), or ecologically destructive where restoration of the ecological function is of greater benefit.

The removal of dams may result in considerable temporary and long-term changes in terms of flows, sediment transport, water quality, and stream habitat and usually requires a substantial environmental and public safety review before implementation. For example, in addition to safety concerns with the dam breaching, a consideration for dam removal is the quantity and quality of the sediments behind the dam. Dams act as sediment traps so there may be substantial accumulations of sediments that must be removed to preclude them from being washed downstream once the dam is removed. In addition, the sediments may potentially be contaminated, impacting their removal and disposal. An evaluation is required to determine the most appropriate methods for the removal and/or transport of the sediments. There are often considerable structural materials that must be removed and disposed of, such as dam materials, gates, and other hardware such as utility connections. The construction of temporary access roads may be required. Following the removal of the dam, restoration of the former reservoir basin as well as tailwater areas is usually required.

While dam removal is complex, and often controversial, it is an important alternative for river restoration. Restoration of the unregulated flow and sediment transport regimes following dam removal have resulted in increased biotic diversity through the enhancement of preferred spawning grounds or other habitats and the reappearance of riffle–pool sequences, gravel, and cobble areas. Fish passage has been another benefit of dam removal (Bednarek 2001). Many of the impacts of dams and the factors associated with their removal are summarized in a special issue of *BioScience* on "Dam removal and river restoration" (Hart and Poff 2002).

8.4.7 DEFLECTORS

Deflectors are one of several instream structures that have been used extensively to create instream habitats for fish as part of restoration projects (Shields 1983; Biron et al. 2004, 2009 and others). Deflectors concentrate the flow either vertically or horizontally. They may be used to increase the speed of flow locally to create areas of differential scour and deposition, such as to create or enhance pool–riffle areas, or to deflect the flow away from erodible banks. Deflector types include (Figure 8.19; Heaton et al. 2002; Stormwater Manager's Resource Center http://www.stormwater-center.net/):

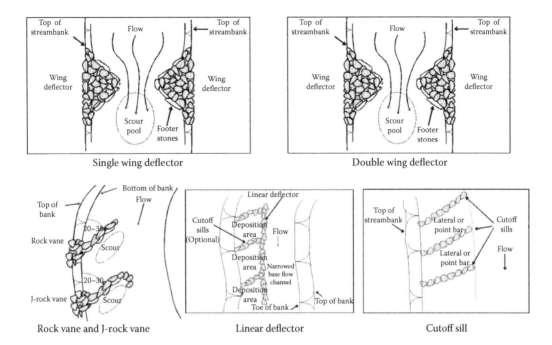

FIGURE 8.19 Deflector types. (From Stormwater Manager's Resource Center, Stream restoration: Flow deflection/concentration practices. Available at: http://www.stormwatercenter.net/.)

- Wing Deflectors (single): A single wing deflector is a triangular structure that extends out from the streambank into the stream, with the widest portion along the bank and the point extending into the channel. The purpose of a single wing deflector is to change or (deflect) the direction of the streamflow either to narrow and deepen the base flow channel or to create sinuosity in the channel. When used to narrow and deepen the base flow channel, the single wing deflector can also promote the formation of overhead cover (undercut banks) on the opposite bank.
- Wing Deflectors (double): Double wing deflectors are installed on each bank of a river to locally reduce the width of the channel, constricting the flow of water, creating scour, and deepening the base flow channel. Double wing deflectors also create an area of increased velocity between them, enhancing riffle habitats between and just upstream of the structure.
- Rock and J-Rock Vanes: Vanes are linear structures that extend out from the streambank into the stream channel in an upstream direction and mimic the effect of a tree partially falling into a stream. Vanes reduce erosion along the streambank, and the "hook" in the J-hook vane produces a longer, wider, and deeper pool than that created by vane-only structures.
- Cutoff Sills: Cutoff sills are low rock sills, which are frequently used in conjunction with linear deflectors, extending from the streambank into the stream channel at an angle of 20°–30° from the bank in an upstream direction. Cutoff sills promote deposition and bar formation along the edge of a channel in order to narrow and better define the base flow channel.
- Linear Deflector: A linear deflector consists of a line of boulders placed within the stream channel rather than along the bank, to narrow, deepen, and better define the base flow channel. The top of the deflector is generally below the bank-full elevation.

Additional information on deflectors and guidance on their design are given by Rosgen (2001), Heaton et al. (2002), Shields (1983), Shields et al. (1995), and Biron et al. (2004).

8.4.8 Education

There are a wide variety of educational opportunities associated with stream restoration, from K-12 and college-level education to professional development and general public involvement. Public involvement, acceptance, and, ideally, support are an important component of the success of restoration projects and an educational and stakeholder involvement program should be part of any restoration project.

For professional development, a relatively large number of courses and workshops are available as well as online educational material from a variety of organizations and agencies. Several universities offer programs and certification programs in stream restoration, such as the Restoration Ecology Certificate from the University of Washington and the North Carolina University Stream Restoration Program. A number of private firms and consultants also provide training. As a result of the diversity of courses and professional workshops that are available, the American Society of Civil Engineers-Environmental and Water Resources Institute formed a stream restoration educational materials task committee (TC) of the River Restoration Committee to provide the restoration community with (1) a recommended standard curriculum based on needs (technicians, engineers, planners, ecologists, biologists, etc.); (2) a list of educator traits most properly suited to cover that curriculum (not specific individuals or institutions); and (3) the logistics of where and how to most effectively disseminate information.

A number of programs for the public and K-12 students and teachers are available that provide workshops, field trips, and educational materials on stream restoration. Agencies such as the U.S. EPA through its teaching center (available at: http://www.epa.gov/teachers/index.htm) provide materials for teachers and classrooms. Adopt-a-Stream programs provide firsthand information about the importance of restoration and gain support for restoration. Public involvement may also be through cleanup efforts and monitoring programs.

8.4.9 Eradication of Weeds/Nonnative Plants

A goal of restoration projects is to restore a sustainable, healthy, and diverse native plant community, where, in many cases, undesirable or invasive species dominate the existing conditions. A variety of techniques are available for weed abatement and the removal of invasive species, including physical, chemical, biological, and cultural methods. Mechanical methods include pulling, cutting, mowing, or burning, while chemical control typically involves the use of pesticides. Biological control involves the introduction of a predator, pathogen, insect, or herbivore that feeds on the invasive species to keep the size of the population of that species in check. Cultural control can be used to prevent reintroduction, such as through transport via boats or the improper disposal of aquarium plants, and also involves public involvement through removal programs, such as "invasive pulling parties."

8.4.10 Fishways

Fishways are structures or modifications to a natural or artificial structure for the purpose of fish passage (WDFW 2000a). Fishway components usually include: attraction features, a barrier dam, entrances, auxiliary water systems, collection and transportation channels, a fish ladder, an exit, and operating and maintenance standards (WDFW 2000b). Some of the design considerations and steps in the design of fishways include (NRCS 2007b):

- Target species for fish passage or screening
- Migratory timing and life history stage at migration
- Physical limitations on fish passage (swimming speed, jumping ability)

- Environmental attractors and stressors (flow volumes, flow velocity, water temperature, seasonal timing)
- Relevant behavioral characteristics of the target species that could affect fish passage (water temperature preferences and avoidances)

The NRCS (2007b) tabulates information on swimming speeds, etc., that can aid in design, as illustrated in Table 8.5.

Fishways can be formal concrete structures or roughened channels, as illustrated in Figure 8.18. Improperly designed culverts are also common impediments to fish movement, and culvert redesign and replacement to provide fishways are a common practice.

8.4.10.1 Concrete Fishways and Ladders

Concrete fish ladders and fishways are structures that are designed to allow fish the opportunity to migrate upstream, over, around, or through a barrier to fish movement. Common examples include pool-weir or vertical-slot fishways. Figure 8.20 illustrates a general configuration and examples are illustrated in Figure 8.21. Pool-weir fishways, the most common type of fishway in the Pacific Northwest (WDFW 2000a), have distinct pools (steps) separated by weirs. The weirs may be with or without orifices (Figure 8.20). Vertical-slot ladders are also commonly used and have a large narrow slot to control the water flow and the depths in the pools between slots. The pools should provide adequate capacity and depth to dissipate hydraulic energy, maintain a stable flow, and provide room for fish to accelerate and jump (NRCS 2007b) according to their maximum swimming speeds and jumping capacity (see Table 8.5).

8.4.10.2 Roughened Channel Fishways

Roughened channels are chutes or flumes with roughness, which are designed to reduce velocity, allowing fish passage. Examples of designed roughened channels include the Denil-style fishways. Denil fishways are rectangular channels fitted with symmetrical, closely spaced baffles that redirect the flowing water and allow fish to swim around or over a barrier (NRCS 2007b; Figures 8.22 and 8.23). Denil fishways have been used in Alaska and the Pacific Northwest to pass migrating adult salmonids over small barriers (Slatick and Basham 1985). Used on large and medium-sized rivers, particularly those with highly variable flows, they are typically constructed as three-sided, sloping,

---

TABLE 8.5

Example of Maximum Swimming Speeds and Maximum Jumping Heights for Selected Adult Salmonids

Salmonid Species	Sustained Speed ft. s⁻¹	Sustained Speed m s⁻¹	Cruising Speed ft. s⁻¹	Cruising Speed m s⁻¹	Burst Speed ft. s⁻¹	Burst Speed m s⁻¹	Jump Height ft.	Jump Height m
Steelhead	4.6	1.4	13.7	4.18	26.5	8.08	11.2	3.4
Chinook	3.4	1.04	10.8	3.29	22.4	6.83	7.8	2.4
Coho	3.4	1.04	10.6	3.23	21.5	6.55	7.2	2.2
Cutthroat	2	0.61	6.4	1.95	13.5	4.11	2.8	0.9
Chum	1.6	0.49	5.2	1.58	10.6	3.23	1.7	0.5
Sockeye	3.2	0.98	10.2	3.11	20.6	6.28	6.9	2.1

Source: NRCS, Fish passage and screening design, Technical supplement 14N (210–VI–NEH), *Part 654 National Engineering Handbook*, Natural Resources Conservation Service, 2007b.

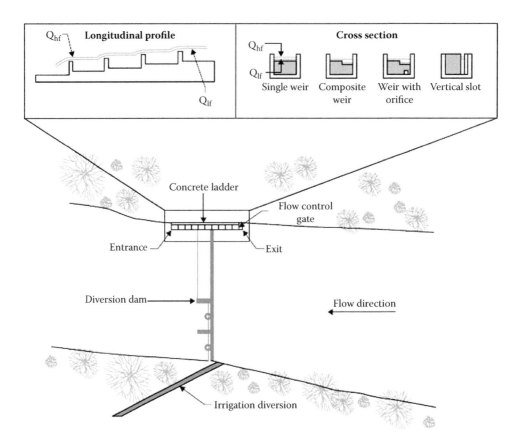

FIGURE 8.20 Plan view of a generalized concrete ladder fish passage facility. (From NRCS, Fish passage and screening design, Technical supplement 14N (210–VI–NEH), *Part 654 National Engineering Handbook*, Natural Resources Conservation Service, 2007b.)

poured-concrete flumes with a series of wooden baffles shaped like window frames installed perpendicular to the water flow. These fishways may be more than 100 ft. long, 10 ft. wide, and 8–10 ft. high. Denil fishways typically require a high degree of operational supervision and maintenance (CDOT 2007) to keep them free of debris.

8.4.10.3 Culvert Modification and Design

Culverts are among the most common barriers to fish movement. Some of the common ways, as listed by the NRCS (2007b), that culverts act as barriers include:

- High velocities or sudden velocity changes at the inlet or outlet or inside the culvert barrel
- Inadequate flow depth in the culvert barrel during critical migration periods
- Excessive length without adequate resting areas
- Significant drop at the culvert outlet
- Debris accumulation at the culvert inlet, outlet, or inside the culvert barrel
- Excessive turbulence inside the culvert or at its outlet or inlet

In many cases, the optimal solution is to replace the culvert, as illustrated by Figure 8.24. The NRCS (2007b) describes a stream simulation method that could be used to design culverts for fish passage. The USDA Forest Service (2006) also distributes FishXing (pronounced fish crossing), which is designed to assist in the evaluation and design of culverts for fish passage.

Pool-weir fishway at Berrien Springs Dam, St. Joseph River, Berrien County, MI

Vertical-slot fishway at Niles Dam, St. Joseph River, Berrien County, MI

FIGURE 8.21 Examples of pool-weir and vertical-slot fishways. (From Michigan Department of Natural Resources. n.d. Available at: http://www.michigan.gov/dnr/0,4570,7-153-10364_52259_19092-46291--,00.html.)

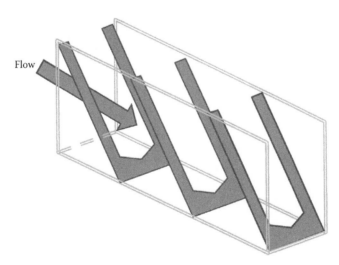

Flow

FIGURE 8.22 Denil fishway. (From California Department of Transportation 2007.)

8.4.10.4 Fish Screens Installed

One solution to prevent fish from passing through or being entrained in structures, such as water intakes for water supply, cooling, hydropower, etc., involves physically stopping them by using screens at water intakes. The screens have to guide fish toward a bypass, generally accomplished by placing them diagonally to the flow, with the bypass in the downstream part of the screen. Uniform velocities and eddy-free currents upstream of the screens must be provided to effectively guide fish toward the bypass (ASCE 1995; Larinier and Travade 1999).

FIGURE 8.23 Denil fishway at Trout Creek Dam, Trout Creek, Ontonagon County, MI. (From Michigan Department of Natural Resources. n.d. Available at: http://www.michigan.gov/dnr/0,4570,7-153-10364_52259_19092-46291--,00.html. With permission.)

FIGURE 8.24 Undersized perched culvert (*left*) replaced with larger pipe designed using stream simulation. (From NRCS, Fish passage and screening design, Technical supplement 14N (210–VI–NEH), *Part 654 National Engineering Handbook*, Natural Resources Conservation Service, 2007b.)

Two common types of fish screens are positive barriers that prevent fish from passing and behavioral barriers that encourage fish to swim away. The Washington Department of Fish and Wildlife (WDFW 2000a) describes and provides design considerations for the following screen types:

- Rotary drum screens
- Vertical fixed-plate screens
- Vertical traveling screens (panel and belt types)
- Nonvertical fixed-plate screens
- Pump screens
- End-of-pipe screens
- Infiltration galleries
- Portable screens

Examples of end-of-pipe screens are illustrated in Figure 8.25 and construction guidance is provided by the Ontario Department of Fisheries and Oceans (ODFO 1995). In addition, visual, auditory, electrical, and hydrodynamic stimuli have resulted in a large number of experimental barriers, such as bubble screens, electrical screens, and hydrodynamic (louver) screens. The hydrodynamic or louver screen consists of an array of vertical slats aligned across a canal intake at a specified angle to the flow direction (ASCE 1995) and guides fish toward a bypass, applicable to sites with relatively high approach velocities, a uniform flow, and relatively shallow depths. The efficiency

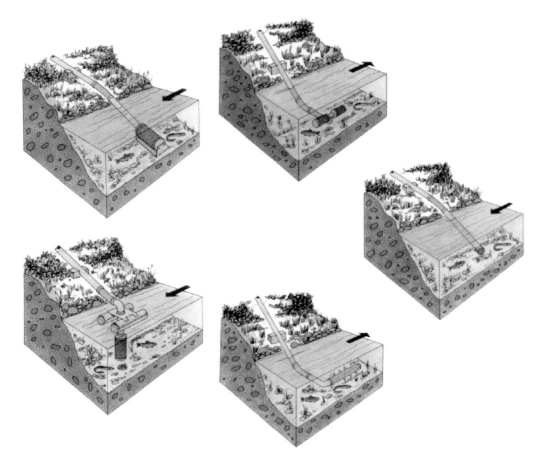

FIGURE 8.25 Examples of end-of-pipe screens. (From ODFO, *Freshwater Intake End-of-Pipe Fish Screen Guideline*, Ontario Department of Fisheries and Oceans, Ottawa, Ontario, 1995.)

of the hydrodynamic or louver screen is highly dependent on the flow pattern in the canal intake (ASCE 1995). Research and studies performed on these systems in the past decade have shown large improvements.

8.4.11 Flow Regime Enhancement

A goal of restoration is to restore or create a self-sustaining, functional flow regime in the stream or river and a system that does not require periodic human intervention. The characterization and impacts of flow were discussed in Chapter 4. Flow, or hydraulic alteration, is one of the most common causes of stream impairment and flows directly impact or drive other factors impacting stream health, such as sedimentation and aquatic habitats. A common goal of restoration projects is to restore the natural flow of rivers, but this is often not achievable since the changes in flow reflect the cumulative impact of all the changes in the watershed, from withdrawals to changes in runoff such as due to increasing impervious areas due to development. These changes may impact both the magnitude of base flows and the timing and duration of stormwater flows (see Chapter 4).

A variety of methods are available to enhance the flow regime. The following list of suggestions was taken directly from the Stream Habitat Restoration Guidelines of the Washington State Aquatic Habitat Guidelines Program (Saldi-Caromile et al. 2004).

1. Techniques to increase the base flow
   a. Remove dams, modify dam impoundments, or modify the water release management plan
   b. Reduce water withdrawal/diversion
      i Reduce water consumption
         A. Reduce irrigation needs by replacing traditional crops and landscapes that require large amounts of supplemental water with ones whose needs more closely match natural precipitation patterns (including the use of native plants)
         B. Improve irrigation practices and systems to maximize their efficiency
         C. Decrease energy demands (Washington is primarily dependent on hydroelectric power) and use alternative energy sources
         D. Use water-efficient appliances and reduce nonessential water use
      ii. Improve the efficiency of water delivery systems (e.g., fixing leaks and using systems that minimize loss of water to evaporation and infiltration)
   c. Increase stormwater retention and groundwater recharge
      i. Improve stormwater management
      ii. Reduce and limit the amount of impervious surfaces in the watershed
         A. Change land use practices and zoning regulations to limit the allowable percentage of impervious surface in the watershed
         B. Decommission roads
         C. Use pervious pavement alternatives where feasible
      iii. Minimize the extent and degree of soil compaction
      iv. Restore stream connectivity to floodplains (see channel modification, levee removal and modification, and dedicating land to the preservation, enhancement, and restoration of stream habitat techniques)
      v. Revegetate denuded areas within the watershed
      vi. Protect, restore, and create wetlands and other infiltration areas
2. Techniques to restore the magnitude and frequency of peak flow events
   a. Remove dams, modify dam impoundments, or modify the water release management plan
   b. Increase stormwater retention and groundwater recharge (as outlined earlier)

3. Techniques to restore the natural flow regime (distribution of flow over time)
   a. Remove dam, modify dam impoundments, or modify the water release management plan
   b. Restore base flow (as outlined earlier)
   c. Restore peak flow magnitude and frequency (as outlined earlier)

8.4.12 Grading-Banks

Grading refers to the balance of cut and fills on-site to avoid off-site stockpiling and further damage to the natural topography. Surface grading is critical to establishing a long-term, stable slope. Grading of the slope surface will enhance its aesthetics and will also improve its ability to establish and maintain good vegetative cover. The cover will reduce the concentration of runoff on slopes and promote sheet flow, which is less erosive and enhances the infiltration of water needed for plant growth. Bank-grading techniques include terracing, cutting and/or filling, keying, and counter weighting (see Figure 8.26; NRCS 1996a).

8.4.13 Irrigation (Increased Efficiency)

Increased irrigation efficiency, as defined based on the ratio of water in the crop root zone to that added, is one of a number of methods commonly used to reduce usage and aid in meeting instream flow goals. Examples of the methods used to improve efficiency include using irrigation ditch piping (rather than earthen ditches), irrigation ditch lining, and efficient scheduling and monitoring.

8.4.14 Land Acquisition or Purchase

The effectiveness of restoration efforts is often impeded by fragmented public/private ownership patterns. Therefore, land acquisition is often required for restoration projects to proceed. This practice refers to obtaining the legal rights to land, by (but not limited to) leases, fee simple purchase, or conservation easements, usually for streamside or watershed management. The land is usually acquired for the purposes of the preservation or protection of critical habitats, the removal of causes of degradation, or to facilitate restoration projects. Some funding sources, such as North Carolina's Clean Water Management Trust Fund, require as a condition a recorded conservation easement or a recorded option agreement to restrict uses and activities in buffer areas. However, while stream restoration is motivated by environmental degradation, often the selection of stream reaches as targets for restoration is based on their availability, rather than the potential benefits of restoration (Sudduth et al. 2007).

8.4.15 Livestock Exclusions and Grazing Controls

One tool for the protection of riparian zones is through control of their use by livestock (see Figure 8.27). One alternative is to use one of a variety of grazing alternatives, such as described in "Riparian area management, grazing management processes and strategies for riparian-wetland areas" (Wyman et al. 2006). Another alternative is to exclude livestock or inhibit their use of riparian zones. For example, fencing may be used to protect riparian zones from livestock, and to keep livestock from wading or using streams and rivers (turning creeks into cricks, Chapter 2). The fencing will not only protect riparian plants and allow their recovery, but it will also reduce creek down-cutting, eroded banks, trampled and hummocky areas, and increase the overall stability of the creek banks and channel (see Figures 8.27 and 8.28). A consideration

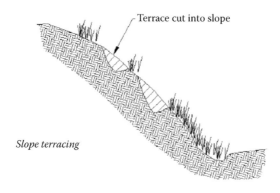

Slope terracing

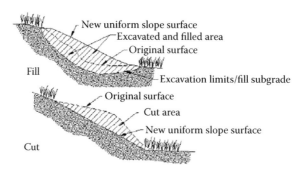

Slope shaping by cutting and/or filling

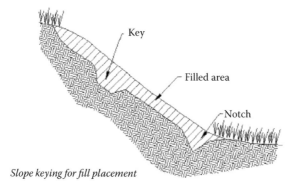

Slope keying for fill placement

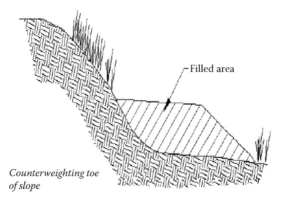

Counterweighting toe of slope

FIGURE 8.26   Bank-grading. (From NRCS, Streambank and shoreline protection. Chapter 16 in *Engineering Field Handbook*, Natural Resources Conservation Service, p. 55, 1996a.)

Fencing to protect riparian zones

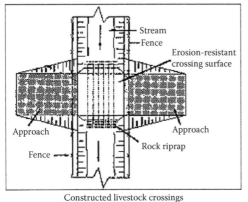

Constructed livestock crossings

FIGURE 8.27   Typical fencing and stream-crossing strategies. (Courtesy of the NRCS.)

FIGURE 8.28   Off-stream watering as a method. (Courtesy of the NRCS.)

for livestock is that the fencing may remove a water source that may need to be replaced. Other alternatives include:

- Using barriers (boulders, logs, etc.) that may inhibit the use of riparian zones.
- Providing hardened crossings and water access at designated points using, for example, coarse gravel pads on a gentle grade may help reduce erosion.
- Locating bedding grounds and livestock handling facilities away from riparian zones.

- Placing livestock away from riparian zones when they are moved to new grazing areas (new turnout locations).
- Installing drift fencing, which is open-ended fencing that is used to retard or alter the natural movement of livestock, in conjunction with natural barriers to regulate natural trailing or loafing by livestock in some riparian areas.
- Frequent range riding and herding can effectively control livestock distribution in many situations.

One difficulty with the protection of riparian zones is the costs and the potential loss of revenue to landowners. However, there are a variety of federal, state, and private programs that may aid landowners in protecting riparian zones. The U.S. Department of Agriculture (USDA) has designated riparian zones as eligible for inclusion in the U.S. Conservation Reserve Program (CRP), in which farmers can be paid for not farming certain environmentally sensitive areas. The USDA also added the Conservation Restoration and Enhancement Program (CREP), which provides funding to enhance and protect riparian areas, including fencing off riparian areas from livestock and replanting native trees and grasses. The Environmental Quality Incentives Program (EQIP), reauthorized in the Farm Security and Rural Investment Act of 2002 (Farm Bill) and administered by the NRCS, also provides a voluntary conservation program for farmers and ranchers that promotes agricultural production and environmental quality as compatible national goals. The EQIP offers financial, educational, and technical help to install practices on croplands or livestock areas to improve and maintain the health of natural resources, such as riparian zones. The Partners for Wildlife program administered by the U.S. Fish and Wildlife Service (USFWS) also provides support for establishing buffer strips, which may be used to aid in riparian management. Conservation easements offer one of the best permanent land protection strategies for riparian landowners. Easements can be donated by the landowner, providing tax benefits, or easements may be purchased by a qualifying organization. The easements may be for an entire property or they may be restricted to a riparian zone.

Regulatory protection, such as through prohibitions against grazing on the riparian zones of some wilderness areas, regulations and standards for grazing on the Bureau of Land Management (BLM) lands, and other actions may also be used for the protection of riparian zones. A variety of federal laws impact grazing activities in riparian zones, such as the federal CWA, restrictions on draining and filling wetlands, the NEPA, and the Endangered Species Act (ESA). Feller (1998) discussed the recent development of laws impacting grazing in western riparian areas.

8.4.16 LARGE WOODY DEBRIS: LWD ADDED

Large woody debris (LWD) provides a variety of stream functions, including providing shelter and low-velocity refuge for fish, pool creation, sediment storage (such as retaining spawning gravel), storage of nutrients and organic matter, bank stabilization, and other functions that have been discussed in previous chapters. LWD removal was common historically, but today the restoration of LWD is recognized as a critical component in the creation and maintenance of suitable aquatic habitats in some systems, particularly in the Pacific Northwest. LWD generally has the greatest influence and range of functions on moderate-slope (0.01–0.03) alluvial channels classified morphologically as pool-riffle or plane-bed (Montgomery et al. 1995, 2003; Montgomery and Buffington 1993; Beechie and Sibley 1997).

The design considerations for LWD include obstruction width, log spacing and key piece frequency, log loading and position. The LWD may be based on native LWD, preferably slowly decaying, or engineered materials. The design concepts for LWD structures are provided by Edminster et al. (1949), Mott (1994), Abbe et al. (1997), Derrick (1997), Shields (2001), and others.

A variety of methods have been used to add LWD to rivers and streams. One method involves cutting and felling trees from the streambank and then stabilizing the LWD by, for example, cabling the logs to stumps. Logs may also be placed in channels using heavy equipment and then secured in place. Individual logs with attached rootwads are less likely to move than other wood pieces during a bank-full flow. These logs and rootwads often make up "key pieces," or pieces of LWD defined as being independently stable within the bank-full channel (i.e., not held or trapped by other material) and retaining or having the ability to retain other LWD (WFPB 1997). A variety of configurations have been used for placing logs, five of which are illustrated in Figures 8.29 and 8.30.

8.4.17 LWD Removed

LWD is a necessary component of many healthy stream systems, and human interference has often led to a decline in LWD. Adding LWD is a common component of restoration projects. However, in

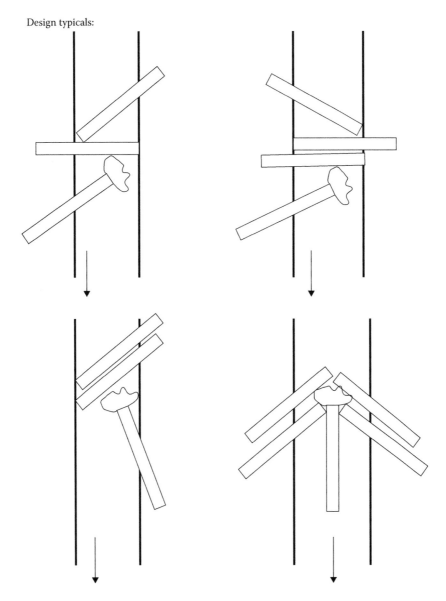

FIGURE 8.29 Typical strategies for wood placement design from the CRW aquatic restoration LWD project plan: Rock Creek above the 40 road. (From Bohle, T., *CRW Aquatic Restoration LWD Project Plan: Rock Creek above the 40 Road*, Seattle Public Utilities, Cedar River Watershed, 2005.)

Preproject   Postproject

FIGURE 8.30 Preproject and postproject conditions for the Cedar River Watershed aquatic restoration LWD project plan: Rock Creek above the 40 road. (From Bohle, T., *CRW Aquatic Restoration LWD Project Plan: Rock Creek above the 40 Road*, Seattle Public Utilities, Cedar River Watershed, 2005.)

some cases where jams of LWD create a blockage to fish passage, some removal may be necessary. This should be done with caution and often only selective removal, or removal of small amounts of the LWD, is necessary to remove the blockage.

8.4.18 MEANDER CREATION

8.4.18.1 Causes and Considerations

A major cause of altered sinuosity is, of course, channel "realignment," which is the intentional straightening of rivers and streams for navigation or other purposes. However, any project changes that tend to alter the channel width, mainly increased channel-forming discharges, tend also to alter the meander dimensions in the course of time (USACE 1994).

Natural rivers are so rarely straight that straight natural rivers are essentially nonexistent and straight reaches of rivers rarely exceed 10 times the channel width (Leopold and Wolman 1957). Even in a straight stretch of river channel, the river thalweg (main channel) is seldom straight, having more commonly a sinuous path between areas of aggradation and degradation. Therefore, river restoration commonly involves putting the "curves" back into a river, such as for those previously "realigned." As discussed previously, one common measure of the "curvature" is the channel sinuosity, defined by the channel length divided by the valley length. Leopold et al. (1964) designated single channels with sinuosities greater than 1.50 as meandering where an absolutely straight channel would have a sinuosity of 1.0, so that restoration involves increasing sinuosity and meanders.

Another consideration for restoration design is that natural rivers are often in a "dynamic equilibrium." Channels commonly change due to, for example, time-varying responses to flows and sediment loads. So, while conditions may change (be dynamic), the river may still be geomorphically stable where its properties (depth, slope, etc.) do not change on average over long periods (such as 10 years or more; Shields et al. 2003b). Such dynamic changes may result in, or in some cases be required for, a healthy ecosystem. However, while restoring that dynamic nature may be desirable from an ecological perspective, it may not be a socially acceptable outcome, particularly in urban environments or where threats are posed to riparian resources or infrastructure (Shields et al. 2003b) due to erosion and flooding. Therefore, stability is often a key component in restoration design.

Putting curves back into rivers is complicated by the fact that the curves, or meanders, not only result from the flow but also the sediment transport characteristics. For example, Ackers (1982) demonstrated that meandering was more common in channels with high sediment concentrations than in those with low concentrations, implying that the restoration of meandering channels must also incorporate sediment transport to maintain channel stability. The relationships between sediment transport and channel geometry are highly variable due to, for example, differences between materials in the bed and banks. However, to be successful, the design of the restored channel should be stable, minimizing aggradation and degradation by reestablishing equilibrium between the supply and transport capacity. The design is commonly based on sediment transport analyses in order to evaluate the relationships between the sediment supply and the sediment transport capacity. Rivers with erodible boundaries not only tend to meander, they also tend to migrate. To be successful, a restoration project must also typically prevent channel migration.

It should also be recognized that all meanders are not the same. That is, for example, meandering rivers vary in their meander geometry and bank materials, such as in the formation of point bars and chutes. Recall that point bars consist of relatively coarse-grained materials (silts and sands) laid down on the inside (convex) bend of a migrating stream channel, while chutes form a new channel across the base of a meander. Based on the classification schemes (see Chapter 2) of Schumm (1963, 1977), Rosgen (1994, 1996), and Brice (1975, 1984), Soar and Thorne (2001) described three of the most common types of meander bends found in stable single-thread channels as:

- Equiwidth Meandering (Schumm Type 3a; Brice Type A/B; Rosgen Type E): Equiwidth indicates that there is only minor variability in the channel width around meander bends. These channels are generally characterized by: low width/depth ratios; erosion-resistant banks; fine-grained bed material (sand or silt); low bed material load; low velocities; and low stream power. Channel migration rates are relatively low because the banks are naturally stable.
- Meandering with Point Bars (Schumm Type 3b; Brice Type C; Rosgen Type C): Meandering with point bars refers to channels that are significantly wider at bendways than at crossings, with well-developed point bars but few chute channels. Point bars consist of relatively coarse-grained materials (silts and sands) laid down on the inside (convex) bend of a migrating stream channel. These channels are generally characterized by: intermediate width/depth ratios; moderately erosion-resistant banks; medium-grained bed material (sand or gravel); medium bed material load; medium velocities; and medium stream power. Channel migration rates are likely to be moderate unless the banks are stabilized.
- Meandering with Point Bars and Chute Channels (Schumm Type 4; Brice Type D; Rosgen Type C/D): Meandering with point bars and chute channels refers to channels that are very much wider at bendways than at crossings, with well-developed point bars and frequent chute channels. These channels are generally characterized by: moderate-to-high width/depth ratios; highly erodible banks; medium- to coarse-grained bed material (sand, gravel, and/or cobbles); heavy bed material load; moderate-to-high velocities; and moderate-to-high stream power. Channel migration rates are likely to be moderate to high unless the banks are stabilized.

Soar and Thorne (2001) indicated that other types (such as in the Rosgen classification system Aa+, A and B, D, DA, F and H) would, for various reasons, not present realistic or attractive targets for a restoration scheme.

8.4.18.2 Restoration Techniques

One approach to restoring a river to "a close approximation of its remaining natural potential" is to base the restoration design on its predisturbed condition, the "carbon copy" approach. Determining and mimicking the "natural" condition, where possible, is perhaps the best method for restoring the sinuosity of channels. The analysis may be based on historical records, such as maps, aerial photographs, or soil surveys. However, in many cases, the natural condition is unknown. Also, disturbances in the river's watershed, resulting in changes in flow and sediment loads, may make what was once a stable condition, unstable under present conditions. Thus, past conditions are often not good estimators for future impacts.

Another approach is the so-called natural approach. This approach involves simply creating a valley and letting the channel form naturally. Due to the instability of the system, this approach is often unsatisfactory. The best design procedure is a systems approach, based on an analysis of the disturbed site by meander analysis and consideration of the undisturbed sites nearby (Hasfurther 1985; Gore et al. 1995).

The initial steps in the restoration project are to (FISRWG 1998):

- Describe the physical aspects of the watershed and characterize its hydrologic response.
- Select a preliminary right-of-way for the restored stream channel corridor and compute the valley length and valley slope.
- Determine the approximate bed material size distribution for the new channel.
- Conduct a hydrologic and hydraulic analysis to select a design discharge or a range of discharges.
- Predict a stable planform type (straight, meandering, or braided).

The last step, predicting a stable channel type, may be based on a sediment transport analysis or an analysis of similar (reference) stable systems. Figure 8.31 provides a definition sketch of cross section geometries and dimensions through a meander.

Following the preliminary steps, the systems approach to design (Hasfurther 1985) is based on performing a basinwide analysis of the stream channel to determine the meander characteristics, such as the meander wavelength (L_m), the belt-width or amplitude (A_m), the radius of curvature (R_c), and the channel's alignment (see Figure 8.31). The sinuosity is computed from the channel length and slope (channel length = sinuosity × valley length; channel slope = valley slope/sinuosity), and a preliminary design meander and sinuosity are determined.

An alternative to the systems approach is to base the meander design on empirical equations that relate meander properties (e.g., belt-width, wavelength, and radius of curvature) to other properties, such as flow, width, radius of curvature, sediment load, and other factors (Rosgen 1994). Generally, the most effective are those that are related to the discharge or the bank-full width (USACE 1994; Copeland et al. 2001). For example, formulations are often of the form:

$$A_m = a\,W^b$$

$$L_m = c\,W^d$$

$$R_c = e\,W^f$$

where a to f are coefficients. For example, in Leopold and Wolman (1966)

$$A_m = 2.7\,W^{1.1}$$

$$L_m = 10.9\,W^{1.01}$$

$$R_c = 2.4\,W^{1.0}$$

in units of feet. Leopold et al. (1964) suggested that meander wavelengths are most commonly in the range of 10–14 times the channel width ($c = 10$–14). Soar and Thorne (2001) indicated that an unbiased morphological expression for the meander wavelength within 95% confidence limits on the mean response suitable for engineering design is given by meander wavelengths within 11.26–12.47 times the channel width ($L_m = 11.16$–$12.47 * W$, in meters). However, the ranges are

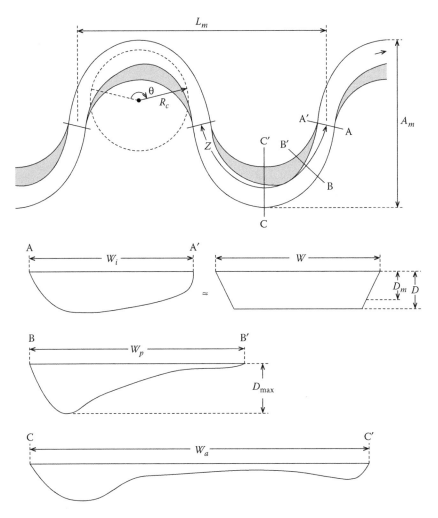

FIGURE 8.31 Meander planform and cross section dimensions for restoration design. (From Soar, P.J. and Thorne, C.R., *Channel Restoration Design for Meandering Rivers*, U.S. Army Corps of Engineers Engineering Research and Development Center, Vicksburg, MS, 2001.) *Note:* Point bars are defined by shaded regions; L_m = meander wavelength, Z = meander arc length (riffle spacing); A_m = meander belt width, R_c = radius of curvature; θ = meander arc angle; W = reach average bank-full width; D = depth of trapezoidal cross section; D_m = mean depth (cross-sectional area/W); D_{max} = maximum scour depth in bendway pool; W_i = width at meander inflexion point; W_p = width at maximum scour location; W_a = width at meander bend apex.

variable as illustrated by Rinaldi and Johnson (1997a,b), who, based on a sample of small Maryland streams, indicated the measured range of meander wavelengths to be between 2.9 and 7.7 times the channel width (in meters). They recommended that designers should ensure that the meander characteristics of the river to be restored be within the range of the data used to develop the regression equations, and that regional differences and fluvial processes be considered when selecting the meander characteristics.

Generally, the ratio of the radius of curvature to the channel width in well-developed meander bends is generally in the range of 1.5–4.5, and commonly in the range of 2–3. While the amplitude of meander systems is variable, the ratio of the amplitude to the wavelength is commonly in the range of 0.5–1.5 (USACE 1994).

A third alternative for sinuosity and meander design (FISRWG 1998; Soar and Thorne 2001) is to first:

- Calculate a stable channel slope analytically on the basis of the sediment transport
- Compute the sinuosity and channel length
- Compute the meander wavelength

For example, a common engineering approach would be to use hydraulic and sediment transport models to estimate the hydraulic characteristics (such as depths, velocities, etc.) and sediment loads based on the design flows. From the hydraulic characteristics, shear stresses could be computed for comparison with an allowable shear stress (one that would not cause degradation). The process would iterate until an allowable channel design (e.g., channel slope) was determined (computed shear stress < allowable). The sinuosity is then given as the ratio of the measured valley slope to the channel slope and is, therefore, directly related to the sediment regime, which controls channel stability. The channel length can then be computed from the product of the sinuosity and valley length. Once sinuosity has been derived, only the wavelength is required to determine the target planform geometry. Meander arc lengths vary from four to nine channel widths (FISRWG 1998).

Once the channel meander design and design sinuosity are selected, the remaining steps include · computing the hydraulic characteristics (depth, etc.) at the design discharge, and then computing the local morphological conditions, such as the pool and riffle locations, depths, and offsets (FISRWG 1998). See FISRWG (1998) and Copeland et al. (2001) for examples and equations.

Rivers with erodible boundaries not only tend to meander, but they also tend to migrate, and the development and control of meanders are a major consideration in restoration projects. Meander migration is not only a problem in areas with high slopes and velocities. Streams with flat slopes and relatively low velocities also exhibit active meander formation and migration, such as in backwater areas and upstream of confluences and reservoirs (USACE 1994).

Protection against migration is typically based on some form of bank protection. This may be continuous forms of protection such as revetment, or discontinuous forms such as groins. Intermittent groins are usually more economical than continuous revetment and while they may be unacceptable for some projects, such as where navigation is an issue, they may also have other benefits such as for fisheries. Short lengths of continuous revetment at points of active river attack are not usually effective in the long term since the attack usually shifts to other points and tends to outflank the short revetments. Bank vegetation and root systems may also provide effective protection (USACE 1994).

8.4.19 Monitoring Biota and/or Flow

Monitoring data are critical to any restoration project for both evaluation and planning purposes, particularly for projects dealing with flow modifications or improvements of instream habitats or biota. Monitoring is also needed to assess the effectiveness of the project (postaudit studies) over time and under variable environmental conditions, as with postaudit monitoring, to determine why and how techniques and practices work, and, equally important, why some fail (Saldi-Caromile et al. 2004).

The monitoring of biota is needed to evaluate sites, plan for restoration, and evaluate the effectiveness of restoration. For example, biota data can be used to evaluate the health or "biotic integrity" of a system, as discussed in Chapter 7. Flow data are critical to restoration projects, both base flow and flow variations, such as over seasons or during events. Flow data are also used (with hydraulic models) to estimate the variations in the hydraulic characteristics (depths, velocities, etc.) affecting aquatic habitats as well as the relationships between flows and vegetation. The streamflow also directly impacts water quality, such as the concentration of dissolved materials. Often, historical data are required to estimate the frequencies of flows, such as for use with the indicators of hydrologic alteration (IHA) software or for determining the probability distribution for flood flows.

For larger systems, flow data may often be obtained from USGS gaging stations. However, gaging stations may not be at the required location, or they may not be available. Flows may be estimated from point velocity measurements along a stream cross section, using acoustic Doppler current meters, or other techniques (see Chapter 4). Monitoring design and issues are discussed by FISRWG (1998), NRCS (2007a), and Saldi-Caromile et al. (2004).

8.4.20 Native Species Protection/Reintroduction

As a general rule for all practices, native plant species are chosen or favored over introduced species (NRCS 2007a). Projects typically include a revegetation plan compatible with native plants, soil, and site conditions or plans for the protection of existing vegetation. In addition to designs that protect existing native vegetation, restoration should be designed to result in a contiguous and connected stream corridor (FISRWG 1998). A variety of guides are available from local agencies and organizations on native plant species, and Dorner (2002) summarized some of the steps and considerations in using native plants in restoration projects. As indicated by Dorner (2002), in order to be successful, careful thought and planning should be used in the selection of plant species (e.g., evaluating the site characteristics and choosing the appropriate species), the installation of those species (selecting seeds of plants, preparing the site, etc.), and the maintenance of those species until they are established (watering, erosion control, pest management, monitoring, etc.).

8.4.21 Pools Created

The creation of instream pool areas in rivers and streams is a common restoration practice. For example, channel rock vanes and LWD are used to create scour pools and control structures are used to create backwater pools for instream habitats. An additional area where pool creation is commonly used is in the restoration or creation of vernal pools (Figure 8.32). Vernal pools are seasonally flooded wetlands and natural vernal pools have substantially declined as a result of urban and agricultural expansion.

8.4.22 Reinstating/Maintaining Hydraulic Connections

The maintenance or reinstatement of hydraulic connections refers to the realignment of stream channels so that ecological functions are maintained in channel bendways and other off-channel

FIGURE 8.32 Vernal pool enhancement at Battelle-Darby Creek Metro Park, Columbus and Franklin County Metro Park, Franklin County, OH. (From Ohio Vernal Pool Partnership, Available at http:\\www.OVPP.org.)

waterbodies. For example, a stream may not have sufficient flows to maintain satisfactory hydraulic connections so that portions of the old channel are no longer connected (Figures 8.33 and 8.34); therefore, either flow or water surface elevation management is required. Sedimentation can require extensive maintenance (FISRWG 1998).

8.4.23 RESEARCH

Research is fundamental to improving the development of methods and tools for stream restoration, both for site modifications and for education. In addition, while there are numerous restoration projects completed or underway, research and monitoring are needed to determine the effectiveness of those projects (postaudit studies) over time and under variable environmental conditions. For example, as discussed previously, postaudit monitoring and research can provide critical information for understanding why techniques and practices work, and, equally important, why some fail (Saldi-Caromile et al. 2004).

8.4.24 REVEGETATION

Revegetation is the reestablishment of plant cover by means of seeding or transplanting on a site disturbed by natural or human-caused actions.

8.4.24.1 Revegetation: Seedlings/Saplings

This technique of erosion control is based on the assumption that soil can be kept in place with a vegetative cover (Long et al. 1982). The revegetation of seedlings is a long-term measurement that is appropriate for exposed soils that have the potential to erode sediments.

8.4.24.2 Revegetation: Live Stake

This technique involves inserting dormant cuttings directly into the ground, which root and have the ability to grow and act as stakes (Heaton et al. 2002). This system of stakes creates a living root mat that stabilizes the soil by reinforcing and binding soil particles together and by extracting excess soil moisture. Live stakes must be cut from dormant materials. Mature stems with diameters of over 3 cm work best. The live stake technique is recommended for streambank protection where the site conditions are uncomplicated, construction time is limited, and an inexpensive method is needed (Figure 8.35; NRCS 1996b).

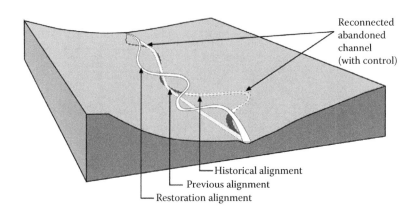

FIGURE 8.33 Maintenance of hydraulic connections. (From FISRWG, *Stream Corridor Restoration: Principles, Processes, and Practices*, Federal Interagency Stream Restoration Working Group, 1998.)

Soil preparation along the shoreline

Placement of live fascines on contour with erosion control fabric

Vegetated shoreline

FIGURE 8.34 Restoration of Jacques Cartier Park along the banks of the Ottawa River in Hull, Quebec, Canada. (From Caulk, A.D., Gannon, J.E., Shaw, J.R., and Hartig, J.H., Best management practices for soft engineering of shorelines, Greater Detroit American Heritage River Initiative, Detroit, Michigan, 2000. With permission.)

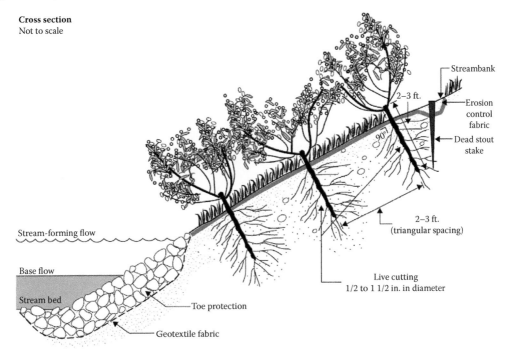

FIGURE 8.35 Live staking. (From NRCS, Bank stabilization. Chapter 5 in *Engineering Field Handbook*, Natural Resources Conservation Service. 1996b.)

8.4.24.3   Other Revegetation Techniques

A number of other revegetation techniques can be used for bank stabilization, such as bundles/fascines, live siltation, brush mat, hedge-brush layering, vegetated cribbing, and grass rolls.

The vegetated geotextiles consist of alternating layers of live branch cuttings and compacted soil with geotextiles (natural or synthetic; Figure 8.36). Geosynthetics are used as reinforcement to strengthen the soil mass by interacting with the soil, creating friction or adhesion, trapping sediments, and rebuilding streambanks. Their effectiveness increases with time, as the vegetation becomes permanently established. Geosynthetics include (NRCS 2007c; Figure 8.37):

- Geotextiles: Are permeable geosynthetics comprised solely of textiles. They may be woven or nonwoven and may be composed of monofilament yarns or monofilament plastic.
- Geogrids: Are geosynthetics formed by a regular network of integrally connected elements with apertures greater than one-fourth of an inch to allow interlocking with the surrounding soil, rock, earth, and other materials to function primarily as reinforcement.
- Geonets: Are geosynthetics consisting of integrally connected parallel sets of ribs overlying similar sets at various angles.
- Geocells: Are composed of polyethylene strips, connected by a series of offset, full-depth welds to form a three-dimensional honeycomb system.

A brush mattress is a protective mat of cuttings placed on the streambank and staked sufficiently to hold it in place; it is the soil bioengineering version of rock riprap (Heaton et al. 2002). The mat provides 100% coverage in the area in which it is placed and cuttings will root along the entire length of the structure, providing long-term protection (Figure 8.38).

Live fascines are long bundles of branch cuttings, commonly willows, bound together in cylindrical structures. They are placed in shallow contour trenches on dry slopes and at an angle on wet slopes to reduce erosion and shallow sliding (Figure 8.39; NRCS 1996b). They principally protect the bank toe and face, but they may also be used as current deflectors. Fascines can, as they grow roots and cover, improve erosion control, infiltration and riparian zone functions (Sotir and Fischenich 2001), and fisheries habitats. Facines are simple and effective, requiring little time to build and can be installed with little site disturbance (Heaton et al. 2002).

FIGURE 8.36   Geocell earth-retaining structure. (From NRCS, Stream restoration design, *National Engineering Handbook*, Department of Agriculture, Natural Resources Conservation Service, 2007a.)

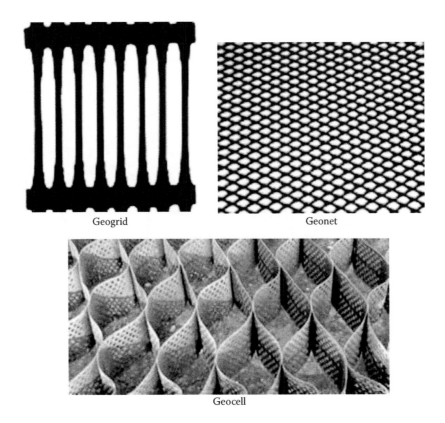

Geogrid

Geonet

Geocell

FIGURE 8.37 Illustration of geotextile types. (From NRCS, Geosynthetics in stream restoration, Technical supplement 14D, *National Engineering Handbook*, Natural Resources Conservation Service, 2007c.)

8.4.25 RIPRAP

Riprap (graded stone or crushed rock) is one of the most commonly used streambank stabilization techniques. Riprap is a permanent, erosion-resistant ground cover of large, loose, angular stone, as illustrated in Figure 8.40. The purpose of this technique is to protect the soil surface from the erosive forces of concentrated runoff, slowing the velocity of the runoff while enhancing the potential for filtration (RI DEM 2007). Riprap is generally used on streambanks at the toe (bottom) of the slope, with other structures placed upslope to prevent soil movement (NRCS 1996a) (Figures 8.40 and 8.41). There are a variety of riprap types, including rock riprap, rubble riprap, gabions, preformed blocks, grouted rock, and paved linings. Riprap design is included in engineering manuals from a variety of federal agencies, including the Corps (USACE 1994), the Federal Highway Administration (FHWA 1989), and the NRCS (1996a), as well as state agencies (e.g., RI DEM 2007).

Riprap also has some moderate benefit for stream habitats and for some organisms. The primary benefit is the stabilization of sediments, reducing the sediment loads and allowing the reestablishment of riparian vegetation. Hehnke and Stone (1978) showed that bird species diversity and density were significantly lower on the riprapped banks than on the unaltered sites. Fischenich (2002) indicated that systems with excessive erosion due to anthropogenic causes are most likely to benefit ecologically from riprap while those with healthy riparian vegetation communities would most likely be harmed ecologically by the addition of riprap structures.

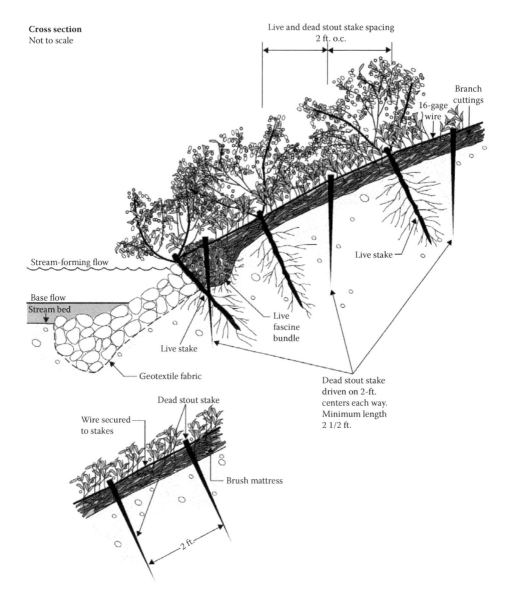

FIGURE 8.38   Brush mattress installation. (From NRCS, Streambank and shoreline protection. Chapter 16 in *Engineering Field Handbook*, Natural Resources Conservation Service. 1996a.) *Note:* Rooted/leafed condition of the living plant material is not representative at the time of installation.

8.4.26   Riparian Buffer Creation/Maintenance

The creation and subsequent maintenance of riparian buffers are often required where the riparian zones have been degraded or eliminated, usually for protection via the removal of nutrients and sediment loads (Mayer et al. 2005). Similarly, vegetated filter strips are a commonly used BMP to reduce nutrient and sediment loads to waterbodies. However, the riparian buffers provide a physical separation between adjacent areas, such as between areas of urban or agricultural development and a river or a stream, and typically consist of higher-level plants (e.g., trees and shrubs) as well as legumes or grasses. The riparian buffer system should typically include a multispecies buffer area established on land next to a stream and plantings that stabilize the streambank and wetlands constructed to absorb storm runoff.

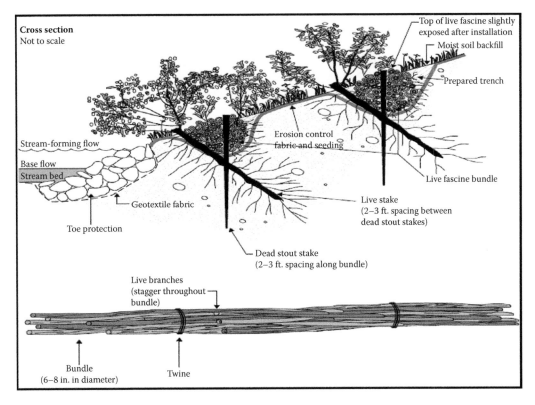

FIGURE 8.39 Live facines. (From NRCS, Streambank and shoreline protection. Chapter 16 in *Engineering Field Handbook*, Natural Resources Conservation Service. 1996a.) *Note:* Rooted/leafed condition of the living plant material is not representative of the time of installation.

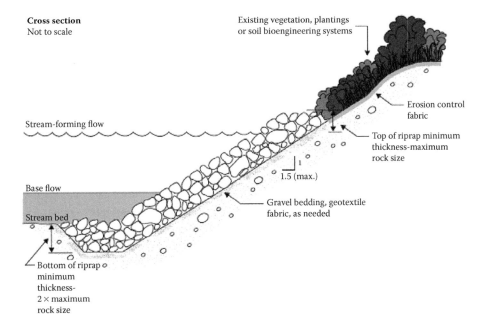

FIGURE 8.40 Rock riprap details. (From NRCS, Streambank and shoreline protection. Chapter 16 in *Engineering Field Handbook*, Natural Resources Conservation Service, 1996a.)

FIGURE 8.41   Riprap for restoration of Pike Creek. (From Delaware Department of Natural Resources and Environmental Control. With permission.)

The design of riparian buffers is usually based on the "three-zone" concept, where (BWM 2006):

- Zone 1: This zone is adjacent to the stream and is heavily vegetated under ideal conditions. This zone is referred to as the "undisturbed forest," where trees help stabilize banks, shade the stream, and provide aquatic food sources. Roots, fallen logs, and other vegetative debris in this zone slow the streamflow velocity, creating pools and habitats for macroinvertebrates, in turn enhancing biodiversity. Decaying debris provides an additional food source for stream-dwelling organisms. The tree canopy, particularly in first-order streams, shades and cools the water temperature, which is critical to sustaining certain macroinvertebrates, as well as critical diatoms, which are essential to support high-quality species/cold-water species.
- Zone 2: This zone is landward of Zone 1 and varies in width. It provides extensive water quality improvement. This zone is considered the "managed forest," since often the trees are managed and harvested. This zone removes, transforms, and stores nutrients, sediments, and other pollutants flowing as sheet flow as well as shallow subsurface flow.
- Zone 3: This zone is landward of Zone 2, and typically consists of grass and lower-order vegetation. This zone provides the first stage in managing upslope runoff so that runoff flows are slowed and evenly dispersed into Zone 2.

These three zones can be managed individually or concurrently (Figure 8.42). However, without concurrent management, such as grazing management in agricultural areas, to go along with attempts to reestablish vegetative communities, many planting operations will fail.

An initial step in the establishment of a riparian forest buffer is to determine what types of plants will be used. A starting point is to determine what is already growing in the vicinity, where trees growing nearby will reveal the parent material of the area and indicate what trees grow naturally on that site (Palone and Todd 1997). Native riparian tree species are preferable because they coevolved with the stream's inhabitants. The soil structure and texture, water table, soil preparation, and other factors affecting planting should be considered in the design as well as the method of planting (e.g., using cuttings, transplanting dry root stock, and seeding) as with any horticultural activity. The plantings in Zone 1 should be protected from disturbance. Zone 2 typically consists of larger trees with an understory of smaller trees and shrubs. This zone can tolerate some disturbance and, where conditions exist, it may be planted with commercially viable species such as for

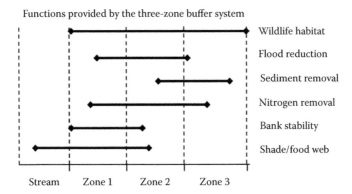

FIGURE 8.42   Zones and functions of a riparian forest buffer. (Redrawn from Palone, R.S. and Todd, A.H. (eds)., *Chesapeake Bay Riparian Handbook: A Guide for Establishing and Maintaining Riparian Forest Buffers*, USDA Forest Service, Radnor, PA, 1997.)

future logging. In Zone 3, the transition zone with adjacent land (such as urban areas or agricultural fields), a dense herbaceous cover with no trees or shrubs works well to slow and filter runoff (Tjaden and Weber 1997).

A second step in the buffer design is the determination of the design width. The width is based, in part, on the designated function of the buffer, with the recommended minimum widths illustrated in Figure 8.43. The NRCS program and criteria for "Riparian forest buffer—terrestrial and aquatic wildlife habitat" (NRCS 2009) call for the width of forested riparian zones to be at least 2.5 times the width of the stream channel. Most commonly, for use in water quality and habitat maintenance, buffer widths are approximately 35–100 ft., where buffers of less than 35 ft. cannot sustain long-term protection of aquatic resources (Tjaden and Weber 1997). The efficiency of buffers, such as in nutrient removal, is variable and is dependent on conditions such as where groundwater maintains saturated soils promoting anaerobic denitrification, so that, in some cases, a narrow strip may be more effective than a wider strip. However, wider strips tend to be more effective; federally recommended buffer widths vary from about 7 to 100 m (Mayer et al. 2005).

The maintenance and management of buffers must also be included in the design. Weed control is essential for the survival and rapid growth of trees and shrubs in a buffer (Tjaden and Weber 1997). To maintain the buffer's integrity, it should be protected against (Mayer et al. 2005):

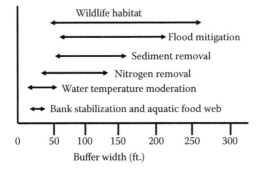

FIGURE 8.43   Range of minimum widths for meeting specific buffer objectives. (Redrawn from Palone, R.S. and Todd, A.H. (eds)., *Chesapeake Bay Riparian Handbook: A Guide for Establishing and Maintaining Riparian Forest Buffers*, USDA Forest Service, Radnor, PA, 1997.)

- Soil compaction from vehicles, livestock, and impervious surfaces (e.g., pavements) that might inhibit infiltration or disrupt water flow patterns.
- Excessive leaf litter removal or the alteration of the natural plant community (e.g., raking, tree thinning, and the introduction of invasive species) that might reduce carbon-rich organic matter from reaching the stream.
- Urbanization and other practices that might disconnect the stream channel from the floodplain (i.e., channelization, bank erosion, stream incision, and drain tiles) and thereby reduce the spatial and temporal extent of soil saturation.

Buffers must be monitored and managed to maintain their maximum water quality and wildlife habitat benefits. Periodic (at least yearly) inspections should be planned and provisions made for repairs where needed (Mayer et al. 2005) (Figure 8.44).

8.4.27 ROCK VANES

Vanes are linear structures that extend from the streambank into the stream channel in an upstream direction and mimic the effect of a tree partially falling into a stream. The vane serves to reduce streambank erosion by redirecting the streamflow toward the center of the stream. In addition, they tend to create scour pools on the downstream side. Rock vanes are typically linear, while a J-rock vane is the same as a rock vane but with the end of the vane curling in the shape of a "J," which tends to enhance downstream scour pool formation (Center for Watershed Protection Inc. 2000). The J-rock vane (or J-hook or fishhook) vane reduces near-bank stress to buy time for root development (NRCS 2007a) (Figure 8.45).

8.4.28 ROOT WADS

Roots wads can be used to deflect currents away from erodible banks. Root wads can also be used to provide complex instream habitats for fish and substrates for aquatic macroinvertebrates (FISRWG 1998; Figure 8.46).

FIGURE 8.44 Forested buffer zone. (Courtesy of NRCS.)

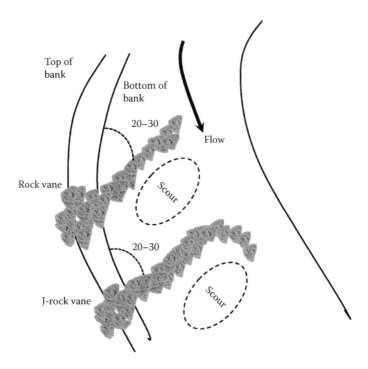

FIGURE 8.45 Plan view of rock vane and J-rock vane. (Redrawn from Center for Watershed Protection, Inc., Fact Sheet: Stream Restoration: Flow Deflection/Concentration Practices, Stormwater Manager's Resource Center (SMRC), Ellicott City, MD, 2000. Available at http:\\www.stormwatercenter.net.)

8.4.29 SAND TRAPS

Sand traps are commonly designed to remove excess sediment from streams, such as for the protection of spawning gravels (Steen 2003), and are often used in conjunction with LWD placement. A depression is typically dug in the channel to collect the sediments, and the deposited sediment is subsequently removed. Sand traps are one of the most common methods for habitat improvement listed in the NRRSS database, and the most common technique in the Midwest (Alexander and Allan 2006).

8.4.30 URBAN BMPS

Urban BMPs are generally accepted as practices that are intended to reduce the peak volume of runoff or the pollutant content of nonpoint source discharges, such as pathogens or indicator species (e.g., coliform bacteria), pesticides, organic materials, sediments, and nutrients. BMPs may require some structural modifications or they may be nonstructural, such as education and public involvement.

Many of the methods used for stream restoration discussed elsewhere in this chapter may also serve to reduce peak flows or pollutant loads, such as restoration of riparian zones. Thus, while the focus of the projects differ, the practices for pollutant load reduction and stream restoration produce synergistic benefits.

This section will focus on urban BMPs, primarily those designed to address stormwater runoff. Although agricultural BMPs were discussed separately, many of the BMPs are applicable to both environments.

Urban BMPs have historically been primarily used to reduce stormwater runoff from urban environments to prevent flooding. Generally, with urbanization the magnitude of stormwater runoff increases,

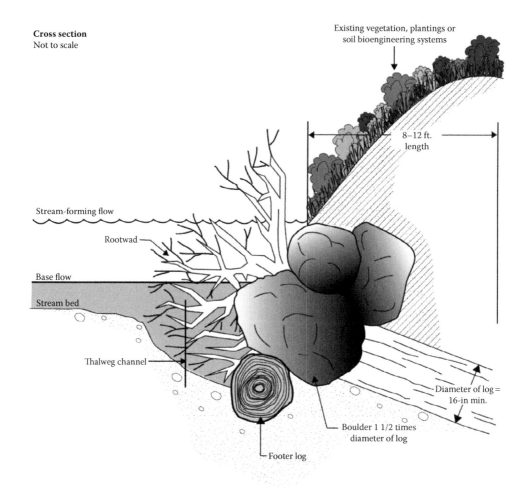

FIGURE 8.46 Root wads. (From NRCS, Bank stabilization. Chapter 5 in *Engineering Field Handbook*, Natural Resources Conservation Service. 1996b.)

while the flow duration (i.e., time of concentration) decreases (Figure 8.47). Consequently, laws and ordinances have focused on the prevention of flooding by requiring the peak flows following construction to be reduced to those prior to construction using some stormwater management system. Traditionally, this system has consisted of structures (curbs, gutters, stormwater drains, etc.) to move stormwater to some central location and off-site as quickly and effectively as possible. Then, centralized structures often route the stormwater to facilities (such as detention ponds) to temporarily store the water and reduce the peak flows. While they have been effective in reducing peak flows, these structures alter the volume and timing of runoff, which often destabilizes the sediments in receiving waters, and they have very limited capabilities in meeting evolving environmental and ecological requirements.

A recent and ongoing paradigm shift is away from the traditional stormwater methods of collection, routing, and storing, to maintaining functional relationships between terrestrial and aquatic ecosystems by keeping water where it falls and using decentralized source control as well as focusing on the quality as well as the quantity of the runoff. Examples include recent trends for low-impact development (LID) and Leadership in Energy and Environmental Design (LEED)-certified green buildings, increasingly encouraged or required in new or revised stormwater ordinances and regulations.

A wide variety of urban BMPs have been developed that are applicable to LID, some of which are illustrated in Figure 8.48. A complete discussion of urban BMPs is beyond the context of this text, but additional information can be obtained from the following selected sources:

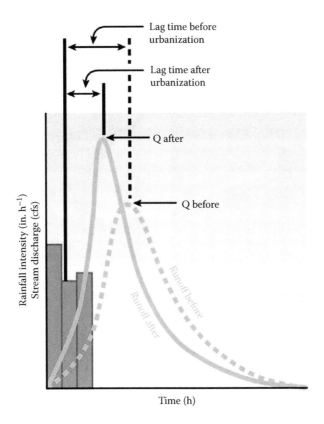

Lag time before urbanization

Lag time after urbanization

Q after

Q before

Runoff before

Runoff after

Rainfall intensity (in. h^{-1})
Stream discharge (cfs)

Time (h)

FIGURE 8.47 Comparison of runoff hydrographs before and after urbanization. (From FISRWG, *Stream Corridor Restoration: Principles, Processes, and Practices*, Federal Interagency Stream Restoration Working Group, 1998.)

- EPA's National Menu of Stormwater BMPs (http://cfpub.epa.gov/npdes/stormwater/menu-ofbmps/index.cfm).
- EPA (2004). Stormwater best management practice design guide, EPA/600/R-04/121.
- FHWA: Stormwater BMPs in an Ultra-Urban Setting: Selection and Monitoring (http://www.fhwa.dot.gov/environment/ultraurb).
- FHWA (2002). *Highway Hydrology. Hydraulic Design Series No. 2, Second Edition.* Publication No. FHWA-NHI-02-001, U.S. Department of Transportation, Federal Highway Administration, Washington DC.
- The Georgia Stormwater Manual.
- ARC (2001a). *Georgia Stormwater Management Manual Volume 1: Stormwater Policy Guidebook*, Atlanta Regional Council, Atlanta, GA.
- ARC (2001b) *Georgia Stormwater Management Manual Volume 2: Technical Handbook*, Atlanta Regional Council, Atlanta, GA.
- USGBC (2009). *LEED Reference Guide for Green Building Design and Construction*, U.S. Green Building Council, Washington, DC.

8.4.31 WATER LEVEL CONTROL/MAINTENANCE

Water level control refers to managing the water levels within a channel and the adjoining riparian zone to control aquatic plants and restore desired functions, including aquatic habitats (FISRWG

Green roofs

Urban forest

Infiltration basins

Bioswales

Porous concrete and pavement

FIGURE 8.48 Examples of BMPs for low-impact development. (From Presentation by Landscape Architecture Department, Mississippi State University.)

1998). The water level is typically maintained or controlled using some structure. Examples include berms, dams, dikes, and levees, which are constructed to contain water, and are often used to increase water levels in a wetland or maintain water levels in a stream. Water control structures such as spillways, pipes with drop inlets, and stoplog water controls are intended to control flows, and also to control water levels. For example, a spillway, a low point in a berm, provides an escape for excess water above the designed level. Typically, the water level requirements vary by season so that flexible control devices are required (FISRWG 1998).

8.4.32 WEIRS (ROCK VORTEX)

A rock vortex weir is designed to serve as a grade control, primarily preventing changes in grade. Rock vortex weirs also introduce variable flow velocities. During base flows, the flows are directed around and through the structure, while during high flows, the structure acts like a weir and tends to form scour pools downstream of the structure while still maintaining the bed load sediment transport regime of the stream. Rock vortex weirs are typically V-shaped, as illustrated in Figure 8.49, with the point upstream and the legs at a 15°–30° angle relative to the streambank.

8.4.33 WETLAND CONSERVATION (AND RESTORATION)

Wetland conservation typically refers to the protection and conservation of wetlands that are maintaining wetland functions, while restoration refers to attempts to restore those functions in wetlands

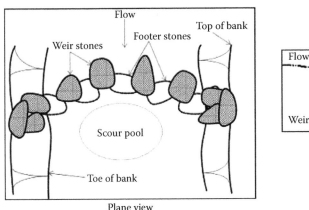
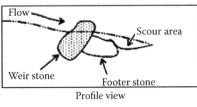
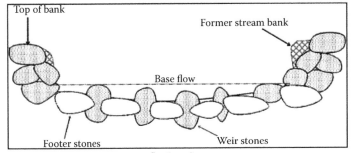

FIGURE 8.49 Rock vortex weir. (From Stormwater Manager's Resource Center. Stream restoration: Flow deflection/concentration practices. Available at: http://www.stormwatercenter.net/.)

that are degraded. Steps for the conservation of wetlands typically proceed first through identification and then through protection, by law or action. For example, wetlands directly connected to navigable waterways are protected under Section 404 of the CWA, while the wetland conservation compliance provisions (Swampbuster) in the 1985 Farm Bill (with amendments) provide protection by removing certain incentives to produce agricultural commodities on converted wetlands. Much of the loss of wetlands in the last century has occurred due to agricultural conversion.

Restoring riparian wetlands is often a critical component of river or stream restoration projects. Riparian wetlands are not only important habitats in themselves, but they also affect stream channel morphology and flows by providing temporary storage and buffering against the physical effects of high flows. Generally, productivity is greatest in riparian wetlands with alternating periods of inundation and drying (NRCS 2007a).

Two common methods for riparian wetland restoration are the "passive" and "active" approaches (USEPA 2003a). The passive method involves identifying and then removing the factors causing wetland loss. An example would be fencing to protect the wetland if cattle usage was identified as a contributing cause. Active methods would involve structural methods, such as restoring connectivity. A variety of causes for wetland degradation and corrective actions are listed in Table 8.6 from the U.S. EPA (USEPA 2003a).

8.4.34 WETLAND CONSTRUCTION

Wetland conservation (or restoration), as discussed earlier, attempts to reestablish ecological processes in damaged natural wetlands, while wetland construction is targeted toward the initiation of wetland processes, or building new wetlands where none previously existed. One of the most common examples of constructed wetlands is that for wastewater or stormwater management. Wetland construction as a BMP is an increasingly popular method for the treatment of stormwater control (see the references in Sections 8.4.1 and 8.4.30 for guidance in wetland construction). Wetland construction is also common as a mitigation measure to compensate for wetlands lost or destroyed, such as due to construction projects. Wetlands are also commonly constructed for wildlife habitats, by government agencies or private organizations (e.g., Ducks Unlimited).

Wetlands have long been known to naturally filter the water of unwanted products. They can be used just as effectively as treatment plants at treating water without the cost and maintenance. A wetland filters water by virtue of aquatic and nonaquatic plants, the absorptive capacity of the soil, and by the different organisms and microorganisms found in water. Artificially constructed wetlands are particularly versatile as a treatment option because they can be customized for a specific set of criteria. They can be designed for any configuration, sediment load, pollutant load, or nutrient load. They can also be designed much the same way as retention ponds, in the respect that the water can be held for a short or long period of time depending on the level of treatment required. These wetlands can also be of some scenic and aesthetic value.

Wetlands constructed for use as a stormwater management tool incorporate wetland plants in shallow pools. Pollutants are removed from stormwater runoff by settling and biological uptake as it flows through the wetland. Wetlands are among the most effective stormwater practices in terms of pollutant removal, and also offer aesthetic value. Stormwater wetlands are fundamentally different from natural wetlands in that they are designed specifically for the purpose of treating stormwater runoff and usually have less biodiversity than natural wetlands. Wetlands are widely applicable stormwater treatment practices. While they have limited applicability in highly urbanized settings and in arid climates, they have few other restrictions (California Stormwater Quality Association 2003).

A constructed wetland is any area designed on a site that would not naturally support wetland plants (Figure 8.50). Constructed wetlands can be used as the basic level of biological treatment for stormwater. Wetland plants allow oxygen to develop at the plants root structure, allowing the water to be oxygen rich. With oxygen-rich water, the hydraulic loading of nutrients creates the proper

TABLE 8.6

Common Wetland Problems and Corrective Measures

Wetland Damage	Reason for Damage	Suggested Correction	Considerations
Hydrology			
Water quality impairment	Excess sediment or nutrients in runoff from adjacent area	Work to change local land use practices; install vegetated buffers/swales/constructed treatment wetlands; install sediment traps	Sediment traps will need periodic cleaning; an expert may be needed to design buffers and swales
Water quality impairment	Excess sediments from eroding slopes	Stabilize slopes with vegetation/biodegradable structures	Many corrective methods exist; look for most sustainable and effective methods
Altered hydrology (drained)	Ditching or tile drains	Fill or plug ditches or drains; break tiles	Organic soil may have decomposed so that the elevation of the site is lower than it used to be
Altered hydrology (constrained)	Road crossing with undersized culvert	Replace with properly sized culvert or with a bridge	Hydrologic expert needed to correct this
Altered hydrology (drained)	Former wetland diked off from its water sources	Remove/breach dikes or install water control structures	Substrate elevation may not be correct for vegetation; add soil or control water level with low maintenance structures
Soils			
Raised elevation	Soil dumping or fill	Remove material	Fill may have compressed soil to lower than initial elevation; take steps to avoid erosion
Subsidence	Soil removal; oxidation of organics; groundwater removal	Add fill; allow natural sedimentation	Fill must support target wetland; test fill for toxic compounds
Toxic soils	By-product of on-site or off-site industrial process; dumping; leaching and concentration of natural compounds	Treatment systems or methods appropriate to the soil/ pollutants; remove material; cover with appropriate soil	Work with experts to choose treatment methods that cause the least amount of indirect damage; choose a different site to avoid serious toxin problems
Biota			
Loss of biodiversity	Change in original habitat	Restore native plant and animal communities using natural processes	Allow species to colonize naturally; import species as appropriate
Loss of native plant species	Invasive and/or nonnative plants; change in hydrology; change in land use	Remove invasive, nonnative plants (allow native plants to recolonize): try to reverse changes in hydrology	Pick lowest impact removal method; repeat removal as nonnatives reinvade; alter conditions to discourage nonnative species

Source: USEPA, *An Introduction and User's Guide to Wetland Restoration, Creation, and Enhancement*, by the U.S. Environmental Protection Agency and developed by the Interagency Workgroup on Wetland Restoration, 2003a.

environment for bacteria to thrive, allowing the biological cycles to begin. The biological processes of a constructed wetland are proven to reduce the levels of biochemical oxygen demand (BOD) and nitrates through nitrification and denitrification (Cathcart 2008; Wetzel 2001).

Constructed or stormwater wetlands can affect restoration efforts in addition to wetlands contiguous to rivers and streams, or riparian wetlands.

FIGURE 8.50   Constructed wetland. (From Cathcart, T., ABE 4313: NPS pollution, agricultural and biological engineering, constructed wetlands, Mississippi State University, Mississippi State, MS, 2008.)

REFERENCES

Abbe, T.B., D.R. Montgomery, and C. Petroff. 1997. Design of stable in-channel wood debris structures for bank protection and habitat restoration: An example from the Cowlitz River, WA. *Proceedings from the Conference on Management of Landscapes Disturbed by Channel Incision*, pp. 809–815.

Ackers, P. 1982. Meandering channels and the influence of bed material. In R.D. Hey, J.C. Bathurst, and C.R. Thorne (eds), *Gravel-Bed Rivers*. Wiley, Chichester, pp. 389–421.

AFS. 2004. Policy statement #32: Study report on dam removal for the AFS resource policy, Committee, American Fisheries Society.

Alexander, G.G. and J.D. Allan. 2006. Stream restoration in the upper midwest, USA. *Restoration Ecology* 14 (4), 595–604.

American Rivers. 2009. River facts. Available at: http://www.americanrivers.org/library/river-facts/river-facts. html.

American Rivers. 2011. 60 dams removed to restore rivers in 2010. Available at: www.AmericanRivers. org/2010DamRemovals.

ARC. 2001a. *Georgia Stormwater Management Manual Volume 1: Stormwater Policy Guidebook*. Atlanta Regional Council, Atlanta, GA.

ARC. 2001b. *Georgia Stormwater Management Manual Volume 2: Technical Handbook*. Atlanta Regional Council, Atlanta, GA.

ASCE. 1995. *Guidelines for Design of Intakes for Hydroelectric Plants*. ASCE Publishers, New York.

Audubon. 2005. Lake Okeechobee: A synthesis of information and recommendations for its restoration. Appendix A, Kissimmee River Restoration, Audubon of Florida, Miami, FL.

Barton, J.R. and F.W. Cron. 1979. *Restoration of Fish Habitat in Relocated Streams*. U.S. Department of Transportation, Federal Highway Administration, FHWA-IP-79-3, Washington, DC, p. 63.

Bednarek, A.T. 2001. Undamming rivers: A review of the ecological impacts of dam removal. *Environmental Management* 27 (6), 803–814.

Beechie, T.J. and T.H. Sibley. 1997. Relationships between channel characteristics, woody debris, and fish habitat in northwestern Washington streams. *Transactions of the American Fisheries Society* 126, 217–229.

Bernhardt, E.S., M.A. Palmer, J.D. Allan, G. Alexander, K. Barnas, S. Brooks, J. Carr, et al. 2005. Synthesizing U.S. river restoration efforts. *Science* 308 (5722), 636–637.

Bigford, T.E. 2004. AFS draft study report and policy statement on dam removal. *Fisheries* 29 (7), 34–35.

Biron, P.M., D.M. Carré, and S.J. Gaskin. 2009. Hydraulics of stream deflectors used in fish-habitat restoration schemes. In C.A. Brebbia (ed.), *River Basin Management V*. WIT Transactions on Ecology and the Environment, 124, pp. 305–319.

Biron, P.M., R. Colleen, M.F. Lapointe, and S.J. Gaskin. 2004. Deflector designs for fish habitat restoration. *Environmental Management* 33 (1), 25–35.

Bohle, T. 2005. *CRW Aquatic Restoration LWD Project Plan: Rock Creek above the 40 Road.* Seattle Public Utilities, Cedar River Watershed.

Boon, P.J. 1992. *River Conservation and Management.* Wiley, Chichester, UK.

Brice, J.C. 1975. Airphoto interpretation of the form and behavior of alluvial rivers. Final Report to the United States Army Research Office Durham under Grant Number DA-ARD-D-31-124-70-G89, Washington University, St Louis, MO, p. 10.

Brice, J.C. 1984. Planform properties of meandering rivers. In C.M. Elliott (ed.), *River Meandering,* Proceedings of the '83 Rivers Conference. American Society of Civil Engineers, New Orleans, LA, October 24–26, 1983, pp. 1–15.

Brookes, A. and F.D. Shields Jr. 1996. *River Channel Restoration: Guiding Principles for Sustainable Projects.* Wiley, Chichester, UK.

BWM. 2006. Pennsylvania stormwater best management practices manual. Pennsylvania Department of Environmental Protection, Bureau of Watershed Management, Document Number: 363-0300-002.

Cairns, Jr. J. 1991. The status of the theoretical and applied science of restoration ecology. *The Environmental Professional* 13, 186–194.

California Stormwater Quality Association. 2003. *Stormwater Best Management Practice (BMP) Handbooks.* California Stormwater Quality Association, Menlo Park, CA.

Cathcart, T. 2008. ABE 4313: NPS pollution. Agricultural and biological engineering, constructed wetlands. Mississippi State University, Mississippi State, MS.

Caulk, A.D., J.E. Gannon, J.R. Shaw, and J.H. Hartig. 2000. Best management practices for soft engineering of shorelines. Greater Detroit American Heritage River Initiative, Detroit, Michigan. Permission is granted to site portions of this report with proper citation.

CDOT. 2007. Fish passage design for road crossings, an Engineering Document, Providing Fish Passage Design Guidance for Caltrans Projects. California Department of Transportation May 2007.

Center for Watershed Protection, Inc. 2000. Fact sheet: Stream restoration—Flow deflection/concentration practices. Stormwater Manager's Resource Center (SMRC), Ellicott City, MD. Available at: www.stormwatercenter.net.

Clark, B. 2007. River restoration through dam removal in the American West: An examination into the variation in magnitude of policy change. Paper presented at the annual meeting of the Western Political Science Association, La Riviera Hotel, Las Vegas, NV.

Copeland, R.R., D.N. McComas, C.R. Thorned, P.J. Soar, M.M. Jonas, and J.B. Fripp. 2001. Hydraulic design of stream restoration projects. ERDC/CHL TR-01-28. U.S. Army Corps of Engineers Engineering Research and Development Center, Vicksburg, MS.

Derrick, D.L. 1997. Harland Creek bendway weir/willow post bank stabilization demonstration project. In S.S.Y. Wang, E.J. Langendoen, and F.D. Shields Jr. (eds), *Management of Landscapes Disturbed by Channel Incision.* University of Mississippi, Oxford, pp. 351–356.

Dorner, J. 2002. An introduction to using native plants in restoration projects; Prepared for the Plant Conservation Alliance, Bureau of Land Management. U.S. Department of Interior, U.S. Environmental Protection Agency by the Center for Urban Horticulture, University of Washington.

Echeverria, J.D., B. Pope, and R.-C. Richard. 1989. *Rivers at Risk: The Concerned Citizen's Guide to Hydropower.* Island Press, Washington, DC.

Edminster, F.C., W.S. Atkinson, and A.C. McIntyre. 1949. Streambank erosion control on the Winooski River, Vermont. Report No. 837, pp. 1–54.

Feller, J.M. 1998. Recent developments in the law affecting livestock grazing on western riparian areas. *Wetlands* 18 (4), 646–657.

FHWA. 1989. Design of Riprap Revetment, HEC-11. Hydraulic circular 11, FHWA-IP-89-016, Federal Highway Administration.

FHWA. 2002. Highway Hydrology. Hydraulic Design Series No. 2, 2nd ed. Publication No. FHWA-NHI-02-001, U.S. Department of Transportation, Federal Highway Administration, Washington, DC.

Fischenich, C. 2002. Overview of Stream Restoration Research in the USACE Presentation, Environmental Laboratory. U.S. Army Corps of Engineers Engineer Research and Development Center, Vicksburg, MS.

FISRWG. 1998. *Stream Corridor Restoration: Principles, Processes, and Practices,* Federal Interagency Stream Restoration Working Group. GPO Item No. 0120-A; SuDocs No. A 57.6/2:EN3/PT.653.

GASWCC. 2007. *Best Management Practices for Georgia Agriculture.* The Georgia Soil and Water Conservation Commission, Athens, GA.

Gore, J.A., F.D. Bryant, and D.J. Crawford. 1995. River and stream rehabilitation. In J.J. Cairns (ed.), *Rehabilitating Damaged Ecosystems.* Lewis Publishers, Boca Raton, FL, pp. 245–275.

Hart, D.D. and N.L. Poff. 2002. A special section on dam removal and river restoration. *BioScience* 52 (8), 653–655.

Hasfurther, V.R. 1985. The use of meander parameters in restoring hydrologic balance to reclaimed stream beds. In J.A. Gore (ed.), *The Restoration of Rivers and Streams—Theories and Experience*. Butterworth Publishers, Boston, MA, pp. 21–40.

Heaton, M.G., R. Grillmayer, and J.G. Imhof. 2002. *Ontario's Stream Rehabilitation Manual*. Ontario Streams, Belfountain, Ontario.

Hehnke, M. and C.P. Stone. 1978. Value of riparian vegetation to avian populations along the Sacramento River system. In R.R. Johnson and J.F. McCormick (eds), *Strategies for Protection and Management of Floodplains, Wetlands, and Other Riparian Ecosystems*. U.S. Forest Service, Washington, DC. GTR-WO-12. Cited in: Guidance Specifying Management Measures for Sources of Nonpoint Pollution in Coastal Waters, EPA 840-B-92-002, January, 1993.

Hirsch, R.M. 2006. USGS science in support of water resources management. Presented at the Mississippi Water Resources Conference, April 25–26, 2006, Jackson, MS.

Johnston Associates. 1989. A status report on the nation's floodplain management activity: An interim report prepared for interagency task force on floodplain Management. U.S. Government Printing Office, Washington, DC.

King County. 2010. *King County Manual of Best Management Practices for Maintenance of Agricultural Waterways*, 1st ed. Washington State University, Seattle, WA.

Lake, P.S., N. Bond, and P. Reich. 2007. Linking ecological theory with stream restoration. *Freshwater Biology* 2, 597–615.

Larinier, M. and F. Travade. 1999. La dévalaison des migrateurs: Problèmes et dispositifs. *Bulletin Français de Pisciculture* 353/354, 181–210.

Leopold, L.B. and M.G. Wolman. 1957. *River Channel Patterns: Braided, Meandering, and Straight*. USGS, Reston, VA.

Leopold, L.B. and M.G. Wolman. 1966. River meanders. *Bulletin of the Geological Society of America* 71, 769–794.

Leopold, L.B., M.G. Wolman, and J.P. Miller. 1964. *Fluvial Processes in Geomorphology*. W.H. Freeman and Co., San Francisco, CA.

Long, S.G., et al. (Preparers). 1982. Manual of revegetation techniques. Equipment Development Center. USDA Forest Service, Missoula, Montana, p. 145, as cited in Darris, Dale. 2001. Plant Materials Technical Note No. 3, Tips for planting trees and shrubs: Revegetation and landscaping technical notes, U.S. Department of Agriculture Natural Resources Conservation Service, Portland, OR.

Marlin, J.C. 2001. Potential use of innovative dredge technology and beneficial use of sediment for river restoration. Waste Management and Research Center, Illinois Department of Natural Resources, Champaign, IL.

Mayer, P.M., S.K. Reynolds Jr., and T.J. Canfield. 2005. Riparian buffer width, vegetative cover, and nitrogen removal effectiveness: A review of current science and regulations. EPA/600/R-05/118. U.S. Environmental Protection Agency, Office of Research and Development National Risk Management Research Laboratory, Ada, Oklahoma.

MDAFRR. 2007. Manual of best management practices for Maine agriculture. Maine Department of Agriculture, Food and Rural Resources, Division of Animal Health and Industry.

Michigan Department of Natural Resources. n.d. What is a fish ladder and weir? Available at: http://www.michigan.gov/dnr/0,4570,7-153-10364_52259_19092-46291--,00.html

Montgomery, D.R. and J.M. Buffington. 1993. Channel classification, prediction of channel response, and assessment of channel condition. Report TFW-SH10-93-002. Washington Department of Ecology, Olympia, Washington.

Montgomery, D.R., J.M. Buffington, R.D. Smith, K.M. Schmidt, and G. Pess. 1995. Pool spacing in forest channels. *Water Resources Research* 31 (4), 1097–1105.

Montgomery, D.R., B.D. Collins, J.M. Buffington, and T.B. Abbe. 2003. Geomorphic effects of wood in rivers. In S.V. Gregory, K.L. Boyer, and A.M. Gurnell (eds), *The Ecology and Management of Wood in World Rivers*. American Fisheries Society, Symposium 37, Bethesda, MD, pp. 21–47.

Mott, D.N. 1994. Streambank stabilization/riparian restoration action plan: Buffalo National River, Arkansas. (unpublished report) National Park Service, Harrison, Arkansas, pp. 1–83.

National Research Council. 1992. *Restoration of Aquatic Ecosystems: Science, Technology and Public Policy*. National Academy Press, Washington, DC

NHDAMF. 2008. Manual of best management practices (BMPs) for agriculture in New Hampshire, New Hampshire Department of Agriculture, Markets, and Food, Concord, NH.

NRCS. 1996a. Streambank and shoreline protection. Chapter 16 in *Engineering Field Handbook*. Natural Resources Conservation Service.

NRCS. 1996b. Bank stabilization. Chapter 5 in *Engineering Field Handbook*. Natural Resources Conservation Service.

NRCS. 2007a. Stream restoration design, *Part 654, National Engineering Handbook*, Department of Agriculture, Natural Resources Conservation Service.

NRCS. 2007b. Fish passage and screening design, Technical supplement 14N (210 -VI -NEH, August 2007), *Part 654 National Engineering Handbook*, Natural Resources Conservation Service.

NRCS. 2007c. Geosynthetics in stream restoration, Technical supplement 14D, In *Part 654, National Engineering Handbook*, Natural Resources Conservation Service.

NRCS. 2009. Enhancement activity—Riparian forest buffer—Terrestrial and aquatic wildlife habitat, animal enhancement activity—ANM 14, United States Department of Agriculture Natural Resources Conservation Service, January 30, 2009.

ODFO. 1995. *Freshwater Intake End-of-Pipe Fish Screen Guideline*. Ontario Department of Fisheries and Oceans, Ottawa, Ontario.

Palmer, M.A., E.S. Bernhardt, J.D. Allan, P.S. Lake, G. Alexander, S. Brooks, J. Carr, et al. 2005. Standards for ecologically successful river restoration. *Journal of Applied Ecology* 42, 208–217.

Palmer, M.A., J.D. Allan, and J.L. Meyer. 2007. River restoration in the United States in the 21st century. *Restoration Ecology*, 15, 472–481.

Palone, R.S. and A.H. Todd (eds). 1997. Chesapeake Bay riparian handbook: A guide for establishing and maintaining riparian forest buffers. USDA Forest Service. NA-TP-02-97. Radnor, PA.

RI DEM. 2007. Rhode Island soil erosion and sediment control handbook, Rhode Island Department of Environmental Management.

Rinaldi, M. and P.A. Johnson. 1997a. Characterization of stream meanders for stream restoration. *Journal of Hydraulic Engineering, ASCE* 123 (6), 567–570.

Rinaldi, M. and P.A. Johnson. 1997b. Stream meander restoration. *Journal of the American Water Resources Association* 33 (4), 867–878.

Rosgen, D.L. 1994. A classification of natural rivers. *Catena* 22, 169–199.

Rosgen, D.L. 1996. *Applied River Morphology*. Wildland Hydrology Books, Pagosa Springs, CO.

Rosgen, D.L. 1997. A geomorphological approach to restoration of incised rivers. In S.S.Y. Wang, E.J. Langendoen, and F.D. Shields Jr. (eds), *Proceedings of the Conference on Management of Landscapes Disturbed by Channel Incision*. University of Mississippi, Oxford, pp. 12–22.

Rosgen, D.L. 2001. *The Cross-Vane, W-Weir and J-Hook Vane Structures. Their Description, Design and Application for Stream Stabilization and River Restoration*, Wildland Hydrology, Pagosa Springs, CO.

Saldi-Caromile, K., K. Bates, P. Skidmore, J. Barenti, and D. Pineo. 2004. *Stream Habitat Restoration Guidelines: Final Draft*. Co-published by the Washington Departments of Fish and Wildlife and Ecology and the U.S. Fish and Wildlife Service, Olympia, Washington.

Shields, Jr. F.D. 1983. Design of habitat structures for open channels. *Journal of Water Resources Planning and Management* 109 (4), 331–344.

Shields, Jr. F.D. 2001. Design of large woody debris structures for channel rehabilitation. USDA-ARS-National Sedimentation Laboratory, Oxford Mississippi.

Shields, Jr. F.D., C.M. Cooper, and S.S. Knight. 1995. Experiment in stream restoration. *Journal of Hydraulic Engineering ASCE* 121, 494–502.

Shields, F.D., C.M. Cooper Jr., S.S. Knight, and M.T. Moore. 2003a. Stream corridor restoration research: A long and winding road. *Ecological Engineering* 20 (5), 441–454.

Shields, Jr. F.D., R.R. Copeland, P.C. Klingeman, M.W. Doyle, and A. Simon. 2003b. Design for stream restoration. *Journal of Hydraulic Engineering* 129 (8), 575–584.

Shields, Jr. F.D., R. Copeland, P.C. Klingeman, M.W. Doyle, and A. Simon. 2008. Stream restoration. Chapter 9 In M.H. Garcia (ed.), *Sedimentation Engineering*. American Society of Civil Engineers, Reston, VA, pp. 461–503.

Schumm, S.A. 1963. Sinuosity of alluvial rivers on the great plains. *Bulletin of the Geological Society of America* 74, 1089–1100.

Schumm, S.A. 1977. *The Fluvial System*. Wiley, New York.

Slatick, E. and L.R. Basham. 1985. The effect of denil fishway length on passage of some nonsalmonid fishes. *Marine Fisheries Review* 47 (1), 83.

Soar, P.J. and C.R. Thorne. 2001. Channel restoration design for meandering rivers. ERDC/CHL CR-01-1, U.S. Army Corps of Engineers Engineering Research and Development Center, Vicksburg, MS.

Society for Ecological Restoration Science and Policy Working Group. 2002. The SER primer on ecological restoration. Available at: http://www.ser.org/content/ecological_restoration_primer.asp.

Sotir, R.B. and C. Fischenich. 2001. Live and inert fascine streambank erosion control. Technical Notes, ERDC TN-EMRRP-SR-31, U.S. Army Corps of Engineers Engineering Research and Development Center, Vicksburg, MS.

Steen, P.J. 2003. History and inventory of stream habitat improvements for the state of Michigan. M.S. thesis. University of Michigan, Ann Arbor, MI.

Stormwater Manager's Resource Center. Stream restoration: Flow deflection/concentration practices. Available at: http://www.stormwatercenter.net/.

Sudduth, E.B., J.L. Meyer, and E.S. Bernhardt. 2007. Stream restoration practices in the southeastern United States. *Restoration Ecology* 15 (3), 573–583.

Tjaden, R.L. and G.M. Weber. 1997. Riparian buffer management riparian forest buffer design, establishment, and maintenance. Fact Sheet 725, Maryland Cooperative Extension Service, University of Maryland.

TWDB. 2005. Water conservation best management practices guide. Texas Water Development Board, Report 362, Austin, TX.

USACE. 1991. Kissimmee River Florida final feasibility report and environmental impact statement: Environmental restoration of the Kissimmee River, Florida. Jacksonville District, Jacksonville, FL.

USACE. 1994. Channel stability assessment for flood control projects. EM 1110-2-1418, United States Army Corps of Engineers, Washington, DC.

USACE. 2008. Kissimmee River restoration. U.S. Army Corps of Engineers Jacksonville District, Everglades Division, Upper East Coast and Kissimmee/Lake Okeechobee Section, Jacksonville, Florida. Available at: http://www.saj.usace.army.mil/Divisions/Everglades/Branches/ProjectExe/Sections/UECKLO/KRR.htm.

USDA Forest Service. 2006. FishXing, Version 3 user manual and reference. USDA Forest Service, San Dimas Technology Development Center, San Dimas, CA.

USEPA. 2000. Principles for the ecological restoration of aquatic resources. EPA841-F-00-003. Office of Water (4501F); United States Environmental Protection Agency, Washington, DC, p. 4.

USEPA. 2003a. *An Introduction and User's Guide to Wetland Restoration, Creation, and Enhancement*, by the U.S. Environmental Protection Agency and developed by the Interagency Workgroup on Wetland Restoration.

USEPA. 2003b. National management measures to control nonpoint source pollution from agriculture, EPA 841-B-03-004, July 2003.

USEPA. 2010. Agricultural management practices for water quality protection. Available at: http://www.epa.gov/owow/watershed/wacademy/acad2000/agmodule/agbmp4.htm, Accessed 30 Oct. 2010.

USGBC. 2009. LEED reference guide for green building design and construction. U.S. Green Building Council, Washington, DC.

WDFW. 2000a. Fishway guidelines for Washington State. Washington Department of Fish and Wildlife, April 25, 2000.

WDFW. 2000b. Fish protection screen guidelines for Washington State. Washington Department of Fish and Wildlife, April 25, 2000.

Wesche, T.A. 1985. Stream channel modifications and reclamation structures to enhance fish habitat. In J.A. Gore (ed.), *Chapter 5 in The Restoration of Rivers and Streams*. Butterworth, Boston (cited in Federal Interagency Stream Restoration Working Group).

Wetzel, R.G. 2001. *Limnology: Lake and River Ecosystems*, 3rd ed. Academic Press, San Diego, CA.

WFPB. 1997. Board manual: Standard methodology for conducting watershed analysis. Version 4.0, November, Washington Forest Practices Board.

White, L.J., I.D. Rutherfurd, and R.E. Hardie. 1999. On the cost of stream management and rehabilitation in Australia. In I.D. Rutherfurd and R. Bartley (eds), *Proceedings of the Second Stream Management Conference, Adelaide, South Australia*. Cooperative Research Centre for Catchment Hydrology, Clayton, Victoria, Australia, pp. 697–703.

Woodworth, P., J. Galster, J. Wyrick, and J. Helminiak. 2010. Dam removal: Adaptive management and bed sediment monitoring before and after. *Proceedings of the World Environmental and Water Resources Congress 2010: Hydraulics and Waterways Council*, pp. 1734–1736.

Wyman, S., D. Bailey, M. Borman, S. Cote, J. Eisner, W. Elmore, B. Leinard, et al. 2006. Riparian area management: Grazing management processes and strategies for riparian-wetland areas. Technical Reference 1737-20. BLM/ST/ST-06/002+1737. U.S. Department of the Interior, Bureau of Land Management, National Science and Technology Center, Denver, CO, pp. 105.

Part II

Lakes and Reservoirs

9 Introduction to Lakes and Reservoirs

Geomorphology and Classification

A lake is the landscape's most beautiful and expressive feature. It is the earth's eye looking into which the beholder measures the depth of his own nature.

Henry David Thoreau

9.1 WHAT IS A LAKE, OR A RESERVOIR?

A lake (see Figure 9.1) is generally taken to be a body of water surrounded by land, but with a free surface (open to the atmosphere). But, what is a lake, versus a reservoir, or versus a pond? Or, for that matter, what is a lake versus a wetland or a river? If a river flows into a reservoir, where does the river end and the reservoir begin? And, does this matter from a regulatory perspective? Or from an ecological perspective? How about if you wanted to name waterbodies or complete a survey of them, how would you determine which is which?

One commonly used distinction between a lake and a reservoir is that a lake is a "natural" body of water, while a reservoir is man-made. A large number of natural lakes are derived from a variety of natural processes (e.g., wind and glacial erosion, as discussed in a later section) and their source or origin often has a profound impact on their characteristics. Similarly, reservoirs are constructed for a variety of purposes (recreation, hydropower, etc.), which also has a profound impact on their characteristics. The origin of lakes (and reservoirs) is of more than casual interest (Wetzel 2001), since it affects their physical, chemical, and biological characteristics.

What is the difference between a lake (or a reservoir) and a pond (natural or man-made)? Often, the distinction is based on their size and/or depth, but there is no precise and accepted distinction. From a regulatory perspective, theoretically it may not make any difference since they are both "waters of the United States" and are thus subject to the same water quality criteria. In practice though, the distinction may be important. Rivers are often distinguished from lakes and reservoirs in that they are vertically well mixed. But, as discussed later, that distinction is not always true. Similarly, wetlands and lakes may have many common characteristics. So, as with rivers and streams (and brooks and becks, creeks and cricks, see Chapter 2), distinctions are typically somewhat arbitrary and no precise definition is available. However, while that may be true, there is a difference from a limnological or ecological perspective and that difference is important.

9.2 GENERAL CHARACTERISTICS

As noted previously, a primary difference between rivers and lakes or reservoirs is in their speed of water movement, or currents. Rivers and streams are *lotic* (from *lotus*, meaning washed). Lotic systems are characterized by running water, as opposed to lakes and reservoirs, which are *lentic* (from *lenis*, meaning calm) and are characterized by standing water. While the flowing nature of rivers dominates their transport and mixing processes, the standing nature of lakes and reservoirs results in other processes assuming an increased importance. The flowing nature of rivers normally

FIGURE 9.1	Lake Louise, Alberta, Canada. (Photograph by J.L. Martin.)

causes complete mixing over their depth and width, and rapid downstream transport. The deeper standing water in lakes tends to move water quality constituents and contaminants more slowly and to stratify vertically, which retards vertical mixing.

Lakes and reservoirs tend to store water for relatively long periods. The length of time is often quantified using the hydraulic residence (or retention) time, or the average time that a water particle resides within a lake or impoundment (one definition of which is volume divided by flow; Martin and McCutcheon 1999). Hydraulic residence times can vary from days to weeks in run-of-the-river reservoirs in which outflows are not controlled and be up to years in larger reservoirs. The run-of-the-river reservoir is a relatively shallow reservoir on a larger river in which velocities are higher than in most reservoirs. In contrast, mean residence times in the Great Lakes range from 3 years for Lake Erie to 180 years for Lake Superior (Chapra 2008).

The depth and hydraulic residence times largely influence the mixing and water quality characteristics of lakes and reservoirs. An increased storage or residence time allows for internal cycling and matter originating within the lake or reservoir (autochthonous materials) to have increased importance relative to materials originating outside and carried into the lake or reservoir (allochthonous materials). Internal productivity and decay often vary over seasons or years, so that variations in transport and mixing occurring over similar time periods become important. Variations with depth are also often more important in lakes and reservoirs than in rivers. Light does not penetrate to the bottom of many lakes and reservoirs so that heat exchange and productivity are limited to the surface layers. Many lakes and reservoirs do not completely mix vertically during periods of the year and large vertical gradients in the temperature, density, and water quality often result.

For rivers, transport is usually dominated by longitudinal advection, and material introduced into them is often rapidly mixed over their width and depth. In contrast, lakes and reservoirs may or may not have well-defined currents or a single direction of flow. The pattern of flows and mixing in a lake or a reservoir is affected by bathymetry, thermal structure, inflows, outflows, and wind mixing. Unlike in rivers, horizontal transport and mixing are often more rapid than those occurring vertically.

9.3 BRIEF HISTORY OF LIMNOLOGY

The study of inland waters, such as lakes and reservoirs, is referred to as limnology, from *limnee* meaning pool, marsh, or lake, so limnology is the study of lakes. This field of study often encompasses rivers and streams as well, but it commonly focuses on lakes.

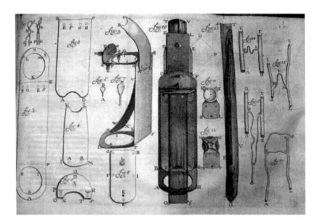

FIGURE 9.2   A 1756 illustration by Henry Baker of van Leeuwenhoek's microscopes.

Limnology has a long and rich history. In the 1600s and early 1700s, lakes in Germany and the Netherlands were categorized based on the presence or absence of stream inflows. Also, in the late 1600s, van Leeuwenhoek developed the first compound microscope and noticed organisms in the water (Figure 9.2). In the 1700s, Horace-Bénédict de Saussure, a Swiss aristocrat, physicist, and Alpine traveler, developed a modified thermometer and determined that the deep waters in some Swiss lakes were much colder than their surface waters all year round.

In the 1800s, further steps in the advancement of limnology were the introduction of the plankton townet by Johannes Muller in 1846 and the Secchi disk to measure water transparency by Pietro Angelo Secchi in 1865. Francois-Alphonse Forel, while at the University of Lausanne, Switzerland, published a monograph on studies of Lac Leman and then published the first limnology text,

FIGURE 9.3   *Handbuch der Seenkunde: allgemeine Limnologie* (Handbook of Limnology) published in 1901; the first limnology-focused text book.

Handbuch der Seenkunde, in 1901 (Figure 9.3). He is widely considered to be the father of limnology and is the first person to have used the term *limnology* in a publication (CGER 1996). His contributions included a study of the hitherto unexplained movement of lake waters, known as seiches, and the discovery of density currents.

In 1871, the U.S. Commission of Fish and Fisheries was established in the United States. In 1887, Forbes published *The Lake as a Microcosm*, describing a lake as an ecological unit in balance with its surroundings. In 1906, Birge and Juday began studies on Wisconsin lakes (Lake Mendota) (Figure 9.4).

In the 1920s, Thienemann and Naumann classified lakes based on their chemistry, production, and fauna as oligotrophic, eutrophic, and dystrophic, respectively. To honor their contributions to the field, each year the International Society of Limnology bestows the Einar Naumann–August Thienemann Medal, which is considered the highest honor that can be bestowed internationally for outstanding scientific contributions to limnology (Figure 9.5).

In 1928, G. Evelyn Hutchinson joined the faculty of Yale University, where he was a professor for 43 years (Figure 9.6). Hutchinson was born in Cambridge, UK, in 1903. He studied at Cambridge College and became a professor at the University of Witwatersrand, Johannesburg, in 1925. Subsequently, he went to Yale University. While he never earned a PhD, over a 36-year period he published the four-volume (with more than 3500 combined pages) *Treatise of Limnology*, considered one of the most comprehensive studies of limnology in existence (Hutchinson 1957, 1667, 1975, 1993).

G.E. Hutchinson was responsible for the introduction of numerous ecological and limnological concepts, such as the "paradox of the plankton" (Hutchinson 1961), which will be discussed in a later chapter. His seminal paper "Homage to Santa Rosalia or why are there so many animals?" (Hutchinson 1959) is considered the first attempt to deal with modern concepts of biological diversity. He is generally considered to be the "father of modern limnology" and has also been recognized as the "father of ecology." His contributions are documented in the 2010 book *G. Evelyn Hutchinson and the Invention of Modern Ecology* (Slack 2010).

FIGURE 9.4 Edward A. Birge and Chancey Juday with a plankton trap circa 1917. (From Wisconsin Historical Society, WHS-3176. With permission.)

FIGURE 9.5 Einar Naumann and August Thienemann. (From the Association for the Sciences of Limnology and Oceanography, ASLO. With permission; Photographs from an old ASLO slide collection submitted by W. Lewis; scanned by K.L. Schulz.)

FIGURE 9.6 G. Evelyn Hutchinson (1903–1991), with a 3-month-old potto in New Haven, CT, 1971. (Photograph by William K. Sacco. With permission.)

Many of the modern and perhaps most cited references in this field were developed by Robert G. Wetzel (Figure 9.7). His textbooks and writing, along with his many awards, such as the G. Evelyn Hutchinson Medal of the American Society of Limnology, made him perhaps the best-known modern limnologist. His contributions included his three editions of *Limnology*, culminating with his 2001 *Limnology, Third Edition: Lake and River Ecosystems*, completed 4 years before his death in 2005.

FIGURE 9.7 Dr. Robert G. Wetzel. (From Soil & Water Conservation Society of Metro Halifax, Nova Scotia, Canada, 1998. Available at http://lakes.chebucto.org/PEOPLE/wetzel.html. With permission.)

9.3.1 LAKES

As with rivers and streams, a variety of classification systems has been developed for lake basins. One of the most common classification systems is based on their origin, and some of the more common forms are discussed here.

9.3.1.1 Tectonic Basins

Tectonic basins result from a diastrophic (movement) event within the earth's crust, such as earthquakes, tilting of faults, depressions between faults, or upheavals forming basins without drainage. Tectonic lakes usually have elongated shapes, steep sides, and great depths. Well-known examples include the Dead Sea and Lake Tiberias in Israel (Bowen 1982). Other examples include Lake Baikal in Russia, Lake Tahoe in California, Pyramid Lake in Nevada, the Great Salt Lake in Utah, and Reelfoot Lake in Tennessee. Somewhat surprisingly, Lake Okeechobee in Florida is also of tectonic origin, formed by a minor depression in the seafloor uplifted during the formation of the Florida Peninsula (Wetzel 2001). Lake Okeechobee, although shallow (its depth is usually less than 4.5 m), has a surface area of 1880 km^2 and is the largest freshwater body in the United States south of the Great Lakes. Both the Great Salt Lakes and Lake Okeechobee are so large that they can be seen from space (Figure 9.8).

9.3.1.2 Volcanic Basins

Volcanic basins may form behind lava-flow dams, in the craters of extinct volcanoes or in cavities formed by subsidence (Bowen 1982). These lakes often have relatively low nutrient contents and

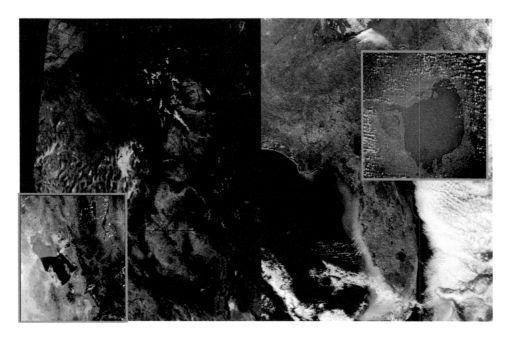

FIGURE 9.8 Utah and the Great Salt Lake (*left*) and Florida and Lake Okeechobee (*right*) as seen from space. (Courtesy of NASA.)

limited productivity. Caldera basins, such as Crater Lake in Oregon, are formed by the collapse of a volcanic cone (Figure 9.9). Crater Lake is the seventh deepest lake in the world, with a depth of 608 m (Wetzel 2001). An example of a lake formed behind volcanic debris flows is Spirit Lake in Washington, which was formed behind debris flows from Mount St. Helens.

9.3.1.3 Glacial Basins

These basins include moraines, formed by the debris from glaciers; cirques, formed in the eroded head of a glacial valley; and tarns, formed by freezing and thawing. The Laurentian Great Lakes are

FIGURE 9.9 Crater Lake. (Courtesy of USGS Crater Lake Data Clearinghouse.)

FIGURE 9.10   The Laurentian Great Lakes. (Courtesy of NASA.)

TABLE 9.1
Physical Features of the Great Lakes

Feature	Units	Superior	Michigan	Huron	Erie	Ontario
Average depth[a]	Feet	483	279	195	62	283
	Meters	147	85	59	19	86
Maximum depth[a]	Feet	1,332	925	750	210	802
	Meters	406	282	229	64	244
Volume[a]	Miles^3	2,900	1,180	850	116	393
	Kilometers^3	12,100	4,920	3,540	484	1,640
Water area	Miles^2	31,700	22,300	23,000	9,910	7,340
	Kilometers^2	82,100	57,800	59,600	25,700	18,960
Land drainage area[b]	Miles^2	49,300	45,600	51,700	30,140	24,720
	Kilometers^2	127,700	118,000	134,100	78,000	64,030
Shoreline length[c]	Miles	2,726	1,638	3,827	871	712
	Kilometers	4,385	2,633	6,157	1,402	1,146
Retention time	Years	191	99	22	2.6	6

Source: The Great Lakes Atlas. Available at: http://www.epa.gov/glnpo/atlas/. Last updated 6/25/2012.

[a] Measured at low-water datum.

[b] Land drainage area for Lake Huron includes St. Marys River. Lake Erie includes the St. Clair-Detroit system. Lake Ontario includes the Niagara River.

[c] Including islands.

glacial lakes, formed by the extensive scouring of continental ice sheets (Figure 9.10). In size and volume, the Great Lakes form the largest group of freshwater lakes in the world (Table 9.1).

The majority of lakes in the northern regions of the United States and Canada are glacier related, and are formed by ice barriers, glacial erosion, glacial deposition, and similar processes. Examples include kettle lakes, formed by melting ice chinks; paternoster lakes, chains of lakes in a glacial valley; and tarns, glacial lakes formed by scour.

FIGURE 9.11 Playa Lake in the Texas Panhandle. (Photograph courtesy of Texas Parks and Wildlife Department © 2013.)

9.3.1.4 Solution Basins

These basins are formed when solution caves collapse. Solution lakes are very common in limestone regions, as illustrated by the many solution lakes in Indiana, Kentucky, Tennessee, Florida (Wetzel 2001), and the Yucatan Peninsula in Mexico. In North America, Florida has the largest concentration of solution lakes. Examples include doline lakes, which are collapse sinkholes, and karst lakes, which are limestone sinks.

9.3.1.5 Wind Basins

These basins are located in arid regions where the soil is fine and the wind creates a depression. Playa lakes, common in the arid regions of the United States and Kinjiang Province in Western China, result from wind erosion as it forms large pans on nearly level soils. Playa lakes are common in the U.S. Great Plains, and are often one of the few sources of water there. They are heavily used by migrating waterfowl. According to the Texas Parks and Wildlife Commission, there are over 19,000 playa lakes in the Texas Panhandle, averaging slightly more than 15 acres in size (Figure 9.11). The lakes often go through extended wet and dry cycles.

9.3.1.6 Stream Basins

These basins include fluvial lakes, which form as a result of stream deposition or erosion, as with the simple expansion in the width of a river, behind deposits or obstructions in streams, or when two rivers of different velocities meet and have a buildup of materials at their confluence. Plunge-pool lakes form from erosion below waterfalls, such as the Grand Coulee Lake in Washington. Oxbow lakes are formed with the cutoff of a stream meander and are common along many of our major river systems, such as the lower Mississippi River (e.g., Figure 9.12).

There are many other types of natural lake basins. For example, G.E. Hutchinson (1957) listed 76 types of lakes based on their geomorphological characteristics. However, the classification provided in this section some indication of the diversity of lakes that may occur.

9.3.2 Reservoirs

In addition to naturally occurring basins, there are man-made basins. Reservoirs are mainly formed by damming a steep-sided valley with a concrete structure, such as Hoover Dam (Figure 9.13), or an earthen embankment. In the southern United States, man-made reservoirs are much more common than lakes. The majority of reservoirs generally have depths of less than 20 m and are relatively shallow in comparison to lakes. Their relatively large flow to volume ratios generally result in their retention times being shorter than those for lakes. The ratio of drainage area to surface area is much

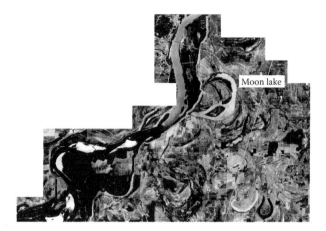

FIGURE 9.12 Aerial photograph of Moon Lake, Mississippi in Coahoma County, showing the oxbow lake, the Mississippi River meanders, and old meander scars.

FIGURE 9.13 Hoover Dam. (Courtesy of USBR.)

larger for reservoirs than for natural lakes, usually resulting in a greater influence of the character of the impounded river basin. Some of the major differences between lakes and reservoirs are listed in Figure 9.14 from Loucks and van Beek (2005).

Reservoirs are constructed for a variety of purposes. Of the dams included in the National Inventory of Dams (NID) 2002–2004 by the U.S. Army Corps of Engineers (authorized by the National Dam Safety and Security Act of 2002), the majority were constructed for recreation (65%) and are privately owned (Figure 9.15). The structure and function of reservoirs will be discussed in more detail in Chapter 10.

Lake	Reservoir
Especially abundant in glaciated areas; orogenic areas are characterized by deep, ancient lakes; riverine and coastal plains are characterized by shallow lakes and lagoons	Located worldwide in most landscapes, including tropical forests, tundra and arid plains; often abundant in areas with a scarcity of natural lakes
Generally circular water basin	Elongated and dendritic water basin
Drainage: Surface area ratio usually <10:1	Drainage: Surface area ratio usually >10:1
Stable shoreline (except for shallow, lakes in semiarid zones)	Shoreline can change because of ability to artificially regulate water level
Water level fluctuation generally small (except for shallow lakes in semiarid zones)	Water level fluctuation can be great
Long water flushing time in deeper lakes	Water flushing time often short for their depth
Rate of sediment deposition in water basin is usually slow under natural conditions	Rate of sediment deposition often rapid
Variable nutrient loading	Usually large nutrient loading their depth
Slow ecosystem succession	Ecosystem succession often rapid
Stable flora and fauna (often includes endemic species under undisturbed conditions)	Variable flora and fauna
Water outlet is at surface	Water outlet is variable, but often at some depth in water column
Water inflow typically from multiple, small tributaries	Water inflow typically from one or more large rivers

FIGURE 9.14   General characteristics of lakes and reservoirs. (Modified from Loucks and van Beek, *Water Resources Systems Planning and Management,* Appendix A: Natural system processes and interactions, United Nations Educational, Scientific and Cultural Organization, Delft, 2005. With permission.)

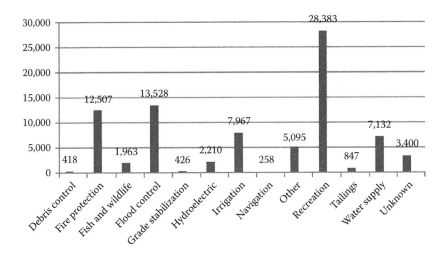

FIGURE 9.15   Project purposes of dams in the U.S. Army Corps of Engineers National Inventory of Dams 2009 database (the *y*-axis is the number of dams in the database).

9.4 OVERVIEW OF ORIGIN AND SIZE

Lakes and reservoirs vary widely in size and number. Inland lakes comprise only approximately 0.02% of the water in the hydrosphere (in contrast to 0.00008% in rivers; Schwoerbel 1987). In contrast, 1.6% is immobilized as ice. However, because of the accessibility of lakes, and their frequent replenishment, they are one of the most important sources of water for humans and wildlife.

TABLE 9.2
Examples of Lakes Greater than 6,000 km²

Name	Continent	Surface Area (km²)	Maximum Depth (m)	Volume (km³)
Caspian Sea	Asia	436,400	946	79,319
L. Superior	N. America	83,300	307	12,000
L. Victoria	Africa	68,800	79	2,700
Aral Sea	Asia	62,000	68	970
L. Huron	N. America	59,510	223	4,600
L. Michigan	N. America	57,850	265	5,760
L. Tanganyika	Africa	34,000	1,470	23,100
L. Baikal	Asia	31,500	1,620	23,000
L. Malawi	Africa	30,800	758	8,400
Gt. Slave Lake	N. America	30,000	614	7,000
Gt. Bear Lake	N. America	29,500	137	
L. Erie	N. America	25,300	64	470
L. Winnipeg	N. America	24,530	19	31,000
L. Ontario	N. America	18,760	225	1,720
L. Ladoga	Europe	18,734	250	920
L. Balkhash	Asia	17,575	27	112
L. Chad	Africa	16,500	12	24
L. Onega	Europe	9,836	124	300
L. Rudolf	Africa	8,600	73	360
L. Nicaragua	C. America	8,400	91	80
L. Volta	Africa	8,300	74	165
L. Titicaca	S. America	8,400	281	893
L. Issyk Kul	Asia	6,200	702	1,732
L. Vanern	Europe	5,570	89	180
L. Kariba	Africa	5,364	93	167
L. Albert	Africa	5,340	48	140
L. Peipus	Europe	2,276	13.4	2.1
L. Kivu	Africa	2,700	489	538
L. Vattern	Europe	1,898	119	72
L. Malaren	Europe	1,163	64	10
Total volume of all lakes			280,000	
Freshwater lakes			150,000	
Reservoirs			5,000	
Salt lakes			125,000	

Source: Martin, J.L. and McCutcheon, S.C., *Hydrodynamics and Transport for Water Quality Modeling*, Lewis Publishers, Boca Raton, FL, 1999. Modified from Loucks and van Beek, *Water Resources Systems Planning and Management*, Appendix A: Natural system processes and interactions, United Nations Educational, Scientific and Cultural Organization, Delft, 2005; and Schwoerbel, J., *Handbook of Limnology*, Ellis Horwood, Chichester, UK, 1987.

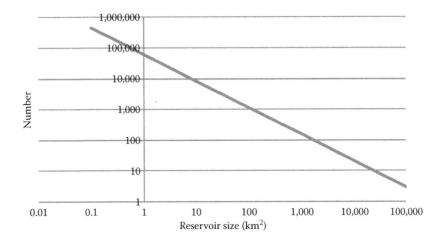

FIGURE 9.16 Comparison of the number of reservoirs worldwide vs. their size. (Based on data from Downing, J.A., Prairie, Y.T., Cole, J.J., Duarte, C.M., Tranvik, L., Striegl, R., McDowell, W.H., et al., *Limnology and Oceanography* 51, 2388–2397, 2006.)

Much of the total inland water is contained within a relatively few large lakes. The surface areas, the maximum depths, and the volumes of lakes over 1000 km² in surface area are provided in Table 9.2. Of these, the Caspian Sea, a large salt lake, is by far the largest. The Black Sea is an inland waterbody connected to the Atlantic Ocean. Downing et al. (2006), based on a global model, estimated that there were approximately 304 million lakes in the world, with a total area of 4.2 million km² and dominated in area by millions of waterbodies smaller than 1 km².

In North America, the Laurentian Great Lakes contain the largest continuous mass of freshwater on earth (Wetzel 2001) with a combined volume of 24,620 km³ making up approximately 20% of the total volume of all freshwater lakes, and a surface area representing approximately 10% of that of all inland waters combined (Table 9.1).

There are many reservoirs on planet Earth. Based on a global model, Downing et al. (2006) estimated that there are more than 500,000 reservoirs, with an average surface area of 0.5 km² (Figure 9.16), covering approximately 0.26 million km². Of the world's reservoirs, as compiled by Downing et al. (2006), 24 exceed 1000 km² in area. For example, the Three Gorges Dam in China slightly exceeds 1000 km² in size, compared to Lake Volta, formed by the Akosombo Dam in the West African nation of Ghana, which covers an area greater than 8500 km².

In the United States as well, small reservoirs (and ponds) dominate, with their numbers difficult to estimate. Using geographic information systems (GIS), Smith et al. (2002) estimated that the area of farm ponds in the United States was approximately 21,000 km². For larger reservoirs, information is available from the NID by the U.S. Army Corps of Engineers. The inventory includes all significant hazard potential classification dams and low hazard or undetermined potential classification dams which equal or exceed 25 ft. in height and which exceed 15 acre-ft. in storage (1 acre-ft. = volume represented by 1 acre in area 1 ft. deep), or equal or exceed 50 acre-ft. in storage and exceed 6 ft. in height. The 2009 NID databases included 82,842 dams, of which the majority (73,423) are earthen dams. Of the dams in the NID, 66% were built prior to 1970, and only 2290 have been built since 2000.

REFERENCES

Bowen, R. 1982. *Surface Water*. Allied Science, London.

CGER. 1996. *Freshwater Ecosystems: Revitalizing Educational Programs in Limnology*. Commission on Geosciences, Environment, and Resources, National Academy Press, Washington, DC.

Chapra, S.C. 2008. *Surface Water Quality Modeling*. Waveland Press, Long Grove, IL.

Downing, J.A., Y.T. Prairie, J.J. Cole, C.M. Duarte, L. Tranvik, R. Striegl, W.H. McDowell, et al. 2006. The global abundance and size distribution of lakes, ponds, and impoundments. *Limnology and Oceanography* 51 (5), 2388–2397.

Hutchinson, G.E. 1957. *A Treatise on Limnology, v. 1. Geography, Physics and Chemistry*. Wiley, New York.

Hutchinson, G.E. 1959. Homage to Santa Rosalia or why are there so many kinds of animals? *American Naturalist* XCIII (870), 145–159.

Hutchinson, G.E. 1961. The paradox of the plankton. *American Naturalist* 95, 137–145.

Hutchinson, G E. 1967. *A Treatise on Limnology, v. 2. Introduction to Lake Biology and the Limnoplankton*. Wiley, New York.

Hutchinson, G.E. 1975. *A Treatise on Limnology, v. 3. Limnological Botany*. Wiley, New York.

Hutchinson, G.E. 1993. *A Treatise on Limnology, v. 4. The Zoobenthos*. Y.H. Edmondson (ed.). Wiley, New York.

Loucks, D.P. and E. van Beek. 2005. *Water Resources Systems Planning and Management*. Appendix A: Natural system processes and interactions. United Nations Educational, Scientific and Cultural Organization, Delft.

Martin, J.L. and S.C. McCutcheon. 1999. *Hydrodynamics and Transport for Water Quality Modeling*. Lewis Publishers, Boca Raton, FL.

Schwoerbel, J. 1987. *Handbook of Limnology*, Ellis Horwood, Chichester, UK.

Slack, N.G. 2010. *G. Evelyn Hutchinson and the Invention of Modern Ecology*. Yale University Press, New Haven, CT.

Smith, S.V., W.H. Renwick, J.D. Bartley, and R.W. Buddemeier. 2002. Distribution and significance of small, artificial water bodies across the United States landscape. *Science of the Total Environment* 299, 21–36.

Wetzel, R.G. 2001. *Limnology: Lake and River Ecosystems*. Academic Press, San Diego, CA.

10 Those Dammed Lakes

10.1 RESERVOIRS (DAMMED RIVERS)

As discussed in Chapter 9, G.E. Hutchinson (1903–1991) is considered to be the father of modern limnology. Among his contributions, he authored the four-volume (with more than 3500 combined pages) *Treatise of Limnology*, which is considered to be one of the most comprehensive studies of limnology in existence. In his 1957 *A Treatise on Limnology. Volume 1. Geography, Physics and Chemistry* (Hutchinson 1957), he defined 76 different types of lakes based on their formation. However, in this over 1000-page long treatise, Hutchinson wrote only two paragraphs about man-made lakes (as lake type 73, Threlkeld 1990).

While lakes are common and undeniably important, reservoirs (Figure 10.1) are common and numerous. Reservoirs typically result from impounding some body of water. Reservoirs are most numerous in regions with few natural lakes, such as the nonglaciated parts of North America (except Florida), which have the largest numbers of reservoirs (Thornton 1990). There are over 80,000 large dams listed in the National Inventory of Dams (NID) and an undetermined total number of dams, probably exceeding 2,500,000 structures nationwide (NRC 1992). Reservoirs are constructed for some design purpose (recreation, hydropower, water supply, etc.). Therefore, it is the design purpose and operation that largely impacts the physical, chemical, and biological characteristics of a reservoir (Martin and McCutcheon 1999). In this chapter, we will review some of the project types and structures as they impact the characteristics of reservoirs.

10.2 PROJECT PURPOSES: STORAGE AND POOL LEVEL CONTROL

Reservoirs are constructed for a variety of purposes. The specific purpose for which a reservoir is constructed dictates its design and operation, or the principle of "form follows function."

Of the 82,642 dams in the NID (https://nid.usace.army.mil), by far the most common project purpose is recreation (Figure 10.2). Other purposes include flood control, fire protection, irrigation, water supply, hydroelectric power, fish and wildlife, tailing control, grade stabilization, debris control, and navigation. Note that this inventory is limited to significant hazard potential classification dams and low hazard or undetermined potential classification dams, which equal or exceed 25 ft. in height and exceed 15 acre-feet in storage, or equal or exceed 50 acre-feet storage and exceed 6 ft. in height. Therefore, many of the smaller reservoirs (and ponds) are not included.

There are a large number of reservoirs, particularly smaller reservoirs, with uncontrolled or unmanaged releases. That is, the releases are through control structures (standpipes, weir overflows, etc.), but the release structure cannot be manipulated to control the releases, such as through increasing gate openings, etc.

10.2.1 Storage

The operation of a reservoir, particularly a large and managed reservoir with a release control, is related to the manner in which its storage is allocated. Typically, reservoir storage is subdivided into specific zones or pools (Wurbs and James 2002) such as those shown in Figure 10.3.

The inactive zone is often not available for use; it is the "dead" zone. Some reservoirs may have a bottom release structure to aid in draining the reservoir or for water quality purposes. A special

FIGURE 10.1   Pomme de Terre Dam and Reservoir. (Courtesy of the U.S. Army Corps of Engineers, Kansas City District.)

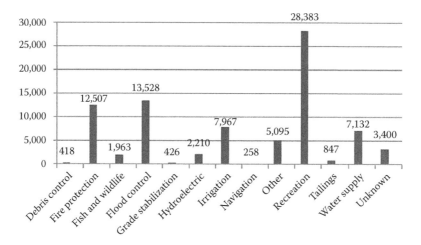

FIGURE 10.2   Primary project purposes of dams in the U.S. Army Corps of Engineers National Inventory of Dams, 2009 database (the *y*-axis is the number of dams in the 2009 database).

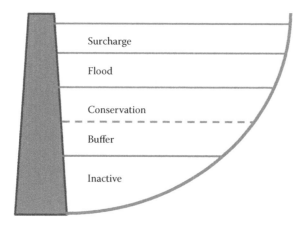

FIGURE 10.3   Reservoir storage zones. (Redrawn from USACE, HEC-5: Simulation of flood control and conservation systems, User's manual, Version 8.0, U.S., Army Corps of Engineers, Washington, DC, 1998.)

case for the conservation pool is a buffer zone that contains storage (stored water) to be used to meet essential demands in the event of a drought. The conservation pool includes usable storage, such as for water supply, and hydropower. The flood control pool allows for the safe storage of water when its release would cause flood damage downstream. If the reservoir does not have flood control storage, this could be included in the conservation pool. Above the flood control storage is the surcharge zone where the water level is above the emergency spillway but below the maximum dam height, above which the dam would be overtopped. An example of the storage allocation for Everett Jordan Lake, a U.S. Army Corps of Engineers multipurpose reservoir in North Carolina, is illustrated in Figure 10.4.

10.2.2 POOL LEVEL CONTROL

For reservoirs that have controlled releases, lake levels (i.e., storage) are varied to meet the reservoir design purpose. For example, if the reservoir is designed for flood control, then one would want the maximum available storage during the wet season, to accommodate or retain the flood. If the reservoir is used for conservation purposes, the conservation pool, then one may want the maximum storage during dry seasons.

For managed reservoirs, those with controlled releases, the design purpose (or purposes) is typically incorporated into the management strategy in a reservoir operations manual, developed when the reservoir is constructed. For federal facilities, those authorized by Congress, the water management strategy is based on the specific provisions of project-authorizing legislation as well as water management criteria defined during the project planning and all applicable laws and regulations (e.g., the Fish and Wildlife Coordination Act, the Federal Water Project Regulation Act—Uniform Policies, the National Environmental Policy Act, and the Clean Water Act) as described in the USACE's (1987) "Management of water control systems." Justification, site selection, design, and design purposes, as incorporated into reservoir operations guidelines, are also based on benefit–cost ratios.

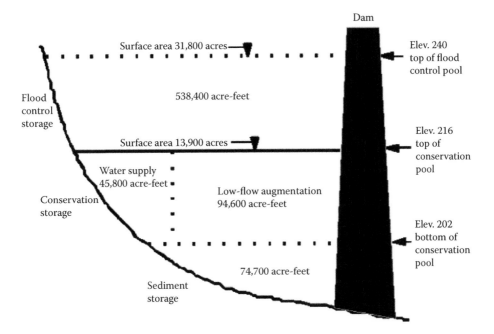

FIGURE 10.4 Example of storage allocation for Lake Jordan, North Carolina. (From NC DENR, Attachment A, B. Everett Jordan Lake water supply allocation request; Division of Water Resources Summary and Recommendation, North Carolina Department Environment and Natural Resources, Raleigh, NC, 2009.)

Reservoir operation, or water level management, strategies are typically incorporated into guide or rule curves, water control diagrams, with the goal of managing the reservoir to follow, as closely as practicable, these guide curves. The guide or rule curve is a seasonally varying plot of water-surface elevations (which translates to storage), depicted as either a line or an envelope, developed to reflect how the reservoir is to be managed. Figure 10.5 illustrates a guide curve for Pickwick Lake. The shaded area is the normal operation (for power production and summer mosquito control), the top of which may be exceeded for the regulation of flood flows.

Kennedy et al. (2000) conducted a categorical evaluation of the guide curves used in the U.S. Army Corps of Engineers reservoirs, as illustrated in Figure 10.6. Case (a) is typical of reservoirs operated for navigational purposes, where a constant water-surface elevation is desirable. Examples include the reservoirs on the Tennessee Tombigbee Waterway and Old Hickory on the Cumberland River. Cases (b) and (c) illustrate a rule or guide curve commonly associated with a hydropower project (e.g., see Figure 10.5), where there is an elevated summer conservation or hydropower pool and the project is operated between an upper and a lower limit. Cases (d) and (e) are commonly associated with flood control (Figure 10.6). Case (f) is where the pool could be depleted during portions of the year due to demands such as irrigation, maintaining minimum releases, water supply, and other consumptive uses (Kennedy et al. 2000). Cases (g) and (h) are for "dry dams" to

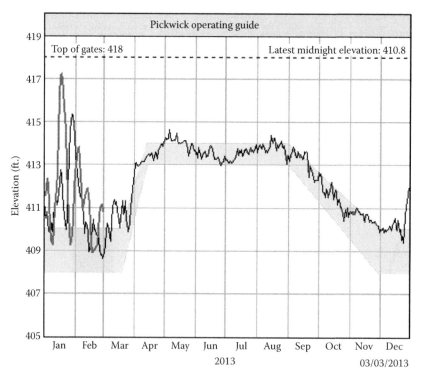

FIGURE 10.5 Guide curve for Pickwick Lake comparing the desired operating envelope (gray area) to the actual lake-water surface elevations for 2012 (black line) and portions of 2013 (red line). (From Tennessee Valley Authority. Pickwick operating guide. Knoxville, TN. Available at: http://www.tva.gov/river/lakeinfo/op_guides/pickwick.htm.)

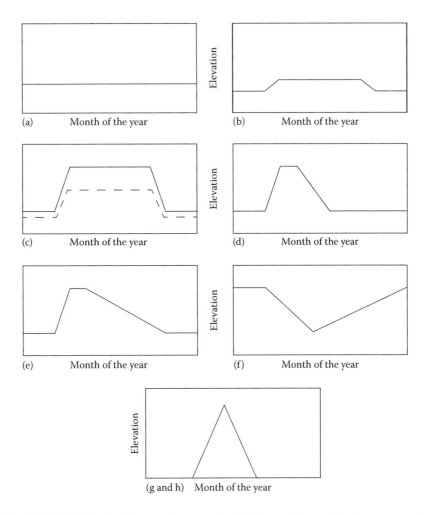

FIGURE 10.6 (a–h) Idealized guide or rule curves for U.S. Army Corps of Engineers reservoirs. (From Kennedy, R., Clarke, J., Boyd, W., and Cole, T., Characterization of U.S. Army Corps of Engineers reservoirs: Design and operational considerations, U.S. Army Engineers Engineering Research and Development Center, Vicksburg, MS, 2000.)

accommodate flooding where for (g) only a small minimum pool is maintained for much of the year (e.g., for fish and wildlife), while for (h) there may not be a permanent pool.

Reservoirs may be designed for a single purpose, such as those illustrated in Figure 10.7. More commonly, reservoirs are operated for multiple purposes. For such cases, the conservation pool (Figure 10.3) is shared and subdivided based on the allocation of water. The conservation pool is allocated in such a way that competing water users are able to get their equitable share based on an agreed policy, such as the amount allocated to each user. This sharing is then incorporated into the rule or guide curve, used for the operation of reservoirs.

One difficulty is that the rule curve is derived from historical data on river flows and water demands (USACE 1987). Usually, the rule curve is followed at all times except in cases of extreme drought or for public safety. However, changes in climate and hydrology can impact those flows and water demands, which, in turn, can impact reservoir operations. For example, in June 2006, the Western Governors Association (WGA 2006) released a report titled "Water needs and strategies for a sustainable future," in which they listed climate change as one of the challenges facing the western states, such as through potential reductions in snowpack, which makes up a

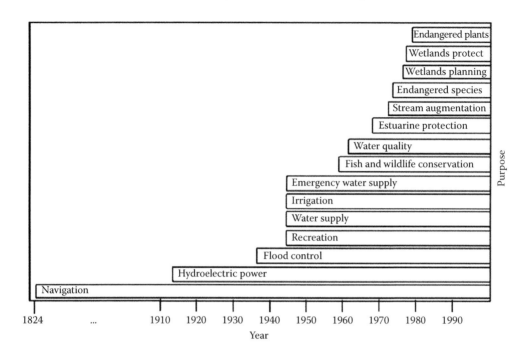

FIGURE 10.7 Chronology of acceptable reservoir purposes and programs as authorized by Congress. (From USACE, Engineering and design—Hydrologic engineering requirements for reservoirs, U.S. Army Corps of Engineers, Davis, CA, 1997.)

significant portion of the total water storage in the West. The report indicated that as a result of climate change, water managers may be forced to make changes in reservoir operations and rule curves.

A second problem is that these multiple purposes are often in conflict with each other. For example, the manner in which a reservoir would be operated for flood control may be different from the manner in which it would be operated for fisheries management. For in-pool wildlife and fisheries management in fluctuating warmwater and cool-water reservoirs, the water level is usually increased during the spring to enhance spawning and the survival of young predators and it is then lowered during the summer to permit regrowth of vegetation in the fluctuation zone. In addition, the pool levels are often increased during autumn for waterfowl management (USACE 1997). Conversely, spring and fall drawdowns are common for flood control management, allowing increased storage during wet seasons. Also, the manner in which a reservoir may be operated for fisheries may differ from the management of in-reservoir fisheries to those in the reservoir tailwaters. As an example, in many hydropower reservoirs in the southern United States, the bottom, cold releases allow the supporting of cold-water (trout) fisheries downstream. The manner in which the reservoir operates clearly impacts the sustainability of that tailwater fishery. So, the management of multipurpose reservoirs is often a problem in optimization.

In addition, needs or conditions change so that it may be necessary to update the reservoir operations to reflect new or changing demands. Examples include a desired shift in operations in response to increased water supply needs, or to a need to increase releases to improve water quality as part of effluent trading, or to respond to the potential impacts of climate change. In addition, there may be a desire to change the reservoir operation to reflect some purpose that previously could not be used to justify the reservoir. For example, it is only since the 1960s that fish and wildlife and environmental issues have been an accepted reservoir design purpose (Figure 10.7). Therefore, federal reservoirs constructed prior to that time and authorized by Congress were precluded from including fish and wildlife and environmental considerations in their operations guide or rule curves.

For federal facilities, the extent of allowable modifications depends on the original congressional authorization. Major changes may require congressional reauthorization. Therefore, legal requirements are a major consideration if there is a desired change in operations. One example that will be discussed in later sections is what was required to alter the releases from reservoirs in northwest Arkansas (White River, etc.) to protect the downstream (tailwater) fisheries.

10.3 TYPES OF DAMS

One of the primary factors affecting the physical, chemical, and biological characteristics of reservoirs is the design and operation of dams. A dam is designed to hold or impound water and resist the forces acting on it, including the force of water (weight and pressure), waves, gravity, and seismic forces.

Dam design considerations include:

- Project purpose
- Material availability and life expectancy (design life)
- Foundations and materials (bearing, sliding, seepage, liquefaction settlement, seepage, leakage, piping, etc.)
- Stability (overturning and slope stability; seismic resistance, water barriers, and drains)

Basically, dams can be subdivided into three types: gravity structures, fill dams, and structural dams. Of the dams included in the 2009 NID, by far the majority were earth-filled dams (Figure 10.8).

10.3.1 GRAVITY STRUCTURES

Gravity structures rely on their mass for stability. Typical gravity structures are composed of concrete, masonry, or roller-compacted concrete (RCC). Concrete gravity dams often run in a straight line across a valley, as illustrated by Grand Coulee Dam (Figure 10.9).

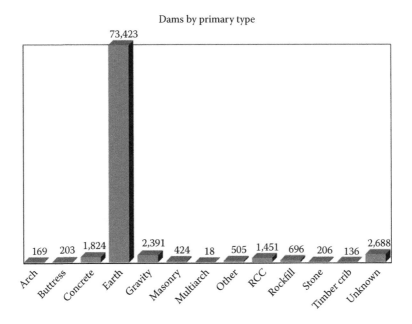

FIGURE 10.8 U.S. Dams in the 2009 National Inventory of Dams database. (From U.S. Army Corps of Engineers National Inventory of Dams, National Data available at: http://geo.usace.army.mil/pgis/f?p=397:5:0::NO.)

FIGURE 10.9   Aerial view of the partially completed Grand Coulee Dam taken on June 15, 1941. (Courtesy of the U.S. Bureau of Reclamation.)

10.3.2   FILL DAMS

Fill dams depend on the frictional resistance of materials and mass to resist the forces acting on them. Examples include earth fill, rock fill, hydraulic fill, and levees. Typically, fill dams are made entirely of an impervious material (are homogeneous) or they are layered or zoned with a waterproof core, such as clay, protected by a filter and drain. The core is then covered with earth or rock fill. While the core materials may not be completely impervious to water, they provide sufficient control of seepage, along with the filter and drain to remove excess water, as illustrated in Figure 10.10. One common type of dam is the concrete-faced, rock-fill dam, which is constructed of permeable rock fill, the impermeable membrane being a concrete slab constructed on the upstream face of the dam wall.

As with gravity dams, fill dams rely only on their weight and friction to resist the forces acting on them. Depending on the availability of materials, earth-fill dams are the least expensive to build and are the most common dam type (Figure 10.8). An example is Fishtrap Dam in Kentucky, illustrated in Figure 10.11.

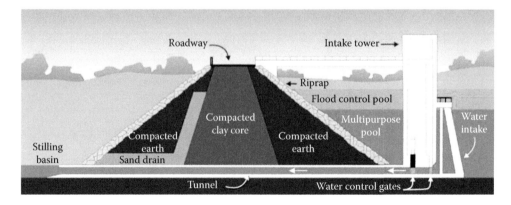

FIGURE 10.10   Structure and outlet works of Melvern Lake, Kansas; a rolled earth-fill structure with an emergency uncontrolled chute spillway and outlet works consisting of a single horseshoe conduit and two hydraulically operated gates. (Courtesy of the U.S. Army Corps of Engineers, Kansas City District.)

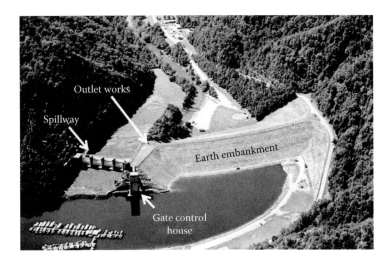

FIGURE 10.11 Fishtrap Lake, Kentucky. (Courtesy of the U.S. Army Corps of Engineers, Huntington District.)

10.3.3 STRUCTURAL DAMS

Structural dams rely on their structural configuration and include arch dams, buttresses, and flood walls. Arch dams are curved and depend on the arch action, as well as the weight of the dam, for their strength. Arch dams are considerably thinner than gravity dams and require much less material, so they are often more cost effective. An example is Hoover Dam (Figure 10.12).

The face of buttress dams is held up by a series of supports and may be flat or curved. These dams are usually made of concrete with steel reinforcement and may be solid or hollow. An example is the Daniel-Johnson Dam in Quebec (Figure 10.13).

FIGURE 10.12 Hoover Dam. (Courtesy of the Bureau of Reclamation—Lower Colorado Region.)

FIGURE 10.13 Daniel-Johnson Dam in Quebec. (Courtesy of Wikimedia Commons. 2012. Available at: http://commons.wikimedia.org/wiki/Category:Daniel-Johnson_Dam.)

10.4 CONVEYANCE STRUCTURES

Conveyance structures convey water over, under, around, or through a dam. Dams are typically designed so that water should never flow over the top of the dam, because overtopping, such as occurs during extreme floods, is a common cause of dam failures. Instead, conveyance structures are designed to allow the release of excess water, such as the surcharge. The surcharge is that water above the flood storage where the water level is above the emergency spillway but below the maximum dam height (Figure 10.14). Other conveyance structures vary with the type of dam and its intended (authorized) purpose. The conveyance structure, location, and design, as well as the manner in which the water is released impact not only the physical, chemical, and biological characteristics of the reservoir, but the downstream waters (tailwaters) as well.

10.4.1 Spillways

Spillways are intended to pass excess water from a reservoir. They include service, auxiliary, and emergency spillways (USBR 2010). Service spillways provide continuous or frequent releases, and may be regulated (controlled) or unregulated (uncontrolled), as illustrated in Figures 10.15 and 10.16, respectively. Auxiliary spillways are infrequently used and may be a secondary spillway, as illustrated in Figure 10.17. Emergency spillways are intended to pass the surcharge storage and prevent the dam from being overtopped (see Figure 10.18). Examples of uncontrolled spillways are illustrated in Figures 10.19 and 10.20.

For gated spillways, the common gate types are tainter (radial) gates, vertical lift gates, and drum gates (USACE 1987). Tainter gates are basically a segment of a cylinder mounted on radial arms that rotate on anchored trunnions (Figures 10.21 and 10.22). They are simple, relatively light, and easy to use and are one of the commonly used gate designs for navigation projects. The vertical lift gate, with wheels (rollers) at each end, moves vertically in slots formed in the pier and consists of a plate and horizontal girders that transmit the water load into the pier (Figures 10.23 and 10.24). Vertical lift gates are most commonly used on low-head dams (USACE 1987). Drum gates are designed to float on water in a chamber located in the spillway crest. A watertight sill chamber is provided for the gate and to raise the gate, upper pool pressures are introduced into the chamber (Figure 10.25).

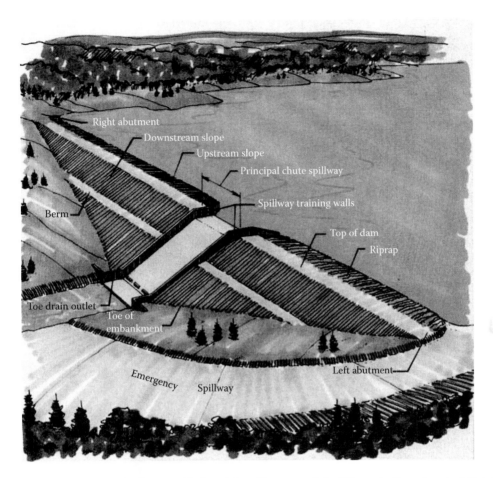

FIGURE 10.14 Earth embankment with drop inlet. (Courtesy of the Wisconsin Department of Natural Resources.)

FIGURE 10.15 Service spillway (gated), Jackson Lake Dam, Wyoming. (From USBR, Design standards No. 14, appurtenant structures for dams (spillway and outlet works), design standards, U.S. Department of the Interior, Bureau of Reclamation, 2010.)

FIGURE 10.16   Service spillway (ogee crest), Crystal Dam, Colorado. (From USBR, Design standards No. 14, appurtenant structures for dams (spillway and outlet works), design standards, U.S. Department of the Interior, Bureau of Reclamation, 2010.)

FIGURE 10.17   Auxiliary spillway (gated) in foreground and service spillway (gated) in background, Stewart Mountain Dam, Arizona. (From USBR, Design standards No. 14, appurtenant structures for dams (spillway and outlet works), design standards, U.S. Department of the Interior, Bureau of Reclamation, 2010.)

FIGURE 10.18   Emergency spillway (fuse plug) in foreground and auxiliary spillway (ogee crest) in background, New Waddell Dam, Arizona. (From USBR, Design standards No. 14, appurtenant structures for dams (spillway and outlet works), design standards, U.S. Department of the Interior, Bureau of Reclamation, 2010.)

FIGURE 10.19 Ogee crest (uncontrolled) spillway, Bumping Lake Dam, Washington. (From USBR, Design standards No. 14, appurtenant structures for dams (spillway and outlet works), design standards, U.S. Department of the Interior, Bureau of Reclamation, 2010.)

FIGURE 10.20 Outlet of Ladybower Reservoir, Peak District, Derbyshire, UK. (Courtesy of Wikimedia Commons.)

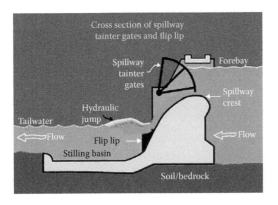

George W. Andrews Lock and Dam

FIGURE 10.21 Structure of tainter gates. (Courtesy of the U.S. Amy Corps of Engineers Institute for Water Resources.)

FIGURE 10.22 Tainter gates and sluices at Folsom Dam. (Courtesy of the U.S. Bureau of Reclamation.)

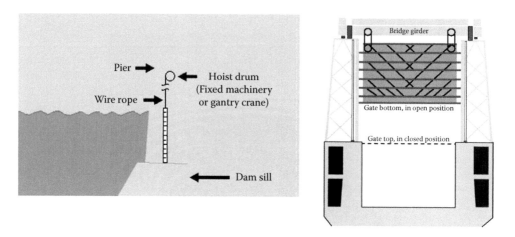

FIGURE 10.23 Vertical lift gates. (Courtesy of the U.S. Army Corps of Engineers Institute for Water Resources.)

The common features associated with spillways are illustrated in Figure 10.26 and include (USBR 2010):

- An approach channel and a safety/debris/log boom
- A control structure, such as a crest structure or grade sill, and gates
- Bulkheads, stop logs, along with associated operating equipment
- Conveyance features, such as a chute floor and walls and/or conduit(s)/tunnel(s)
- A terminal structure, such as a hydraulic jump stilling basin, a flip bucket, and a plunge pool
- A downstream channel

10.4.2 OUTLET WORKS

The outlet works consist of a combination of features and operating equipment for the safe and controlled release of water to meet project purposes. Examples of outlet works include an intake structure and conveyance features such as conduits, a control structure, etc. (USBR 2010). The common features associated with outlet works include (USBR 2010; Figure 10.27):

- Intake structures, trashracks, gates/valves, and bulkheads (if appropriate)
- Conveyance features, such as conduit(s)/tunnel(s)
- A control structure, such as a gate chamber, gates/valves, an access shaft/adit/conduit, along with operating equipment
- A terminal structure, such as a hydraulic jump stilling basin, an impact structure, and a plunge pool
- A downstream channel

FIGURE 10.24 Downstream navigation gate construction of the vertical lift gates at Lower Monumental Lock and Dam, among the largest in the world, near Kahlouts, WA, in 2011. (*Middle*) Project manager Steve Hartman displays the mechanics and rope wires used to lift the 1.5 million pound gate. (From USACE, *Intercom* 40, 11, 2013. U.S. Army Corps of Engineers, Walla Walla District.)

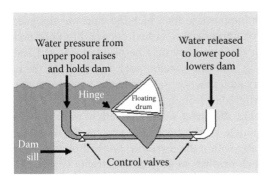

FIGURE 10.25 Drum gate (*left*) and drum gate on a diversion dam (*right*). (Courtesy of the U.S. Army Corps of Engineers Institute for Water Resources.)

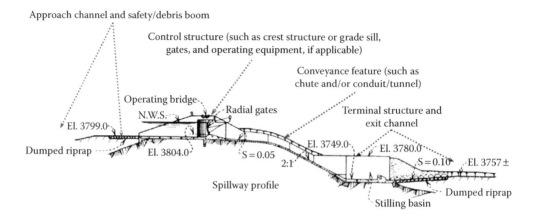

FIGURE 10.26 Common features of spillways. (From USBR, Design standards No. 14, appurtenant structures for dams (spillway and outlet works), design standards, U.S. Department of the Interior, Bureau of Reclamation, 2010.)

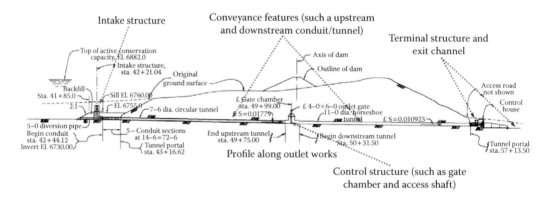

FIGURE 10.27 Common features of outlet works. (From USBR, Design standards No. 14, appurtenant structures for dams (spillway and outlet works), design standards, U.S. Department of the Interior, Bureau of Reclamation, 2010.)

10.4.2.1 Functional Requirements

The outlet works of a dam depend largely upon the purpose for which the dam was designed, or its authorized purpose, for example, flood control, conservation, or hydropower. For flood control, the outlet structures are designed to allow high outflows; they may be gated or ungated and are commonly surface releases. For conservation (e.g., the conservation pool, Figure 10.3), the outlet works may vary with the specific type of conservation practice, such as navigation, irrigation, and water supply and may be from the surface (overflows) or from below the water surface. For meeting downstream water quality targets, for example, the outlet works may allow water to be withdrawn from different vertical levels, or from multiple levels, during the year, as the water quality and the temperature vary seasonally and vertically within the reservoir. For hydropower, water is discharged through a penstock to a turbine, which revolves to generate electricity (Figure 10.28). Since the water-surface elevation provides the potential energy, the water is typically withdrawn from the lower levels of the reservoir, frequently resulting in releases of cold(er) water often with low dissolved-oxygen concentrations.

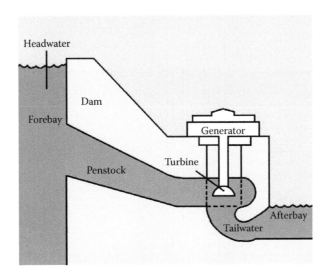

FIGURE 10.28 Typical hydropower section. (Courtesy of the U.S. Bureau of Reclamation.)

10.4.2.2 Intake Structure

Intake structures for embankment dams are typically gated towers, multilevel, uncontrolled two-way risers, and/or a combination of both (USACE 1980). The lake level at which the intake structure is placed depends on the design purpose. Dams that store water for irrigation, domestic use, or other conservation purposes must have the outlet works intake structure low enough to be able to draw the water down to the bottom of the allocated storage space (USBR 1987; USACE 1997). Similarly, for hydropower the intake structure is at lower reservoir levels. While there may a "dead" or inactive storage zone (Figure 10.3), the intake structure is usually placed as low as practicable to allow for drainage, inspection, and repair.

As will be discussed later, reservoirs commonly thermally stratify for much of the year, resulting in vertical variations in temperatures and water quality. For example, oxygen is often depleted in the lower levels of many reservoirs during summer and fall. As a result, for water supply or for the management of tailwater and downstream systems, it is often necessary to withdraw water at some level other than at the bottom. Since the "best" water elevation varies seasonally, some reservoirs are equipped with selective withdrawal structures so that the elevation of the intake can be varied (Figure 10.29).

10.4.2.3 Conduits

Conduits convey water through, under, or around embankment dams, and may be under pressure (penstocks), or to hydropower turbines, where the outlet works are used as the penstock for power plants (USBR 1987). Often, there are several conduits, allowing one to be taken out of service for maintenance and repair. Conduits that pass water through or under embankment dams are most susceptible to failure; where possible, tunnels that are not in direct contact with the dam embankment are used. Conduits may be designed in a variety of shapes. Shape A, illustrated in Figure 10.30, is most commonly used for pressurized conduits (penstocks), while shapes B and C are commonly used for nonpressurized conduits, and may be larger and used for downstream access.

10.4.2.4 Control Structure

Outlet works are usually regulated by gates and valves, typically motor operated, hydraulically operated, or manually operated (USACE 1997), with emergency backup systems. The regulating gates and valves may be placed at the beginning or at the end of the conduit, or at an intermediate location (Figure 10.31).

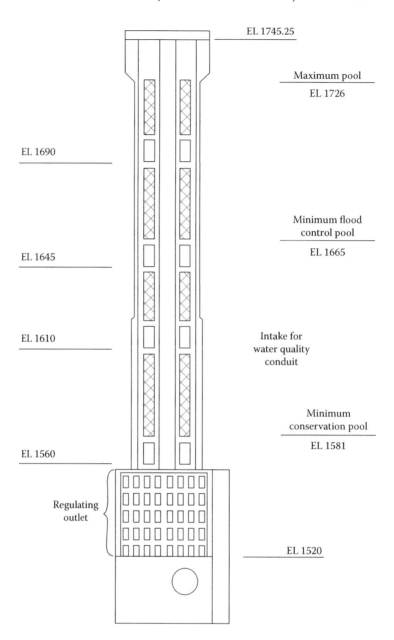

EL 1745.25

Maximum pool
EL 1726

EL 1690

Minimum flood
control pool
EL 1665

EL 1645

EL 1610

Intake for
water quality
conduit

Minimum
conservation pool
EL 1581

EL 1560

Regulating
outlet

EL 1520

FIGURE 10.29 Diagram of a selective withdrawal structure for Elk Creek Dam. (From U.S. Army Engineer District. Rogue River Basin Project. Portland, OR, 1983. Howington, S.E. Intake structure operations study, Elk Creek Dam, Oregon. Technical Report HL-90-16, US Army Engineer Waterways Experiment Station, Vicksburg, MS, 1990.)

10.4.2.5 Energy Dissipation

Often, reservoir releases have considerable potentially destructive energy associated with them, which, if not dissipated, can result in damage to the structure, scour below the dam, and other undesirable impacts. An example is when water is passed through a spillway and down steep chutes, where the water is basically in free fall. If that energy is not dissipated, it may often result in scour. For example, the spillway releases between 1962 and 1982 from the Kariba Dam (Zambia–Zimbabwe),

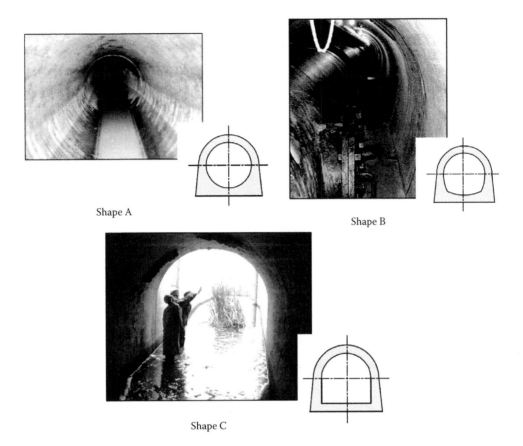

Shape A

Shape B

Shape C

FIGURE 10.30 Examples of reservoir conduits and shapes. (From FEMA, Technical manual: Conduits through embankment dams best practices for design, construction, problem identification and evaluation, inspection, maintenance, renovation, and repair, Federal Emergency Management Authority, Washington, DC, 2005.)

a 128 m-high concrete arch dam on the Zambezi River resulted in an 80 m deep (below the initial riverbed) downstream scour hole (Mason and Arumugam 1985).

Examples of energy dissipation designs (stilling basins) for spillways are illustrated in Figure 10.32. These devices are designed to introduce a hydraulic jump, which results from a change in velocities from supercritical to subcritical flows. Hydraulic jumps result in large and rapid changes in the water-surface elevation, and a subsequent large decrease in kinetic energy. Examples of hydraulic jumps as a function of the Froude number are illustrated in Figure 10.33 and dissipation structures are illustrated in Figure 10.34. For supercritical flows, the Froude number is greater than one and the larger the Froude number the higher the spillway velocity in relation to the speed (velocity) of a gravity wave.

In addition to dissipating energy, hydraulic jumps also aid in increasing oxygen concentrations and enhancing mixing.

In addition to dissipating energy by inducing a hydraulic jump, there are a variety of other energy-dissipating devices or structures. For example, jet releases from pressure conduits are often directed or deflected onto impact-type dissipaters. Impact dissipaters direct the water into an obstruction that diverts the flow and generates turbulence (Wei and Lindell 2004). Examples of jet releases are illustrated in Figures 10.35 and 10.36. Strontia Springs Dam, illustrated in Figure 10.35, is owned and operated by Denver Water and was put into service in 1986. It is a 292 ft. high, double-curvature, thin-arch, concrete dam with a maximum structural height of 292 ft.

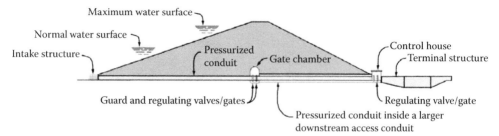

(a) Pressurized outlet works with the control feature intermediate along the conduit, with downstream access

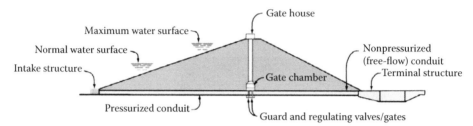

(b) Pressurized outlet works with control feature intermediate along the conduit, without downstream access

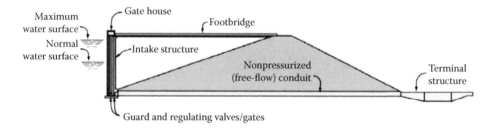

(c) Nonpressurized outlet works with the control feature at the upstream end of the conduit

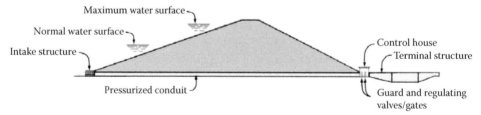

(d) Pressurized outlet works with control feature at the downstream end of the conduit

FIGURE 10.31 (a–d) Common outlet works and conduits for embankment dams. (From FEMA, Technical manual: Conduits through embankment dams best practices for design, construction, problem identification and evaluation, inspection, maintenance, renovation, and repair, Federal Emergency Management Authority, Washington, DC, 2005.)

10.4.2.6 Structures for Water Quality and Fisheries Management

Dams and their operations have large impacts, both positive and negative, on fisheries and water quality both within and below the impoundment. Often, additional structural or operational controls are implemented to enhance the water quality and fisheries. Examples include the devices discussed in the previous section (rivers and streams) for fish passage, since dams acting as barriers to migration is a major issue for many fish species. There are a variety of other fisheries and water quality

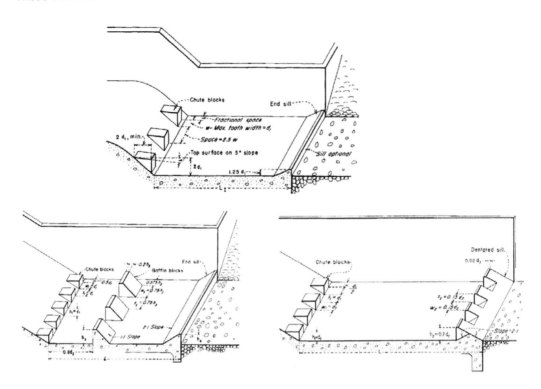

FIGURE 10.32 Examples of spillway energy dissipation designs. (From USBR, *Design of Small Dams*, 3rd ed., U.S. Government Printing Office, Washington, DC, 1987.)

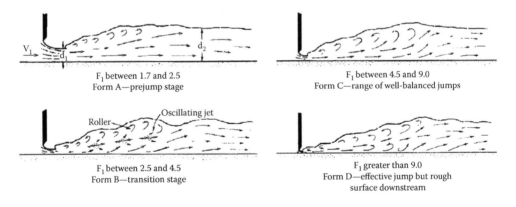

FIGURE 10.33 Characteristic forms of hydraulic jumps as a function of the Froude number. (From USBR, *Design of Small Dams*, 3rd ed., U.S. Government Printing Office, Washington, DC, 1987.)

issues, which will be discussed in later chapters. One example is dissolved gases, either too much or too little.

One problem in many areas is total dissolved gas supersaturation (TDGS). The dissolved gas concentrations in water are a function of pressure (as well as temperature, etc.), such as the partial pressure of the individual gases in the atmosphere. The concentration at which water is in equilibrium with the atmosphere is the saturation concentration. When air is entrained and the pressure increases, such as when water falls over a dam into a plunge pool or a jet of water contacts a surface, the concentration can increase well beyond that saturation concentration, or become

FIGURE 10.34 Spillway chute and energy dissipation structure on Willow Lake, Illinois. (Photograph from willow-lake.org.)

FIGURE 10.35 Strontia Springs Reservoir. (Photograph courtesy of Denver Water. With permission.)

FIGURE 10.36 Jet tube release from Glen Canyon Dam during a 2008 high-flow experiment. (Courtesy of the U.S. Bureau of Reclamation.)

supersaturated. Exposure to excess or supersaturated total dissolved gas can result in "gas bubble disease" or "gas trauma" in fish and other aquatic organisms, which can be lethal. The condition is similar to the "bends" that divers encounter. Beeman et al. (2003), for example, discuss TDGS and gas bubble disease below Grand Coulee Dam. Gas bubble disease will be discussed in greater detail in Chapters 14 and 18.

Often too, the water released from lower levels of the reservoir pool is devoid of oxygen and has high concentrations of reduced materials (e.g., methane, hydrogen sulfide, and ammonia). A variety of methods have been used to improve the quality of reservoir releases by adding oxygen, from oxygen injection above the dam to hydropower autoventing turbine technologies, and others that will be discussed in Chapters 17 and 18.

10.5 PRETTY DAMMED OLD (DAM FAILURES AND DAM SAFETY)

Any man-made structure can fail, and dams are no exception. However, the impact of a dam failure can be catastrophic, often resulting in considerable loss of life and property. One of the most catastrophic dam failures in U.S. history was the Johnstown Flood, which is the subject of a highly recommended book by David McCullough (1987). The dam was a 72 ft.-high embankment dam, created as a reservoir for the canal basin in Johnstown, PA, and built between 1838 and 1852. On May 31, 1889, the dam failed, largely as a result of mismanagement and neglect. Its failure resulted in a flood wave reported to be 36 in. high and a half mile wide passing downstream through Johnstown, resulting in an official death toll of 2209, but hundreds more were lost (Figure 10.37). Other major historical U.S. dam failures include:

- St. Francis Dam, California, failed in 1928, 450 killed
- Baldwin Hills Dam, California, failed in 1963, 5 killed
- Buffalo Creek Dam, West Virginia, failed in 1972, 125 killed
- Teton Dam, a new Bureau of Reclamation dam in Idaho, failed during its first filling in June 1976, killing 11 people and causing over $1 billion in damages
- Laurel Run Dam, Pennsylvania, failed in July 1977, killing 40 people and causing over $5.3 million in damages
- Kelly Barnes Dam, Georgia, failed in November 1977, killing 39 students in a bible college downstream

FIGURE 10.37   Debris in Johnstown, PA, following the dam's failure. (From Wikimedia Commons.)

Figures 10.38 and 10.39 illustrate, respectively, the 1995 gate failure of Folsom Dam and the 2010 Delhi Dam failure in Iowa, neither of which resulted in loss of life.

Major causes of dam failure include (NRC 1983):

- Overtopping—34% of all failures (nationally)
  - Inadequate spillway design
  - Debris blockage of spillway
  - Settlement of dam crest
- Foundation defects—30% of all failures
  - Differential settlement
  - Sliding and slope instability
  - High uplift pressures
  - Uncontrolled foundation seepage
- Piping and seepage—20%
  - Internal erosion through dam caused by seepage—"piping"
  - Seepage and erosion along hydraulic structures such as outlets
  - Conduits or spillways, or leakage through animal burrows
  - Cracks in the dam
- Conduits and valves—10%
  - Piping of embankment material into conduit through joints or cracks
- Other—6%

FIGURE 10.38 The 1995 Folsom Dam radial gate failure. (From USBR, *Best Practices and Risk Methodology*, Chapter 16, U.S. Department of the Interior, Bureau of Reclamation, 2009.)

FIGURE 10.39 The 2010 Delhi Dam failure in eastern Iowa. (Republished from KRCG.com. © 2010 SourceMedia Group, Cedar Rapids, IA. With permission.)

Of the 84,130 dams in the NID 2009 database hazard rating, 13,991 are listed as high-hazard dams and an additional 12,662 as intermediate-hazard dams. A high-hazard dam is one where there is a high probability of loss of human life if the dam fails. The failure of an intermediate-hazard dam may result in loss of human life but will more likely result in significant property or environmental destruction.

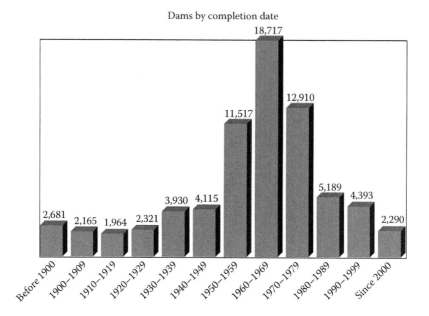

FIGURE 10.40 U.S. dams by their completion date. (From U.S. Army Corps of Engineers National Inventory of Dams, National data available at: http://geo.usace.army.mil/pgis/f?p=397:5:0::NO.)

With age, dams like any other infrastructure deteriorate. As illustrated in Figure 10.40, many U.S. dams were constructed between 1950 and 1980, so the average age of U.S. dams is 51 years. A commonly used design life expectancy for dams in the United States is 50 years. As a result, a large number of U.S. dams have exceeded, or will soon exceed, their design life. FEMA (2001, cited by Conyngham et al. 2006) estimates that 58,000 large dams (85% of dams in the NID) will have exceeded their design lifespan by 2020.

Only 11% of dams are federally owned, thus the responsibility for ensuring the safety of the remaining 89% of all dams falls on state agencies, which often do not have the resources to adequately inspect dams within their states and take appropriate corrective actions (ASCE 2009). For example, in 2009, Texas had only seven engineers and an annual budget of $435,000 to regulate more than 7400 dams, so that each inspector was responsible for more than 1050 dams (ASCE 2009). Since 1998, the American Society of Civil Engineers (ASCE) has issued report cards on the condition of America's infrastructure, and America's dams have consistently received a D grade. The number of dams in the United States that are deficient, in terms of age or structure, is increasing.

10.6 DECOMMISSIONING/REMOVING DAMS

A recent trend related to large dams in the United States and elsewhere has been to decommission dams that no longer serve a useful purpose, are too expensive to maintain safely, or have unacceptable levels of impact (World Commission on Dams 2000). This trend is particularly common in the United States where over 500 dams have been decommissioned and the rate for decommissioning large dams has overtaken the rate of dam construction (World Commission on Dams 2000).

Dam removal may have positive environmental and social impacts, such as the removal of barriers to migrating fish as part of river restoration projects or in reducing hazards. Dam removal may have negative environmental and economic impacts, so the process must be evaluated in terms of the risk and hazards as well as the cost and benefits of the dam and its removal.

Dam removal is a major engineering activity and there are a variety of potential impacts to consider if a dam is to be removed, including (Conyngham et al. 2006):

- Release of excessive sediments: reservoirs store large quantities of sediments, which may be released or must be removed as part of dam removal projects.
- Release of toxic sediments: sediments deposited behind dams may be contaminated and may require remediation prior to or following removal.
- Release of nutrients: sediments also store nutrients, which may be released following dam removal causing downstream increases in aquatic plants, such as algae.
- Undesirable vegetation response: plant growth downstream may be adversely impacted.
- Physical instability and bank erosion: as discussed for rivers and streams, taking a system out of sediment equilibrium may cause channel degradation, such as bank collapse, channel incision, and other impacts.
- Risk of downstream ice damming: dam removals in ice-prone rivers have been observed to cause an increased risk of ice jamming and damming (USACE 2001; White and Moore 2002, cited by Conyngham et al. 2006).
- Mobility of invasive organisms: some dams act as barriers to invasive species (such as the sea lamprey) or to separate introduced and native species; thus, the removal of that barrier may negatively impact the river or stream.

A variety of legal and regulatory issues are associated with dam removal. For example, federal, state, and local permits are required for dam removal. However, the laws and regulations on which these permits are based are often designed for public safety and environmental protection, rather than restoration, which may complicate the dam removal process. Bowman (2002) provides a review of the laws and regulations impacting dam removal. Dam removal as a restoration process will be discussed in greater detail in Chapter 18 on restoration and management.

REFERENCES

ASCE. 2009. Report card for America's infrastructure. American Society of Civil Engineers, Reston, VA.

Beeman, J.W., D.A. Venditti, R.G. Morris, D.M. Gadomski, B.J. Adams, S.P. VanderKooi, T.C. Robinson, and A.G. Maule. 2003. Gas bubble disease in resident fish below Grand Coulee Dam. U.S. Department of the Interior and U.S. Geological Survey, Cook, WA.

Bowman, M.B. 2002. Legal perspectives on dam removal. *BioScience* 52 (8), 739–747.

Conyngham, J., J.C. Fischenich, and K.D. White. 2006. Engineering and ecological aspects of dam removal— An overview. Environmental Laboratory and Cold Regions Research and Engineering Laboratory, ERDC TN-EMRRP-SR-80. Engineer Research and Development Center, U.S. Army Corps of Engineers, Vicksburg, MS.

FEMA. 2001. National dam safety program. Federal Emergency Management Agency, Washington, DC. Available at: http://www.fema.gov/fima/damsafe/.

FEMA. 2005. Technical manual: Conduits through embankment dams best practices for design, construction, problem identification and evaluation, inspection, maintenance, renovation, and repair. Federal Emergency Management Authority, Washington, DC.

Howington, S.E. 1990. Intake structure operations study, Elk Creek Dam, Oregon. Technical Report HL-90-16, U.S. Army Engineer Waterways Experiment Station, Vicksburg, MS.

Hutchinson, G.E. 1957. *A Treatise on Limnology. Volume 1. Geography, Physics and Chemistry.* Wiley, New York.

Kennedy, R., J. Clarke, W. Boyd, and T. Cole. 2000. Characterization of U.S. Army Corps of Engineers reservoirs: Design and operational considerations, ERDC WQTN-MS-05. U.S. Army Engineers Engineering Research and Development Center (ERDC), Vicksburg, MS.

Martin, J.L. and S.C. McCutcheon. 1999. *Hydrodynamics and Transport for Water Quality Modeling.* Lewis Publishers/CRC Press, Boca Raton, FL.

Mason, P.J. and K. Arumugam. 1985. A review of 20 years of scour development at Kariba Dam. International Conference on the Hydraulics of Floods and Flood Control, Cambridge, UK.

McCullough, D. 1987. *The Johnstown Flood*. Simon & Schuster, New York.

NC DENR. 2009. Attachment A, B. Everett Jordan Lake water supply allocation request; Division of Water Resources Summary and Recommendation. North Carolina Department of Environment and Natural Resources, Raleigh, NC.

NRC. 1983. *Safety of Existing Dams, Evaluation and Improvement*. National Academy Press, Washington, DC.

NRC. 1992. *Restoration of Aquatic Ecosystems: Science, Technology, and Public Policy*. National Academy Press, Washington, DC.

Tennessee Valley Authority. Pickwick operating guide. Knoxville, TN. Available at: http://www.tva.gov/river/lakeinfo/op_guides/pickwick.htm.

Thornton, K.W. 1990. Perspectives on limnology. In K.W. Thornton, B.L. Kimmel, and F.E. Payne (eds), *Reservoir Limnology: Ecological Perspectives*, Chapter 1. Wiley, New York.

Threlkeld, S.T. 1990. Book review: *Reservoir Limnology* by Thornton et al. 1990. *Limnology and Oceanography* 35 (6), 1412–1413.

USACE. 1980. Engineering and design: Hydraulic design of reservoir outlet works, EM 1110-2-1602. U.S. Army Corps of Engineers, Washington, DC.

USACE. 1987. Management of water control systems, EM 1110-2-3600, CECW-EH-W. U.S. Army Corps of Engineers, Washington, DC.

USACE. 1997. Engineering and design—Hydrologic engineering requirements for reservoirs, EM 1110-2-1420. U.S. Army Corps of Engineers, Washington, DC.

USACE. 1998. HEC-5: Simulation of flood control and conservation systems. User's manual. Version 8.0. U.S. Army Corps of Engineers Hydrologic Engineering Center, Davis, CA.

USACE. 2001. Considerations for dam removal in ice-affected rivers, Ice Engineering, #27. Cold Regions Research and Engineering Laboratory, U.S. Army Corps of Engineers, Hanover, NH.

USACE. 2013. District wins 2012 Chief of Engineers Award for Excellence, *Intercom* 40, 11. U.S. Army Corps of Engineers Walla Walla District.

U.S. Army Engineer District. 1983. Rogue River Basin project. Portland, OR.

USBR. 1987. *Design of Small Dams*, 3rd ed. U.S. Government Printing Office, Washington, DC.

USBR. 2009. Trunnion friction radial gate failure. Chapter 16 in *Best Practices and Risk Methodology*. U.S. Department of the Interior, Bureau of Reclamation, Washington, D.C.

USBR. 2010. Design standards No. 14, appurtenant structures for dams (spillway and outlet works), Design Standards. U.S. Department of the Interior, Bureau of Reclamation.

Wei, C.Y. and J.E. Lindell. 2004. Hydraulic design of stilling basins and energy dissipaters. In L.W. Mays (ed.), *Hydraulic Design Handbook*, Chapter 18. McGraw-Hill, New York.

WGA. 2006. Water needs and strategies for a sustainable future. Western Governors Association, Denver, CO.

White, K.D. and J.N. Moore. 2002. Impacts of dam removal on riverine ice regime. *Journal of Cold Regions Engineering* 16 (1), 3–16.

World Commission on Dams. 2000. Dams and development: A new framework. Report of The World Commission on Dams. Earthscan Publications Ltd, London and Sterling.

Wurbs, R.A. and W.P. James. 2002. *Water Resources Engineering*. Prentice-Hall, Upper Saddle River, NJ.

11 Zones and Shapes in Lakes and Reservoirs

11.1 INTRODUCTION

Lakes and reservoirs vary widely in shape, size, and origin. Their size is often characterized by their depth, surface area, and volume. However, these metrics are not constant, but can vary both spatially and temporally, particularly for some reservoirs. For reservoirs, the dam and dam design, as translated into reservoir operations, impact the characteristics of, and transport within, the reservoir. Other metrics are needed to characterize the physical differences between and within lakes and reservoirs as they impact their physical, biological, and chemical characteristics.

11.2 LAKE ZONATION AND NOMENCLATURE

The size and physical characteristics of lakes and reservoirs impact their chemical and biological characteristics. Depth, for example, can influence a variety of other characteristics, such as the propensity to mix as a function of wind (Figure 11.1). Light and light availability is another characteristic that can be used to establish zones. Light striking the water surface decreases exponentially with depth as a function of the properties of light, water, and the materials dissolved or suspended in it (see Chapter 12). The zone that light penetrates is referred to as the *photic* zone, the bottom of which is usually taken as the depth at which the light is 1% of that striking the surface. The zone below the photic zone is referred to as the *aphotic* zone. The photic zone is obviously the zone in which plants can grow, and at the bottom of the photic zone is the compensation point, below which respiration dominates. So, in the dark and usually colder zone, also called the *profundal* zone, decomposition processes dominate, which result in reductions in oxygen and increases in reduced materials (e.g., reduced forms of iron and other metals).

If the depth of a lake or a reservoir, or portions of a lake or a reservoir, is shallow enough that the light penetrates to the bottom (the photic zone extends to the bottom), then rooted aquatic vegetation or bottom algae may survive. The zone in which this occurs is the *littoral* zone. The extent and the impact of the littoral zone vary widely. In many small and shallow lakes and reservoirs, the littoral zone represents a large fraction of their surface area. Steep-sided lakes, such as those formed in caldera, or reservoirs created in steep valleys may have little or no littoral zone. Littoral currents are usually parallel to the shoreline and are strongly influenced by the shoreline shape. The shallow depth and wind mixing often cause complete vertical mixing of the littoral zone. Aquatic plants may also impact transport. Transpiration by aquatic plants may represent a large water loss. Synthesis by aquatic plants often serves as a major source of organic materials. The littoral zone often provides an essential habitat for maintaining productive fisheries.

The offshore region where emergent plants cannot grow is the *pelagic* or *limnetic* zone. This is the zone where plants are dominated by floating or planktonic species and the bottom is free of vegetation. In this zone, there are also often vertical variations in temperatures and water quality.

Since light (heat) transfer is restricted to the surface (photic) zone, often a temperature gradient is established between a well-mixed, warm, productive, surface zone (the *epilimnion*) and the dark, colder *hypolimnion*. The zone of transition separating these two zones is the *metalimnion*.

None of these zones are static, and may vary seasonally and spatially. For example, neither the factors impacting light penetration nor the depth is constant. In natural lakes, the depth can vary

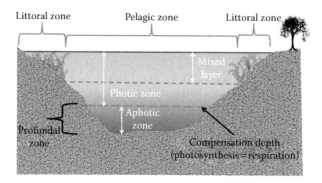

FIGURE 11.1   Lake zones.

with the season. In reservoirs, the regulation of the depth is part of the management process. For example, for flood control reservoirs, the water volume (and depth) may be reduced in the spring and fall in order to accommodate floodwaters. As a result, the location and the extent of the zones discussed previously will vary.

There are strong longitudinal as well as horizontal and vertical variations in the physical, chemical, and biological characteristics of reservoirs, particularly those constructed in drowned river valleys (Figure 11.2).

The riverine zone is dominated by the advective flow of a river, and its characteristics are closer to those of a river than a reservoir (Figure 11.2). This zone is relatively narrow and shallow and has a shorter water residence time. The riverine zone is characterized by higher levels of suspended solids and available nutrients, derived from upstream sources or from outside of the reservoir (allochthonous). This zone is usually well mixed and aerobic (Wetzel 2001). High particulate turbidity commonly reduces light penetration and limits primary production in this zone (Kirk 1985). The spatial extent of this zone is not constant, but varies with the magnitude of the inflows (Martin and McCutcheon 1999), so that for high inflow events the zone could extend far into the reservoir.

In the transition zone, the reservoir becomes wider and deeper, as compared to the riverine zone, and the velocities drop (Wetzel 2001). This is often the most dynamic region of a reservoir (Kennedy et al. 1985). Suspended solids often settle in this zone and productivity increases as decreased turbidity results in enhanced light penetration. In this zone, a shift occurs to an increasing percentage of total organic matter loading from phytoplankton and rooted vascular plants (autochthonous production; Kennedy et al. 1985). In many reservoirs, due to the increased productivity and enriched bottom sediments, seasonal anoxia begins in this zone (Martin and McCutcheon 1999).

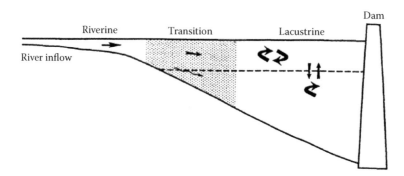

FIGURE 11.2   Longitudinal zones of a reservoir. (From Thornton, K.W., et al., *Proceedings of the Symposium on Surface Water Impoundments*, Volume 1, American Society of Civil Engineers, New York, 1981; From Wetzel, R.G., *Limnology*, 3rd ed., Academic Press, San Diego, CA, 2001. With permission.)

The "true lake" (or reservoir) zone is the lacustrine zone. This zone is wider and deeper and has a much longer residence time. The velocities are small and impacted by the dam operation, and wind mixing dominates the surface layers. The reservoir often becomes thermally stratified due to the density differences between the warm upper layer (epilimnion) and the colder bottom waters (hypolimnion). Productivity is usually limited to the surface photic zone (or well-mixed epilimnion) and is dominated by planktonic algae. Internal productivity (autochthonous) dominates and nutrient concentrations in the photic zone become depleted. In the deeper, profundal zone, oxygen is often seasonally depleted and there are high concentrations of reduced materials. As a result, the characteristics of the lacustrine zone become more similar to natural lake ecosystems (Wetzel 2001).

11.3 LAKE BASIN AND CHARACTERISTICS

Lake basin morphology influences lake hydrodynamics and lake responses to pollution. Some of the more commonly used metrics to assess lake basin morphological metrics are discussed next.

11.3.1 DEPTH AND ELEVATIONS

Depth impacts the lake or reservoir in terms of mixing, light penetration, and the propensity to stratify. Shallow lakes and reservoirs are also commonly more productive. In deep lakes, it is less likely that the bottom sediments would be impacted by mixing due to waves.

Some of the common depth metrics include the mean depth, the maximum depth, and the relative depth. The mean depth is typically estimated by the lake volume divided by the lake surface area, or by just averaging the depth measurements. The relative depth is the maximum depth divided by the mean diameter (Hutchinson 1957). All of these spatial averages are also commonly taken as time-averaged values, such as annual averaged mean depth.

Depths vary spatially and temporally. For a natural lake the maximum depth may be located near the lake's center, while for reservoirs the greatest depth is near the dam (Figure 11.2), or in "borrow pits" or areas near the dam from which earth was "borrowed" in the dam's construction.

Rather than the depth, the water-surface elevation, referenced from some datum, is most commonly measured, such as near the dam or reservoir operations. This measurement is also commonly referred to as the "gage height" (obtained from a gage) or the "stage height." The stage height is usually measured with a water-stage recorder, encased in a stilling well to reduce the impact of waves and other local disturbances on the measured stage or water-surface elevation.

In rivers and streams, there is typically some slope to the water surface. For some special systems of a constant shape (prismatic), the water-surface slope is the same as the bottom slope (steady-uniform flow).

In reservoirs, while the bottom (such as an old inundated river channel or "thalweg") slopes toward the dam (Figure 11.2), the water-surface elevation (or stage) becomes nearly constant. That is, the water surface becomes relatively flat. This is referred to as a backwater impact, and in hydraulics, the curve of the water surface is referred to as an "M1" curve (Martin and McCutcheon 1999; Sturm 2001). The surface of lakes is also relatively flat. Flat is, of course, a relative term, since the water surface varies as a function of waves, seiches, and other variations, as will be discussed in Chapter 14.

11.3.2 FLOWS AND VELOCITIES

Rivers are advective systems and their flows are commonly estimated using a rating curve, which is the relationship between water-surface elevations (and/or depths) and flows (see Chapter 4). Rating curves are developed at points in a river called control points where there is a unique relationship between depths and flows. These relationships are used by agencies such as the U.S. Geological Survey (USGS) to convert measured gage heights to flows, which are then reported.

However, in the backwater area of a reservoir, there is typically no such relationship. That is, there is no unique relationship between water-surface elevations and flows. For stations located within this backwater zone, agencies such as the USGS only report elevations. Since rating curves are not applicable and velocities are small and variable, measuring the flows within a reservoir is often problematic.

Upstream inflows to reservoirs are commonly measured, such as at the USGS gauging stations. However, rarely are all of the tributaries into a lake or reservoir monitored. In addition to the tributaries, there are often ungaged nonpoint source inflows, as well as precipitation falling directly onto the lake or reservoir. Since large portions of the inflow are commonly unknown, a common practice is to use the known outflow (discharge from the dam) and the known increase or decrease in the amount of water stored in the reservoir (based on changes in pool elevation) to back compute the inflow from the continuity equation (see Equation 11.1). Where possible, actual measurements of the upstream flows are used to verify the validity of the calculated reservoir inflow.

Discharges from reservoirs may be monitored, providing a detailed record of outflows. For lakes, the outflows are uncontrolled. However, for lakes modified by structures, the flow may be monitored at the structure or through ungaged weirs or spillways. In large reservoirs, such as those used for hydropower, there may be multiple outlet structures, such as multiple penstocks and turbines for hydropower, multiple-gated spillways for flood releases as well as emergency spillways, which may or may not be monitored. The discharges, such as releases through gates, may be based on measured water-surface elevations in comparison with the operational guide or rule curve for the reservoir, as discussed in Chapter 10.

11.3.3 BATHYMETRY

The depth of a lake or a reservoir is not constant, but varies spatially and temporally. The depth and its variations are important for navigation, fisheries management, recreation, and other purposes and are commonly incorporated into bathymetric maps. Bathymetric maps are analogous to terrestrial topographic maps, which show contours of equal elevations. But, bathymetric maps are typically based on contours of equal depths. As with topographic maps, the closer the lines are, the more rapid the changes that occur (in depth or elevation). Bathymetric maps are typically estimated from sedimentation or sounding (bathymetry or hydrographic) surveys. The U.S. Army Corps of Engineers (the Corps) (USACE 1995, 2002) provides guidance on methods, accuracy standards, and quality control criteria. These maps are available from a variety of agencies and organizations. Bathymetric maps for many Florida lakes (e.g., Figure 11.3) are available from Florida Lake Watch (University of Florida; http://lakewatch.ifas.ufl.edu/).

11.3.4 SURFACE AREA AND HYPSOGRAPHIC CURVES

Lake surface areas are used not only to indicate the size of a lake or a waterbody, but also to estimate the potential impacts of wind, waves, evaporation, precipitation, gas exchange, solar heating, and other impacts that vary as a function of the surface area. Changes in the surface area are also important in lake management, such as to control littoral areas.

The surface area varies as a function of depth (or surface elevation) and the relationship is commonly expressed using a hypsographic curve or a plot of the elevation or depth versus the surface area. Note that, depending on the purpose, the area may be a discrete surface area; that is, the plan area at a particular depth, such as to estimate wind effects, etc. Alternatively, the area may represent the benthic area. The two areas may be similar, but the differences are a function of the bottom slope. Particularly for lakes, the curve may also be expressed as a percentage of the mean surface area (Wetzel 2001).

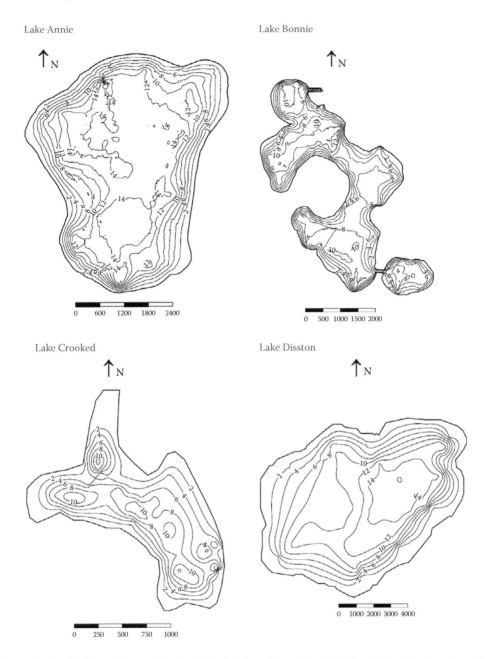

FIGURE 11.3 Bathymetric maps for four Florida lakes. (From Florida LakeWatch, A beginner's guide to water management—Lake morphometry, Information Circular 104, Florida LakeWatch, University of Florida, Gainesville, FL, 2001.)

The shape of the hypsographic curve can also be used for a comparison of lakes, such as to aid in determining the dominant processes impacting their biological characteristics. Of course, in a lake that is perfectly rectangular, the surface area would be constant. As illustrated in Figure 11.4, the hypsographic curve for a cone would be convex, and for a truncated cone, it would be relatively flat in comparison with a circular cone. In contrast, a hemispherical or a bowl-shaped lake would be expected to have a more convex hypsographic curve.

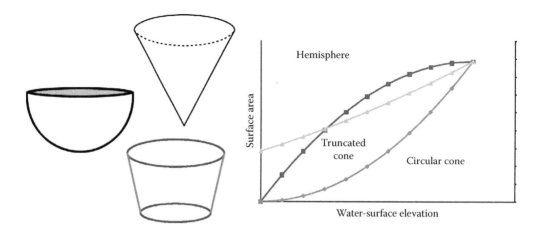

FIGURE 11.4 Examples of the shapes of hypsographic curves, surface areas versus depth or elevation, for representative geometric shapes.

11.3.5 Volume–Depth Curves

The volume of a lake or a reservoir is also a basic distinguishing characteristic that affects the quantities of materials in the lake, the impacts of sedimentation, the residence time, and the water available for given management purposes.

Similar to area, volume is a function of depth and the relationship is expressed as a depth–volume curve, similar to a hypsographic curve. The shape of the curve can provide information about the general shape of the lake or reservoir, as illustrated in Figure 11.5.

The management of reservoirs is usually based on their storage capacity and inflows in order to manage outflows, based on the relationship:

$$\frac{\Delta S}{\Delta t} = I - O \tag{11.1}$$

where
  S indicates storage (volume)
  t is time
  I is the sum of the inflows
  O is the sum of the outflows

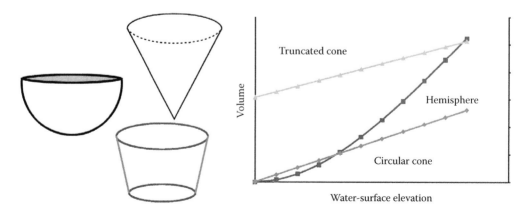

FIGURE 11.5 Comparison of depth–volume curves for three geometric shapes.

The outflows are typically measured and may be expressed as a function of the water level (h) over some structure ($O = f(h)$). The storage is expressed as a function of the depth or water-surface elevation ($S = f(h)$), as in the depth–volume curve. Measured changes in the water-surface elevation over time can then be translated to changes in the volume over time using the depth–volume curve. Given inflows such as during a flood and the depth–volume curve, the relationship can also be used to estimate the outflows needed to obtain a target water-surface elevation, such as specified in the reservoir rule or the guide curve (see Chapter 11). The depth–volume curves are also known as storage capacity versus elevation curves, an example of which is provided in Figure 11.6.

11.3.6 Shoreline Development Ratio

Another commonly used physical metric is the shoreline development ratio, which provides an index for the potential interactions between the shoreline, the shoreline development, and the lake or reservoir.

The shoreline development ratio (D_L) is the ratio of the existing shoreline length to the circumference of a circle of area equal to that of the lake:

$$D_L = \frac{L}{2\sqrt{\pi A}} \tag{11.2}$$

so that if $D_L = 1$, the lake is a circle. Figure 11.7 illustrates the impact of the shoreline development ratio, where, obviously, Lake B has a greater shoreline length. Both lakes have the same surface area, so that alone would not be enough to differentiate between the two lakes. However, Lake B, with the larger L (and $D_L > 1$) would have more of an interface between the water and the surrounding land (i.e., coves and peninsulas), which often translates into more habitats for fish, birds, and other wildlife (Florida LakeWatch 2001). An increased shoreline is also a factor impacting human population distributions around lakes and reservoirs, and human disturbance. The population distribution around the Great Lakes is illustrated in Figure 11.8.

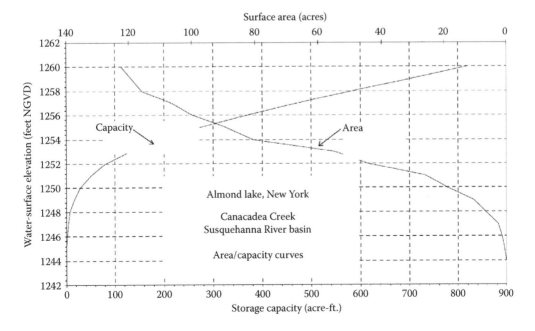

FIGURE 11.6   Capacity curves for Almond Lake, New York. (Courtesy of the U.S. Army Corps of Engineers.)

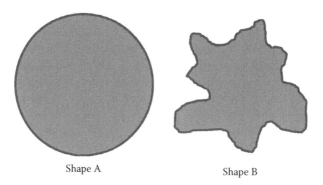

Shape A   Shape B

FIGURE 11.7 Comparison of lake shoreline indices. (Modified from Florida LakeWatch, A beginner's guide to water management—Lake morphometry, Information Circular 104, Florida LakeWatch, University of Florida, Gainesville, FL, 2001.)

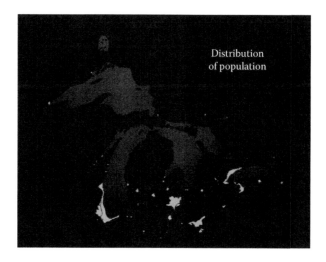

FIGURE 11.8 Population distribution in the Great Lakes. (Courtesy of the USEPA, *Great Lakes Atlas*.)

11.3.7 PERCENT LITTORAL ZONE

Generally, the greater the percentage of the lake that is within the littoral zone, the greater the influence of that zone on the lake. The percentage of the littoral zone impacts the abundance and diversity of plants, benthic organisms, and fisheries. Some lakes and reservoirs are almost 100% littoral (they lack a lacustrine zone).

11.3.8 SURFACE AREA AND WIND EXPOSURE (FETCH)

The impact of wind varies with the unobstructed distance that wind can travel over water before intersecting a landmass, which is known as fetch. The fetch can impact wave formation and wind mixing, and seich formation. The greater the fetch, usually the greater the wave height and the deeper the wave energy and the resulting mixing extend below the water surface. Impacts include recreational uses (e.g., boating), sediments, and sediment resuspension. The fetch, and its impact on waves, can also impact the distribution of plants and benthic organisms. For example, a long fetch and the resulting increased wave energy may preclude emergent vegetation. Fetch is a common metric used to evaluate shoreline stability, as described in the Corps' "Shore protection manual" (CERC 1984).

11.3.9 SEDIMENT TRAP EFFICIENCY

Lakes and, in particular, reservoirs tend to trap sediments. There is usually some storage allowance for sedimentation, and surveys are carried out periodically to estimate how much sedimentation has occurred (see USACE 1995). The sediment trap efficiency is the difference in the sediment yield (sediment weight) of the reservoir inflow minus the yield of the outflow, divided by the yield of the inflow. The greater the trap efficiency, the greater the sedimentation.

Reservoirs are typically efficient sediment traps, and trap efficiency is a design consideration. The trap efficiency of reservoirs typically decreases with increasing age, as the capacity of the reservoir decreases. The trap efficiency is impacted by the type of sediments, the hydraulic characteristics of the reservoir, and how the reservoir is operated (USACE 1995).

The U.S. Bureau of Reclamation (USBR 2006) describes a variety of methods for estimating sediment yields and trap efficiency. Basically, reservoirs will eventually fill and if the inflow yield is large, the useful life of a reservoir may be short. For example, Linsley and Franzini (1979) described a small reservoir on the Solomon River near Osborne, Kansas, that filled with sediment during the first year of operation. The accumulated sediments not only impact the life of the reservoir, but they also pose problems for the restoration or decommissioning of the reservoir. Methods are available to reduce the rate of reservoir sedimentation, to increase the life of the reservoir and prolong its use, as discussed by USBR (2006).

11.3.10 RETENTION OR RESIDENCE TIME

Another characteristic or metric used to characterize lakes and reservoirs is their retention or residence time, basically how long the water stays in the waterbody (on average).

The residence time is often estimated by the average volume divided by the average outflow. Lakes or reservoirs with long residence times tend to trap sediments and contaminants more efficiently, have greater internal (autochthonous) productivity, and are slower to respond to changes in their management or operations.

Lakes tend to have longer residence times than most reservoirs (Figure 11.9) and, in some cases, these residence times may be very long. For example, the residence time of Lake Superior is approximately 191 years and that of Lake Michigan is 99 years. Lake Erie, on the other hand, has a relatively short residence time of approximately 2.6 years. According to the Tahoe Environmental Research Center (TERC) at the University of California, Davis, Lake Tahoe has a maximum depth of greater than 500 m and a relatively small watershed with an area of 800 km^2 as compared to a surface area of 500 km^2. With a large volume and relatively small flows, Lake Tahoe has a remarkable residence time of 650–700 years.

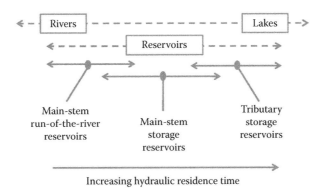

FIGURE 11.9 Comparison of hydraulic residence times. (Modified from Kimmel, B.L. and Groeger, A.W., *Lake and Reservoir Management*, U.S. Environmental Protection Agency, Washington, DC, 1984.)

TABLE 11.1

Comparison of Three Corps of Engineers Reservoirs

Characteristics	DeGray Lake	Red Rock Lake	West Point Lake
Impoundment type	Tributary	Main stem	Main stem
Major tributary	Caddo River	Des Moines River	Chattahoochee River
Volume (10^6 m^3)	808	78	746
Surface area (km^2)	54	26	105
Length (km)	32	12	53
Mean depth (m)	14.9	3.1	7.1
Maximum depth (m)	60	11	31
Annual hydraulic residence time (year)	2.06	0.05	0.13
High-flow hydraulic residence time (year)	1.11	0.02	0.09
Low-flow hydraulic residence time (year)	3.39	0.12	0.18

Source: Modified from Kennedy, R.H., *Lake and Reservoir Management*, U.S. Environmental Protection Agency, Washington, DC, 1984.

The residence time of reservoirs is variable. Main-stem, run-of-the-river reservoirs have little storage and the residence time may be days to weeks. Main-stem storage reservoirs have residence times that are often weeks to months, while larger reservoirs may have a residence time from weeks to months to years. For example, Kennedy (1984) compared three Corps' reservoirs, as shown in Table 11.1. The residence time of the tributary reservoir (DeGray Lake) averaged 2.1 years (range of 1.1–3.4 years), while the main-stem reservoirs were typically much less than 1 year.

The residence or retention times are not temporally constant, but vary with volumes and flows. A common practice is to use mathematical models of hydrodynamics and transport in lakes or reservoirs to estimate the residence time (Martin and McCutcheon 1999). Two common methods are simulating a dye tracer and simulating the water age. For the dye tracer, commonly a concentration is specified initially everywhere within the reservoir with all inflows having concentrations of zero. The residence time is computed by how long it takes for the dye to be flushed out, or reduced to some fraction of its initial concentration. For the water age, a nonconservative tracer is simulated, which accumulates at a positive (zeroth-order) rate of 1.0 day^{-1}. In essence, the water ages at that rate unless it is flushed out. The water age allows the visualization of areas of a lake with little flushing, density inflows, and other features that may then impact the reservoir's quality and management. Loga-Karpinska et al. (2003) and Camacho and Martin (2013) provide discussions of the methods used to estimate residence times.

11.3.11 DRAINAGE AREAS

The characteristics of lakes and reservoirs are also largely impacted by their watersheds. Thornton (1984) compared the Corps' reservoirs with natural lakes that were surveyed as part of the U.S. EPA's national eutrophication survey (1972–1975). This comparison indicated that, in general, reservoirs have greater drainage and surface areas, drainage area/surface area ratios, mean and maximum depths, shoreline development ratios, and aerial water loads (Table 11.2).

The characteristics of the watershed also impact the characteristics of the lake or reservoir. One example is the differences in lakes in urban as opposed to rural environments. Schueler and Simpson (2002) used the following criteria to distinguish urban lakes:

- They tend to be rather small, and generally have a surface area of 10 mi^2 or less (this excludes larger lakes).
- They tend to be shallow, with an average depth of 20 ft. or less.

TABLE 11.2

Comparison of Lakes and Reservoirs

Characteristics	Natural Lakes (N = 309)	CE Reservoirs (N = 107)	Probability Means Are Equal
Drainage area (km^2)	222	3228	<0.0001
Surface area (km^2)	5.6	34.5	<0.0001
Drainage/surface area (DA/SA)	33	93	<0.0001
Mean depth (m)	4.5	6.9	<0.0001
Maximum depth (m)	10.7	19.8	<0.0001
Shoreline development ratio	2.9 (N = 34)[a]	9.0 (N = 179)[b]	<0.001
Areal water load (m yr^{-1})	6.5	19	<0.0001
Mean hydraulic residence time (year)	0.74	0.37	<0.0001

Source: Taken from Thornton, K.W., *Lake and Reservoir Management*, U.S. Environmental Protection Agency, Washington, DC, 1984; (a) Hutchinson, G.E., *A Treatise on Limnology*, Volume 1, Geography, Physics and Chemistry, Wiley, New York, 1957; (b) Leidy, G.R. and Jenkins, R.M., The development of fishery compartments and population rate coefficients for use in reservoir ecosystem modeling, prepared for Office, Chief of Engineers, U.S. Army, Washington, DC, 1977.

- They have a watershed area/drainage area ratio of at least 10:1, meaning that their watersheds exert a strong influence on the lake.
- The lake watershed must contain at least 5% impervious cover as an overall index of development.
- Whether natural or man-made, the lake must be managed for recreation, water supply, flood control, or some other direct human use.
- Several types of lakes with a unique hydrology or nutrient cycling are excluded, such as solution lakes that are strongly influenced by groundwater, the rare nitrogen-limited lakes, saline lakes, and playa lakes.

Schueler and Simpson (2002) suggested that because urban lakes are sufficiently different, and the impact of watershed development on lake quality so pervasive, these lakes should be treated as a separate group from other lakes. For example, urban lakes receive higher phosphorus loads, resulting in higher rates of algal growth bringing about cultural eutrophication or extreme eutrophication (hypereutrophication). As an example, Schueler and Simpson (2002, cited in USEPA 1986) reported that about half of all U.S. lakes are eutrophic or hypereutrophic. In contrast, in a survey of 3700 urban lakes evaluated by the U.S. EPA (USEPA 1980), the percentage of eutrophic or hypereutrophic lakes exceeded 80%.

REFERENCES

Bonvechio, T.F. 2003. Relations between hydrological variables and year-class strength of sportfish in eight Florida waterbodies. MS thesis, University of Florida.

Camacho, R.A. and J.L. Martin. 2013. Hydrodynamic modeling of first order transport time scales in the St. Louis Bay estuary, Mississippi. *ASCE Journal of Environmental Engineering* 139 (3), 317–331.

Coastal Engineering Research Center (CERC). 1984. Shore protection manual. U.S. Army Corps of Engineers, Washington, DC.

Florida LakeWatch. 2001. A beginner's guide to water management—Lake morphometry, Information Circular 104. Florida LakeWatch, University of Florida, Gainesville, FL.

Hutchinson, G.E. 1957. *A Treatise on Limnology.* Volume 1. Geography, Physics and Chemistry. Wiley, New York.

Kennedy, R.H. 1984. Lake–river interactions: Implications for nutrient dynamics in reservoirs. In *Lake and Reservoir Management*, EPA 440/5/84-001. U.S. Environmental Protection Agency, Washington, DC, pp. 266–271.

Kennedy, R.H., K.W. Thornton, and D.E. Ford. 1985. Characterization of the reservoir ecosystem. In G. Gunnison (ed.), *Microbial Processes in Reservoirs*. Dr. W. Junk Publishers, Boston, MA, pp. 27–38.

Kimmel, B.L. and A.W. Groeger. 1984. Factors controlling primary productivity in lakes and reservoirs: A perspective. In *Lake and Reservoir Management*, EPA 440/5/84-001. U.S. Environmental Protection Agency, Washington, DC, pp. 277–281.

Kirk, J.T.O. 1985. Effects of suspensoids (turbidity) on penetration of solar radiation in aquatic ecosystems. *Hydrobiologia* 125, 195–208.

Leidy, G.R. and R.M. Jenkins. 1977. The development of fishery compartments and population rate coefficients for use in reservoir ecosystem modeling. Contract Report Y-77-1, prepared for Office, Chief of Engineers, U.S. Army, Washington, DC.

Linsley, R.K. and J.B. Franzini (1979). *Water-Resources Engineering*. McGraw-Hill, New York.

Loga-Karpinska, M., K. Duwe, C. Guilbaud, M. O'Hare, U. Lemmin, L. Umlauf, E. Hollan, et al. 2003. D24: Realistic residence times studies. Integrated water resource management for important deep European lakes and their catchment areas. Eurolakes. Technical Report. FP5-Contract No.: EVK1-CT1999-00004.

Martin, J.L. and S.C. McCutcheon. 1999. *Hydrodynamics and Transport for Water Quality Modeling*. Lewis Publishers, Boca Raton, FL.

Schueler, T. and J. Simpson. 2002. Why urban lakes are different. Courtesy of the Center for Watershed Protection, as part of a Storm Water Design to Protect Watersheds Workshop held by EPA Region 10 and CWP in Coeur d'Alene, Idaho.

Sturm, T.W. 2001. *Open Channel Hydraulics*. McGraw-Hill, New York.

Thornton, K.W., R. H. Kennedy, J. H. Carroll, W. W. Walker, R. C. Gunkel, and S. Ashby. 1981. Reservoir sedimentation and water quality—A heuristic model. In H. Stefan (ed.), *Proceedings of the Symposium on Surface Water Impoundments*, Volume 1. American Society of Civil Engineers, New York, pp. 654–664.

Thornton, K.W. 1984. Regional comparisons of lakes and reservoirs: Geology, climatology and morphology. In *Lake and Reservoir Management*, EPA 440/5/84-001. U.S. Environmental Protection Agency, Washington, DC, pp. 261–265.

Thornton, K.W., B.L. Kimmel, and F.E. Payne. 1990. *Reservoir Limnology: Ecological Perspectives by Basin Effects*. Wiley-Interscience, New York.

USACE. 1995 (revised). Sedimentation investigations for rivers and reservoirs, EM 1110-2-4000. U.S. Army Corps of Engineers, Washington, DC.

USACE. 2002. Engineering and design—Hydrographic surveying, EM 1110-2-1003. U.S. Army Corps of Engineers, Washington, DC.

USBR. 2006. Erosion and sedimentation manual. U.S. Department of the Interior Bureau of Reclamation, Denver, CO.

USEPA. 1980. Our nation's lakes, Office of Water, EPA-440-5-80-009. U.S. Environmental Protection Agency, Washington, DC.

USEPA. 1986. Quality criteria for water—1986, Office of Water, EPA-440-5-86-001. U.S. Environmental Protection Agency, Washington, DC.

USEPA. 1990. *The Lake and Reservoir Restoration Guidance Manual*, 2nd ed., EPS-440/4-90-006. U.S. Environmental Protection Agency, Washington, DC.

Wetzel, R.G. 2001. *Limnology*, 3rd ed. Academic Press, San Diego, CA.

12 Light and Heat in Lakes and Reservoirs

12.1 DISTRIBUTION OF LIGHT AND HEAT EXCHANGE

Our primary source of light is radiation, which comes from the sun. At specific locations on the earth, the magnitude of the incoming solar radiation varies with the altitude of the sun, which varies daily and seasonally (Figures 12.1 and 12.2). The summer solstice or midsummer occurs on the longest day and shortest night of the year, when the earth's axial tilt is most inclined toward the sun. The summer solstice occurs in June in the Northern Hemisphere and in December in the Southern Hemisphere. Conversely, the winter solstice occurs when the earth's axial tilt is most inclined away from the sun, and corresponds to the shortest day and longest night of the year. The range of seasonal variations in solar radiation varies with the earth's latitude.

Solar radiation also varies daily (Figure 12.3). The magnitude of peak solar radiation as well as day length varies seasonally (Figure 12.4). The magnitude of the radiation reaching the water's surface at a particular location on the earth and the time of year is also reduced by the atmospheric absorption and reflection from clouds, and, of the remaining radiation, some is reflected from the water's surface (Figures 12.5 and 12.6).

The intensity of solar radiation striking the earth's surface varies in wavelengths from ultraviolet (UV) through the visible to the infrared. Some of the lower UV wavelengths are completely absorbed by the atmospheric ozone layer, so organisms on Earth have developed no tolerance to it. Similarly, the atmosphere absorbs some of the infrared wavelengths but it is more transparent to visible light (wavelengths from 380 and 760 nm; Figure 12.7).

Of the visible light striking the water's surface, some of it penetrates into the water and some of it is reflected. The wave energy of the reflected light typically decreases, becoming longer-wavelength thermal energy. This longwave radiation can be reflected back to the water's surface by clouds.

Both longwave and shortwave radiation are important contributors to the heating of lakes and reservoirs (see Martin and McCutcheon 1999). Also important is the portion of the shortwave, visible spectrum that photosynthetic organisms are able to use in the process of photosynthesis, referred to as photosynthetically active radiation or photosynthetically available radiation (PAR). Plant pigments such as chlorophyll or accessory pigments such as carotenoids absorb the radiation. Chlorophyll, for example, is most efficient in capturing red and blue light (at 665 and 465 nm; Figure 12.8).

12.2 LIGHT AND WATER INTERACTIONS

Depending on the color and turbidity of the water, the visible light striking the water's surface can penetrate to considerable depths. The depth to which light penetrates the water affects the thermal characteristics of lakes and reservoirs as well as the water quality, for example, by influencing the distribution of aquatic plants. An understanding of the light environment and the factors affecting it is a critical component of lake management.

FIGURE 12.1   Sunrise at Hammond Bay, Lake Huron, MI. (Courtesy of the USEPA; photograph by Carol Y. Swinehart, Michigan Sea Grant Extension.)

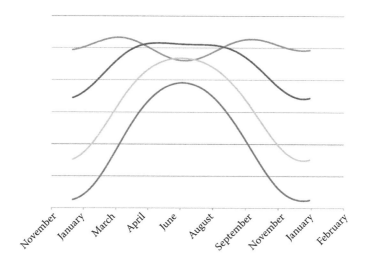

FIGURE 12.2   Annual variations in solar radiation.

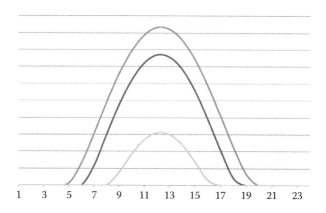

FIGURE 12.3   Daily solar radiation (at latitude 45° north).

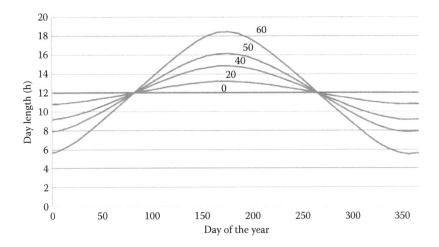

FIGURE 12.4 Seasonal variations in day length as a function of latitude (degrees).

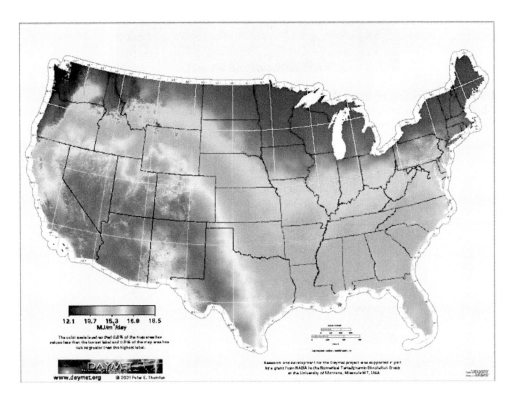

FIGURE 12.5 Eighteen-year annual mean daily total shortwave radiation for the United States (MJ m^{-2} day^{-1}). (Courtesy of the National Center for Atmospheric Research, DAYMET program.)

The relationship between light penetration and water quality has been known for a long time. In 1865, Angelo Secchi, an Italian astrophysicist commissioned to measure water transparency in the Mediterranean Sea, developed a method to measure light penetration, or the transparency of water. He invented a device that consists of a disk with alternating black and white quadrants (Figure 12.9). The disk is mounted on a pole or a line and is then lowered through the water column until the pattern on the disk is no longer visible. The depth at which this occurs is known as the Secchi depth. This device is still in common use today.

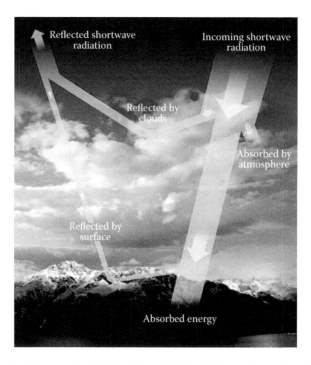

FIGURE 12.6 Incoming (shortwave) radiation. (From NASA, "The earth's radiation budget" science mission directorate, National Aeronautics and Space Administration, 2010. Available at http://missionscience. nasa.gov/ems/13_radiationbudget.html. With permission.)

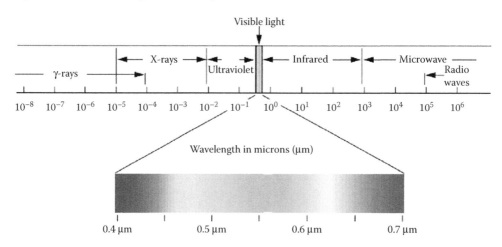

FIGURE 12.7 Solar radiation spectrum. (From Malm, W.C., Introduction to visibility, National Park Service Cooperative Institute for Research in the Atmosphere (CIRA), Colorado State University, 1999. With permission.)

The distribution of light in water is also measured using underwater photometers (or light meters). These devices measure light at the surface and at depth as they are lowered through the water column.

Light typically decreases exponentially with depth. The relationship of light at the surface (H_0) to light remaining at a given depth (H) can be described using Beer's law, also known as the Beer–Lambert law or Lambert–Beer law, named after Johann Heinrich Lambert (1728–1777) and August Beer (1825–1863), and is expressed as:

$$H = H_0\, e^{-K_e Z} \tag{12.1}$$

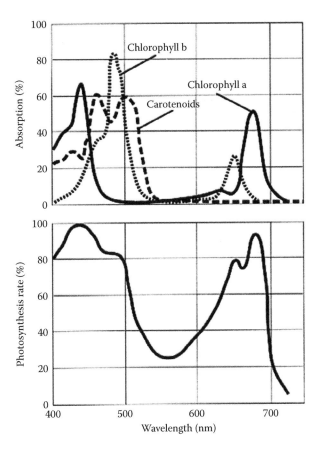

FIGURE 12.8 Variation in absorption and the rates of photosynthesis as a function of the wavelength of light. (From Wikimedia Commons.)

FIGURE 12.9 Pietro Angelo Secchi, 1818–1878 and his Secchi disk. (From Wikimedia Commons.)

where Z is the downward distance from the water surface and K_e is the light extinction coefficient. The light extinction coefficient can be calculated from the measured light over depth based on Beer's law. Typical extinction coefficients, as well as Secchi depths, for a number of lakes are listed in Table 12.1.

The extinction coefficient for reservoirs is frequently greater than that for natural lakes. For example, the light extinction coefficients for Lake Houston, Texas, range from about 2.5 to 8.0 m^{-1} (Lee and Rast 1997) compared to 0.06–0.12 m^{-1} for Crater Lake, Oregon (Table 12.1).

The penetration of light through the water column is affected by the materials dissolved or suspended in it. Pure water has a light extinction coefficient of about 0.03–0.04 m^{-1} (Lorenzen 1972; Verdiun et al. 1976). Other factors that may increase light extinction include pigments or organic acids dissolved in water, suspended solids, and phytoplankton (Martin and McCutcheon 1999).

The greater the extinction coefficient or the lower the Secchi depth, the less the light will penetrate. Higher extinction coefficients may be due to the naturally higher color or turbidity associated with southern waterbodies, but they may also be the result of pollution. For example, excess nutrient loadings result in excess growth of aquatic plants.

As such, the light extinction impacts the depth of the euphotic zone, or the zone in which plants can live. Recall from Chapter 11 that the bottom of the photic zone is usually taken to be that depth where the radiation is 1% of that at the surface ($H/H_0 = 0.01$), so that the depth of the photic zone will be $Z = 4.61/K_e$. Although the Secchi depth is not a direct measure of the photic zone depth, a common rule of thumb is that the depth of the photic zone is about twice the Secchi depth. This is

TABLE 12.1
Representative Values of Extinction Coefficients, Secchi Depths, and Photic Zone Depths

Lake	Ke (m⁻¹)	Secchi Depth (m)	Depth of Photic Zone (m)	Description
Crater Lake (OR)	0.06–0.12	25–45	>120	Clear, sky blue ultraoligotrophic lake
Lake Tahoe (CA/NV)	0.12	40	90–136	As above but with decreasing clarity since 1960s due to watershed overdevelopment
Lake Superior (blue water)	0.13	15–20	46–60	Ultraoligotrophic, most oligotrophic of the Laurentian Great Lakes
Lake Superior (green water near Duluth)	0.3	5–12	20–30	Western arm near Duluth and St. Louis River and harbor inputs
St. Louis River (Duluth-Superior Harbor)	4.21	0.7	>5	Brown (bog) stained from river plus high suspended sediments
Lake Michigan	0.19–0.24	?	19–31	Mesooligotrophic
Lake Huron	0.1–0.5	?	25–31	Mesooligotrophic
Lake Erie	0.2–1.2	2–10	12–26	Eutrophic (clarity improving recently due to zebra mussels)
Lake Ontario	0.15–1.2	?	12–29	Mesotrophic
Lake Baikal (Siberia)	0.2	5–40	15–75	Oligotrophic
Grindstone Lake (Pine County, MN)	0.82	3–6	8–20	Mesotrophic, water is fairly stained or colored
Ice Lake (Itasca County, MN)	0.83	2–5	6–15	Mesotrophic
Lake Minnetonka				
Halsted Bay (Hennepin County, MN)	2.9	0.5	<2	Eutrophic

Source: USEPA, Watershed Academy Web, Understanding lake ecology, Available at: http://cfpub.epa.gov/watertrain.

based on the assumption that, at the Secchi depth, light is about 10% of the light at the surface $(\ln(0.1)/\ln(0.01) = 2)$. Representative photic zone depths are illustrated in Table 12.1.

Since the light extinction coefficient increases and the depth of the photic zone decreases as a result of excess plant growth such as by phytoplankton, the light extinction coefficient has commonly been used to characterize the trophic status of lakes and reservoirs (see Table 12.1). Carlson (1977) used the Secchi depth as a basis for his trophic status of lakes, as will be discussed in Chapter 16.

Common trophic levels include eutrophy, mesotropy, and oligotrophy. Eutrophic lakes and reservoirs are those that receive excess nutrients and hence are highly productive and have excess plant growth. Oligotrophic lakes are very low in nutrients and productivity. Mesotrophic lakes are in between.

The light extinction coefficient ranges for lakes of varying trophic status provided by Likens (1975) are ultraoligotrophic (0.03–0.8 m^{-1}), oligotrophic (0.05–1.0 m^{-1}), mesotrophic (0.1–2.0 m^{-1}), and eutrophic (0.5–4.0 m^{-1}). Forsberg and Ryding (1980) based their classification on the Secchi depth, with oligotrophic lakes having Secchi depths greater than 13 ft., mesotrophic lakes having Secchi depths ranging from 8 to 13 ft., eutrophic lakes having Secchi depths from 3 to 8 ft., and hypereutrophic lakes having Secchi depths less than 3 ft.

The light extinctions illustrated in Table 12.1 refer to total light. However, the penetration of light through the water column varies with the wavelength so that the spectral distribution varies with the depth. Light underwater is diminished by absorption and scattering (Kirk 1994). While absorption removes light, scattering increases the probability that light will be absorbed by increasing the path length. Since different wavelengths are absorbed differently, each wavelength would have its own extinction coefficient. The sorption of longwave or thermal (infrared) radiation is very rapid, and about 53% of the total light energy is converted to that in the first meter of water (Wetzel and Likens 2000). Of the visible light, depending on the amount of scattering and absorption, the longer wavelengths, such as red, yellow, and orange, are absorbed first (resulting in large extinction coefficients) while the shorter wavelengths (violet, blue, and green) can penetrate further (smaller coefficients) with blue penetrating the deepest. Chlorophyll, a photosynthetic plant pigment, is most efficient in capturing red and blue light (at 665 and 465 nm, Figure 12.4). Relatively pure water will appear blue, but depending on the dissolved and particulate materials in the water affecting scattering and sorption, water may appear more green or brown in color.

12.3 SURFACE HEAT BALANCE

The exchange of heat between the water surface and the overlying atmosphere (surface heat exchange) is primarily the result of the five processes depicted in Figure 12.10. The first two processes are the absorption of the longwave and shortwave radiation discussed earlier. Solar energy may be absorbed in the water column. If light penetrates to the bottom, that energy may then be absorbed by the bottom substrate and later reemitted. The remaining processes include longwave back radiation from the water's surface, conduction and convection, and evaporation, all of which are impacted by the water temperature.

The longwave radiation from the water is basically heat emission from a warm object. The radiation emitted is proportional to the absolute temperature of water to the fourth power (blackbody radiation).

Conduction is the transfer of heat due to the differences in temperatures between the water and the overlying air, and convection is the exchange of heat due to the movement of fluids. For surface exchanges, the magnitude of both is proportional to the temperature gradient and, at equilibrium, the water temperature would be equal to the air temperature (no gradient). The transfer of heat is also enhanced by wind. Convection and diffusion can also transport heat from the warmed surface layer to lower layers. Where light penetrates to the bottom, the heat may be absorbed and reemitted.

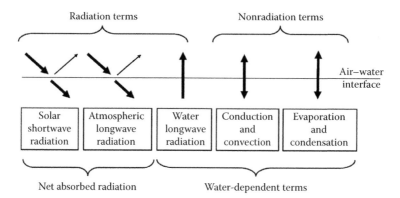

FIGURE 12.10 Components of the heat balance. (From Chapra, S.C., Pelletier, G.J., and Tao, H., QUAL2K: A modeling framework for simulating river and stream water quality, Civil and Environmental Engineering Department, Tufts University, Medford, MA, 2007. With permission.)

Similarly, evaporation is based on the gradient of the water vapor pressure between water and air. So, if there is no gradient (the air is saturated, such as at 100% relative humidity), there is no heat transfer. Evaporation is also enhanced by wind (Martin and McCutcheon 1999). The typical magnitudes of each of these processes at midlatitudes are illustrated in Figure 12.11.

12.4 WATER DENSITY

One of the basic properties of water is its density. Because of its unique properties, water is most dense at 4°C and then decreases as it heats or cools (Figure 12.12). The density of water is also influenced by the materials dissolved or suspended in it. For example, seawater is about 2.5% denser than freshwater.

12.5 LAKE STRATIFICATION

Water quality is profoundly affected by the tendency of many lakes not to mix completely in the vertical dimension, at least during some portions of the year. The degree of mixing is often based on

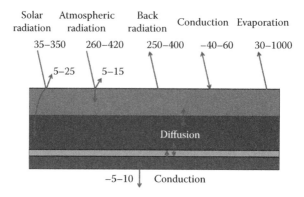

FIGURE 12.11 Heat exchange fluxes, typical midlatitude daily average (W m^{-2}). (From Wool, T.A., Ambrose, R.B., and Martin, J.L., WASP7 temperature and fecal coliform: Model theory and user's guide; supplement to water quality analysis simulation program (WASP) user documentation, U.S. EPA, Region 4 Water Management Division, Atlanta, GA, 2008; USEPA WASP Workshop.)

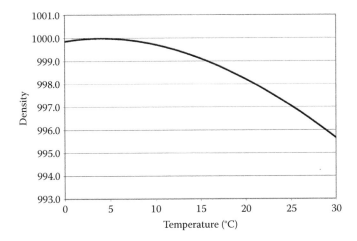

FIGURE 12.12 Water density variations with temperature.

the available energy for mixing, such as from wind, and vertical variations in density. These density differences result in the vertical stratification of water. The density differences provide a barrier that resists mixing, thus the greater the density differences the more difficult it is to mix the strata.

In freshwater the density differences are primarily impacted by the water temperatures (see Figure 12.20). In saline lakes, differences in both the temperature and dissolved solids cause stratification. Density differences can also result from differences in the concentration of suspended solids.

An idealized stratification cycle begins when heat enters the water surface faster than it is mixed over the depth, usually beginning in spring or early summer. As the surface layer heats, it becomes less dense than the waters below it. As the surface layer continues to heat, the density difference becomes greater, forming a barrier that further inhibits vertical mixing. Wind mixing is usually sufficient to mix the upper portion of the water column, forming an upper, warm, and well-mixed layer, the *epilimnion* (Figure 12.13). This layer overlays a deeper, colder, and denser layer, the *hypolimnion*. These two layers are separated by a zone of transition, the *metalimnion*. In the metalimnion

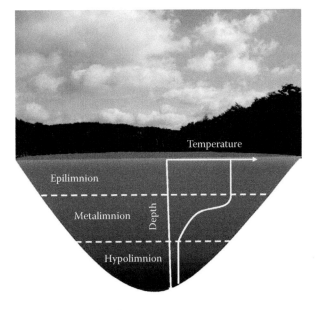

FIGURE 12.13 Idealized summer stratification.

or *thermocline*, the maximum vertical temperature gradients occur. The thermocline is commonly defined where a minimum of a 1°C temperature change per meter of depth occurs.

The depth of the thermocline at any time is controlled by the seasonal climate, the occurrence of storms, water temperature, water depth, lake bathymetry, the strength of inflow and outflow currents, and other factors covered in more detail by Hutchinson (1957), Wetzel (2001), Chapra (2008), and Ford and Johnson (1981).

In the fall, changes in the surface heat budget (a reduction in solar heating and air temperatures, etc.) result in cooling of the surface waters. As the epilimnion cools, the density gradient lessens and wind mixing penetrates further down the water column. Also, as the surface water cools and becomes denser than the waters immediately beneath it, the water sinks causing convective mixing. The thermocline erodes and becomes deeper and ultimately the entire lake mixes (Figure 12.14). This is referred to as the fall overturn. Following overturn, the entire lake or reservoir may continue to cool.

As cooling continues during the winter months, the water temperature may approach 4°C and then cool beyond that temperature. As the temperature of the surface water drops to less than 4°C and becomes less dense, inverse stratification occurs. In inverse stratification, the temperature of the surface water is colder but less dense than the waters below it, which remain at 4°C on the bottom of the lake. Ice cover can contribute further once the lake freezes. The ice cover prevents wind mixing and erosion of the density differences. The surface layers next to the ice remain near 0°C to maximize the mild inverse density difference. Inverse stratification is so mild that often a distinct thermocline does not form and the epilimnion and hypolimnion are not well defined. Inverse stratification persists until spring warming heats the surface layer. As the temperature of the surface layer approaches the temperature of the hypolimnion, which is usually 4°C, spring overturn occurs. Then, with springtime solar heating, the stratification cycle is started anew (Figure 12.15).

Some lakes and reservoirs, especially run-of-the-river reservoirs, do not follow the general stratification pattern. Shallow lakes and reservoirs only weakly stratify due to relatively high flows or the dominance of wind mixing. The thermocline may be difficult to define. Complete mixing may occur during the summer stratification period as a result of wind or runoff events. Fall overturn occurs earlier in shallow lakes than in deeper lakes. In warmer climates, such as much of the Southern United States,

FIGURE 12.14   Period of complete mixing (e.g., spring and fall).

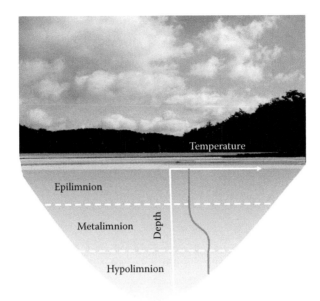

FIGURE 12.15 Idealized winter stratification.

water does not cool beyond 4°C and lakes and reservoirs remain unstratified during winter. That is, rather than mixing twice (*dimictic*), they only mix once a year (one mixing or *warm monomictic*).

The 1980 thermal structure of DeGray Lake (reservoir) in Arkansas is used here as an example of the stratification cycle of a warm monomictic lake. During January and February, the lake is unstratified (completely mixed). During March, stratification begins (Figure 12.16). The stratification continues through the year with increasing temperatures in the epilimnion and the establishment of a strong metalimnion. While mixing across the metalimnion is reduced by the density gradient, the limited mixing remains sufficient to cause some slight seasonal increases in hypolimnetic temperatures. Then, in late fall, the surface cools, ultimately resulting in complete mixing during December through the following spring.

The vertical and temporal temperature variations for DeGray Lake are also illustrated using the results of the application of the CE-QUAL-W2 two-dimensional model to the reservoir (Martin 1988;

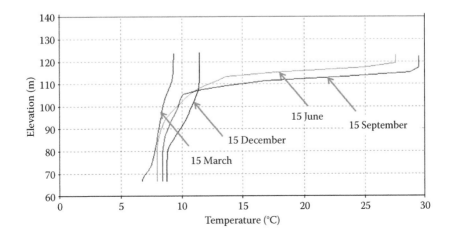

FIGURE 12.16 Selected 1980 vertical temperature profiles for DeGray Lake, Arkansas.

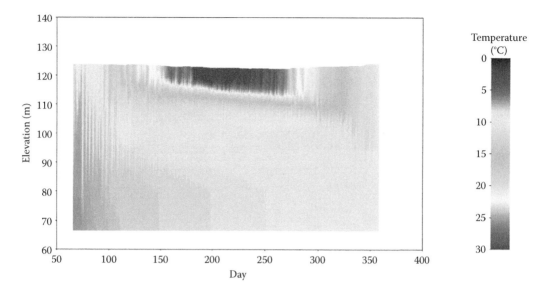

FIGURE 12.17 1980 temperature time-vertical contours for DeGray Reservoir from near the dam.

Martin and McCutcheon 1999). In these plots (Figures 12.17 and 12.18), the seasonal heating of the epilimnion can be seen more clearly, along with periodic intervals of convective mixing, such as at nighttime. This is also a peaking hydropower reservoir and the operation of the reservoir can impact mixing as well. The longitudinal–vertical profile plots (Figure 12.18) show that water temperatures are usually reasonably uniform horizontally, as would be expected since changes are largely due to surface transfers. Some variations can be seen, such as those due to inflow temperatures and other factors impacting transport, which will be discussed in Chapter 13. The differences are often sufficient to result in strong longitudinal gradients in other water quality parameters.

A second example is illustrated in Figure 12.19 for Folsom Lake, California (Bender et al. 2007), also using the results from the CE-QUAL-W2 model. Folsom Lake becomes thermally stratified each spring and maintains a separation between the warmer epilimnetic waters and the cold-water pool comprising the hypolimnion. Thermal stratification usually begins in April and continues through the summer and into November when winds and inflow begin to mix the top and bottom layers.

12.6 CLASSIFICATION BASED ON MIXING

Lakes and reservoirs are often classified based on stratification patterns, whether they mix completely during the year, and if so, how many times. The number of mixing events and the degree to which mixing occurs often has a direct impact on water quality.

The general mixing characteristics of lakes may often be related to their location (e.g., latitude and elevation), as summarized in Figure 12.20, while the mixing characteristics of reservoirs often depend on their structure and operation. One distinction is based on whether the entire lake or reservoir mixes during a year, or only part thereof:

- Holomictic: refers to lakes and reservoirs in which the entire water column completely mixes during the year.
- Meromictic: refers to those relatively rare lakes and reservoirs that do not mix completely. They often have a deep stratum that is completely stagnant and often anaerobic. Meromixis can result, in part, due to lake bathymetry, for example, where it minimizes the impacts of wind mixing. Meromixes can also result, in whole or in part, from chemical, biological, or physical effects, and are subsequently referred to as (Hutchinson 1937):

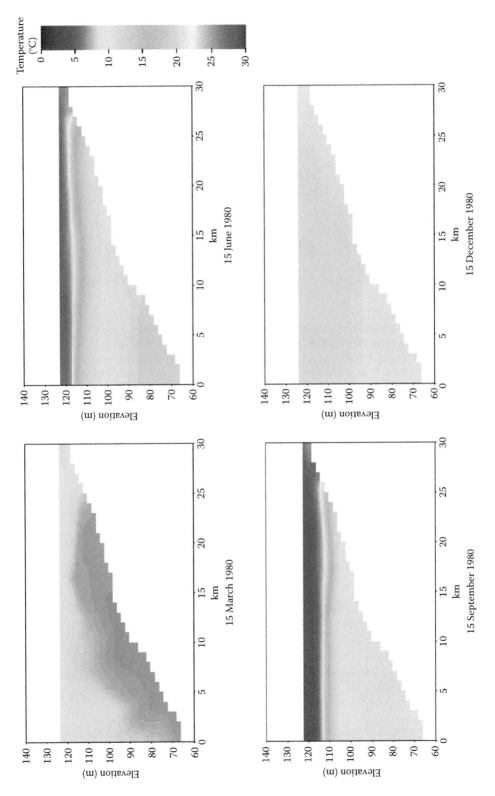

FIGURE 12.18 1980 longitudinal–vertical temperature contours for DeGray Reservoir extending from the dam (left) up-reservoir.

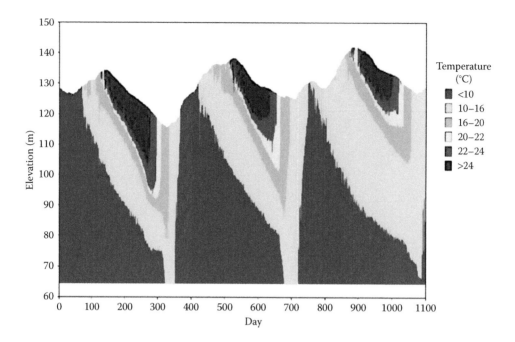

FIGURE 12.19 Modeled temperature profile contours (vertical profiles vs. time) at the Folsom Dam fore-bay (just upstream of the dam) during 2001–2003. (From Bender, M.D., Kibitschek, J.P., and Vermeyen, T.B., Temperature modeling of Folsom Lake, Lake Natoma and the Lower American River. Special Report, Sacramento Water Forum and Bureau of Reclamation, Sacramento County, CA, 2007. With permission.)

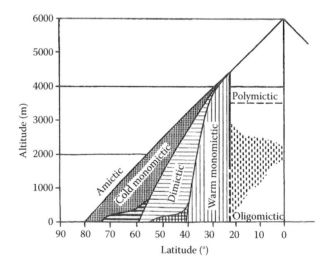

FIGURE 12.20 Lake types vs. latitude and altitude. (Reprinted from *Limnology: Lake and River Ecosystems*, 3rd ed., Wetzel, R.G., Copyright 2001, with permission from Elsevier.)

- Biogenic meromixis: results from a biological activity, such as where the electrolyte concentration increases in bottom waters due to the decomposition of organic materials (Hutchinson 1937).
- Ectogenic meromixis: is due to something originating or produced by some source outside of the lake and reservoir. An example would be where denser saltwater enters a

lake or reservoir creating a density barrier. Ectogenic meromixis often refers to a transient event, following which a lake returns to holomixis (Hutchinson 1937). The Great Salt Lake is an extreme case of ectogenic meromixis caused by diking (Lina 1976).

- Crenogenic meromixis: refers to a condition where the bottom water brings denser water into the lake or reservoir.

Meromixes may also occur as one stage, often an early stage, in the evolution of a lake (Hakala 2004; see Chapter 16).

For lakes and reservoirs that do mix completely (are holomictic), a second distinction is based on the number of annual mixing events, which often varies as a function of latitude and altitude (Figure 12.20):

- Monomictic: refers to lakes and reservoirs that mix only once a year.
- Cold monomictic: are lakes or reservoirs typically at high elevations, or at high latitudes, where the lakes' temperatures do not exceed 4°C and may only mix once during the summer months.
- Warm monomictic: are lakes at lower elevations, or lower latitudes, which do not reach temperatures below 4°C and mix only once a year during the winter.
- Dimictic: refers to lakes and reservoirs that mix twice a year.
- Polymictic: refers to lakes and reservoirs that mix frequently or continuously.
- Oligomictic: refers to lakes and reservoirs that rarely mix, and are often continuously stratified. Oligomictic lakes are more prevalent in tropical regions.
- Amictic: refers to lakes and reservoirs that do not mix at all, usually as a result of densities due to high dissolved-solids concentrations.

The number and extent of mixing events has a profound impact on the water quality of lakes and reservoirs and their releases. Many reservoirs in the Southern United States, particularly deeper ones such as DeGray Reservoir in Arkansas, are warm monomictic. A few may be dimictic, cold monomictic, or oligomictic. Many shallow run-of-the-river reservoirs or regulation reservoirs are polymictic.

12.7 ICE FORMATION AND COVER

During the winter months, ice covers many lakes and reservoirs in North America. Its extent is illustrated in Figure 12.21, which provides isopleths of the mean air freezing index values (cumulative degree-days below 32°F, based on mean air temperatures). Ashton (1982) suggested that stable ice cover may form on lakes and reservoirs located in areas with a freezing index above 100°F days. As illustrated in Figure 12.21, this encompassed much of North America in 1984.

12.7.1 Ice Formation

Ice formation begins when water is supercooled to slightly (much less than 0.1°C) below 0°C. Supercooling takes place in lakes and rivers when the air temperature is significantly less than 0°C and usually an air temperature of −8°C or lower is required (USACE 2002). The density difference is sufficient to form a thin film in which ice crystals may form (Ashton 1986).

In flowing waterbodies, such as rivers and streams, where vertical mixing is sufficient to transport supercooled water beneath the surface, ice crystals may be at depth. The resulting ice crystals are referred to as frazil ice. Frazil ice consists of small platelike crystals that are produced throughout the flow (Ashton 1982) (Figure 12.22).

Rather than frazil ice, sheet ice formation is associated with the lower velocities found in lakes and reservoirs (Ashton 1982). In slow-flowing waterbodies such as lakes, vertical mixing is not

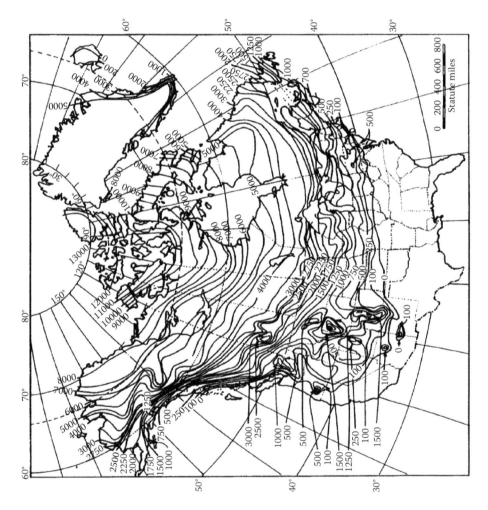

FIGURE 12.21 Distribution of mean air-freezing index values in North America. Design freezing index values are cumulative degree-days of air temperatures below 32°F for the coldest year in a 10-year cycle or the average of the 3 coldest years in a 30-year cycle. (From USACE, Pavement criteria for seasonal frost conditions—Mobilization construction, U.S. Army Corps of Engineers, 1984.)

FIGURE 12.22 Frazil ice in Yosemite Creek. (Courtesy of National Park Service and Wikimedia Commons.)

sufficient to carry supercooled water, or ice, to any appreciable depth, and a surface film or sheet ice forms. Ice formation usually begins in open water on calm days or along the shore (Ashton 1986). The formation of ice normally occurs first at night, thaws the following day until finally enough ice forms to prevent that complete thawing, or the daytime temperatures remain low enough for a continuous ice cover to form (Ashton 1979). Generally, for a surface ice cover to become firmly established, the mean (depth-averaged) temperature of water must be less than 2°C, the daily average air temperature must be less than –5°C, and the wind speed must be less than 5 m s^{-1} (Ashton 1982). Once formed, ice grows at the ice–water interface as a result of heat transfer upward from the interface, through the ice, to the atmosphere. In the absence of snow, the ice thickness is proportional to the square root of the number of degree-days of freezing (1 day of –10°C would be 10 degree-days below freezing; Ashton 1986).

12.7.2 Light Penetration through Ice and Snow

Light penetration is retarded by snow and ice, affecting both the heat balance and the survival of aquatic organisms. Light penetrating through the snow and ice acts to both heat the water and provide radiation for primary production. Where sufficient light penetrates the ice and snow cover, algal blooms often occur. In particularly productive waterbodies during prolonged periods of ice and snow cover, the retardation of light penetration causes dissolved oxygen depletion and winter fish kills. Snow cover is usually the most significant factor affecting the retardation of light penetration. Bolsenga (1977) found that 3 cm of wind-packed snow over 28 cm of clear ice reduced radiation transmittance from the clear ice to the snow-covered ice by up to 90%. Snow clearing has been used in some areas to aid in the prevention of winter fish kills. For example, Barica et al. (1983) investigated the feasibility of snow clearing to prevent winter fish kills in Rock Lake, southern Manitoba.

12.7.3 Lake Ice Decay

During periods of warming, lake ice will begin to decay. Ice decay usually begins when the snow cover melts, forming pools and decreasing the albedo of the ice surface, or how the light is reflected (Ashton 1986). As the reflectance decreases, the heat absorbed increases, and as the ice cover reaches and exceeds 0°C, the ice melts at the grain boundaries and becomes weak and unstable. As the ice begins to break up, the open water, which has a lower albedo than ice (less reflective), will melt and the process continues. Shen and Chiang (1984) found that, in the Saint Lawrence River, thermal dissipation, or the mixing of warmer waters from shallower areas to open waters, was the main factor affecting the breakup.

Ice may melt both at its bottom, due to increased water temperatures and at its surface, due to increased air temperatures. The dissipation of the ice cover occurs not only by direct melting at the top and bottom of the ice but also by a phenomenon that Ashton (1982) refers to as deterioration. Once the ice cover warms to near the melting point, its mechanical integrity is lost by a process called rotting or candling. Rotting results from melting at the ice grain boundaries. This rotting or candling makes the ice more susceptible to break up by wind and currents.

REFERENCES

Ashton, G.D. 1979. River ice. *American Scientist* 67 (1), 39–45.

Ashton, G.D. 1982. Theory of thermal control and prevention of ice cover in rivers and lakes. In V.T. Chow (ed.), *Advances in Hydroscience*, Vol 13. Academic Press, New York, pp. 132–185.

Ashton, G.D. 1986. *River Lake Ice Engineering*. Water Resources Publications, Highlands Ranch, CO.

Barica, J., J. Gibson, and W. Howard. 1983. Feasibility of snow clearing to improve dissolved oxygen conditions in a winterkill lake. *Canadian Journal of Fish Aquatic Science* 40, 1526–1531.

Bender, M.D., J.P. Kibitschek, and T.B. Vermeyen. 2007. Temperature modeling of Folsom Lake, Lake Natoma and the Lower American River. Special Report, Sacramento Water Forum and Bureau of Reclamation, Sacramento County, CA.

Bolsenga, S.J. 1977. Preliminary observations on the daily variation of ice albedo. *Journal of Glaciology* 18 (80), 517–521.

Carlson, R.E. 1977. A trophic state index for lakes. *Limnology and Oceanography* 22 (2), 361–369.

Chapra, S.C. 2008. *Surface Water-Quality Modeling*. Waveland Press Inc., Long Grove, IL.

Chapra, S.C., G.J. Pelletier, and H. Tao. 2007. QUAL2K: A modeling framework for simulating river and stream water quality, Version 2.07: Documentation and users' manual. Civil and Environmental Engineering Department, Tufts University, Medford, MA.

Ford, D.E. and M.C. Johnson. 1981. Field observations of density currents in impoundments. In H.G. Stefan (ed.), *Proceedings of the Symposium on Surface Water Impoundments*. American Society of Civil Engineers, New York, pp. 1239–1248.

Forsberg, C. and S.O. Ryding. 1980. Eutrophication parameters and trophic state indices in 30 Swedish waste-receiving lakes. *Archives of Hydrobiology* 89, 189–207.

Hakala, A. 2004. Meromixes as part of lake evolution: Observations and a revised classification of true meromictic lakes in Finland. *Boreal Environment Research* 9, 37–53.

Hutchinson, G.E. 1937. A contribution to the limnology of arid regions. *Transactions of the Connecticut Academy of Arts and Sciences* 33, 47–132.

Hutchinson, G.E. 1957. *A Treatise on Limnology*. Volume 1. Geography, Physics and Chemistry. Wiley, New York.

Kirk, J.T.O. 1994. *Light and Photosynthesis in Aquatic Ecosystems*, 2nd ed. Cambridge University Press, Cambridge.

Lee, R.W. and W. Rast. 1997. Light attenuation in a shallow, turbid reservoir, Lake Houston, Texas. Water-Resources Investigations Report 97–4064, U.S. Geological Survey, Reston, VA.

Likens, G.E. 1975. Primary productivity of inland aquatic ecosystems. In H. Leth and R.H. Whittaker (eds), *Primary Productivity of the Biosphere*. Springer-Verlag, New. York, pp. 185–202.

Lina, A. 1976. The meromictic Great Salt Lake. *Journal of Great Lakes Research* 2 (2), 374–383.

Lorenzen, C.J. 1972. Extinction of light in the ocean by phytoplankton. *Journal du Conseil* 34, 262–267.

Malm, W.C. 1999. Introduction to visibility. National Park Service Cooperative Institute for Research in the Atmosphere (CIRA), Colorado State University.

Martin, J.L. 1988. Application of a two-dimensional model to DeGray Lake, Arkansas. *ASCE Journal of the Environmental Engineering* 114 (2), 317–336.

Martin, J.L. and S.C. McCutcheon. 1999. *Hydrodynamics and Transport for Water Quality Modeling*. CRC/Lewis Publishers, Boca Raton, FL.

NASA. 2010. "The earth's radiation budget" science mission directorate. National Aeronautics and Space Administration. March 19, 2011. Available at: http://missionscience.nasa.gov/ems/13_radiationbudget.html.

Shen, H.T. and L.A. Chiang. 1984. Simulation of growth and decay of river ice cover. *ASCE Journal of Hydraulics* 110 (7), 958–971.

USACE. 1984. Pavement criteria for seasonal frost conditions—Mobilization construction, EM 1110-3-138 CEMP-ET. U.S. Army Corps of Engineers.

USACE. 2002. Engineering and design, ice engineering, EM 1110-2-1612. U.S. Army Corps of Engineers.

Verdiun, J., L.R. Williams, and V.W. Lambou. 1976. Components contributing to light extinction in natural waters: Method for isolation. Working Paper 369. U.S. Environmental Protection Agency.

Wetzel, R.G. 2001. *Limnology: Lake and River Ecosystems*, 3rd ed. Elsevier Academic Press, San Diego, CA.

Wetzel, R.G. and G.E. Likens. 2000. *Limnological Analysis*, 3rd ed. Springer-Verlag, New York.

Wool, T.A., R.B. Ambrose, and J.L. Martin. 2008. WASP7 temperature and fecal coliform: Model theory and user's guide; supplement to water quality analysis simulation program (WASP) user documentation. U.S. EPA, Region 4 Water Management Division, Atlanta, GA.

13 Transport and Mixing Processes in Lakes and Reservoirs

13.1 INTRODUCTION

The processes impacting physical transport in lakes and reservoirs largely impact their quality, so an understanding of those physical transport processes is fundamental to identifying the causes of variations in the water quality and in the management of lakes and reservoirs. Transport processes in lakes and reservoirs are controlled by natural variations in inflows, outflows, and meteorological forcings (see e.g., Figure 13.1) as they are impacted by reservoir bathymetry, heat exchange, and other factors. In man-made reservoirs, the transport and mixing processes are also controlled, to a large degree, by the manner in which they are regulated by the dam(s) that formed them.

13.2 WATER MOVEMENT: WAVES, CURRENTS, AND INFLOWS

One common and very noticeable physical transport and mixing process in lakes and reservoirs, particularly on windy days, is due to surface waves. Surface waves are a highly visible manifestation of the impact of wind shear over the interface between air and water or the shear of a lighter layer over a denser layer (Smith 1975; Turner 1973). Waves can also be internal, such as on the interface between layers of different densities but within the lake or reservoir. Internal waves are caused by shearing currents set up by both the wind and other currents. Although not as obvious as surface waves, internal waves can be larger and more effective in causing mixing. The intensity of wave mixing and turbulence is a direct result of wind energy or the energy in other shearing currents.

13.2.1 PROGRESSIVE SURFACE WAVES

The basic characteristics of waves include their amplitude (height between trough and crest) and wavelength, or the distance between crests, as shown in Figure 13.2. The time required for successive waves to pass a given point is the wave period.

Waves can be in motion horizontally, and those that move with respect to a fixed point of observation are progressive waves. In contrast, some waves remain in a constant horizontal position, and are referred to as standing waves. One important form of a standing wave is a seiche.

The wave height and wave period are related to the wind speed, duration, and fetch. Wind blowing over water results in a shear stress so that the work being done by the wind results from a frictional drag that transfers energy from the wind to the water. The greater the wind speed is, the greater the energy transfer will be. Similarly, the greater the distance over which the wind can blow undisturbed, the greater the energy transfer will be.

Fetch is the unobstructed distance over which the wind can travel over water in a constant direction. Therefore, the effect of fetch will vary as a function of both the morphometry of a lake or reservoir and the wind direction, and together with the wind speed, it impacts the size and power of waves.

FIGURE 13.1 Storm surge on Lake Superior. (Photo from Great Lakes Storms Photo Gallery, October 25–27, 2001, Lake Superior Storm Surge Photos.)

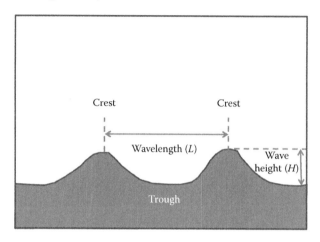

FIGURE 13.2 Wave characteristics.

As fetch increases, the wavelength increases. Long wavelengths are only produced in the presence of a long fetch. In addition, with greater fetch, larger waves are produced that can, in turn, increase shoreline erosion and sediment resuspension. The height of the highest wave is proportional to the square root of the fetch (Wetzel 2001). For example, Wetzel (2001) showed that the maximum wave height is about 1% of the square root of the fetch (F, in kilometers):

$$H_{max} = 0.01\sqrt{F} \tag{13.1}$$

As an example, for Lake Superior the observed and estimated maximum wave height using this relationship is about 6.9 m for a fetch of 482 km.

Fetch also impacts habitats. For example, Prince (2011) found significant relationships between fetch and water depth and the occurrence of Eurasian milfoil. Fetch may also have other impacts. In Lake Michigan, a longer fetch results in stronger lake-effect snow bands due to more warmth and moisture being added to the air as it crosses the lake. Fetch length is commonly determined using methods outlined in the U.S. Army Corps of Engineers (the Corps) shore protection manual (USACE 1984). Rohweder et al. (2008) incorporated the Corps' methods (USACE 1984) along with a wind wave model, into an ArcGIS platform for assessing the impacts of fetch on habitat rehabilitation and enhancement projects.

The shortest wavelengths require only limited contact between wind and water. Waves with a wavelength less than 2π cm (6.28 cm) are capillary waves. The more important gravity waves have wavelengths longer than 2π cm. The two types of gravity waves are short waves and long waves, distinguished by the interaction with the lake bottom. The wavelength of short waves is much less than the water depth (depth $< L/2$, where L is the wavelength; Figure 13.3a), thus they are not affected by shear at the water bottom. The waves seen by eye on lake and reservoir surfaces are typically short waves. Long waves, such as the lake seiche, do interact with and are influenced by bottom friction (Figure 13.3b). The distinction between gravity waves is important because of the differences in the resulting mixing.

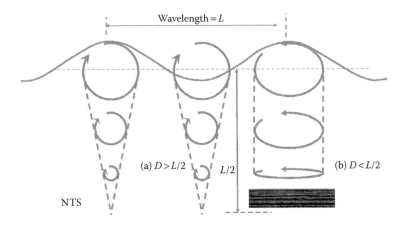

FIGURE 13.3 Relationship of orbital motion wave with wave height and wavelength.

The ratio of the wave height to the wavelength is highly variable, in a range of 1:100–1:10 (Wetzel 2001). Above a ratio of about 1:10, the wave becomes unstable and the peak collapses, forming whitecaps.

Observations of short waves would seem to suggest that both the wave and the associated water mass are traveling at the wave speed, but this is misleading. If that were the case, then a floating object would be expected to move at or close to the wave speed (with some reduction due to drag) in the direction of the wave. However, a closer observation over a short period indicates that the floating object simply bobs up and down without moving with the wave. While, over a longer period of observation, there would be some residual circulation or movement of the floating object, this may not occur over a shorter period. The reason that causal observation indicates that water is moving at the speed of the wave is the orbital motion caused by wave action. Water at the surface does not simply move vertically up and down with the rising and falling water surface as the wave passes. Instead, the motion is circular in a vertical plane, making a complete revolution as each successive wave passes, as illustrated in Figure 13.3. Thus, an important distinction for mixing is that wave momentum and wave mass move in very different ways that tend to mix surface layers or layers at an interface.

The wave-induced orbit will be greatest at the surface, where the radius is one-half of the wave height (H). Below the water surface, the orbital radius of particles decreases (Figure 13.3). Since short waves result in orbital moment with no net advection of water, the overall effect is to induce mixing and not horizontal mass transport.

There is no appreciable orbital motion below a depth of approximately one-half of the wavelength in an unstratified flow. This depth is often referred to as the wind-mixed depth (Figure 13.3). The wind-mixed depth increases with fetch since the wave height and the wavelength increase with increasing fetch. For example, Ragotzkie (1978) developed a relationship for lakes in Wisconsin and Central Canada in which the mean depth of the thermocline (D_{th}) is proportional to four times the square root of the fetch (in kilometers):

$$D_{th} = 4\sqrt{F} \tag{13.2}$$

As the wavelength becomes longer in relation to the depth or as the water becomes shallower, the wave orbits become increasingly flatter or elliptical, as illustrated in Figure 13.3. As the orbits flatten, the motion of the water becomes essentially a horizontal oscillation (Smith 1975) so that the water motion due to long wavelengths is more organized rather than dispersive. Motion of this type is characteristic of long waves, which have wavelengths that are much greater than the water depth.

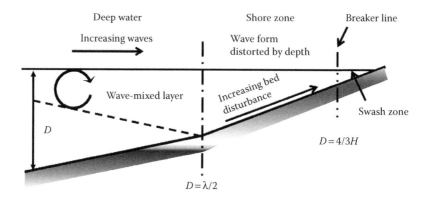

FIGURE 13.4 Schematic of surface waves. (Redrawn from Ford, D.E. and Johnson, L.S., An assessment of reservoir mixing processes, U.S. Army Engineers Waterways Experiment Station, Vicksburg, MS, 1986.)

Both short and long waves are gravity waves whose wave speed or celerity is dependent upon gravitational acceleration. The short wave speed or celerity depends on the wavelength, while for long waves, the wave speed or celerity is a function of the water depth. For example, the celerity (c) of a long wave may be estimated from gravitational acceleration (g) and depth (D):

$$c = \sqrt{g\,D} \qquad\qquad (13.3)$$

13.2.2 BREAKING WAVES

As short waves enter shallow water, interaction with the bottom begins to affect the orbital motion, as shown in Figure 13.4. The bottom interaction begins when the depth decreases to less than one-half of the wavelength (L in Figure 13.3 or λ in Figure 13.4). From this point inland to the breaker line is the shore zone where the depth is less than $\lambda/2$. In the shore zone, the wave velocity decreases with the square root of the depth, which results in a corresponding increase in the wave height. Waves distort as water at the crest moves faster than the wave, creating instability. These unstable waves may eventually collapse, forming breakers or white caps, as illustrated in Figure 13.4. The type of breaker that is formed depends on the wave steepness, the wind speed and direction, the direction of the wave, and the shape and roughness of the bottom. A spilling breaker (Figure 13.5) tends to form over a gradually shoaling bottom and tends to break over long distances with the wave collapsing downward at the front of the wave. Plunging breakers (Figure 13.5) occur when the bottom shoals rapidly or when the wind direction opposes the wave. The plunging breaker begins to curl and then collapses before the curl is complete. A plunging or surging breaker does not actually break or collapse but forms a steep peak as the wave moves up the beach. This type of breaking wave and the associated energy controls beach erosion, aquatic plant growth, surf zone mixing, and the exchange of contaminants between the surface water and the groundwater.

After breaking, waves continue to move up a gradually sloping beach until the force of gravity forces the water back. The extent to which the water runs up the beach is called the swash zone. The movement of the swash up the beach may result in the deposition of particles and debris, resulting in swash marks at the highest point of the zone.

13.2.3 LANGMUIR CIRCULATION

In very large lakes, seas, and oceans with a very long fetch, a series of parallel pairs of large vertical vortices or circulatory cells, known as Langmuir circulation, develop in the general direction of

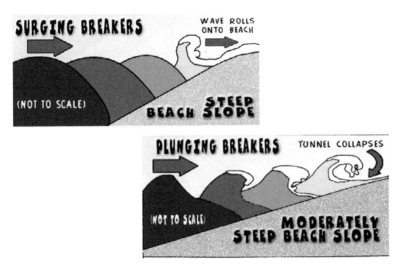

FIGURE 13.5 Types of breaking waves. (From the Office of Naval Research's Science and Technology Focus Site for Students and Teachers.)

a sustained wind when the wave and current conditions are favorable (Figure 13.6; Leibovic 1983, cited by Ford and Johnson 1986). Actually, these long vortices develop side by side over the entire surface of the deeper water but at a slight angle with the wind direction. As shown in Figure 13.6, the pairs of longitudinal vortices or Langmuir cells form at approximately 15° clockwise from the path of the wind. The depth of the vortices depends on stratification and there may be a relationship with the internal waves formed on the thermocline. Langmuir cells may be deeper over the troughs of internal waves. Where the counterrotating Langmuir cells converge, visible streaks or bands form on the surface that tends to accumulate floating debris. In the convergence zone, downward velocities carry surface waters toward the thermocline. These downward currents move in a circular fashion and turn upward into a divergence zone midway between the Langmuir streaks. The water near the thermocline may move to a zone nearby, resulting in an upwelling of water typically higher in nutrients. As first proposed by Langmuir (1938), this type of large-scale circulation also contributes to the vertical mixing of the epilimnion.

13.2.4 STANDING WAVES, SURFACE SEICHE

One of the more important standing waves is the seiche. A seiche refers to a "sloshing" of water around some point or node of oscillation. The term *seiche* was coined by the Swiss hydrologist François-Alphonse Forel, based on his observations of Lake Geneva, Switzerland, where the term is derived from a French dialect and means "to sway back and forth." Surface seiches are periodic oscillations of the water surface resulting from displacements. Displacements are typically caused by large-scale wind events. They can also be caused by seismic activity or large reservoir withdrawals. During hydropower operations or reservoir releases, there is a net flow of water toward the dam.

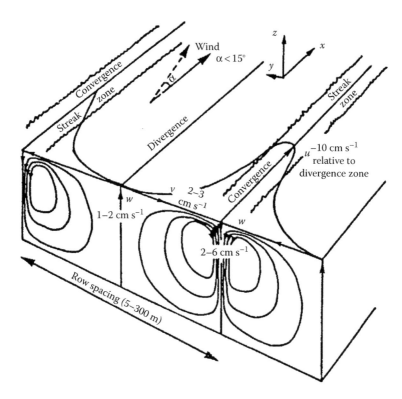

FIGURE 13.6 Langmuir circulation cell. (From Ford, D.E. and Johnson, L.S., An assessment of reservoir mixing processes, U.S. Army Engineers Waterways Experiment Station, Vicksburg, MS, 1986; After Pollard, R.T., Rhines, P.B., and Thompson, R.O.R.Y., *Geophysical Fluid Dynamics* 3, 381–404, 1973.)

Once hydropower withdrawals cease, momentum will cause the water to pile up at the dam and to ultimately form a seiche (Martin and McCutcheon 1999).

For example, wind blowing across a lake for a sustained period will result in water piling up on the far shore, such as a storm surge. The increase in the water-surface elevation at the downwind lake boundary is wind set-up, while on the opposite side the water-surface elevation is depressed (wind set-down). Figure 13.7 illustrates a 7-ft. seiche on Lake Erie, observed during 2003, with the wind set-up on the eastern shore and the wind set-down on the western shore. After the wind subsides, the water-surface tilt or displacement results in a sloshing motion, or a seiche of the lake surface. As the rocking or sloshing motion takes place, potential energy is converted to kinetic energy and dissipated (Figure 13.8). Over time, the wave magnitude or the wave height decreases or decays as bottom friction causes potential energy to be converted into kinetic energy and dissipated.

The seiche can oscillate around a single point, called a uninodal seiche. Where pressure is exerted on the center of the lake and then released, a bimodal seiche is generated (Figure 13.8). Multimodal seiches have also been observed with up to 17 nodes (Wetzel 2001).

Surface seiches are often relatively small in magnitude. However, for lakes with large fetch, seiches can be quite large. For example, Lake Erie is relatively long (maximum of 210 mi.) and narrow (57 mi.), so that wind blowing along the length of the lake can result in large variations in water-surface elevations and large seiches. Martin and McCutcheon (1999) demonstrated that a set-up of 3 ft. for Lake Erie requires a wind speed of about 30 mph. A 7-ft. seiche is illustrated in Figure 13.7, but surface seiches of up to 16 ft. have been observed. These elevation changes, associated with storm surges, are in addition to surface waves, which can add an additional 10 ft. or more to the total elevation change.

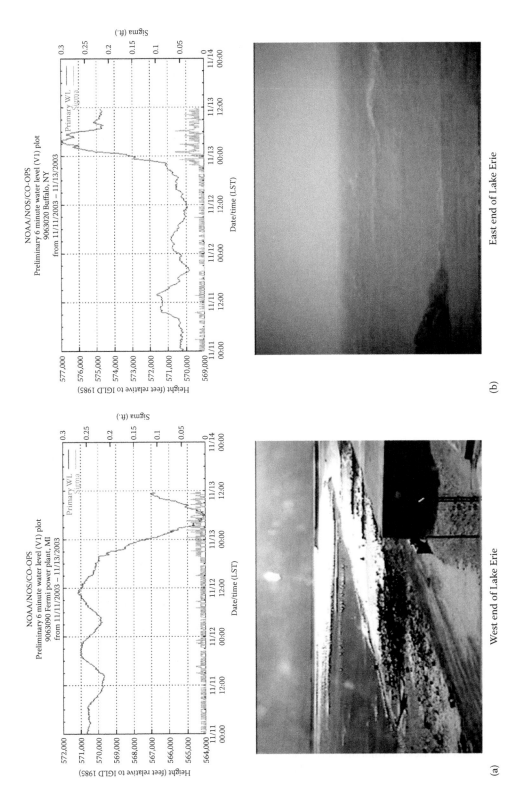

West end of Lake Erie

(a)

East end of Lake Erie

(b)

FIGURE 13.7 (a,b) Seven-feet seiche on Lake Erie, 2003. (Courtesy of NOAA Great Lakes Environmental Research Laboratory.)

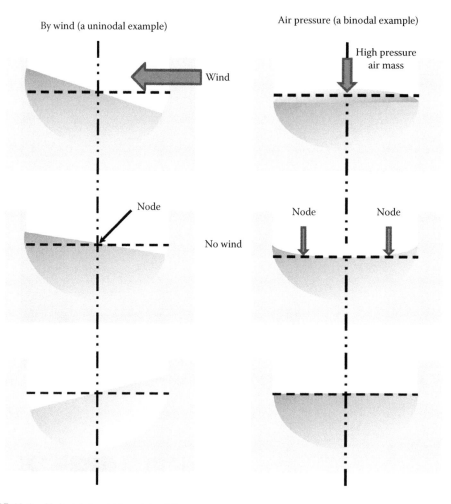

FIGURE 13.8   Unimodal and bimodal seiches.

13.2.5   INTERNAL WAVES, UPWELLING

Internal waves are more difficult to observe than surface waves but they may be more important in lake and reservoir mixing. Smaller density differences within the waterbody tend to make internal waves travel more slowly than surface waves but achieve greater wave heights. At the water surface, the tilt or slope is affected by the large density difference between water and air. However, across the interface of a two-layered stratified lake, the density differences are relatively small. Therefore, the magnitude of the tilt of the interface can be many times that of the surface tilt. For example, Cole (1975) described a remarkable internal wave in Lake Baikal, with a period of 38 days and an amplitude of 75 m. Theoretically, a surface tilt on the order of 10 cm could have energized that wave (Martin and McCutcheon 1999). Internal seiches with magnitudes of >10 m are common in Lake Michigan (Wetzel 2001).

Internal waves include standing waves like internal seiches (Mortimer 1974) and internal hydraulic jumps (French 1985), but most are progressive waves that radiate energy away from the point of disturbance where the waves were generated (Ford and Johnson 1986). Wind shear, water withdrawals, hydropower releases, thermal discharges, as well as local disturbances produce internal waves. Internal waves can propagate and break similar to surface waves (Wetzel 2001).

Internal seiches can be detected by changes in the temperature-depth profiles where the depth of the thermocline moves upward or downward in the water column. If the interface tilt is large enough, the interface may temporarily contact the surface, creating an upwelling. Upwellings commonly result in locally colder surface water temperatures, in comparison with other surface waters. Upwellings can also introduce additional nutrients to the surface waters from the colder and often more nutrient-rich metalimnion. Oxygen concentrations may also be depressed with the introduction of hypoxic subsurface waters. Ford and Johnson (1986) observed that an upwelling following a large storm depressed the surface dissolved oxygen concentrations of C.J. Brown Lake, Ohio, to about 2 mg L^{-1}.

13.2.6 EARTH'S ROTATION—THE CORIOLIS FORCE

The rotation of the earth also impacts water movement. Coriolis forces, due to the rotation of the earth, cause currents in the Northern Hemisphere to be deflected to the right, and to the left in the Southern Hemisphere when an observer looks in the direction of the currents. The affect is typically important for lakes, such as the Great Lakes. Jin et al. (2002) demonstrated the importance of the Coriolis force for Lake Okeechobee, Florida.

13.2.7 PENETRATIVE CONVECTION

Another surface-mixing phenomenon occurs as surface water cools during the nighttime or during seasonal cooling periods. The cooler, denser water overlying the less dense, warmer water causes negative buoyancy or unstable stratification. This density instability causes the cooler surface waters to sink and the underlying warmer waters to rise, mixing the upper layers. This mixing will continue until a stable density profile is once again achieved.

13.2.8 INFLOWS AND CURRENTS

The impact of inflows may be illustrated by longitudinal variations in reservoirs, as discussed in Chapter 12 and illustrated in Figure 13.9. In the riverine zone, the initial momentum of an inflow into a lake or a reservoir pushes the more stagnant lake water ahead of it until the initial momentum is substantially dissipated (Ford and Johnson 1986; Cassidy 1989). Prior to dissipation in this zone, the advective riverine processes dominate the transport within a lake. In the riverine zone, turbulent kinetic energy is usually sufficient to keep the water column well mixed, preventing vertical stratification and the settling of some materials. The mixing between the riverine inflow and the reservoir waters is limited and the quality or identity of the inflow remains similar to that of the riverine water (Martin and McCutcheon 1999).

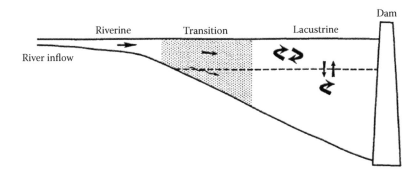

FIGURE 13.9 Longitudinal zones of a reservoir. (From Thornton, K.W., et al., *Proceedings of the Symposium on Surface Water Impoundments*, Volume 1, American Society of Civil Engineers, New York, 1981; Wetzel, R.G., *Limnology*, 3rd ed., Academic Press, San Diego, CA, 2001. With permission.)

The transition zone is characterized by the transition from riverinelike to lakelike characteristics. In this zone, buoyancy forces due to differences in the densities between inflows and lake waters begin to dominate the advective forces of the inflows. With decreased velocities, denser particles settle. As the initial kinetic energy of the inflows is dissipated, the potential energy of the inflows, due to the density differences between the inflows and the lake waters, begins to dominate transport. If the inflow energy is not strong enough to eliminate those density differences (e.g., cause complete mixing), then the inflow will flow to a layer of an equivalent density level and move along that layer. Where density differences result from differences in temperature or dissolved materials, they are referred to as density currents. However, if the density differences result from particulate loads, they are more correctly referred to as turbidity currents (Smith 1975).

If the river water is less dense (more buoyant) than the surface waters of the lake or reservoir, the more buoyant river water will flow over the top of the reservoir water as an overflow (Figure 13.10). Overflows are common in the springtime, for example, when shallower river waters generally heat more rapidly (become less dense) than reservoir waters. Overflows also result from thermal discharges. The underside of the overflow is a surface of discontinuity (due to velocity and density differences), where eddy mixing and the entrainment of reservoir waters occur. The degree of mixing and entrainment is largely controlled by the magnitude of the flows and the density differences.

Overflows may contribute substantially to water quality variations since they add materials directly to the more productive surface zones of the lake or reservoir (Ford and Johnson 1986). Since overflows will tend to move over the top of the reservoir waters, the increased water-surface elevation will result in a hydrostatic instability, which will tend to cause the overflow to quickly spread out as it moves down-reservoir. The actual amount of spreading that may occur will be influenced by processes such as wind mixing, which may tend to mix the overflow with underlying waters or "push" it to one side of the reservoir. Surface winds often cause the overflow to be rapidly dissipated (Smith 1975).

If the density of the river water is greater than the surface water, the negative buoyancy causes the inflows to plunge beneath the lake or reservoir waters, becoming an underflow (Figure 13.10). The higher density may result from cooler inflow temperatures or higher concentrations of dissolved or particulate materials. The point where underflows plunge (the plunge point) or the separation point where overflows detach from the bottom is the transition point between the turbulent open channel or riverine inflows and the stratified flows in lakes and reservoirs (Martin and McCutcheon 1999). Usually, the plunge point is easily located by the foam and floating debris trapped by the converging currents.

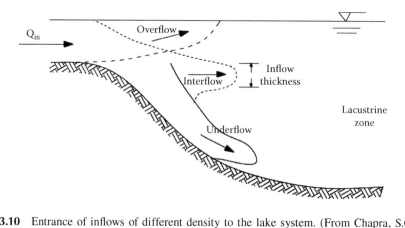

FIGURE 13.10 Entrance of inflows of different density to the lake system. (From Chapra, S.C., Martin, J.L., and Camacho, R.A., LAKE2K: A modeling framework for simulating lake water quality (Version 2.0), Documentation and Users Manual, Tufts University, Medford, MA, 2011.)

Underflows are common in the fall. Shallow riverine waters cool faster than lake waters and fall storms cause higher turbidity inflows. In lakes, underflows will tend to follow the deepest channel, spreading out and dissipating due to the turbulence generated by bottom shear. In reservoirs, underflows tend to follow the thalweg, or drowned river channel. Velocities will decrease as the underflow moves along the bottom, due to the shear between the flow and the bottom. As the current velocities decrease, suspended solids may settle out, causing turbidity currents to lose their identity. However, in reservoirs with relatively steep bottom slopes, turbidity currents may flow downward to the dam, resulting in problems in the design and operation of outlet structures (Smith 1975). Density currents may also reach the dam but generally do not pose problems like the sedimentation that occurs when a turbidity current reaches outlet structures.

If the density of the underflows is greater than the bottom waters, they may remain on the bottom. However, most underflows are less dense than the bottom layer. In such a case, the underflow separates from the bottom as an interflow (either density or turbidity; Figure 13.10). Interflows are not affected by the bed shear, as with underflows.

Density and turbidity currents have been reported for many lakes and reservoirs. For example, Anderson and Pritchard (1951) reported winter underflows and spring overflows, along with summer and fall interflows of the Colorado River into Lake Mead. Lehn (1965, cited by Wetzel 1975, 2001) observed a rapid attenuation of light in the vicinity of the metalimnion due to the intrusion of river water with high suspended loads (turbidity currents) into Lake Bodensee, Germany. Ford and Johnson (1981) reported density and turbidity currents on 10 occasions in DeGray Lake, Arkansas. In many small lakes and reservoirs, the relative importance of the density and turbidity currents is undetectable (Smith 1975).

Underflows, interflows, and overflows will typically move down-reservoir as long as additional riverine inflows of similar characteristics can push it forward. In reservoirs, the formation and movement of underflows and interflows may be enhanced by subsurface withdrawals. Studies of DeGray Lake, Arkansas (Martin 1987), suggested that summer midlayer dam releases enhanced interflows above the thermocline. The movement of materials from anoxic zones at the head of the reservoir along the interflow contributed to a metalimnetic dissolved oxygen minimum and affected the quality of reservoir releases.

In general, however, underflows and interflows add oxygen-consuming materials to the more poorly mixed and less productive regions of the reservoir, thereby contributing to dissolved oxygen depletion in the bottom waters (Ford and Johnson 1986). Density differences result in an effective isolation of the riverine waters so that they may become anoxic and chemically reduced. Interflows and underflows are often a major source of reduced materials to downstream reservoir regions (Nix 1981, 1987; Martin 1987). Alternatively, if the underflows are well oxygenated, they may improve the bottom water quality. Ford and Johnson (1986) found that cold, oxygenated inflows from Lake Ouachota, Arkansas, maintain anoxic hypolimnion in downstream Lake Hamilton, although coves within the reservoir become anoxic.

An estimation of the extent of the initial mixing due to inflows, and of the possible formation of density or turbidity currents, is often necessary to determine the transport of materials through lakes and reservoirs, to compute release concentrations, and to compute the impact of the inflows on ambient quality. Failure to consider the formation of density currents, for example, may result in an overestimation of the dilution of the inflows and their residence time within the lake or reservoir. High turbidity interflows in Lake Cumberland, Kentucky, have been observed to essentially "short circuit" the reservoir. That is, these currents have remained relatively intact through the reservoir so that estimates of release concentrations based on assuming mixing with the reservoir waters would be inaccurate. Similarly, Worden and Pistrang (2003) observed shortcuts in the Wachusett Reservoir, essentially connecting releases from the upstream Quabbin Lake to the outflow in a metalimnetic "short circuit" undergoing minimal mixing with the ambient Wachusett Reservoir water.

13.3 WHAT ABOUT DAMS?

The transport processes in lakes and reservoirs are controlled by natural variations in inflows, outflows, and meteorological forcings. In man-made reservoirs, the transport and mixing are also controlled to a large degree by the manner in which they are regulated by the dam(s) that formed them (Figure 13.11).

13.3.1 PROJECT PURPOSES AND RESERVOIR OPERATIONS

The project design and operation of reservoirs directly impacts the physical transport within the reservoir and the magnitude and quality of the release waters. Therefore, regulation of the water levels within reservoirs affects their quality and productivity. For example, in some flood control reservoirs, seasonal drawdown of the water level provides additional flood storage capacity. However, the drawdown may reduce breeding areas along the shoreline, affecting fisheries and productivity.

The impact of reservoir operations on transport may be illustrated by hydropower reservoirs. Reservoir operations to produce electricity using water releases, or hydropower, affect both mixing and quality characteristics. Hydropower reservoirs are generally of two types: base load and peaking hydropower operations. Base load operations occur over relatively long periods of time during the day. As an example, an increasing number of low-head hydropower facilities produce electricity by continuously releasing water from relatively shallow reservoirs. In contrast, peaking operations release water and generate electricity only during peak demand periods. A typical hydropower dam is illustrated in Figure 13.12. Water may be released for only a few hours during the day and operations may show significant weekly and seasonal variations. These peak demands are typically in the morning and the evening during the week. Heating demands during winter, and cooling during summer, result in seasonal variations. The rapid releases that occur during hydropower operations produce rapid variations in reservoir and downstream water levels, thereby affecting mixing in both reservoirs and streams.

Pump storage hydropower operations produce special mixing and water quality characteristics and differ from conventional operations. In one configuration for pump storage peaking hydropower reservoirs, water is released from an upper reservoir to create energy during peak demand periods and then captured and stored in a lower reservoir. That is, the release water is stored in a typically lower reservoir (the storage reservoir) and then pumped back to the upper reservoir during periods of low demand when power may be obtained at reduced costs. Another configuration is illustrated

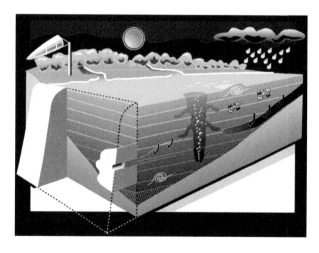

FIGURE 13.11 Reservoir transport and mixing patterns. Simulated in DYRESM model. (Courtesy of Jorg Imberger, Center for Water Research, University of Western Australia.)

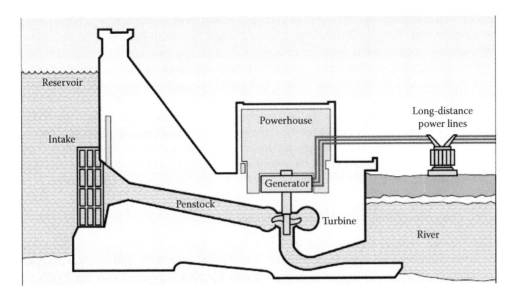

FIGURE 13.12   Schematic of a hydroelectric dam. (Courtesy of the Tennessee Valley Authority.)

by Tennessee Valley Authority's 1530-mW Raccoon Mountain project on the Tennessee River near Chattanooga, Tennessee, the largest federally owned pumped storage project. In this project, during periods of low demand, water is pumped from Nickajack Reservoir at the base of the mountain to the reservoir built at the top. When demand is high, water is released via a tunnel drilled through the center of the mountain to drive generators in the mountain's underground power plant (Figure 13.13). Pumped storage represents about 2.2% of all generation capacity in the United States, 10.2% in Japan, and 18.7% in Austria (Deane et al. 2010).

Reversible turbines usually act as pumps to produce the return flows and then produce energy when water is released. The return flow rates may be appreciable. For example, in Carters Lake, Georgia, a 500-mW pump storage facility, the maximum discharge during peaking hydropower operations is 612 m^3 s^{-1}, while the maximum pumpback flow is 213 m^3 s^{-1}. One issue that has been

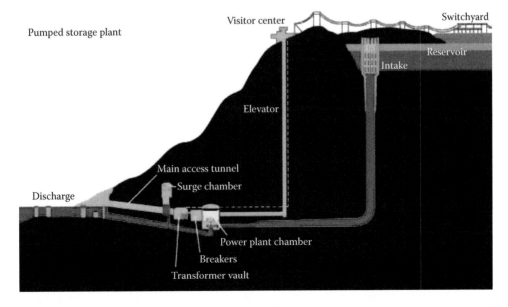

FIGURE 13.13   Diagram of Raccoon Mountain storage design. (Courtesy of the Tennessee Valley Authority.)

associated with pumpback operations is the entrainment and mortality of fish, which has impacted the operation of some facilities. For example, for the Richard B. Russell Reservoir, between Georgia and Alabama, the pumpback is from the reservoir tailwaters and has resulted in fish mortality as fish in the tailwaters become entrained. In order to reduce pumpback mortality, additional fish protection structures and seasonal and daily limits (e.g., nighttime only) on pumpback operations have been imposed.

Pumpback operations impact not only the transport in the storage reservoir, but also the transport in the main reservoir. Since the intake for the turbines is located at an appreciable depth, the return flow usually enters the upper reservoir as a jet (momentum dominates its transport), which entrains ambient water. Density stratification in the reservoir eventually results in jet collapse and density current formation (Roberts and Dortch 1985), which can impact temperature stratification.

13.3.2 Conveyance Structures and Operations

The size, location, type, and operation of these conveyance structures impact not only the transport within a reservoir, but also the transport in the downstream tailwaters. An issue, for example, is that in many stratified reservoirs, the hypolimnetic water quality degrades following stratification. The hypolimnion may become anoxic, or deoxygenated, during the summer months, particularly in highly productive or recently constructed reservoirs. The resulting hypolimnetic concentrations of reduced materials represent a deficit that must be overcome before the hypolimnion can again become oxic. Deoxygenation affects the quality of release water, particularly for those reservoirs that withdraw water from the hypolimnion, such as J. Percy Priest Reservoir in Tennessee, as illustrated in Figure 13.14.

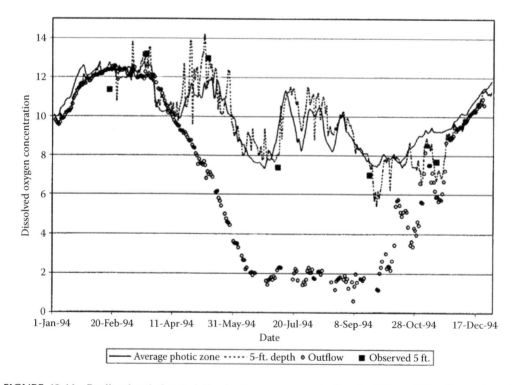

FIGURE 13.14 Predicted and observed dissolved oxygen concentrations in J. Percy Priest Reservoir and reservoir outflows during 1981. (From Martin, J.L. and Cole, T.M., Water quality modeling of J. Percy Priest Reservoir using CE-QUAL-W2, U.S. Army Engineer Research and Development Center, Vicksburg, MS, 1999.)

Artificial reaeration can be used to increase dissolved oxygen concentrations in the hypolimnion, typically using either air injection or oxygen injection. The quantity of air injected is often large and the increased vertical mixing that results may destroy the stratification within a lake or reservoir (Figure 13.11). There are many cases where increased vertical mixing is not desirable. For such cases, dissolved oxygen rather than air is often injected. The quantity of gas required is much reduced so that oxygen injection may often leave the stratification intact. The reader is referred to Gächter and Imboden (1985) and Pastorok et al. (1981) for additional information.

The transport within a reservoir and the characteristics (e.g., temperature and oxygen concentrations) of reservoir releases are impacted by the characteristics of the conveyance structure, the magnitude of the outflow, and the density stratification (Martin and McCutcheon 1999). The vertical zones from which water is withdrawn are referred to as the withdrawal distribution. In the absence of stratification, the withdrawal distribution may be essentially uniform over the depth of the reservoir. However, under stratified conditions, the withdrawal may be limited to a smaller zone within the epilimnion or the hypolimnion or both (Figures 13.15 and 13.16). The withdrawal distribution depends on the degree of stratification, the conveyance structure, and the energy associated with the withdrawal.

Two examples of representative withdrawal distributions (computed using the SELECT model; Davis et al. 1987; Schneider et al. 2004) are illustrated in Figures 13.15 and 13.16, along with the vertical temperatures and oxygen concentrations. For the first example, the withdrawal is from a port 150 ft. above the bottom of the reservoir and located near the thermocline (Figure 13.15). The normalized withdrawal distribution is a relatively narrow, nearly Gaussian, distribution centered on the port location and extending between 123 and 171 ft. This distribution would result in a current along the thermocline. For this stratification pattern, the estimated release temperature (from SELECT) would be 21.7°C and an oxygen concentration of 5.7 mg L^{-1}. In the second example, the port is located 15 ft. from the bottom of the reservoir and the rate of the withdrawal is increased by a factor of 10 (Figure 13.16). The withdrawal zone is larger than the first example, and the lower limit is restricted by the bottom of the reservoir. The estimated release temperature is colder (16.5°C) and the oxygen concentration is lower (3.5 mg L^{-1}).

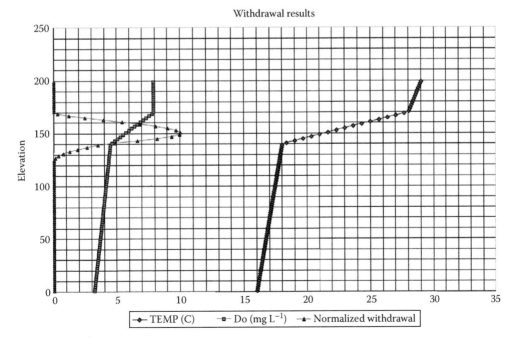

FIGURE 13.15 A representative withdrawal pattern for a point source (port) withdrawal from a stratified reservoir; Case A, with the withdrawal from a port at an elevation of 150 ft.

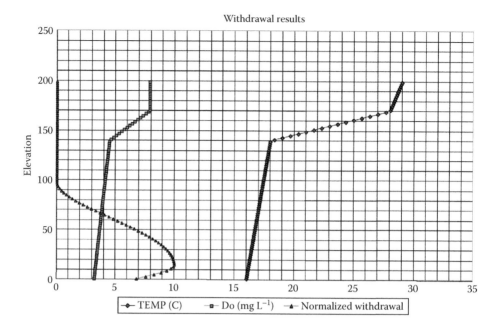

FIGURE 13.16   A representative withdrawal pattern for a point source (port) withdrawal from a stratified reservoir; Case B with the withdrawal from a port at an elevation of 15 ft. and at 10 times the flow rate of Case A.

REFERENCES

Anderson, E.R. and D.W. Pritchard. 1951. Physical limnology of Lake Mead, U.S. Navy Electronics Laboratory Report 258, San Diego, CA.

Cassidy, R.A. 1989. Water temperature, dissolved oxygen, and turbidity control in reservoir releases. In J.A. Gore, and G.E. Petts (eds), *Alternatives in Regulated River Management.* CRC Press, Boca Raton, FL, pp. 27–62.

Chapra, S.C., J.L. Martin, and R.A. Camacho. 2011. LAKE2K: A modeling framework for simulating lake water quality (Version 2.0), Documentation and Users Manual. Tufts University, Medford, MA.

Cole, G.A. 1975. *Textbook of Limnology.* C.V. Mosby Company, Saint Louis, MO.

Davis, J. E., J.P. Holland, M.L. Schneider, and S.C. Wilhelms. 1987. SELECT: A numerical one-dimensional model for selective withdrawal, Instruction Report E-87-2. U.S. Army Engineer Waterways Experiment Station, Vicksburg, MS.

Deane, J.P., B.P. Ó Gallachóir, and E.J. McKeogh. 2010. Techno-economic review of existing and new pumped hydro energy storage plant. *Renewable and Sustainable Energy Reviews* 14 (4), 1293–1302.

Ford, D.E. and M.C. Johnson. 1981. Field observations of density currents in impoundments. In H.G. Stefan (ed.), *Proceedings of the Symposium on Surface Water Impoundments.* American Society of Civil Engineers, New York.

Ford, D.E. and L.S. Johnson. 1986. An assessment of reservoir mixing processes, Technical Report E-86-7. U.S. Army Engineers Waterways Experiment Station, Vicksburg, MS.

French, R.H. 1985. *Open-Channel Hydraulics.* McGraw-Hill, New York.

Gächter, R. and D.M. Imboden. 1985. Lake restoration. In W. Stumm (ed.), *Chemical Processes in Lakes.* John Wiley, New York, pp. 363–388.

Jin, K.R., Z.G. Ji, and J.M. Hamrick. 2002. Modeling winter circulation in Lake Okeechobee, Florida. *Journal of Waterway, Port, Coastal, and Ocean Engineering* 128, 114–125.

Langmuir, I. 1938. Surface motion of water induced by wind. *Science* 87, 119–123.

Lehn, H. 1965. Litorale Aufwuchsalgen im Pelagial des Bodwensee. *Beiträge zur naturkundlichen Forschung in Südwestdeutschland* 27, 97–100.

Leibovic, S. 1983. The form and dynamics of Langmuir circulations. *Annual Review of Fluid Mechanics* 15, 391–427.

Martin, J.L. 1987. Application of a two-dimensional model of hydrodynamics and water quality (CE-QUAL-W2) to DeGray Lake, Arkansas, Technical Report E-87-1. U.S. Army Waterways Experiment Station, Vicksburg, MS.

Martin, J.L. and T.M. Cole. 1999. Water quality modeling of J. Percy Priest Reservoir using CE-QUAL-W2, ERDC/EL SR-00-9. U.S. Army Engineer Research and Development Center, Vicksburg, MS.

Martin, J.L. and S.C. McCutcheon. 1999. *Hydrodynamics and Transport for Water Quality Modeling*. Lewis Publishers, Boca Raton, FL.

Mortimer, C.H. 1974. Lake hydrodynamics. *International Association of Applied Limnology Mitteilungen* 20, 124–197.

Nix, J. 1981. Contribution of hypolimnetic water on metalimnetic dissolved oxygen minima in a reservoir. *Water Resources Research* 17, 329–332.

Nix, J. 1987. Distribution of iron and manganese in DeGray Lake, Arkansas. In R.H. Kennedy and J. Nix (eds), *Proceedings of the DeGray Lake Symposium*. U.S. Army Engineer Waterways Experiment Station, Vicksburg, MS.

Pastorok, R.A., T.C. Ginn, and M.W. Lorenzen. 1981. Evaluation of aeration/circulation as a lake restoration technique, EPA 600/3-81-014. U.S. Environmental Protection Agency, Washington, DC.

Pollard, R.T., P.B. Rhines, and R.O.R.Y. Thompson. 1973. The deepening of the wind mixed layer. *Geophysical Fluid Dynamics* 3, 381–404.

Prince, J.M. 2011. Modeling Eurasian watermilfoil (*Myriophyllum spicatum*) habitat with geographic information systems. PhD dissertation, Mississippi State University.

Ragotzkie, R.A. 1978. Heat budget of lakes. Chapter 1 in A. Lerman (ed.), *Lakes: Chemistry, Geology, Physics*. Springer-Verlag, New York, pp. 1–19.

Roberts, P.J.W. and M.S. Dortch. 1985. Entrainment descriptions for mathematical modeling of pumped-storage in reservoirs, Technical Report E-85-12. U.S. Army Engineers Waterways Experiment Station, Vicksburg, MS.

Rohweder, J., J.T. Rogala, B.L. Johnson, D. Anderson, S. Clark, F. Chamberlin, and K. Runyon. 2008. Application of wind fetch and wave models for habitat rehabilitation and enhancement projects. U.S. Geological Survey Open-file Report 2008–1200.

Schneider, M.L., S.C. Wilhelms, and L.I. Yates. 2004. SELECT Version 1.0 beta: A one-dimensional reservoir selective withdrawal model spreadsheet, ERDC/EL SR-04-1. US Army Corps of Engineers, Water Operations Technical Support Program, Vicksburg, MS.

Smith, I.R. 1975. *Turbulence in Lakes and Rivers*. Scientific Publication 29. Freshwater Biological Association, Ambleside, UK.

Thornton, K.W., R.H. Kennedy, J.H. Carroll, W.W. Walker, R.C. Gunkel, and S. Ashby. 1981. Reservoir sedimentation and water quality—A heuristic model. In H.G. Stefan (ed.), *Proceedings of the Symposium on Surface Water Impoundments*. American Society of Civil Engineers, New York, pp. 654–664.

Turner, J.S. 1973. *Buoyancy Effects in Fluids*. Cambridge University Press, Cambridge, UK.

USACE. 1984. Shore protection manual. Coastal Engineering Research Center, Fort Belvoir, VA.

Wetzel, R.G. 1975. *Limnology*. Saunders College Publishing, Philadelphia, PA.

Wetzel, R.G. 2001. *Limnology*, 3rd ed. Elsevier, New York.

Worden, D. and L. Pistrang. 2003. Nutrient and plankton dynamics in Wachusett Reservoir: Results of the DCR/DWSP's 1998–2002 monitoring program, a review of plankton data from Cosgrove intake, and an evaluation of historical records. Massachusetts Department of Conservation and Recreation, Division of Water Supply Protection, MA.

14 Chemical and Water Quality Kinetic Characteristics and Processes

14.1 DISSOLVED GASES

The solubility of gases in water is often an important consideration in the management of lakes and reservoirs, both from a regulatory and an ecological perspective. Total dissolved gases (TDGs) may be important, particularly when they are in excess, resulting in gas bubble disease. Other dissolved gases commonly of interest include nitrogen, ammonia, oxygen, carbon dioxide, methane, and hydrogen sulfide. Dissolved oxygen (DO) is always a critical water quality parameter. In many lakes that have an anaerobic hypolimnion, methane, hydrogen sulfide, ammonia, and other gases may accumulate, creating problems associated with hypolimnetic releases.

14.1.1 ATMOSPHERIC COMPOSITION

A primary source of gases dissolved in water is the atmosphere and, under equilibrium conditions, the amount of gas dissolved is proportional to the partial pressure of the individual gases. The atmosphere is primarily composed of the gases listed, along with their properties, in Table 14.1 (AFS 1984). These gases comprise approximately 99.996% of the atmosphere and, with the exception of carbon dioxide, are of relatively constant proportions. Other gases include neon (Ne, about 0.0018%) and hydrogen (H_2, about 0.00005%), with the remaining gases being made up of variable quantities of water vapor, helium (He, about 0.0005%), and methane (CH_4, about 0.00017%).

14.1.2 ATMOSPHERIC EXCHANGES

At equilibrium with the atmosphere, the aqueous-phase concentration would be equal to the gas-phase concentration. This equilibrium concentration is also called the saturation concentration (c_s). If the liquid concentration exceeds the saturation concentration ($c > c_s$), the net movement will be out of the liquid, and if the liquid concentration is less ($c < c_s$), the net movement would be into the liquid. That is, the rate of movement (R) is proportional to the gradient, or it is between the saturation concentration (c_s) and the liquid concentration c.

$$R = K_r \left(c_s - c \right) \tag{14.1}$$

The rate constant (K_r) is a transfer rate that is a function of fluid (air and/or water) turbulence. That is, the greater the turbulence is, the higher the rate will be.

14.1.3 OTHER SOURCES

While gas exchange with the atmosphere is a source of dissolved gases, other mechanisms impact the gas concentration and the composition of lakes, including chemical and biological processes.

TABLE 14.1

Characteristics of Selected Gases

	Oxygen	Nitrogen	Argon	Carbon Dioxide
Composition of air (%)	20.946	78.084	0.934	0.032
Atomic weights (g mol^{-1})	31.9988	28.0134	39.948	44.0098
Atomic volume (L mol^{-1})	22.392	22.403	22.39	22.263
K[a]	1.42903	1.25043	1.78419	1.97681
A[b]	0.5318	0.6078	0.426	0.3845

Source: AFS, *Computation of Dissolved Gas Concentrations in Water as Functions of Temperature, Salinity and Pressure*, Special Publication 14, American Fisheries Society, Bethesda, MD, 1984.

[a] mL L^{-1}*K = mg L^{-1}.

[b] Partial pressure (mmHg) = C_i A_i/Beta$_i$, where C_i = concentration in milligrams per liter and Beta = Bunsen coefficient.

Autochthonous sources may include production and respiration by plants, decomposition (aerobic or anaerobic), and chemical reactions. Allochthonous sources of gases in lakes, in addition to atmospheric exchange, may include groundwaters, and volcanic activity. In addition, the artificial injection of air into lakes may be used to improve oxygen conditions or to destratify lakes. The injection of pure oxygen is also used to improve oxygen conditions in lakes.

14.1.4 GAS SOLUBILITY

The relationship between the amount of gas that can be absorbed by water and the partial pressure of the gas over the water is described by Henry's law. Henry's law is named after J. William Henry (1774–1836; Figure 14.1), an English chemist credited with the relationship.

FIGURE 14.1   William Henry and his paper on gas law. (Courtesy of Wikimedia Commons from the trustees of the British Museum and the Royal Society of London.)

Henry's law can be expressed as:

$$H_e = \frac{p}{c_1}; \quad H_e' = \frac{H_e}{R\,T_a} = \frac{c_g}{c_1}$$ (14.2)

where
H_e is Henry's constant (given in concentration basis, in units of atmosphere per cubic meter per mole)
p is the partial pressure
c_1 is the liquid concentration (molar)
Henry's constant can also be written in a nondimensional form by dividing by the universal gas constant (R) and temperature (T_a), so that it represents the ratio of the aqueous-phase concentration (c_1) to the gas-phase concentration (c_g).

Henry's law represents an equilibrium relationship in that it predicts the water concentration, as a function of the partial pressure, when that concentration is in equilibrium with the atmosphere. At equilibrium, the aqueous-phase concentration (c_1) would be equal to the gas-phase concentration (c_g), resulting in the saturation concentration (c_s). As described earlier, the rate of movement (R) is proportional to the gradient between c_1 and c_g, or between the saturation concentration (c_s) and the liquid concentration (c_1) and the fluid (air and/or water) turbulence.

The saturation concentration can be computed from Henry's constant as

$$c_s = \frac{p}{H_e}$$ (14.3)

Henry's constants for selected compounds are provided in Table 14.2. For example, the atmosphere is about 21% oxygen, so the partial pressure of oxygen is 0.21. The molecular weight of oxygen is about 32 g mol^{-1}; therefore, the saturation concentration of oxygen (at 20°C) is 0.21 atm/(7.74E-01 atm m^{-3} mol^{-1}) = 0.27 mol m^{-3} or 8.7 g m^{-3} = 8.7 mg L^{-1}. The oxygen concentration decreases with altitude (decreasing barometric pressure), and decreases with increasing temperature and salinity. For example, the solubility of oxygen in saltwater is, depending on the temperature, 16%–21% less than that in freshwater.

Henry's law accurately describes the behavior of gases dissolving in water (or other liquids) when concentrations and partial pressures are reasonably low. Henry's constant varies with temperature; therefore, as the temperature increases, so does Henry's constant, the result of which is that solubility decreases. Solubility also varies with other materials dissolved in water, and solubility is commonly related to salinity.

TABLE 14.2
Henry's Constant for Selected Compounds

Compound	Formula	Henry's Constant (20°C)	
		(Dimensionless)	(atm m^{-3} mol^{-1})
Methane	CH_4	64.4	1.55E+00
Oxygen	O_2	32.2	7.74E−01
Nitrogen	N_2	28.4	6.84E−01
Carbon dioxide	CO_2	1.13	2.72E−02
Hydrogen sulfide	H_2S	0.386	9.27E−03
Sulfur dioxide	SO_2	0.0284	6.84E−04
Ammonia	NH_3	0.000569	1.37E−05

There are a wide variety of expressions relating changes in solubility to temperature, salinity, and other factors, from modifications to Henry's law constant to experimentally derived equations. For example, a relatively large number of equations have been developed to predict the saturation concentrations of oxygen, since oxygen is a critical environmental and regulatory parameter. Bowie et al. (1985) reviewed a number of these formulations and compared their predictions, noting discrepancies as high as 11% between formulations, particularly for highly saline conditions.

14.1.5 Impacts of Barometric Pressure, Elevation, and Depth

The equilibrium water solubility of gases is a function of pressure. For air solubility, that pressure is due to the weight of the atmospheric gases (the sum of the partial pressures), which is the atmospheric or barometric pressure. The normal sea-level barometric pressure is 760 mmHg (about 15 psi). That atmospheric pressure decreases as a function of altitude or due to changes in meteorological conditions. For example, a storm may change the atmospheric pressure by 20 mmHg. Similarly, the pressure of pure gases also impacts gas solubility, whether in a closed or an open vessel (or at depth).

Altitude decreases the partial pressure by about 1.0%/100 m of elevation above sea level. So, the partial pressure in Denver, Colorado, the "mile high city," would only be about 82% of what it would be at sea level. The relationship is slightly nonlinear; for example, the p_{alt} for Mount Everest (8850 m) is about 31%, rather less than zero, which would result from a constant 1.0%/100 m.

Henry's relationship indicates that liquid saturation concentrations are a function of pressure. Since water is relatively incompressible, the hydrostatic pressure due to the weight of water is a linear function of the specific weight of water and the water depth ($P = \gamma z$, where z is the depth in meters and γ is the specific weight of water, about 9.81 kN m^{-3} at 4°C). The specific weight can also be computed from the product of density and gravitational acceleration.

For example, the hydrostatic pressure at 1 m depth would be 9.81 kN m^{-2}. Standard atmospheric pressure is 1 atm = 101325 Pa = 1.01325 bar = 101.325 kN m^{-2} = 760 mmHg. Therefore, there is an increase in pressure of about 1 atm for each 10.33 m depth of water. For example, at a depth of 200 m, the pressure would be about 20.4 atm.

Some of the consequences of the increases in pressure are well known. One well-known impact occurs for divers who would have elevated gas levels in their blood as a consequence of the increased pressure and depths. If that pressure is reduced too rapidly for the excess gas to be expelled safely, bubble formation would cause the "bends", which could be fatal. Similarly, gas bubble disease can occur in fish below dams where, rather than depth, the momentum of falling water results in pressure increases. The increase in pressure and the resulting gas concentrations are also of importance for some of the practices used to increase the oxygen concentration in reservoir releases, such as through the injection of oxygen into the hypolimnion near the dam. The increased pressure will also cause gases to be rapidly lost to the atmosphere when reservoir releases are from deep in the reservoir, such as hypolimnetic releases from hydropower operations. The hypolimnetic waters are often devoid of oxygen during summer and fall, and have high concentrations of anaerobic materials, such as methane, hydrogen sulfide, and ammonia, which can cause odor or toxicity problems in the tailwaters. One extreme case of degassing from deep gas-saturated waters is known as a limnetic eruption.

The equilibrium solubility of a gas bubble at a specified depth can be computed from the sum of the air and hydrostatic pressures:

$$P_t = BP + k\rho gz \tag{14.4}$$

where
  P_t is the total pressure (mmHg)
  BP is the air (barometric) pressure at the water surface (760 mmHg)

g is the gravitational acceleration ($9.80655 \text{ m}^{-2} \text{ s}^{-1}$)

ρ is the average water density (kilograms per cubic meter)

k is a conversion factor that converts kPa to mmHg ($1 \text{ kPa} = 7.50062 \text{ mmHg}$) (mmHg/kPa)

z is the depth (meters)

14.1.5.1 Bubble Formation

The relationships between depth and solubility discussed earlier were based on the assumption that the water and an air bubble were in equilibrium. Often, equilibrium does not occur (such as due to a pressure drop) and bubbles may form and rise up through water; if they reach the surface, the gases are ventilated to the atmosphere. The rate of bubble rise is dependent on the bubble size, the properties of the fluid (density and viscosity), the presence of surfactants, etc. In addition, mass exchange occurs between the bubble and the water column. Therefore, the rate of bubble rise versus the rate of dissolution impacts whether a bubble would reach the surface.

The characterization and prediction of bubble formation and transport are important in a number of applications, such as in the design of oxygen injection systems. For those systems, if the bubbles that are formed rise up and are ventilated, then their effectiveness is compromised. Examples of studies of bubble formation and its impact on diffuser design include Little and McGinnis (2001), Martin and Cole (2000), and McGinnis and Little (1998, 2002).

14.1.5.2 Limnetic CO_2 Eruptions

Carbon dioxide concentrations may increase at depth due to the dominance of degradation and decomposition reactions in the hypolimnion, which consume electron acceptors (such as O_2, NO_3^-, and SO_4^{2-}) and produce CO_2 and CH_4. Other sources include the dissolution of calcium carbonate and surface transfers. Carbon dioxide may also be introduced into the hypolimnion from groundwaters and, in some lakes, from volcanic origin. The result can be supersaturated carbon dioxide concentrations in the hypolimnion. The catastrophic release of this carbon dioxide, such as that resulting from overturn or other factors bringing this highly supersaturated water to the surface, may result in an explosive eruption, called a limnetic eruption.

In 1986, a tremendous explosion of CO_2 from Lake Nyos, a crater lake west of Cameroon, killed more than 1700 people and livestock up to 25 km away. Lake Nyos has an area of 1.5 km^2, its depth exceeds 200 m, and it is strongly stratified (Figure 14.2). Dissolved CO_2 seeping from springs beneath the lake became trapped by the high hydrostatic pressure. While the exact mechanism is not known, it is speculated that an overturn of the whole lake resulted in exposing the supersaturated waters, with the consequent limnetic eruption of CO_2 on August 21, 1986 (Evans et al. 1993). A CO_2 cloud formed, estimated to be 50 m thick, and since CO_2 is more dense than air, the cloud

FIGURE 14.2 Lake Nyos, Cameroon, following the gas eruption. (Courtesy of Wikimedia Commons.)

FIGURE 14.3   Degassing project for Lake Nyos. (Courtesy of Wikimedia Commons.)

sank downslope, killing people and animals in its path. A similar but smaller event occurred at Lake Monoun, Cameroon, on August 15, 1984 (Sigurdsson et al. 1987). The recharge of Lake Nyos is of concern and was measured by Evans et al. (1993). Degassing methods have been used to aid in preventing a recurrence of the limnetic eruption, and include a pipe from the surface to the lake bottom allowing excess gas to be released to the atmosphere (Figure 14.3).

Conditions similar to those in Lake Nyos, which may potentially result in limnetic eruptions, occur in other lakes, including Lake Kivu. Lake Kivu is located in a heavily populated area situated on the border of Rwanda and the Democratic Republic of the Congo. Both carbon dioxide and methane are problems in Lake Kivu.

14.2   TOTAL DISSOLVED GAS

TDG refers to the total pressure of all gases and is a concern particularly below reservoir spillways where increases in pressure, due to the spillage of water, cause increases in gas concentrations. This is typically a transient condition, therefore as the pressure decreases, the water will degas. Similarly, fish exposed to excessive dissolved gas pressure or tension also have excess gas dissolved in their circulatory system. When the gas comes out of a solution it may form bubbles (emboli), which block the flow of blood through the capillary vessels, causing distress or death; this is called "gas bubble disease" (discussed in more detail in Chapter 18). Water quality criteria to protect aquatic life generally limit the TDG concentration to 110% of the saturation value for gases at the existing atmospheric and hydrostatic pressures.

14.3   OXIC VERSUS ANOXIC RESERVOIR PROCESSES

The water quality of lakes and reservoirs is controlled to a large degree by the transport and distribution of light and heat, in addition to chemical and biological processes (e.g., productivity, respiration, and decomposition). The classical characterization of the seasonal cycle in deep dimictic lakes is that following spring overturn a seasonal stratification pattern develops where a warm, well-oxygenated epilimnion lies over a cold, less oxygenated (and often hypoxic) hypolimnion. For dimictic lakes, there is then a fall overturn followed by an inverse stratification period in the winter months, where due to reduced exchange through ice and snow, deoxygenation can again occur in deeper waters.

The water quality in a stratified reservoir depends on time, temperature, environmental conditions, and terminal electron acceptors (TEAs). TEAs are necessary in the metabolic process of oxidizing organic matter to provide food for aquatic life. When DO is present (oxic conditions), oxygen serves as the TEA, while under anoxic conditions a different set of microorganisms emerge that can use other forms of oxygen (e.g., NO_3, MnO_2, SO_4, and Fe^{2+}).

Autotrophic microorganisms are those that obtain essential "biologically useful energy" (BUE) from light (phototrophs) or inorganic chemical (chemotrophs) reactions. BUE is used by autotrophs to synthesize the biomass from inorganic nutrients (e.g., CO_2, N, P, K, and Si). The process can be represented by the following equation (Gordon and Higgins 2007):

$$CO_2 + H_2X + BUE + NH_3 \rightleftarrows C_5H_7O_2N \text{ (cells)} + X \text{ (unbalanced)} \qquad (14.5)$$

When X indicates oxygen, photosynthesis results in DO production. When X indicates sulfur, non-oxygenic growth results in sulfide production.

Heterotrophic bacterial growth uses organics for both BUE and cell synthesis. Since organics are oxidized for BUE, there must also be a reduction in materials so that the oxidation and reduction are in equilibrium. These types of reactions in which electrons are exchanged are referred to as oxidation–reduction reactions or redox reactions.

In redox reactions, one substance gives up electrons and another receives electrons. These two materials are referred to as a redox pair, where the material that loses electrons is oxidized while material that gains electrons (the TEA) is simultaneously reduced. Since oxygen has one of the highest affinities for electrons, the term *oxidized* is used, but oxygen is not necessary for a redox reaction to occur.

The oxidation–reduction potential (ORP) is a measure of the ability of a substrate to gain or lose electrons, and is typically measured in millivolts (mV). Energy sources that are more highly oxidized have greater ORP values, therefore high and positive values are associated with oxic conditions and negative values are associated with anoxic conditions. The ORP is a commonly measured limnological parameter (where there are sufficient TEAs), particularly where data are being collected to support the computation of metal speciation. The reader is referred to Nordstrom and Wilde (2005) for a discussion of measurement techniques.

TEAs must be available for heterotrophic organisms to utilize organic matter and facilitate growth. The equation representing the process is (Gordon and Higgins 2007)

$$\text{Organics} + \text{TEA} \rightleftarrows CO_2 + EP \text{ (end products)} \qquad (14.6)$$

However, the value of the BUE obtained varies depending on the quality of the food (organics) and the TEA used. Table 14.3 and Figure 14.4 show the common TEAs in an aquatic environment and their end products and the BUE yields as reported by Bouwer (1992) and Fenchel et al. (1998), as cited in Gordon and Higgins (2007).

TABLE 14.3

Terminal Electron Acceptors and Their End Products and BUE

Desirability	Terminal e⁻ Acceptor	End Products	BUE Yield
Most favorable	O_2	H_2O	29.9 kcal eq⁻¹
Less favorable	NO_3	N_2	28.4
Undesirable	MnO_2	Mn^{2+}	23.3
Undesirable	$FeO(OH)$	Fe^{2+}	10.1
Undesirable	SO_4	H_2S	5.9
Unfavorable	Organics	Reduced organics	Na
Least favorable	CO_2	CH_4	5.6

Source: Gordon, J.A. and Higgins, J.M., *Energy Production and Reservoir Water Quality*, American Society of Civil Engineers, Reston, VA, 2007.

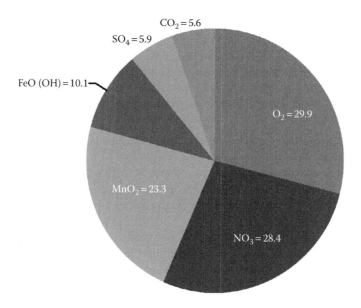

FIGURE 14.4   Relative BUE yield of various oxidants.

Typically in lakes and reservoirs, the ORP would be a high positive value in the oxic epilimnion and would then decrease with depth. In an anoxic hypolimnion, and in the anoxic sediments, the ORP would be relatively high and negative (reducing).

As oxygen (and organics) is consumed under oxic conditions (i.e., oxygen is the TEA), the ORP is typically from about 400 to 300 mV. This oxidation also results in a decrease in pH, typically from neutral to 6.5 (Gordon and Higgins 2007).

Once the oxygen is depleted, the chemical characteristics of the lakes or reservoirs are largely impacted by the availability of TEAs. The end product of these processes may be problematic as these materials are entrained into surface waters or they are released. For example, hypolimnetic releases such as from hydropower facilities may not only be anoxic, but they may also contain high concentrations of reduced materials, such as NH_4, Mn^{2+}, Fe^{2+}, H_2S, and CH_4.

Somewhat surprisingly, the sequence in which the TEAs are utilized generally follows a thermo-dynamically predicted sequence, illustrated by a BUE yield (Table 14.3). As indicated by Di Toro (2001), it is not intuitively obvious why this should be the case, since all of these reactions are bac-terially mediated. But, data, as reported by Di Toro, demonstrate that is the case, with the exception of NO_3, where the predicted route of denitrification is from NO_3 to NH_4^+, where most if not all of the reduced nitrogen appears as N_2 (Table 14.3).

Denitrification poises the ORP at about 200 mV and the pH may drop to 6 (Gordon and Higgins 2007). As will be discussed later, there are usually substantially elevated concentrations of ammonia (NH_4^+) in the anoxic hypolimnion of lakes and reservoirs. However, this may occur due to ammonifica-tion, which is the release of ammonia from decomposing organics. Bacteria first strip off the ammonia from nitrogenous organics before further oxidation; subsequently, the ammonia may build up under anoxic conditions because the only sink is cell synthesis, which is minimal (Gordon and Higgins 2007).

Under anaerobic conditions and following the depletion of nitrate, manganese is the next most thermodynamically preferable TEA. Manganese reduction is primarily associated with bottom sediments, and, as a result, the ORP drops from 200 to 50 mV and the pH remains at about 6–6.5 (Gordon and Higgins 2007).

Next, solid-phase hydrous oxides of iron are used for TEAs with the soluble form released being Fe^{2+}, which accumulates under anaerobic conditions. The ORP must be at about 50 mV for 2 weeks before reduction begins and the pH will be 6–6.5 (Gordon and Higgins 2007).

Following iron, the next most thermodynamically favorable TEA is sulfate (SO_4). Sulfate reduction typically occurs following extended periods of anoxia with the ORP becoming negative at about –150 mV (Gordon and Higgins 2007). Sulfate reduction is recognized as an important TEA in estuarine and marine sediments, but it is often assumed to have a less important role in freshwater systems (Capone and Kiene 1988), partially due to its relatively lower sulfate concentrations. However, perhaps even in cases where the magnitude of the sulfate reduction is small compared to other TEAs, it still remains important because of the end product of that reduction, hydrogen sulfide (H_2S), commonly referred to as "rotten egg" gas.

Sulfate reduction also reduces the pH and the ORP to a point where fermentation begins to occur, where organics are reduced to organic acids and alcohols, which accumulate as end products. These end products produce an immediate oxygen demand when released to an aerobic environment and typically have objectionable odors. These reduced organics are also precursors to methane formation (Gordon and Higgins 2007).

When anaerobic conditions and organic acids are present, CO_2 will serve as a TEA and form methane, which may accumulate in bottom waters or sediments. For methanogenesis, the ORP remains at about –200 mV and the pH increases due to the destruction of organic acids (Gordon and Higgins 2007). The methane is relatively insoluble, forming a gas phase that may escape from the sediments and bottom waters as bubbles (Di Toro 2001).

14.4 OXYGEN

14.4.1 STANDARDS AND CRITERIA

DO is vital to aquatic life (for aerobic organisms) and is used by most regulatory agencies as an indicator of aquatic health and as a surrogate enforcement parameter. The typical minimum DO criteria for the protection of aquatic life are 5 mg L^{-1} for warmwater ecosystems and 6 mg L^{-1} for cold-water ecosystems. However, criteria vary as a function of aquatic life and life stages and the exposure periods (e.g., using instantaneous, daily, weekly, or monthly values). Therefore, the protection of aquatic health in the implementation of the Clean Water Act (e.g., the national pollutant discharge elimination system [NPDES] permitting and total maximum daily load [TMDL] studies) is commonly based, in part, on maintaining adequate DO concentrations in lakes and reservoirs.

While the concept for oxygen criteria in lakes and reservoirs is similar to that used in rivers and streams, its implementation is problematic since, unlike rivers and streams, lakes and reservoirs commonly stratify so that there are often large spatial as well as temporal variations in the DO concentrations. In the case of stratified lakes, hypolimnetic oxygen depletion may be naturally occurring or enhanced due to excess loads of nutrients and organic materials (e.g., eutrophication, which will be discussed in Chapter 17).

In reservoirs, which are not natural systems, stratification is commonly the result of what is called "hydromodification." The U.S. Environmental Protection Agency (USEPA 1993) defined hydromodification as the "alteration of the hydrologic characteristics of coastal and non-coastal waters, which in turn could cause degradation of water resources," which includes the impacts of dams. For example, if a dam were not there then perhaps stratification (and hence hypolimnetic oxygen depletion) would not have occurred. For this reason, the U.S. EPA concluded that TMDL evaluations were not required for systems that are impaired (not meeting water quality standards) due to hydromodification.

The Kansas water quality standards illustrate the dilemma. Kansas established universal (applicable streams, lakes, etc.) numeric criteria of 5 mg L^{-1} for DO and included the following condition: "the concentration of dissolved oxygen in surface waters shall not be lowered by the influence of artificial sources of pollution." However, as noted by the Kansas Department of Health and Environment (KDHE 2011), lakes "with moderate levels of nutrient or algal content have sufficient levels of dissolved oxygen, but see dissolved oxygen diminish at their lower depths," such as Clinton

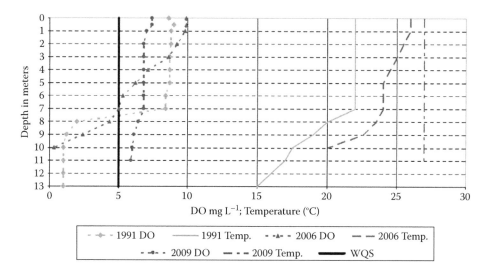

FIGURE 14.5 Oxygen and temperature profiles for Clinton Lake, Kansas, in comparison to the state's water quality standard. (WQS; From KDHE, Water Quality Standards White Paper, Allowances for low dissolved oxygen levels for aquatic life use, Kansas Department of Health and Environment, Bureau of Water, 2011.)

Lake, illustrated in Figure 14.5. Clinton Lake periodically stratifies and, during periods of stratification, oxygen concentrations decline. Thus, during periods of stratification, the state standards are violated. Among the options considered by Kansas (KDHE 2011) for potential exceptions to the state's rule is to "explicitly state allowances for dissolved oxygen lower than 5 mg L^{-1} when caused by documented natural conditions," or to "explicitly exclude applying dissolved oxygen criteria to the lowest portions of a lake (i.e., the hypolimnion)."

A number of other states (e.g., Maryland, MDE 2004; Minnesota, MPCA 2003) do not currently address the effects of stratification on DO concentrations in their water quality regulations. For Louisiana, the state standard is "for a diversified population of fresh warm-water biota including sport fish, the DO concentration (Section 1113 Title 33, Part IX, Subpart 1 May 2007) shall be at or above 5 mg L^{-1}." As illustrated in Figure 14.6 for an oxbow lake in Louisiana, this poses a similar problem to that of Kansas, since the criteria are not met in the typically seasonally anoxic hypolimnion.

Some states address the issue by specifying a particular depth at which the criteria are applied. For example, Mississippi (MDEQ 2007) standards for DO are daily average concentrations of not less than 5.0 mg L^{-1} with an instantaneous minimum of not less than 4.0 mg L^{-1}, and for waters that are thermally stratified, based on samples that are at mid-depth of the epilimnion if the epilimnion depth is 10 ft. or less, or 5 ft. from the water surface if the epilimnion depth is greater than 10 ft.

14.4.2 OXYGEN SATURATION

As discussed in Section 14.1.4, one source of oxygen is the atmosphere, which is approximately 21% oxygen (Table 14.1). Saturation concentrations vary as a function of temperature, salinity, and pressure. The impact of temperature may be estimated from (APHA 1995)

$$\ln o_s(T,0) = -139.34411 + \frac{1.575701 \times 10^5}{T_a} - \frac{6.642308 \times 10^7}{T_a^2}$$

$$+ \frac{1.243800 \times 10^{10}}{T_a^3} - \frac{8.621949 \times 10^{11}}{T_a^4} \tag{14.7}$$

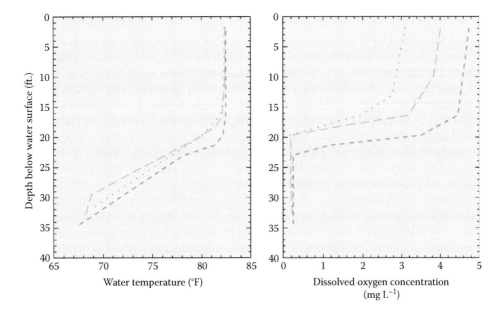

FIGURE 14.6 Variation of water temperature and dissolved oxygen concentration, specific conductance, and pH at Lake Bruin, LA, September 24, 1997. (From Ensminger, P.A., Bathymetric survey and physical and chemical-related properties of Lake Bruin, Louisiana, February and September 1997, from Water-Resources Investigations Report 98-4243, U.S. Geological Survey, Baton Rouge, LA, 1998.)

where $o_s(T, 0)$ is the saturation concentration of DO in freshwater at 1 atm (magnesium peroxide per liter) and T_a is the absolute temperature (kelvin), where $T_a = T + 273.15$. The effect of elevation is accounted for by Equation 14.8:

$$o_s(T, \text{elev}) = e^{\ln o_s (T,0)}(1 - 0.0001148\, H) \tag{14.8}$$

where H is the elevation in meters. The impact of temperature and elevation on saturation concentrations is illustrated in Figure 14.7.

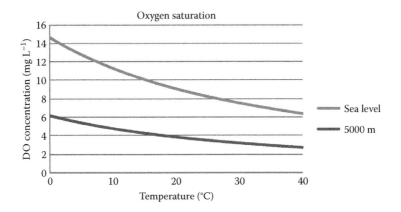

FIGURE 14.7 Oxygen saturation concentrations at sea level and at 5000 m.

14.4.3 Sources and Sinks

The factors affecting DO in lakes and reservoirs include the production and decomposition of organic matter, as illustrated by Figure 14.8. One source (or sink) of oxygen is exchange with the atmosphere. Unlike rivers that are dominated by stream turbulence, most of the exchange in lakes is due to wind and waves. The exchange is based on the rate of turbulence and the difference in the partial pressures of the oxygen (or the differences between the oxygen concentration and the saturation concentration). Depending on the degree of vertical mixing, the exchange may contribute to an oxygenated surface layer (epilimnion) but not to deeper waters. At times of high productivity, the surface waters of some lakes may become supersaturated, resulting in the venting of oxygen to the atmosphere. Oxygen may also be artificially injected into some lakes as a management or restoration practice, which will be discussed in later chapters.

Organic material added to lakes and reservoirs may be derived from external sources (point and nonpoint loads, such as tributaries, waste discharges, and runoff) or internal sources. Lakes and reservoirs often differ from streams and rivers, particularly low-order systems, in that much of the organic material is autochthonous rather than allochthonous. While generalizations are often problematic, the relative importance of the littoral components of autochthonous production increases greatly and changes markedly from nutrient-limited to nutrient-rich lakes (Wetzel 2001).

The stoichiometric relationship between production and respiration or the decomposition of organic matter is commonly based on some assumed representative reaction, such as shown in Equation 14.9 assuming that ammonium is a substrate (Chapra et al. 2008):

$$106CO_2 + 16NH_4^+ + HPO_4^{2-} + 108H_2O \rightleftarrows C_{106}H_{263}O_{110}N_{16}P_1 + 107O_2 + 14H^+ \qquad (14.9)$$

where the forward reaction represents production and the backward reaction represents respiration. If that (Equation 14.9) were the case, then for each gram of carbon degraded or respired, 2.67 g of oxygen (as O_2) would be consumed as computed by

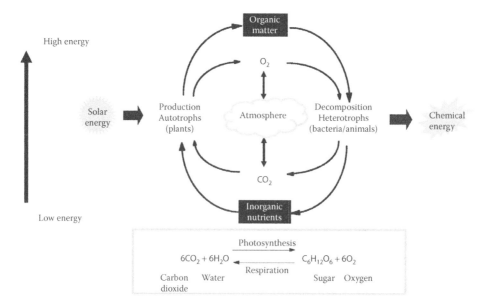

FIGURE 14.8 Natural cycle of organic production and decomposition. (Modified from Chapra, S.C., *Surface Water Quality Modeling*, McGraw-Hill, New York, 1997.)

$$r_{oca} = \frac{106\,\text{mol}\,O_2(32\,g\,O_2/\text{mol}\,O_2)}{106\,\text{mol}\,C(12\,g\,C/\text{mol}\,C)} = 2.67\,\frac{g\,O_2}{g\,C}$$

or conversely for each gram of carbon produced, 2.67 g of O_2 would be produced.

Production is impacted by the presence of light (Figure 14.8) and by temperature, which varies over the course of a day as well as over the seasons. Partially as a result, there are often large temporal as well as spatial variations in the distribution of DO.

14.4.4 DISTRIBUTION OF OXYGEN

14.4.4.1 Diel Variations

As light variations occur over the course of a day, relatively large diel variations in oxygen concentrations may occur. The extent of these variations is a function of a number of environmental factors, including the rates of productivity and respiration, basin morphometry, wind speed and direction, depth, and other factors.

Deep dimictic temperate lakes are likely to undergo only moderate diel variations in one or more physicochemical properties (Hutchinson 1957). In contrast, shallow productive tropical lakes may show less stratification and more extreme diel variations in their physicochemical parameters, as illustrated by diel variations in temperature and DO in Lake Kissimmee, Florida (Dye et al. 1980; Figures 14.9 and 14.10). As illustrated, temperature stratification occurs during daylight hours but it is broken down by nocturnal cooling and wind mixing. Similarly, strong vertical variations in oxygen occur during daytime hours, but they become mixed during the night.

A somewhat similar pattern in diel variations occurs in deeper southern U.S. reservoirs, such as Lake Livingston, Texas (Martin 1984). During July, the reservoir is stably stratified with oxygen depleted in the hypolimnetic waters (Figure 14.11). During daylight hours and under low wind

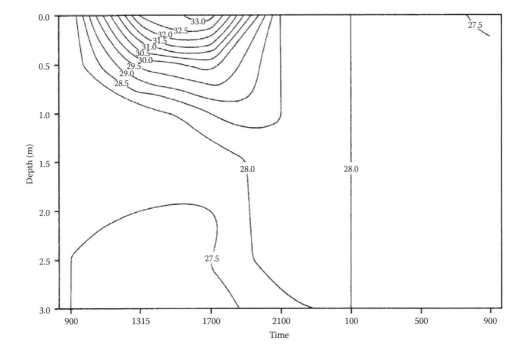

FIGURE 14.9 Variations in water temperatures at Station 22 of Lake Kissimmee, Florida, July 16–17, 1974. (From Dye, C.W., Jones, D.A., Ross, L.T., and Gernert, J.L., *Hydrobiologia* 71, 51–60, 1980. With permission.)

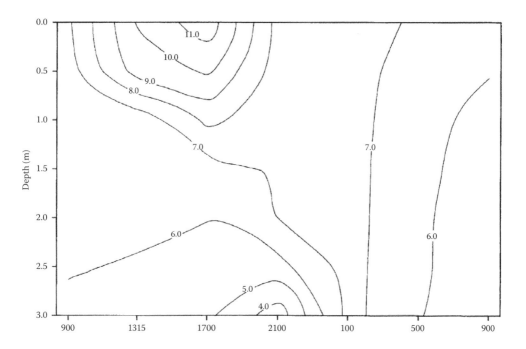

FIGURE 14.10　Variations in dissolved oxygen concentrations at Station 22 of Lake Kissimmee, Florida, July 16–17, 1974. (From Dye, C.W., Jones, D.A., Ross, L.T., and Gernert, J.L., *Hydrobiologia* 71, 51–60, 1980. With permission.)

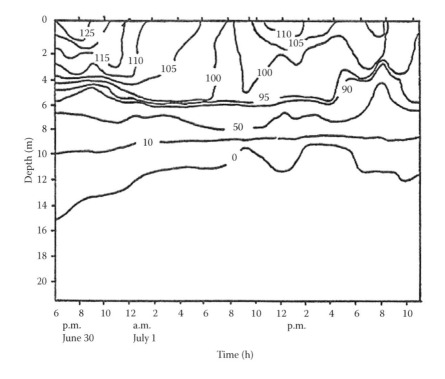

FIGURE 14.11　Measured variations in the percentage of oxygen saturation during July 1983, for Lake Livingston, Texas. (From Martin, J.L., Models of diel variations of water quality in a stratified eutrophic reservoir (Lake Livingston, Texas), PhD. dissertation, Texas A&M University, 1984.)

speeds, epilimnetic DO concentrations in the highly productive reservoir become supersaturated, associated with strong vertical oxygen gradients. However, in the early morning hours with increased wind speeds, mixing substantially reduces the gradient and results in decreased oxygen concentrations (exceeding 4 mg L^{-1}) in surface waters. This is not due to a change in productivity, but rather to a change in the rates of vertical mixing. The impact of wind speed on vertical distribution of oxygen and pH is illustrated in Figure 14.12.

14.4.4.2　Seasonal Distribution: Vertical Distribution

Typically, oxygen concentrations vary over a yearly cycle in lakes and reservoirs as functions of temperature, productivity, and decomposition. In a completely mixed lake with no production, the oxygen concentration would be expected to be a constant value equal to the saturation concentration. That would be the case for a period of complete mixing for either of the vertical profiles for a eutrophic or oligotrophic lake, as illustrated in Figure 14.13.

For an oligotrophic lake (little to no production or resulting decomposition), during periods of summer stratification, the temperature of the surface waters would be warmer resulting in, due to lower saturation concentrations, the oxygen concentrations in the surface water being lower than in the bottom waters, an *orthograde* profile (Wetzel 2001). With inverse stratification during the winter months, and with decreased surface temperatures, the oxygen concentrations in the surface waters may increase, but will still commonly remain lower than that of the hypolimnion.

For a stratified eutrophic lake during the summer months, with increased productivity (limited to the photic zone) and reduced transparency, oxygen concentrations in the warmer epilimnion would be greater than those in the cooler hypolimnion, where decomposition dominates and the oxygen concentration would be reduced as a result. The resulting profile would be a *clinograde*

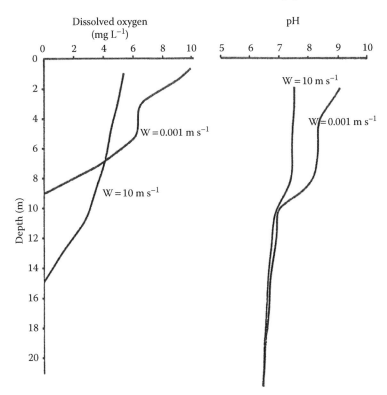

FIGURE 14.12　Predicted vertical variations in oxygen and pH as a function of wind speed. (From Martin, J.L., Models of diel variations of water quality in a stratified eutrophic reservoir (Lake Livingston, Texas), PhD. dissertation, Texas A&M University, 1984.)

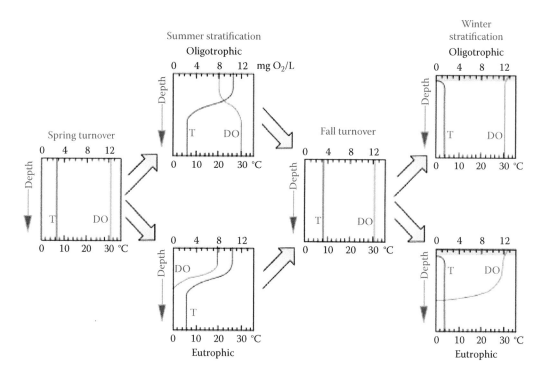

FIGURE 14.13 Idealized vertical distribution of oxygen concentrations and temperature for an oligotrophic and eutrophic dimictic lake. (From Water on the Web, Monitoring Minnesota Lakes on the Internet and training water science technicians for the future—A national on-line curriculum using advanced technologies and real-time data. University of Minnesota, Duluth, MN, 2004. With permission. Reprinted from *Limnology: Lake and River Ecosystems*, 3rd ed., Wetzel, R.G., Copyright 2001, with permission from Elsevier.)

profile (Wetzel 2001; Figure 14.13). During the winter months, in the presence of ice cover, oxygen concentrations may again be reduced with depth as a function of decomposition, reduced surface exchange, and light penetration through the snow and ice.

The amount or proportion of the reservoir that is hypoxic or anoxic, and the rate at which it becomes anoxic, is a commonly used limnological parameter in lake management and in assessing the trophic status of a lake or a reservoir. It may be expressed in various ways, perhaps the most common of which is the areal hypolimnetic oxygen deficit (AHOD). The AHOD is defined as the change in DO per unit area (e.g., square centimeter) of hypolimnetic surface per day (e.g., milligrams of dissolved oxygen per square centimeter per day) over the stratification season. The AHOD was initially proposed by Strom (1931) and Hutchinson (1938) and is intended to remove the influence of lake morphometry, making this metric more comparable between reservoirs.

Another common feature of the vertical structure of many reservoirs is oxygen minima and maxima. That is, the maximum or minimum concentrations occur at some intermediate depth, rather than at the surface or at the bottom. An example of DO maxima is illustrated in Figure 14.14 for Bulls Shoals Reservoir in Arkansas and an example of DO minima and maxima is illustrated in Figure 14.15 for Lake Mead, Arizona and Nevada. Oxygen maxima and minima may be due to a number of factors, such as interflows bringing high or low DO water from upper reservoir regions to material accumulating at the density gradient represented by the thermocline.

14.4.4.3 Seasonal Variations in Longitudinal Distribution

Seasonal variations in the longitudinal distribution of oxygen also typically occur, particularly in deep storage reservoirs, as described by Cole and Hannan (1990). Sedimentation in the transition zone and spring warming often result in hypoxia first developing in this upper zone of a reservoir.

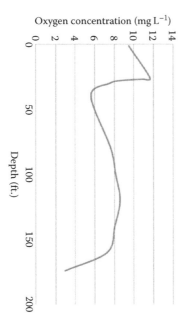

FIGURE 14.14 Vertical oxygen concentration profile for Bull Shoals Reservoir, Arkansas, on June 12, 2012. (Based on online dissolved oxygen and temperature lake reports from the U.S. Army Corps of Engineers, Little Rock District, Available at http://www.swl-wc.usace.army.mil/.)

Then, as the season progresses, and depending on the flow and other conditions, the hypoxic zone extends further down the reservoir toward the dam, as illustrated in Figure 14.16.

This trend is illustrated in Figure 14.17a and b using CE-QUAL-W2 model predictions for Lake Lanier in Georgia by Martin and Hesterlee (1998). Lake Lanier is a warm monomictic reservoir, impounded by Buford Dam on the Chattahoochee River in north-central Georgia (northeast of Atlanta). The reservoir was initially constructed for hydropower, flood control, navigation, and for streamflow regulation. As shown in Figure 14.17a and b, the oxygen depression in the upper portions of the reservoir begins as early as March and then progresses down the reservoir, with much of the deeper portions of the reservoir being anoxic during the summer months. The oxygen distribution is impacted by inflows, wind mixing, and the operation of the hydropower facility. As cooling begins in the late fall, the oxygen concentrations increase in the upper portions of the reservoir and the bottom volume of the hypoxic water diminishes. The reservoir completely mixes usually in January.

Similarly, predicted for July 31, 1995, the spatial distributions of water temperature, water age, chlorophyll-a, and DO concentrations are illustrated in Figures 14.18 through 14.21 for Lake Walter F. George and Lake Allatoona in Georgia, both of which are warm monomictic systems. Lake Walter F. George is located along the Georgia–Alabama border in the southern portion of both states, and is managed for hydropower, navigation, and recreation. Lake Allatoona is one of the most frequently visited U.S. Army Corps of Engineers ("the Corps") lakes in the nation, receiving more than 13 million visitors each year and is located north of Atlanta. The Allatoona Dam is also operated for hydropower.

The predictions for Lake Walter F. George for this particular date demonstrate the impact of a high-flow event moving into the reservoir, so that the riverine zone extends a considerable distance into the reservoir. This can be seen in the contours of the temperature and water quality constituents, but it is most apparent in the contours for water age. Recall that water age is a synthetic (model) parameter that is used to estimate the residence time, which accumulates at a rate of 1.0 day^{-1} with all "new" (boundary) water entering the system having an age of zero. So, by July 31, beginning from January 1, the maximum possible age would be 212 days (365 days for the year). The event illustrated (Figure 14.18b) that the "oldest" water for this date had been pushed to near the dam,

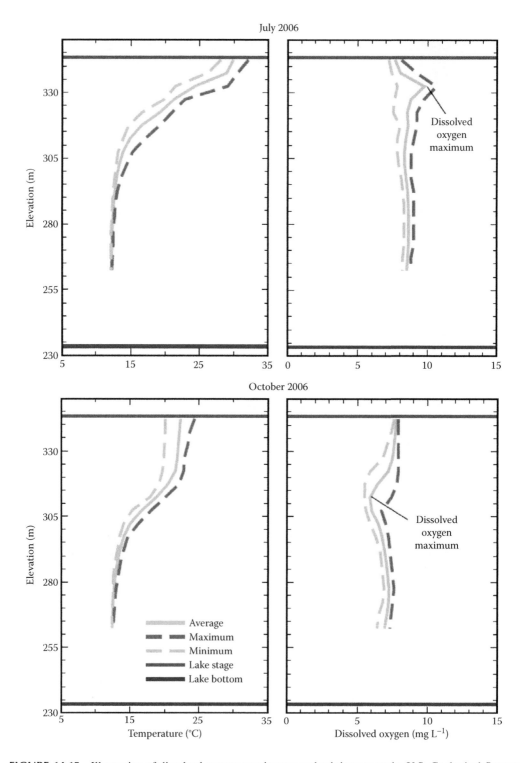

FIGURE 14.15 Illustration of dissolved oxygen maximums and minimums at the U.S. Geological Survey Virgin Basin station on Lake Mead for July 2006 and October 2006. (From Veley, R.J. and Moran, M.J., Evaluating lake stratification and temporal trends by using near-continuous water-quality data from automated profiling systems for water years 2005–2009, Lake Mead, Arizona and Nevada, U.S. Geological Survey, Carson City, NV, 2011.)

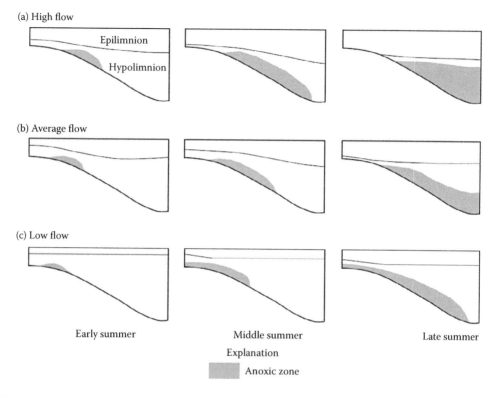

(a) High flow

Epilimnion

Hypolimnion

(b) Average flow

(c) Low flow

Early summer   Middle summer   Late summer

Explanation

Anoxic zone

FIGURE 14.16 (a–c) Generalized longitudinal distribution in hypoxia. (From De Lanois, J.L. and Green, W.R., *Hypolimnetic dissolved-oxygen dynamics within selected White River reservoirs, northern Arkansas– southern Missouri, 1974–2008*, U.S. Geological Survey, Reston, VA, 2011; After Cole, T.M. and Hannan, H.H., *Reservoir Limnology—Ecological Perspectives*, Wiley, New York, 1990.)

while much of the upper reservoir consisted of "young" water. Figure 14.19a also demonstrates a peak chlorophyll-a concentration located midway down the reservoir, illustrating the impact of a local discharge. Even given the impact of the high-flow event, beyond the riverine zone the hypo-limnion of the reservoir is anoxic for this date and much of the year.

For Lake Allatoona, the residence time, as indicated by water age, is greater than it is for Lake Walter F. Georgia and the horizontal gradients are less, but with the highest productivity in the upper zone of the reservoir. For this date, most of the reservoir below the relatively shallow hypo-limnion is anoxic (Figure 14.21).

14.5 NITROGEN

14.5.1 Sources and Sinks of Nitrogen

Nitrogen is important for lakes and reservoirs for a number of reasons, not the least of which is that it is an essential nutrient and an indicator of pollution (e.g., eutrophication).

Nitrogen can exist in a number of oxidation states, as shown in Table 14.4, all of which are of environmental importance. The processes impacting nitrogen are illustrated in Figure 14.22.

14.5.1.1 Atmospheric Sources/Sinks

Nitrogen in the atmosphere typically occurs as molecular nitrogen (N_2), or in dissolved or par-ticulate forms. The molecular form makes up approximately 78% of our atmosphere (Table 14.1). Other forms in the atmosphere include trace amounts of nitrogen oxides, nitric acid vapor, gaseous

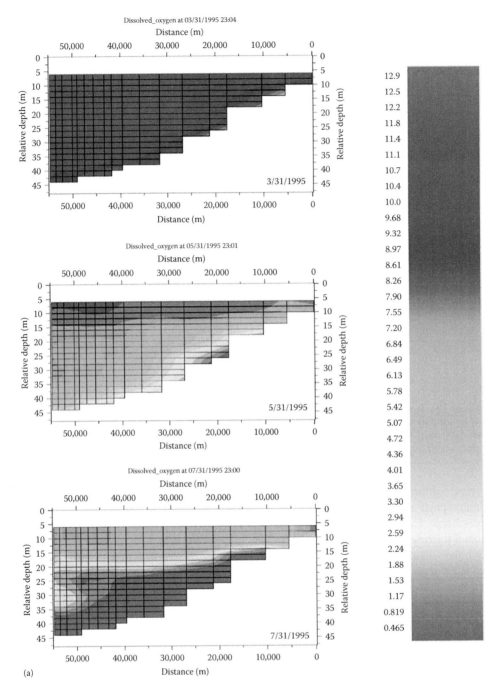

FIGURE 14.17 (a,b) Predicted dissolved oxygen concentrations (mg L^{-1}) in Lake Lanier, Georgia, for selected dates in 1995. (From Martin, J.L. and Hesterlee, C., Detailed reservoir water quality modeling to support environmental impact statements for the evaluation of ACT and ACF river basins water allocation, Contract Report prepared for the U.S. Army Engineer District, Mobile, 1988.)

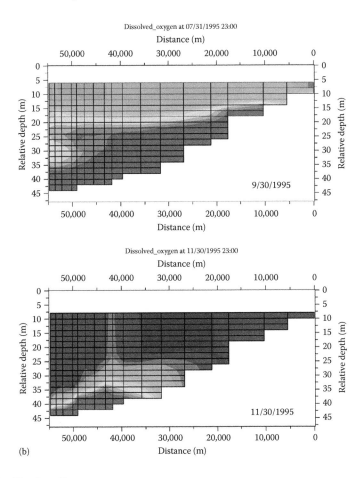

FIGURE 14.17 (Continued)

ammonia, particulate nitrate and ammonium compounds, and organic nitrogen. These forms can serve as sources or sinks for nitrogen in lakes and reservoirs, such as through gas exchange, and wet and dry deposition.

The processes impacting the gas exchange of molecular nitrogen are similar to those impacting the gas exchange of oxygen. The exchange is a function of turbulent mixing at the water surface, such as due to wind and waves (or, as discussed in the chapter on tailwaters, water plunging over the dam) and the concentration gradient between the water column and the atmosphere. As with oxygen, the solubility of nitrogen is a function of the temperature and pressure, and the computation is described in the American Fisheries Society's *Computation of Dissolved Gas in Water as Functions of Temperature Salinity and Pressure* (AFS 1984). The molecular nitrogen in water is also impacted by denitrification, a source of N_2 (Figure 14.22), and can be lost by the uptake of some plankton, such as some cyanobacteria (see Chapter 16), which can fix atmospheric nitrogen.

In addition to gas exchange, atmospheric wet and dry deposition can serve as an important source of nitrogen to lakes and reservoirs (a nutrient) as well as a source of lake acidification (e.g., by nitrous oxides; USEPA 2008, 2009). Particularly for large lakes, the atmosphere can serve as a substantial source of nutrient loads. The National Atmospheric Deposition Program monitors wet atmospheric deposition of nitrogen, as illustrated by the rates of wet deposition for 2006 in Figure 14.23. In addition, in 2010 the Ammonia Monitoring Network (AMoN) was established to monitor ammonia gas concentrations across the United States.

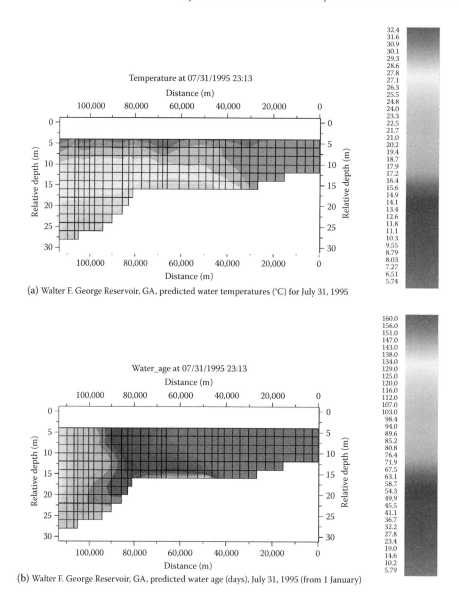

(a) Walter F. George Reservoir, GA, predicted water temperatures (°C) for July 31, 1995

(b) Walter F. George Reservoir, GA, predicted water age (days), July 31, 1995 (from 1 January)

FIGURE 14.18 (a,b) Predicted water temperature (°C) and water age (days from January 1) in Lake Walter F. George, Georgia, for July 31, 1995. (From Martin, J.L. and Hesterlee, C., Detailed reservoir water quality modeling to support environmental impact statements for the evaluation of ACT and ACF river basins water allocation, Contract Report prepared for the U.S. Army Engineer District, Mobile, 1988.)

14.5.1.2 Ammonification

The decomposition of organic forms of nitrogen (organic nitrogen) leads to the release of ammonia, a process referred to as ammonification (Figure 14.22; Equation 14.10):

$$\text{Protein (organic-N)} + \text{bacteria} \rightarrow NH_3$$

(14.10)

Ammonification occurs under either oxic or anoxic conditions. In the photic zone, ammonia may be taken up by plants as an essential nutrient, or it may be produced via excretion. In the aphotic zone, ammonification can result in accumulations of oxygen during periods of stratification.

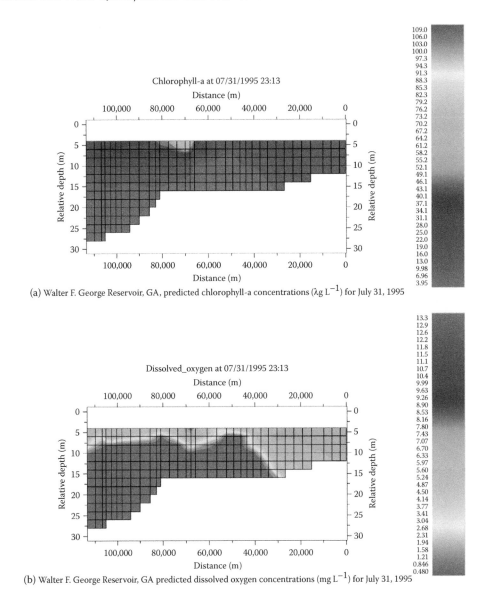

(a) Walter F. George Reservoir, GA, predicted chlorophyll-a concentrations (λg L^{-1}) for July 31, 1995

(b) Walter F. George Reservoir, GA predicted dissolved oxygen concentrations (mg L^{-1}) for July 31, 1995

FIGURE 14.19 (a,b) Predicted chlorophyll-a (ppb) and dissolved oxygen concentrations (ppm) for Lake Walter F. George, GA, for July 31, 1995. (From Martin, J.L. and Hesterlee, C., Detailed reservoir water quality modeling to support environmental impact statements for the evaluation of ACT and ACF river basins water allocation, Contract Report prepared for the U.S. Army Engineer District, Mobile, 1988.)

14.5.1.3 Nitrification

Under oxic conditions, and in the presence of appropriate bacteria, organisms oxidize ammonia, which is referred to as nitrification. This occurs as a two-step process (Figure 14.22), where first the *Nitrosomonas* bacteria convert ammonium (NH_4^+) to nitrite (NO_2^-):

$$NH_4^+ + 1.5O_2 \rightarrow NO_2^- + H_2O + 2H^+ \tag{14.11}$$

and then the *Nitrobacter* bacteria convert nitrite to nitrate:

$$NO_2^- + 0.5O_2 \rightarrow NO_3^- \tag{14.12}$$

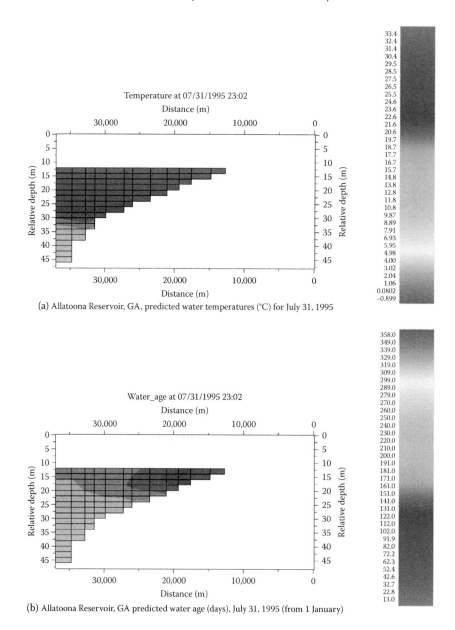

(a) Allatoona Reservoir, GA, predicted water temperatures (°C) for July 31, 1995

(b) Allatoona Reservoir, GA predicted water age (days), July 31, 1995 (from 1 January)

FIGURE 14.20 (a,b) Predicted water temperature (°C) and water age (days) in Lake Allatoona, GA, for July 31, 1995. (From Martin, J.L. and Hesterlee, C., Detailed reservoir water quality modeling to support environmental impact statements for the evaluation of ACT and ACF river basins water allocation, Contract Report prepared for the U.S. Army Engineer District, Mobile, 1988.)

The second reaction is relatively fast compared to the first one (rate limiting), therefore sometimes the two reactions are combined, such as in:

$$NH_4^+ + 2O_2 \rightarrow NO_3^- + H_2O + 2H^+ \tag{14.13}$$

The process of nitrification impacts the form and availability of nitrogen, where ammonia is the form preferred by autotrophs. The process also impacts oxygen. Using the stoichiom-

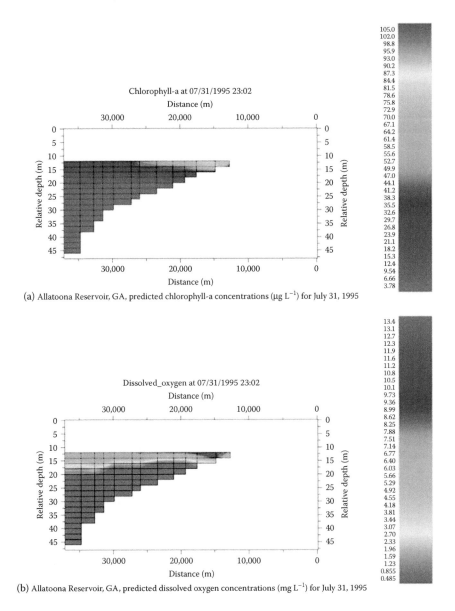

(a) Allatoona Reservoir, GA, predicted chlorophyll-a concentrations (μg L^{-1}) for July 31, 1995

(b) Allatoona Reservoir, GA, predicted dissolved oxygen concentrations (mg L^{-1}) for July 31, 1995

FIGURE 14.21 (a,b) Predicted chlorophyll-a (ppb, μg L^{-1}) and dissolved oxygen concentrations (ppm, mg L^{-1}) for Lake Allatoona, GA, for July 31, 1995. (From Martin, J.L. and Hesterlee, C., Detailed reservoir water quality modeling to support environmental impact statements for the evaluation of ACT and ACF river basins water allocation, Contract Report prepared for the U.S. Army Engineer District, Mobile, 1988.)

etry from Equation 14.13, for each gram of nitrogen nitrified, 4.57 g of oxygen would be consumed.

$$R_{nitrif} = \frac{2\,mol\,O_2\left(32\,g\,O_2/mol\,O_2\right)}{1\,mol\,N\left(14\,g\,N/mol\,N\right)} = 4.57\frac{g\,O_2}{g\,N} \tag{14.14}$$

So, nitrification has a large impact on oxygen concentrations.

TABLE 14.4
Common Oxidation States for Nitrogen

Oxidation State of N	Increasing Levels of Sulfur Oxidation						
Form	-3	0	$+1$	$+2$	$+3$	$+4$	$+5$
Aqueous solution and salts	NH_4^+				NO_2^-		NO_3^-
	NH_3						
Gas phase	NH_3	N_2	N_2O	NO		NO_2	

14.5.1.4 Ammonia Speciation

Ammonia in water may occur in one of two forms, depending on the pH and temperature. The ionization reaction maybe written as (USEPA 1999b)

$$NH_4^+ \rightleftarrows NH_3 + H^+ \tag{14.15}$$

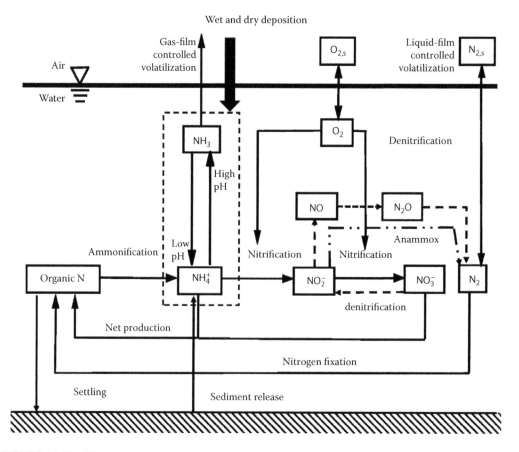

FIGURE 14.22 Nitrogen processes.

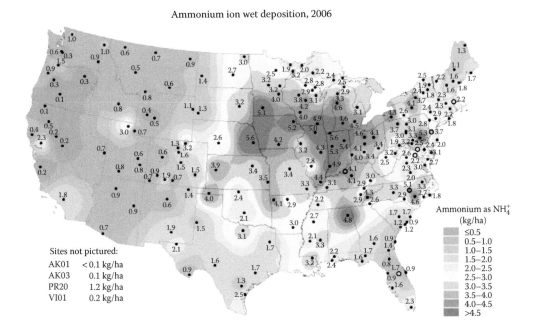

Ammonium ion wet deposition, 2006

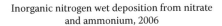

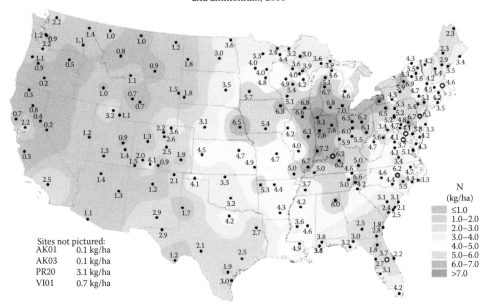

Inorganic nitrogen wet deposition from nitrate
and ammonium, 2006

FIGURE 14.23 Wet deposition nitrogen loads for 2006. (From the National Atmospheric Deposition Program/ National Trends Network. http://nadp.sws.uiuc.edu.)

where the equilibrium relationship is given by

$$K = \frac{\left[NH_3 \right]\left[H^+ \right]}{NH_4^+}$$

$$pK = -\log_{10} K = 0.09018 + \frac{2729.92}{T_a}$$

(14.16)

where K is the equilibrium constant, T_a is the absolute temperature (kelvin), and the concentrations are molar. By substituting the Equation 14.16 into a mass balance for the total ammonia, the fraction unionized as a function of the temperature and the hydrogen ion concentration can be computed from (Figure 14.24):

$$f_u = \frac{1}{1 + \dfrac{\left[H^+ \right]}{K}}$$

(14.17)

The importance of this is that unionized ammonia may be toxic, as will be described in Section 14.4.1.

14.5.1.5 Productivity and Respiration

Nitrogen is also an essential component of biomass. Thus, it is taken up by autotrophs and introduced back into the water column via their excretions and death (Figure 14.22). The preferred form is ammonia, although autotrophs may use any of the dissolved inorganic nitrogen (DIN) forms. Some autotrophs, notably some cyanobacteria, can fix atmospheric nitrogen so that they can use dissolved N_2 as a nitrogen source when other forms are unavailable.

The nitrogen requirements of autotrophs are commonly estimated based on some assumed stoichiometry, or composition of organic matter, such as (Redfield et al. 1963; Chapra 1997)

$$100\,gD{:}40\,gC{:}7200\,mgN{:}1000\,mgP{:}1000\,mgA$$

(14.18)

where gX is the mass of element X (g) and mgY is the mass of element Y (mg). The terms D, C, N, P, and A refer to dry weight, carbon, nitrogen, phosphorus, and chlorophyll-a, respectively.

The general concept is that the ratios of the components can be used to determine which of the components limit growth. For many lakes and reservoirs, the two nutrients that primarily control

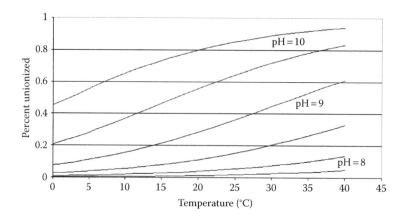

FIGURE 14.24   Fraction of ammonia that is unionized as a function of pH and temperature.

the growth of aquatic plants (e.g., phytoplankton and macrophytes) are nitrogen (N) and phosphorus (P), so N/P ratios are commonly used to estimate which nutrient limits growth. Bergström (2010), for example, proposed the following characterizations of nutrient limitations based on the weight ratios of DIN and total phosphorus (TP):

- Lakes with DIN:TP ratios (by weight) less than 1.5 were characterized as N limited.
- Lakes with DIN:TP ratios (by weight) greater than 4 were characterized as P limited, on the basis of the work.
- Lakes with DIN:TP ratios (by weight) between 1.5 and 4 could not be assigned to a nutrient-limitation class and were characterized as either colimited or limited by something other than N or P.

Often, phosphorus is assumed to be the limiting nutrient in freshwater systems, while nitrogen is the limiting nutrient in estuarine and marine systems. However, often, the limiting nutrient may vary between waterbodies and with the time of the day or year. For example, in a survey of DIN/TP ratios in nearly 30,000 lakes, Baron et al. (2011) demonstrated that while P is most commonly limiting, frequently N may be limiting or both nutrients are colimiting (Table 14.5).

TABLE 14.5

Proportion of Lakes in Three Nutrient Limitation Classes Based on Ratios of Dissolved Inorganic Nitrogen to Total Phosphorus

Region	Total No. of Lakes[a]	N-Limited[b]		P-Limited[c]		Intermediate N–P Ratios[d]	
		No.	%	No.	%	No.	%
Adirondacks	1,290	41	3	1,082	84	137	13
New England	4,361	330	8	2,891	66	1,139	26
Poconos/Catskills	1,506	105	7	899	60	502	33
Southeast	2,424	280	11	1,689	70	455	19
Upper Midwest	8,755	1,890	22	5,433	63	1,251	15
Rockies	6,666	1,641	25	3,668	55	1,357	20
Sierra/Cascades	4,155	934	22	2,687	65	534	13
Total	28,976	5,220	18	18,350	63	5,407	19

Source: Data from the Eastern Lake Survey (Linthurst, R.A., Landers, D.H., Eilers, J.M., Brakke, D.F., Overton, W.S., Meier, E.P., and Crowe, R.E., Characteristics of lakes in the eastern United States, Vol. 1, Population description and physico-chemical relationships, U.S. Environmental Protection Agency, Washington, DC), conducted in the fall of 1984, and the Western Lake Survey (Eilers et al., 1987), conducted in the fall of 1985; Baron, J.S., Driscoll, C.T., Stoddard, J.L., and Richer, E.E., *Bioscience*, 61(8), 602–615, 2011.

[a] Eastern and Western Lake Surveys were stratified random samples of all lakes less than 4 ha in size; the estimates of the number of lakes in each region are based on the target population sizes for each survey.

[b] Lakes with DIN:TP ratios (by weight) less than 1.5 were characterized as N limited, on the basis of the work of Bergström (2010).

[c] Lakes with DIN:TP ratios (by weight) greater than 4 were characterized as P limited, on the basis of the work of Bergström (2010).

[d] Lakes with DIN:TP ratios (by weight) between 1.5 and 4 could not be assigned to a nutrient-limitation class and are characterized as either colimited or limited by something other than N or P.

Note: N, nitrogen; P, phosphorus.

14.5.1.6   Denitrification

Denitrification (Figure 14.22) is the process by which nitrate nitrogen is converted to molecular nitrogen (N_2). As discussed in Section 14.3, once oxygen is depleted, nitrate is the next most thermodynamically favorable oxidant for heterotrophic organisms (Table 14.3). That is, in addition to oxygen, it produces the most energy in the processing of food (organic matter). Since it occurs as oxygen is depleted, it is generally assumed that denitrification only occurs under hypoxic (reduced oxygen) conditions.

Denitrification was long thought to be the only pathway by which the oxidized forms of DIN were converted to N_2. However, a relatively recent process has been discovered, known as anaerobic ammonium oxidation (anammox; Figure 14.22), whereby the anaerobic oxidation of ammonium (NH_4) with nitrite (NO_2) generates N_2 (Thamdrup and Dalsgaard 2002). This process has been found to be very important in marine sediments, but its importance in anaerobic waters or sediments of freshwater systems is not well understood (Hamersley et al. 2009).

14.5.1.7   Sediment Release

Particulate forms of organic nitrogen may settle from the water column to the sediment bed, where they undergo decomposition. This process is known as sediment diagenesis, as described by Di Toro (2001). As a result of the ammonification (diagenesis) of particulate organic carbon, ammonia can be produced and the flux of sediments serves as a source to the water column. This process is of particular importance because of the relatively long timescales associated with sediment diagenesis. For example, the sediments may serve as a source of nutrients to the water column long after the external loads to a lake or reservoir have been reduced.

14.5.2   Nitrogen: Regulatory Environment

Nitrogen has been the subject of regulation for a number of reasons, including:

- Impact on acidification
- Impact on oxygen
- Toxicity
- Impact on plant growth (as a nutrient)

Lake acidification has been a major problem, resulting primarily from emissions of sulfur and nitrogen oxides into the atmosphere and their subsequent introduction into lakes and reservoirs as mist, fog, rain, or snow. Their impact has been greatest in lakes with low alkalinity or buffering capacity. The regulatory focus has been on limiting these emissions.

Nitrification impacts oxygen concentrations, so nitrogen has been a common component of waste load allocations and TMDLs where low DO concentrations are typically the problem. This may be a problem in some lakes and reservoirs, but it is more commonly a regulatory target for streams and rivers.

Similarly, ammonia is regulated as a result of the toxicity of unionized ammonia, as described in the U.S. EPA's ambient water quality criteria for ammonia (USEPA 1999b) (Figure 14.25). Ammonia toxicity may be a problem, particularly in highly eutrophic alkaline lakes.

One of the major regulatory focuses on nutrients (including nitrogen) is on developing nutrient criteria, with the goal of reducing eutrophication at the national scale. The U.S. EPA developed its first guidance document on developing nutrient criteria for lakes and reservoirs in 2000 (USEPA 2000a). The general approach is to develop specific guidance documents for specific ecoregions (14 in total) as illustrated by Figure 14.26 for the contiguous states. The U.S. EPA has provided this guidance as a starting point to states, and all states are presently in the process of developing or finalizing nutrient criteria. For nitrogen, the criteria have typically been developed in terms of total nitrogen concentrations, such as 0.36 mg L^{-1} for ecoregion 9 (USEPA 2000b) (Table 14.6).

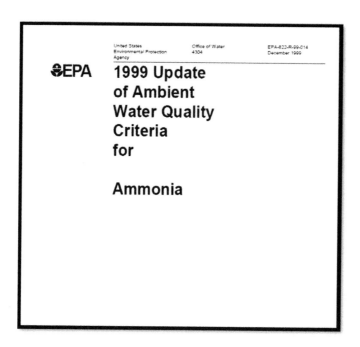

United States
Environmental Protection
Agency

Office of Water
4304

EPA-822-R-99-014
December 1999

$EPA

1999 Update of Ambient Water Quality Criteria for

Ammonia

FIGURE 14.25 U.S. EPA criteria for ammonia.

14.5.3 DISTRIBUTION OF NITROGEN

The seasonal distribution of nitrogen in lakes and reservoirs is highly variable, as described by Wetzel (2001). However, there are some general trends often related to the lake's trophic status. For example, in eutrophic systems, concentrations of nitrate and ammonia are often relatively low in epilimnetic waters, due to their uptake by plants. Below the photic zone, as oxygen is depleted, nitrate is used as an oxidant (TEA) by heterotrophic bacteria (see Table 14.3). Ammonia is used in cell synthesis. However, the rates of ammonification, the release of ammonia from decomposing organics, usually exceed the rate of synthesis, therefore ammonia builds up under anoxic conditions (Gordon and Higgins 2007). The result of the uptake of ammonia in the epilimnion and ammonification in the anoxic hypolimnion often leads to extremes in the vertical distribution of ammonia concentrations.

Seasonal variations in the longitudinal and vertical distribution of nitrate-N and ammonia-N concentrations are illustrated in Figures 14.27 and 14.28 using model predictions for Allatoona Reservoir and Walter F. George Reservoir, Georgia, for July 31, 1995. The concentrations of both nutrients for Allatoona Reservoir are low in surface waters. For Allatoona Reservoir, there are relatively strong predicted longitudinal as well as vertical variations in ammonia concentrations, even though the oxygen concentrations for this date are relatively uniform (Figure 14.21). For Walter F. George Reservoir, a high-flow event had a large impact on the nitrogen concentrations on this date (see also Figures 14.18 and 14.19). These predictions illustrate the complexity in making generalizations regarding concentration distributions for nitrogen and the importance of the physical transport on those distributions.

The seasonal variation in the distribution of ammonia for Lanier Reservoir is illustrated in Figure 14.29a and b. For this reservoir and for this year, ammonia concentrations followed but lagged the trends in the development of the anoxic hypolimnion (Figure 14.17a and b). As the season progressed, ammonia concentrations increased in anoxic waters in the upper reaches of the reservoir, gradually extending down into the reservoir. The greatest concentrations were eventually restricted to the deepest portion of the reservoir during the fall deepening of the epilimnion. The reservoir completely mixes usually in January.

Draft aggregations of level III ecoregions
for the national nutrient strategy

I. Willamette and Central Valleys
II. Western forested mountains
III. Xeric west
IV. Great Plains grass and shrublands
V. South central cultivated Great Plains
VI. Corn belt and northern Great Plains
VII. Mostly glaciated dairy region
VIII. Nutrient poor largely glaciated upper Midwest and Northeast
IX. Southeastern temperate forested plains and hills
X. Texas–Louisiana coastal and Mississippi alluvial plains
XI. Central and eastern forested uplands
XII. Southern coastal plain
XIII. Southern Florida coastal plain
XIV. Eastern coastal plain

FIGURE 14.26   United States ecoregions for contiguous states. (From USEPA, Nutrient criteria technical guidance manual, lakes and reservoirs, 1st ed., U.S. Environmental Protection Agency, Washington, DC, 2000a; updated maps available from http://www.epa.gov/wed/pages/models.htm.)

14.6   PHOSPHORUS

Phosphorus occurs in lakes and reservoirs in organic and inorganic forms. Phosphorus is an essential component of all organisms. The most limiting nutrient is usually considered to be the one that controls growth, and phosphorus is commonly considered to be the most limiting nutrient for most freshwater systems. However, as illustrated in Table 14.5, that is not always the case.

As indicated by Wetzel (2001), no other element in freshwater systems has been studied as extensively as phosphorus. The importance of phosphorus in controlling plant growth is well known, and it is a common component of fertilizers. Understanding the sources and role of phosphorus is critical to the management of lakes and to determining their health.

Much of the effort associated with phosphorus is directed toward quantifying and managing loads. Chapra (1997) described the phosphorus-loading concept and its history of use in lake management. Vollenweider (1968) first introduced the relationships between loadings of phosphorus and the trophic status of lakes, linking the management of loadings to lakes and lake water quality. As described by Chapra (1997), developing phosphorus budget models was the basis for much of the effort to reduce eutrophication in the Great Lakes and elsewhere. Currently, a common target of lake restoration is the reduction of external or internal loadings of phosphorus.

Phosphorus is a component of igneous rock most commonly as the mineral apatite $(Ca_5(PO_4)_3)$ usually combined with either OH^- (hydroxyapatite) or fluoride (fluoroapatite). It is mined in many areas to produce fertilizers, with some of the largest operations in the world in Florida.

TABLE 14.6

Summary of Recommended Criteria for Each of the Aggregate Nutrient Ecoregions for Rivers and Streams and Lakes and Reservoirs

	Rivers and Streams			Lakes and Reservoirs		
Ecoregion	TP (µg L^{-1})	TN (mg L^{-1})	Chl-a (µg L^{-1})[a]	TP (µg L^{-1})	TN (mg L^{-1})	Chl-a (µg L^{-1})[a]
I	47.00	0.31	1.80			
II	10.00	0.12	1.08	8.75	0.10	1.90
III	21.88	0.38	1.78	17.00	0.40	3.40
IV	23.00	0.56	2.40	20.00	0.44	2 (S)[b]
V	67.00	0.88	3.00	33.00	0.56	2.30 (S)[b]
VI	76.25	2.18	2.70	37.50	0.78	8.59 (S)[b]
VII	33.00	0.54	1.50	14.75	0.66	2.63
VIII	10.00	0.38	0.63	8.00	0.24	2.43
IX	36.56	0.69 (S)[b]	0.93	20.00	0.36	4.93
X	128.00	0.76 (S)[b]	2.10			
XI	10.00	0.31 (S)[b]	1.61	8.00	0.46	2.79 (S)[b]
XII	40.00	0.9 (S)[b]	0.4	10.00	0.52	2.60
XIII				17.50	1.27	12.35
XIV	31.25	0.71 (S)b	3.75	8.00	0.32	2.90 (T)[c]

Source: Data from USEPA Summary July 2002.

[a] Chl-a = chlorophyll-a measured by the fluorometric method, unless specified.

[b] S = spectrophotometric method.

[c] T = trichromatic method.

Note: TN, total nitrogen; TP, total phosphorus.

The chemistry of phosphorus is complex. However, for lake management, for assessing nutrient health, and for nutrient criteria, only TP is often considered. Other than TP, the common forms measured and used in lake management include organic phosphorus and (ortho) inorganic phosphorus, either as totals or analytically separated into dissolved and particulate fractions. The dissolved inorganic phosphorus (DIP) is assumed to be the bioavailable form.

Organic phosphorus may include a variety of components, both living and nonliving, reactive (labile) or inert, particulate or dissolved. Nonliving organic phosphorus will mineralize (either slowly or quickly depending on the form) to inorganic forms. This may occur in the water column or sediments (sediment diagenesis). Living plants will uptake phosphorus during growth, excrete phosphorus, or produce nonliving phosphorus as they die.

An additional process impacting phosphorus and phosphorus management is the chemical deactivation of available forms. For example, phosphorus inactivation using alum is a commonly used method in lake restoration. The alum forms a precipitate or floc that scrounges phosphorus (and other materials) as it settles through the water column, and it may also create a "blanket" over the sediment that reduces sediment phosphorus release. It has long been known that there are similar controls for the release of phosphorus from sediments resulting from sediment diagenesis. The internal loads from sediments may be a major contributor to total lake phosphorus loads. Mortimer (1941) postulated that some barrier exists in the aerobic portion of sediments due to the formation of an iron hydroxide precipitate. Soluble forms accumulate in the anaerobic water column and sediments. When those soluble forms are introduced to oxic conditions, they precipitate and form flocs that tie up phosphorus, which may remove it from the water column or inhibit its escape from sediments (Di Toro 2001).

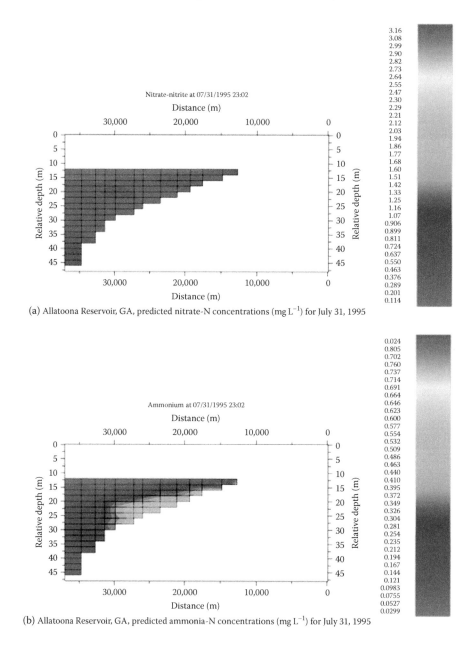

(a) Allatoona Reservoir, GA, predicted nitrate-N concentrations (mg L^{-1}) for July 31, 1995

(b) Allatoona Reservoir, GA, predicted ammonia-N concentrations (mg L^{-1}) for July 31, 1995

FIGURE 14.27 (a,b) Predicted nitrate-N and ammonia-N concentrations in Lake Allatoona, Georgia, for July 31, 1995. (From Martin, J.L. and Hesterlee, C., Detailed reservoir water quality modeling to support environmental impact statements for the evaluation of ACT and ACF river basins water allocation, Contract Report prepared for the U.S. Army Engineer District, Mobile, 1988.)

14.6.1 SEASONAL DISTRIBUTION

The trophic status of lakes and reservoirs is largely a function of the phosphorus concentrations, and the temporal and spatial distribution of phosphorus is largely impacted, in turn, by those trophic conditions. In an oligotrophic lake, the phosphorus concentrations would be relatively constant. For a eutrophic lake, phosphorus would be much more variable and would be expected to be in relatively low concentrations during seasons and in areas of greatest productivity (epilimnion,

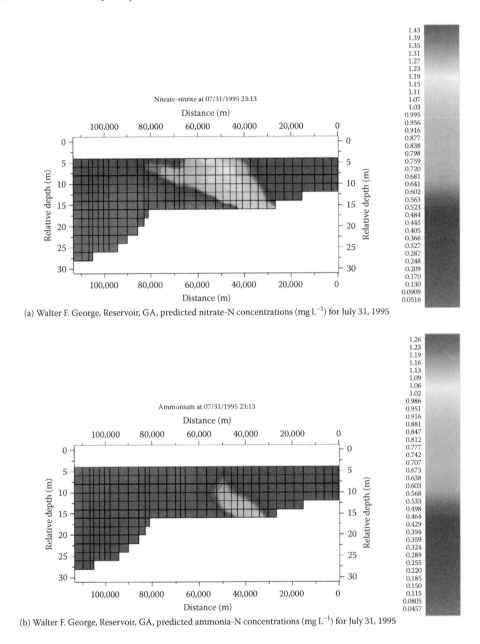

(a) Walter F. George, Reservoir, GA, predicted nitrate-N concentrations (mg L^{-1}) for July 31, 1995

(b) Walter F. George, Reservoir, GA, predicted ammonia-N concentrations (mg L^{-1}) for July 31, 1995

FIGURE 14.28 (a,b) Predicted nitrate-N and ammonia-N concentrations in Lake Walter F. George, Georgia, for July 31, 1995. (From Martin, J.L. and Hesterlee, C., Detailed reservoir water quality modeling to support environmental impact statements for the evaluation of ACT and ACF river basins water allocation, Contract Report prepared for the U.S. Army Engineer District, Mobile, 1988.)

transition zone). In the hypolimnion, and particularly under anaerobic conditions, phosphorus concentrations would increase due to the decay and decomposition of organic matter. With relatively high organic loads to the sediments, sediment diagenesis would result in releases of phosphorus to the overlying water column. Reduced forms of iron and manganese would be common in an anoxic hypolimnion; however, under oxic conditions, they may reduce phosphorus concentrations from the water column.

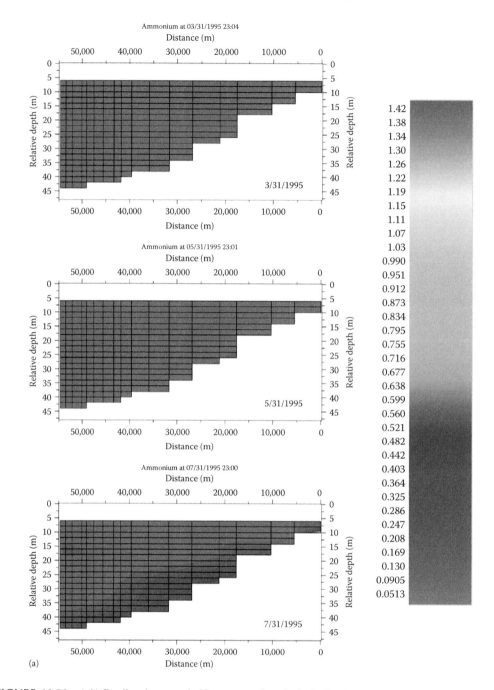

FIGURE 14.29 (a,b) Predicted ammonia-N concentrations in Lake Lanier, Georgia, for selected dates in 1995. (From Martin, J.L. and Hesterlee, C., Detailed reservoir water quality modeling to support environmental impact statements for the evaluation of ACT and ACF river basins water allocation, Contract Report prepared for the U.S. Army Engineer District, Mobile, 1988.)

14.7   pH, ALKALINITY, AND CO$_2$

The pH of lakes is impacted by a number of biological (e.g., autotrophic or heterotrophic productivity and respiration) and chemical processes as well as by external loadings. Changes in the pH have a variety of impacts on the chemical and biological characteristics of lakes. The pH may control

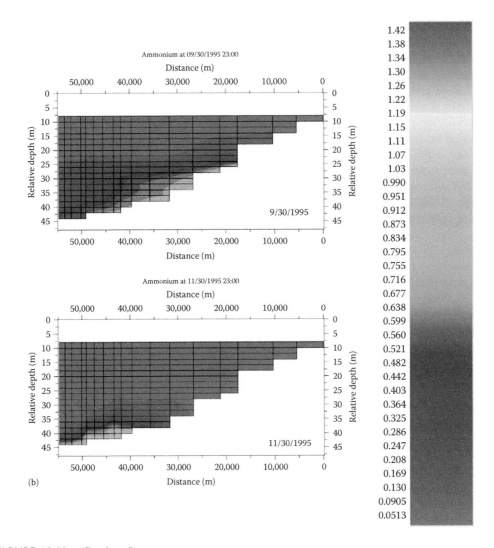

FIGURE 14.29 (Continued)

metal speciation, calcium solubility, and other chemical properties. Lake organisms may have only a limited tolerance to pH changes, with the specific degree of tolerance being a function of the life stage.

Generally, the young of most species are least tolerant. Each state may have differing standards, but the U.S. EPA's national recommended water quality criteria (USEPA 1986) is for a pH in the range of 5–9. As of July 26, 2012, of the 71,577 listed causes of impairment of waters of the United States (from 303(d) lists by the U.S. EPA of waters not meeting water quality standards), 4126 were due to pH, acidity, or caustic conditions. Lake acidification has been a major environmental problem in the United States and Canada.

A number of chemical and biological processes impact the pH in lakes. One of the factors impacting the pH is variations in carbon dioxide, such as those due to biological processes, gas exchange, and external loads.

Biological processes produce or consume carbon dioxide. The addition of carbon dioxide to water can produce a weak acid, carbonic acid:

$$CO_2 + H_2O \rightleftharpoons H_2CO_3 \qquad (14.19)$$

The formation of carbonic acid can reduce the pH. As a result, during daylight hours when net productivity by autotrophs is positive, the pH could increase and then decrease during periods of nighttime respiration. Similarly, heterotrophic production and respiration impact carbon dioxide. Heterotrophic respiration under oxic conditions will produce CO_2 and decrease the pH, typically from neutral to 6.5 (Gordon and Higgins 2007). Under anaerobic conditions, the pH is also impacted by denitrification, the formation of organic acids, and the reduction of nitrate, manganese, iron, and sulfate.

External loads, such as from acidic streams or snowmelt, or from the atmosphere (e.g., nitrous oxides and sulfides), impact the pH of lakes. Other sources include the dissolution of calcium carbonate and the introduction of carbon dioxide from groundwaters and, in some lakes, from volcanic origin. As discussed in Section 14.1.5.2, loads from volcanic sources can produce highly supersaturated carbon dioxide concentrations in the hypolimnion, which have resulted in a catastrophic release of carbon dioxide during overturn (a limnetic eruption).

One of the factors impacting the pH is the exchange of carbon dioxide between water and the atmosphere. As atmospheric carbon dioxide concentrations increase, lake concentrations will increase, with an impact a resulting impact (lowering) on the pH.

The susceptibility of a lake or a reservoir to pH changes is largely reflected in its alkalinity, where alkalinity is a measure of the buffering capacity or the ability to resist pH changes. For example, lakes with low alkalinity and hence a low buffering capacity are much more susceptible to acidification. Alkalinity and pH relationships are discussed in Chapter 5.

14.8 SULFIDES AND SULFATES

Sulfur in freshwaters may exist in a number of forms, as listed in Table 14.7. Sulfur is common in the earth's crust in the form of gypsum ($CaSO_4$) and pyrite (FeS_2). It is common in the atmosphere in the form of sulfur oxides (SO_2, SO_4, such as forming sulfuric acid, H_2SO_4), which can serve as a source of sulfur to lakes and reservoirs in the form of dry or wet deposition, with the greatest rates of deposition in the north-eastern United States (USEPA 1999). Mean annual atmospheric SO_4 concentrations for 2011 are illustrated by Figure 14.30. External loads of sulfur oxides may contribute to lake acidification, particularly in lakes with low alkalinity.

A generalized sulfur cycle is illustrated in Figure 14.31. Sulfur is taken up during the process of building plant and animal cells and is also a component of waste. Therefore, organic forms in the water column and sediment impact sulfur cycling, via production, degradation, and excretion (e.g., of SO_4).

Under aerobic conditions, sulfides may be microbiologically oxidized to form sulfites and sulfates. Under anaerobic conditions, sulfates are utilized as an oxidant (TEA; Table 14.3), ultimately resulting in a reduction to sulfides. Sulfates follow nitrates, iron, and manganese oxides in terms of biologically usable energy yield for oxidation.

TABLE 14.7
Common Oxidation States for Sulfur

Oxidation State of S	Increasing Levels of Sulfur Oxidation $\longrightarrow$				
Form	-2	-1	0	$+4$	$+8$
Aqueous solution and salts	H_2S			H_2SO^3	H_2SO_4
	HS^-			HSO^{3-}	HSO^{4-}
	S^{2-}	S_2^{2-}		SO_3^{2-}	SO_4^{2-}
Gas phase	H_2S			SO_2	SO_3
Molecular solid			S_8		

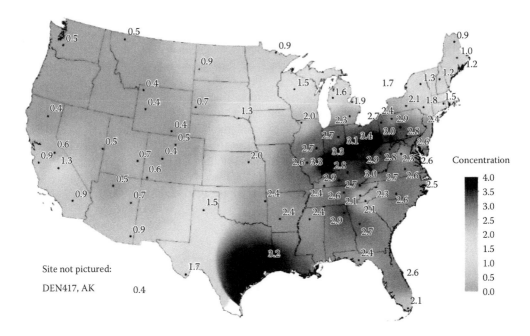

FIGURE 14.30 Annual mean sulfate concentrations for 2011. (From AMEC, Clean air status and trends network annual report, AMEC Environment & Infrastructure, Inc., Prepared for the U.S. Environmental Protection Agency Office of Air and Radiation, Washington, DC, 2011.)

The forms of sulfide may be unionized hydrogen sulfide at lower pH values, or ionized forms (e.g., HS^-) at higher pH values. The presence of sulfide is problematic since even at small concentrations it creates taste and odor (it is the "rotten egg gas") problems. Sulfides are also toxicants, and the tolerance of aquatic organisms varies between species and life stages. The U.S. EPA (USEPA 1986) recommended that an aquatic life criterion for chronic conditions (the criterion continuous

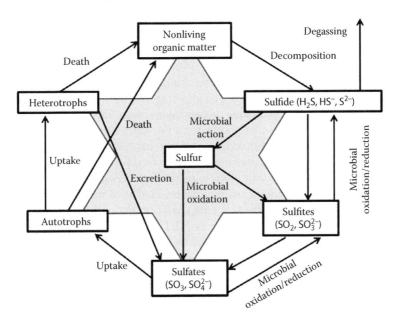

FIGURE 14.31 Generalized sulfur cycle.

concentration or CCC) for sulfides be 2 µg L^{-1}. Excessive concentrations may be a particular problem in reservoir tailwaters receiving untreated outflows from reservoir anaerobic bottom waters.

Unionized hydrogen sulfide (H_2S) is also only slightly soluble in water (Table 14.2). So, hydrogen sulfide may be volatilized from the water surface or from reservoir releases, often creating odor problems in tailwaters.

Sulfides are also important as they impact the bioavailability of many metals, thereby impacting metal toxicity. Sulfides form insoluble precipitates with a number of metals, including mercury. Therefore, the presence of sulfur is often used to assess the toxicity of sediments. A common analytical method is to acidify a sample, convert the sulfur to gaseous hydrogen sulfide (H_2S), which is then purged from the sample, trapped in an aqueous solution and measured; the result is referred to as acid volatile sulfide (AVS; Allen et al. 1993). AVS concentrations are then used to predict the toxicity in the sediments of divalent metals, including copper (Cu), cadmium (Cd), nickels (Ni), lead (Pb), and zinc (Zn) (Di Toro et al. 1992, 2005).

14.9   IRON AND MANGANESE

Oxides of manganese (e.g., MnO_2) follow nitrate in terms of being thermodynamically attractive oxidants under anaerobic conditions, and in terms of producing biologically usable energy. Manganese is followed thermodynamically by iron oxides (e.g., FeO(OH)) as TEAs. The end products of the microbial reduction of oxides of manganese and iron are the ionic forms Mn^{2+} and Fe^{2+} (Table 14.3).

As with other end products of microbial reduction, iron and manganese may accumulate in anaerobic waters and sediments, such as in or below the hypolimnion of lakes and reservoirs. This accumulation is often impacted by pH and other materials, such as sulfides, with which they may form insoluble precipitates. For example, ferric iron may form $Fe(OH)_2$ and $FeCO_3$, which are insoluble, and FeS, which is very insoluble.

When reduced forms of iron and manganese are introduced into oxic waters, such as across a redox boundary, they both form insoluble oxyhydroxides. The redox potential and the oxidation rate of manganese are slower than those of iron, therefore it will remain in solution longer in the oxic epilimnion or in reservoir releases. The oxidation is typically estimated using kinetic rates rather than equilibrium chemistry. For example, Dortch et al. (1992) modeled the oxidation of iron and manganese in reservoir tailwaters using first-order rates.

Iron and manganese are generally not toxic, but are considered nuisance chemicals for water supply. The oxidation of the reduced forms also results in a loss of oxygen.

One of the important aspects of the formation of iron and manganese oxyhydroxides is that they commonly form high surface area flocs to which a number of other chemicals sorb. Generally, iron is considered the more important of the two. The importance of iron was noted by Mortimer (1941), who postulated the role of an iron hydroxide precipitate in controlling releases of phosphorus from sediments. While the scavenging of phosphorus from the water column by iron oxyhydroxides is well known and important, the oxyhydroxides sorb and remove a number of other metals as well. It is generally accepted that the sorption of metals to iron oxyhydroxides is the dominant chemical process regulating the dissolved concentrations of metals in waters and sediments (Horowitz 1991; Dzombak and Morel 1990). Because of its importance, models of metal speciation, for example, equilibrium chemical models, such as the U.S. EPA's MINTEQA2 (Allison et al. 1991), or the metal transport, kinetic, and speciation models, such as META4-WASP (Martin et al. 2012), generally include oxyhydroxide formation and sorption, such as using the hydrous ferric oxide double layer (HFO DLM) sorption model initially of Dzombak and Morel (1990).

14.10   METHANE

Methane in lakes is typically produced by bacteria (methanogens) using one of several metabolic pathways and substrates (e.g., acetate, formate, hydrogen, and carbon dioxide), in a process known as methanogenesis. As indicated in Table 14.3, this is the last most thermodynamically attractive oxidation pathway, so it generally only occurs after other TEAs (oxidants, e.g., O_2, NO, MnO_2, FeO(OH), and SO_4) are depleted and while hydrogen (H_2) and carbon dioxide (CO_2) accumulate.

The general reaction is

$$CO_2 + 4H_2 \rightarrow CH_4 + 2H_2O \tag{14.20}$$

where CO_2 is the TEA (Table 14.3) and methane is the end product. However, methane may also be produced by the destruction of organic carbon (generalized as CH_2O in Equation 14.22) and organic acids, represented by (assuming complete destruction of the organic acids; Di Toro 2001):

$$CH_2O \rightarrow \tfrac{1}{2}CH_4 + \tfrac{1}{2}CO_2 \tag{14.21}$$

Once produced, if the methane enters an oxic environment it can create an oxygen demand, as by

$$\tfrac{1}{2}CH_4 + O_2 \rightarrow + \tfrac{1}{2}CO_2 + H_2O \tag{14.22}$$

Methane is relatively insoluble in water (Figure 14.32), so it may escape from the water column and sediments via degassing from the surface. St. Louis et al. (2000) noted that the global average areal methane flux from the surface of reservoirs to the atmosphere was approximately 120 mg CH_4 m^{-2} day^{-1}, whereas natural lakes averaged only 9 mg CH_4 m^{-2} day^{-1}, suggesting that reservoirs may serve as anthropogenic sources of this greenhouse gas. An alternative pathway is from the ebullition of methane from saturated releases of bottom waters. In their study of a Swiss hydropower facility, Adelsontro et al. (2010) reported that releases through the surface and turbines combined generated total methane emissions on warm summer days of up to 500 mg CH_4 m^{-2} day^{-1}, which averaged >150 mg CH_4 m^{-2} day^{-1}. Adelsontro et al. (2010) further indicated that the source of the methane was almost entirely bubbling from sediments.

Methanogenesis in sediments can also contribute to the methylation of mercury, if present. Methylmercury is a bioaccumulative toxicant and excess concentrations have resulted in fish consumption advisories in a number of systems.

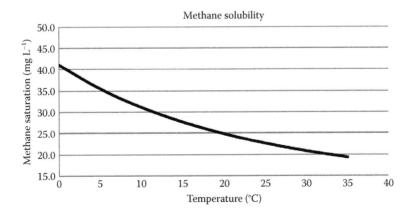

FIGURE 14.32 Methane solubility. (Computed from Yamamoto, S., Alcauskas, J.B., and Crozier, T.E., *Journal of Chemical & Engineering Data*, 21, 78–80, 1976.)

REFERENCES

Adelsontro, T., D.F. McGinnis, S. Sobek, I. Ostrovsky, and B. Wehrli. 2010. Extreme methane emissions from a Swiss hydropower reservoir: Contribution from bubbling sediments. *Environmental Science and Technology* 44, 2419–2425.

AFS. 1984. *Computation of Dissolved Gas Concentrations in Water as Functions of Temperature, Salinity and Pressure*, Special Publication 14. American Fisheries Society, Bethesda, MD.

Allen, H.E., G. Fu, and B. Deng. 1993. Analysis of acid-volatile sulfide (AVS) and simultaneously extracted metals (SEM) for the estimation of potential toxicity in aquatic sediments. *Environmental Toxicology and Chemistry* 12 (8), 1441–1453.

Allison, J.D., D.S. Brown, and K.J. Novo-Gradac. 1991. MINTEQA2/PRODEFA2, a geochemical assessment model for environmental systems: Version 3.0 user's manual, EPA/600/3-91/021. U.S. Environmental Protection Agency, Environmental Research Laboratory, Athens, GA.

AMEC. 2011. Clean air status and trends network annual report, AMEC Environment & Infrastructure, Inc., EPA Contract No. EP-W-09-028. Prepared for the U.S. Environmental Protection Agency Office of Air and Radiation, Washington, DC.

APHA. 1995. *Standard Methods for the Examination of Water and Wastewater*, 19th ed. American Public Health Association, American Water Works Association and Water Environment Federation, Washington, DC.

Baron, J.S., C.T. Driscoll, J.L. Stoddard, and E.E. Richer. 2011. Empirical critical loads of atmospheric nitrogen deposition for nutrient enrichment and acidification of sensitive U.S. lakes. *Bioscience* 61 (8), 602–615.

Bergström, A-K. 2010. The use of TN:TP and DIN:TP ratios as indicators for phytoplankton nutrient limitation in oligotrophic lakes affected by N deposition. *Aquatic Sciences* 72, 277–281.

Bouwer, E.J. 1992. Bioremediation of organic contaminants in the subsurface. In R. Mitchell (ed.), *Environmental Microbiology*. Wiley-Liss, New York, pp. 287–318.

Bowie, G.L., W.B. Mills, D.B. Porcella, C.L. Campbell, J.R. Pagenkopf, G.L. Rupp, K.M. Johnson, P.W.H. Chan, and S.A. Gherini. 1985. Rates, constants, and kinetics formulations in surface water quality modeling (2nd ed.), EPA/600/3-5/040. U.S. Environmental Protection Agency, Athens, GA.

Capone, D.G. and R.P Kiene. 1988. Comparison of microbial dynamics in marine and freshwater sediments: Contrasts in anaerobic carbon catabolism. *Limnology and Oceanography* 33, 725–749.

Chapra, S.C. 1997. *Surface Water Quality Modeling*. McGraw-Hill, New York.

Chapra, S.C., G.J. Pelletier, and H. Tao. 2008. QUAL2K: A modeling framework for simulating river and stream water quality, Version 2.11: Documentation and user's manual. Civil and Environmental Engineering Department, Tufts University, Medford, MA.

Cole, T.M. and H.H. Hannan. 1990. Dissolved oxygen dynamics. In K.W. Thornton, B.L. Kimmel, and F.E. Payne (eds), *Reservoir Limnology—Ecological Perspectives*. Wiley, New York, pp. 71–107.

De Lanois, J.L. and W.R. Green. 2011. Hypolimnetic dissolved-oxygen dynamics within selected White River reservoirs, northern Arkansas–southern Missouri, 1974–2008. U.S. Geological Survey Scientific Investigations Report 2011–5090. U.S. Geological Survey, Reston, VA. (Appendix available at: http://pubs.usgs.gov/sir/2011/5090/.)

Di Toro, D. 2001. *Sediment Flux Modeling*. Wiley Interscience, New York.

Di Toro, D.M., J.D. Mahony, D.J. Hansen, K.J. Scott, A.R. Carlson, and G.T. Ankley. 1992. Acid volatile sulfide predicts the acute toxicity of cadmium and nickel in sediments. *Environmental Science and Technology* 26, 96–101.

Di Toro, D.M., J.A. McGrath, D.J. Hansen, W.J. Berry, P.R. Paquin, R. Mathew, K.B. Wu, and R.C. Santore. 2005. Predicting sediment metal toxicity using a sediment biotic ligand model: Methodology and initial application. *Environmental Toxicology and Chemistry* 24 (10), 2410–2427.

Dortch, M.S., D.H. Tillman, and B.W. Bunch. 1992. Modeling water quality of reservoir tailwaters. Technical Report W-92-1. U.S. Army Corps of Engineers Waterways Experiment Station, Vicksburg, MS.

Dye, C.W., D.A. Jones, L.T. Ross, and J.L. Gernert. 1980. Diel variations of selected physico-chemical parameters in Lake Kissimmee, Florida. *Hydrobiologia* 71, 51–60.

Dzombak, D.A. and F.M.M. Morel. 1990. *Surface Complexation Modeling Hydrous Ferric Oxides*. Wiley, New York.

Eilers, J.M., P. Kanciruk, R.A. McCord, W.S. Overton, L. Hook, D.J. Blick, D.R. Brakke, P.E. Kellar, M.S. DeHaan, M.E. Silverstein, and D.H. Landers. 1987. Western lake survey. EPA/600/3-86/054b. USEPA, Washington, D.C.

Ensminger, P.A. 1998. Bathymetric survey and physical and chemical-related properties of Lake Bruin, Louisiana, February and September 1997, from Water-Resources Investigations Report 98-4243 Bathymetric Survey and Physical and Chemical-Related Properties of Lake Bruin. U.S. Geological Survey, Baton Rouge, LA.

Evans, W.C., G.W. Kling, M.L. Tuttle, G. Tanyileke, and L.D. White. 1993. Gas buildup in Lake Nyos, Cameroon: The recharge process and its consequences. *Applied Geochemistry* 8, 207–221.

Fenchel, T., G.H. King, and T.H. Blackburn. 1998. *Bacterial Biogeochemistry, The Ecophysiology of Mineral Cycling*. Academic Press, San Diego, CA.

Gordon, J.A. and J.M. Higgins. 2007. Fundamental water quality processes. In J.L. Martin, J. Edinger, J. A. Gordon, and J. Higgins (eds), *Energy Production and Reservoir Water Quality*. American Society of Civil Engineers, Reston, VA, pp. 3.1–3.22.

Hamersley, M.R., D. Woebken, B. Boehrer, M. Schultze, G. Lavik, and M.M.M. Kuypers. 2009. Water column anammox and denitrification in a temperate permanently stratified lake (Lake Rassnitzer, Germany). *Systematic and Applied Microbiology* 32, 571–582.

Horowitz, A.J. 1991. *Sediment-Trace Element Chemistry*. Lewis, Chelsea, MI.

Hutchinson, G.E. 1938. On the relation between oxygen deficit and the productivity and typology of lakes. *Internationale Revue gesamten Hydrobiologie und Hydrogrographie* 36, 336–355.

Hutchinson, G.E. 1957. *A Treatise on Limnology,* Vol. 1, Physics and Chemistry. Wiley, New York.

KDHE. 2011. Allowances for low dissolved oxygen levels for aquatic life use. Water Quality Standards White Paper. Kansas Department of Health and Environment, Bureau of Water, Topeka, KS.

Linthurst, R.A., D.H. Landers, J.M. Eilers, D.F. Brakke, W.S. Overton, E.P. Meier, and R.E. Crowe. 1986. Characteristics of lakes in the eastern United States. Vol. 1. Population description and physico-chemical relationships, Report no. EPA/600/4-86/007a. U.S. Environmental Protection Agency, Washington, DC.

Little, J.C. and D.F. McGinnis. 2001. Hypolimnetic oxygenation: Predicting performance using a discrete-bubble model. *Water Science & Technology: Water Supply* 1, 185–191.

Martin, J.L. 1984. Models of diel variations of water quality in a stratified eutrophic reservoir (Lake Livingston, Texas). PhD dissertation, Texas A&M University.

Martin, J.L. and C. Hesterlee. 1998. Detailed reservoir water quality modeling to support environmental impact statements for the evaluation of ACT and ACF River Basins water allocation. Contract Report prepared for the U.S. Army Engineer District, Mobile, March.

Martin, J.L. and T. Cole. 2000. Water quality modeling of J. Percy Priest Reservoir, Technical Report, ERDC/EL SR-00-9. ERDC Waterways Experiment Station, Vicksburg, MS.

Martin, J.L., B.A. Butler, T.A. Wool, and A. Comer. 2012. Model documentation for WASP 8: META4-WASP, a metal exposure and transformation assessment model, model theory and user's guide. U.S. Environmental Protection Agency, Region 4, TMDL, Atlanta, GA.

Maryland Department of the Environment (MDE). 2004. Code of Maryland regulations; Chapter 26.08.02–Water Quality. Available at: http://www.dsd.state.md.us/comar/subtitle_chapters/26_Chapters.htm.

McGinnis, D.F. and J.C. Little. 1998. Bubble dynamics and oxygen transfer in a speece cone. *Water Science & Technology* 37, 285–292.

McGinnis, D.F. and J.C. Little. 2002. Predicting diffused-bubble oxygen transfer rate using the discrete-bubble model. *Water Research* 36, 4627–4635.

MDEQ. 2007. State of Mississippi water quality criteria for intrastate, interstate, and coastal waters. Mississippi Department of Environmental Quality, Office of Pollution Control, Jackson, MS.

Minnesota Pollution Control Agency (MPCA). 2003. Minnesota rules chapter 7050.0220–Specific standards of quality and purity by associated use classes. Available at: http://www.revisor.leg.state.mn.us/arule/7050/0220.html.

Mortimer, C.H. 1941. The exchange of dissolved substances between mud and water in lakes. *Journal of Ecology* 29, 280–329.

Nordstrom, D.K. and F.D. Wilde. 2005. Reduction- 6.5 oxidation potential (electrode method), Chapter A6. *Field Measurements*, U.S. Geological Survey, Reston, VA.

Redfield, A.C., B.H. Ketchum, and F.A. Richards. 1963. The influence of organisms on the composition of seawater. In M.N. Hill (ed.), *The Sea*, Vol. 2, Wiley-Interscience, New York, pp. 27–46.

Sigurdsson, H., J.D. Devine, F.M. Tchoua, T.S. Presser, M.K.W. Pringle, and W.C. Evans. 1987. Origin of the lethal gas burst from Lake Monoun, Cameroon. *Journal of Volcanology and Geothermal Research* 31, 1–16.

St. Louis, V.L., C.A. Kelly, E. Duchemin, J.W.M. Rudd, and D.M. Rosenberg. 2000. Reservoir surfaces as sources of greenhouse gases to the atmosphere: A global estimate. *Bioscience* 50, 766–775.

Strom, K.M. 1931. FETOVATN: A physiographic and biological study of a mountain lake. *Archiv für Hydrobiologie* 22, 491–536.

Thamdrup, B. and T. Dalsgaard. 2002. Production of N2 through anaerobic ammonium oxidation coupled to nitrate reduction in marine sediments. *Applied and Environmental Microbiology* 68, 1312–1318.

USEPA. 1986. Quality criteria for water, EPA 440/5-86-001. Office of Water, Regulations and Standards, U.S. Environmental Protection Agency, Washington, DC.

USEPA. 1993. Guidance specifying management measures for sources of nonpoint pollution in coastal waters, EPA 840-B-92-002B. U.S. Environmental Protection Agency, Washington, DC.

USEPA. 1999a. Atmospheric deposition of sulfur and nitrogen compounds. In *National Air Quality and Emissions Trends Report*, 1999. U.S. Environmental Protection Agency, Washington, DC.

USEPA. 1999b. 1999 update of ambient water criteria for ammonia, EPA 822-99-014. U.S. Environmental Protection Agency, Washington, DC.

USEPA. 2000a. *Nutrient Criteria Technical Guidance Manual, Lakes and Reservoirs*, 1st ed. EPA-822-B00-001. U.S. Environmental Protection Agency, Washington, DC.

USEPA. 2000b. Ambient water quality criteria recommendations information supporting the development of state and tribal nutrient criteria lakes and reservoirs in nutrient ecoregion IX, EPA 822-B-00-011. U.S. Environmental Protection Agency, Washington, DC.

USEPA. 2008. Integrated science assessment (ISA) for oxides of nitrogen and sulfur–ecological criteria (final report), EPA/600/R-08/082. U.S. Environmental Protection Agency, Washington, DC.

USEPA. 2009. Risk and exposure assessment for review of the secondary national ambient air quality standards for oxides of nitrogen and oxides of sulfur (final report), EPA-452/P-09-004a. U.S. Environmental Protection Agency, Washington, DC.

Veley, R.J. and M.J. Moran. 2011. Evaluating lake stratification and temporal trends by using near-continuous water-quality data from automated profiling systems for water years 2005–2009, Lake Mead, Arizona and Nevada. U.S. Geological Survey Scientific Investigation Report 2012–5080. U.S. Geological Survey, Carson City, NV.

Vollenweider, R.A. 1968. The scientific basis of lake and stream eutrophication, with particular reference to phosphorus and nitrogen as eutrophication factors, Technical Report OAS/DSI/68.27. Organization for Economic Cooperation and Development, Paris.

Water on the Web. 2004. Monitoring Minnesota lakes on the internet and training water science technicians for the future—A national on-line curriculum using advanced technologies and real-time data. University of Minnesota, Duluth, MN. Available at: http://WaterOntheWeb.org.

Wetzel, R.G. 2001. *Limnology: Lake and River Ecosystems*, 3rd ed. Academic Press, San Diego, CA.

Yamamoto, S., J.B. Alcauskas, and T.E. Crozier. 1976. Solubility of methane in distilled water and seawater. *Journal of Chemical & Engineering Data* 21 (1), 78–80.

15 Biota of Lakes and Reservoirs

15.1 CLASSIFICATION

The biota of lakes and reservoirs are very taxonomically diverse. However, many of these organisms share common characteristics that are useful in their management, based on where they live, both physically and in their relationship to other organisms. Two of the commonly used classification schemes, as discussed in the following sections, are based on trophic levels and zonation.

15.1.1 TROPHIC LEVEL

The biological communities in lakes and reservoirs are often classified into trophic levels according to their source of energy or organic materials, where trophic derives from the Greek *trophē* referring to food or feeding. One of the systems used to characterize the condition of lakes is based on productivity and is referred to as the trophic condition (e.g., eutrophic and oligotrophic). The organisms that make up that productivity may be divided into trophic levels consisting of primary and secondary consumers and decomposers.

Primary producers are those organisms that synthesize organic material from inorganic materials, such as elemental nutrients and carbon dioxide. These organisms are also the autotrophs, or "self-feeders." Autotrophs require an energy source, which may include light (phototrophs) or chemical reactions (chemoautotrophs). Phototrophs are restricted to the upper levels of lakes and reservoirs and are limited by the availability of light and nutrients. Phototrophs include small floating plants, phytoplankton, the larger macrophytes and their attached periphyton. Chemoautotrophs more commonly live in the harsher anaerobic portions of lakes such as an anoxic hypolimnion and sediments. These include bacteria and archaea that use inorganic energy sources, such as nitrates, oxides of manganese and iron, and sulfates.

Consumers are those organisms that feed on primary producers or other consumers. That is, they cannot fix carbon so they must rely on consuming those organisms that can fix carbon. Primary consumers are those organisms that feed directly on the producers (living or detrital), such as zooplankton, and are the herbivores. Organisms that feed on the primary consumers are the secondary consumers, such as some invertebrates and planktivorous fish (planktivores). And so on, until the tertiary consumers that feed on the piscivorous fish, known as the piscivores, which are at the "top of the food chain."

Decomposers are those organisms that convert organic material back into nutrients, for example, mineralization. Decomposers may be found throughout lakes, but they dominate in the aphotic zone, where, in the absence of light, they feed on the flux of organic material falling from the photic zone.

One consequence of the trophic levels is that since no transfer of energy is 100% efficient, the higher the order is on the food chain, the less biomass that can be supported. This is often depicted as the ecological pyramid (Figure 15.1). Also, note that only the primary producers are limited by the availability of light. The consumers are not restricted to the photic zone.

There are exceptions to the foregoing classification, as with all classifications of biotic assemblages. For example, there are organisms that are facultative heterotrophic phototrophs, or facultative phototrophs. That is, they are generally primary producers, but can also become consumers. They are not "obligate" phototrophs but "facultative" phototrophs, so that primary production is not essential for their growth, allowing them to grow in both the light and the dark.

The relative importance and classification of some organisms are based on what they produce and what they consume. Primary producers consume nutrients during growth and that growth is

The ecological pyramid

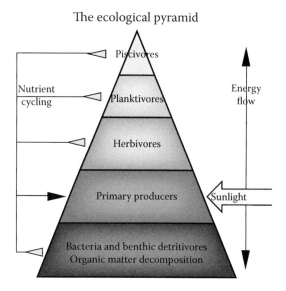

FIGURE 15.1 Food pyramid. (From Water on the Web, Monitoring Minnesota lakes on the Internet and training water science technicians for the future—A national on-line curriculum using advanced technologies and real-time data, University of Minnesota–Duluth, Duluth, MN, 2004, Available at http:// WaterOntheWeb.org.)

limited by the availability of nutrients, principally phosphorus and nitrogen. Those nutrients are subsequently recycled by plant death and excretion. Thus, controlling plant growth is often directed toward controlling nutrients, such as in developing nutrient criteria.

One important example where products are important is the methanogens that produce methane as a metabolic by-product under anaerobic conditions. Another example where both products and uptake are important is the processing of nitrogen. Common forms of nitrogen include that in living organisms, nonliving organic nitrogen, inorganic forms (ammonia, NH_4; nitrite, NO_2; nitrate, NO_3), and dissolved nitrogen gas (N_2). Organic nitrogen typically refers to the nonliving organic material that was once a component of living organisms (e.g., both autotrophs and heterotrophs; detrital or excreted materials). The bacterial breakdown of these organic materials produces inorganic ammonia. In the nitrification process, under aerobic conditions, ammonia is converted to nitrite and ultimately to nitrate (Figure 15.2). This process is important in lake environments since inorganic nitrogen is an essential nutrient for primary production (for the autotrophs). Dissolved nitrogen (N_2) is not biologically available to most autotrophs, but some do have the ability to fix nitrogen, converting N_2 to ammonia (NH_4). Another important group of organisms are those that transform inorganic N to gaseous forms (e.g., ultimately N_2). This transformation is a pivotal sink of inorganic nitrogen in many aquatic environments, including lakes and reservoirs. The processes include:

- Denitrification: The microbial transformation of NO_3 to gaseous nitrogen, which is usually accomplished by facultative heterotrophic chemotrophs. In this case, the bacteria can obtain oxygen from either that dissolved under aerobic conditions or from the nitrate molecule under aerobic conditions. Denitrification is a primary mechanism for the removal of nitrogen from lakes (Saunders and Kalff 2001).
- Anaerobic ammonium oxidation (anammox): In this case, bacteria combine ammonium and nitrite or nitrate to ultimately form nitrogen gas. This process, more recently discovered, has been found to be of particular importance in marine environments, accounting for up to 50% of the total nitrogen turnover (Kuypers et al. 2003, 2005; Dalsgaard et al. 2005). However, the importance of anammox in freshwater lacustrine nitrogen cycling remains to be determined (Rissaned et al. 2011).

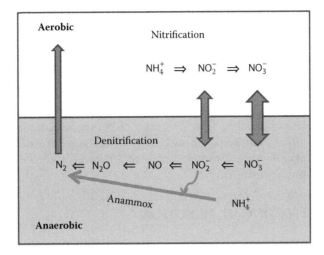

FIGURE 15.2 Nitrogen transformation processes.

One consequence of anaerobic reduction versus aerobic oxidation is that there is a reduction in energy yield. That is, organisms obtain less energy from a chemical reduction in the absence of free oxygen. Under anaerobic conditions, organisms that can reduce nitrate nitrogen get the most "bang for the buck," followed by sulfate reduction and finally methanogenesis. Organisms that reduce sulfate get less than 30% of the energy yield of aerobic respiration, while methanogenesis generates only about 1% of the energy yield of aerobic respiration (Figure 15.3). Therefore, as systems become anaerobic, there is usually a population shift, with populations initially dominated by the nitrifiers, and then the sulfate reducers, and finally the methanogens.

15.1.2 ZONATION

A second classification system is based on where the organisms live. The zones included are

- Psammolittoral zone
- Littoral zone
- Limnetic or pelagic zone
- Benthic zone

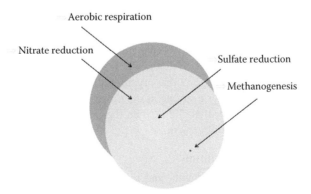

FIGURE 15.3 Relative energy yield (energy produced per mole oxidant) associated with oxidation and reduction processes.

The psammolittoral is the beach zone, and is home to a variety of specialized copepods, rotifers, and nematodes. The littoral zone is the shallow water zone where light may penetrate to the bottom, where macrophytes may be found. The limnetic or pelagic zone is the deep-water zone where phytoplankton dominate primary productivity. The benthic zone is the zone on or in the bottom. The epifauna, such as crayfish and dragonfly larvae, live and move about on the lake bottom, while the infauna, such as nematodes and some insect larvae, live beneath the mud surface.

The relative proportion or abundance of each of these zones directly affects their biological characteristics and is also often taken as an indication of the successional stage. If one considers that as lakes age, they fill in, then the rate of succession would be greater for lakes dominated by a littoral zone than for lakes dominated by a pelagic zone. Or, as lakes age and fill, the littoral zone will become increasingly dominant.

15.2 FACTORS AFFECTING DISTRIBUTION

15.2.1 GENERAL PRINCIPLES

The distribution of organisms is affected by factors such as light, temperature, substrate, and nutrients. For example, phototrophs will be restricted to the photic zone of lakes, while heterotrophs may not be so restricted. Some of the factors affecting the distribution of organisms are discussed in the following sections.

15.2.1.1 Stoichiometry and Redfield Ratios

Autotrophic production takes inorganic materials and energy and produces organic materials from them. The building blocks for this organic material consist of carbon, nitrogen, phosphorus, and other materials. The specific ratios of the materials in organic materials are commonly based on some assumed stoichiometry, such as in the following approximation (Redfield et al. 1963, Chapra et al. 2007):

$$C_{106}H_{263}O_{110}N_{16}P_1$$

which can be used to determine mass ratios, such as

C:N:P
Molar basis
   106:16:1
Mass basis
   106×12: 16×14: $1 \times 31 = 1272$: 224: 31

So, the organic matter of, say, algae is composed of mass of about 40% carbon, 7.2% nitrogen, and 1% phosphorus, or on a molar basis, of 16 times more nitrogen than phosphorus.

These ratios may be used, for example, to determine if phosphorus or nitrogen is in limited supply for plant growth. This equation suggests that organisms require (molar basis) 16 times as much nitrogen as they do phosphorus. So, for example, if the molar ratio of N:P is greater that 16:1 (or the mass concentration ratios are greater than 7.2:1), then there is an excess of nitrogen in relation to phosphorus, and phosphorus may limit growth.

15.2.1.2 Liebig's "Law of the Minimum"

All organisms have requirements for growth, and that growth may be limited by those requirements not being met. A limiting factor to biological activity is that material available in an amount most closely approaching the critical minimum required to sustain that activity (Odum 1971). But, if you have a variety of requirements for growth, how do you determine which is limiting? One common approach assumes that the limiting factor is the one that is in least supply (relative to the growth requirement). That is the basis for Liebig's "law of the minimum." The principle is most commonly illustrated using a wooden barrel, in which each stave represents some growth requirement (Figure 15.4). If water is added to the barrel, the amount of water that the barrel can hold is determined by the shortest stave.

Liebig's law of the minimum in determining growth requirements is similar to the Redfield approach, but different. For example, using the Redfield ratios, if the molar ratio of N:P is greater than 16:1, this implies that phosphorus is limiting, regardless of the molar concentration. The relationship between nutrients and growth is also described using the Michaelis–Menton equation (Chapra 1997), as illustrated by Figure 15.5. That is, if concentrations were very low in relation to the growth requirement, then the relationship between the rates of growth (due to the impact of this limiting nutrient) would be linear. At some point, if the concentrations were high enough, then the growth rate would approach its maximum value (1.0 times the maximum growth rate). The half-saturation concentration (K_s, see Table 15.1) would be that concentration at which the rate of growth would be one-half of its maximum value. Using this model would imply that where N and/or P (and/or silica for diatoms to form their frustules) were in very high concentrations, the

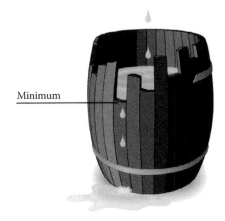

Minimum

FIGURE 15.4 The leaky barrel. (Courtesy of Wikimedia Commons.)

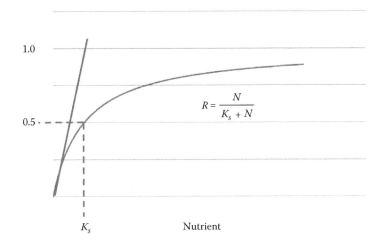

$$R = \frac{N}{K_s + N}$$

K_s Nutrient

FIGURE 15.5 Michaelis–Menton relationship between nutrients and growth.

TABLE 15.1
Half-Saturation Constants for Nutrient
Limitation of Phytoplankton Growth

Nutrient	K_s
Phosphorus	1–5 µg P/L
Nitrogen	5–20 µg N/L
Silica (diatoms)	20–80 µg Si/L

Source: Chapra, S.C., *Surface Water Quality Modeling*,
McGraw-Hill, New York, 1997.

concentrations would not be growth limiting regardless of the ratios between them. An analog would be related to food, where if there is little food then the quantity limits growth, but where if there were surpluses then growth would be limited by the rate at which an organism could eat. A distinction must also be made between productivity or the rate of growth and biomass. For example, under a variety of conditions (e.g., tropical forests), the biomass may be high but the rate of growth may be slow or the opposite.

15.2.1.3 Shelford's Law of Tolerance

Shelford's law of tolerance describes the success or failure of organisms depending on how well a complex set of conditions is satisfied. That is, the absence or failure of an organism can be controlled by the qualitative or quantitative deficiency or excess with respect to any one of several factors (limiting factors) that may approach the limits of tolerance for that organism.

For example, organisms may have a set of optimal conditions where they may be most successful, while as conditions vary above or below those conditions, they may be less successful, or in the extreme, they may die. For example, organisms may have a specific range of temperatures within which they can grow and be successful, while if the temperatures are outside of that range, their growth or success rate would decline. The "range" of optimal temperatures may vary between organisms. Eurythermal organisms, such as carp, can withstand a wide range of temperatures, while stenothermal organisms may survive only under narrow ranges of temperature. Stenothermal fish may, for example, be more common in the hypolimnion of lakes where temperature variations are smaller, while carp and catfish may inhabit the more shallow waters of southern lakes and reservoirs with larger seasonal temperature variations.

The optimal conditions do not apply just to temperature, but to all environmental factors. For example, phototrophs require light, so they are restricted to the photic zone. However, at certain times of the day or year, and also as a function of depth, high levels of light may inhibit photosynthesis. Substrate may also be limiting, as can be oxygen. Hydrodynamics can also impact organisms, so that there may be an optimal range of velocities, above or below which the organism or populations may decline.

An organism may be tolerant of one condition and not of another. The tolerances may also vary with life stages. The U.S. Environmental Protection Agency (U.S. EPA) (USEPA 1974) identified the maximum weekly temperature for the survival of rainbow trout as 24°C. That condition is met in much of the bottom water released from southern reservoirs, allowing a viable downstream trout fisheries. However, while trout may survive in these waters, often the tailwaters are not conducive to natural reproduction due to their relatively high temperatures, the lack of suitable habitats, and other factors. So, while they may survive, only a "put and take" fisheries is sustainable. In addition, in general the limits of tolerance for adult reproductive organisms are lower than those of nonreproductive organisms and are lower for young organism than for adults.

15.3 CHARACTERISTICS OF ORGANISMS BY ZONE

15.3.1 LIMNETIC ZONE

15.3.1.1 Phytoplankton

15.3.1.1.1 Taxonomic Classification

Phytoplankton in general refers to the assemblage of plants ("phyto") or autotrophic organisms, typically microscopic, which are "planktonic" or float, drift, or weakly swim in the water column (including the pelagic zone). Various general schemes have been used to classify phytoplankton, one example of which is that some models subdivide phytoplankton into blue-green algae, green algae, brown algae, and diatoms. To understand the implications of this classification, let us first revisit the system of the taxonomic classification of organisms.

There are a variety of variations in taxonomic classification systems, but most modern systems are based on that first established by Carl Linnaeus in his *Systema Naturae* (1735; Figure 15.6). Today, most taxonomists organize organisms into ranks, based on similarities in their structure. The highest rank is typically the domain, which is then further subdivided into lower ranks down

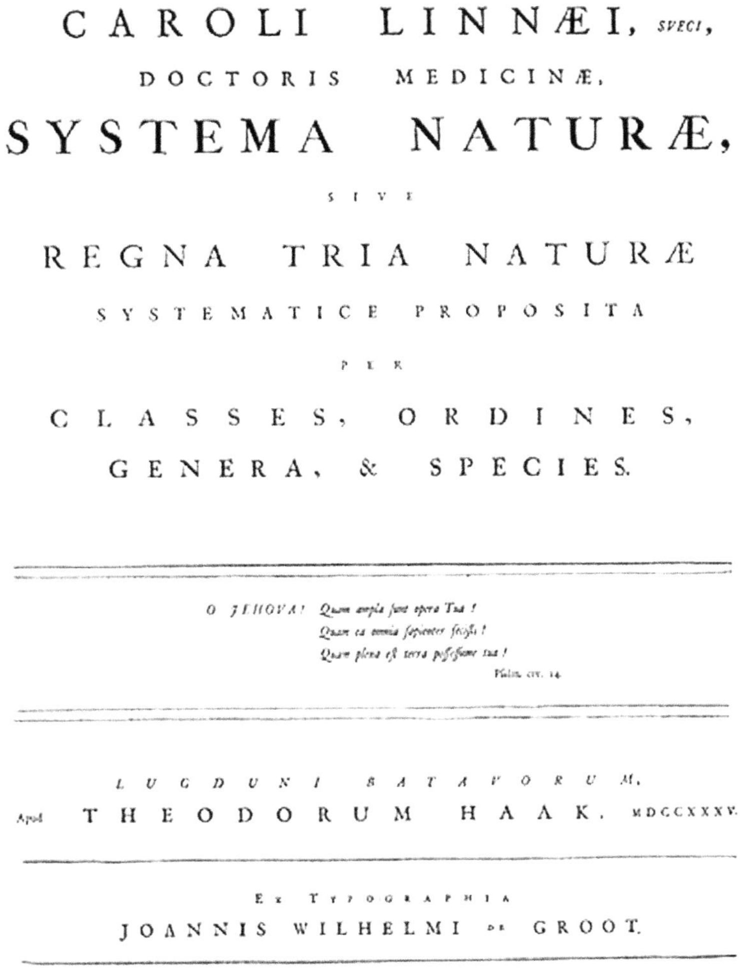

CAROLI LINNÆI, *sveci,*

DOCTORIS MEDICINÆ,

SYSTEMA NATURÆ,

sive

REGNA TRIA NATURÆ

SYSTEMATICE PROPOSITA

per

CLASSES, ORDINES,

GENERA, & SPECIES.

O JEHOVA! *Quam ampla sunt opera Tua!*
Quam ea omnia sapienter fecisti!
Quam plena est terra possessione tua!

Psalm. civ. 14.

LUGDUNI BATAVORUM,

Apud THEODORUM HAAK, MDCCXXXV.

Ex Typographia
JOANNIS WILHELMI DE GROOT.

FIGURE 15.6 *Systema Naturae* by Carl Linnaeus (1735).

to the basic taxonomic rank, the species. The classification may be illustrated by that for the *Paramecium*:

- Domain Eukaryotic
  - Kingdom Animalia
- Phylum Ciliophora
  - Class Ciliatea
  - Order Hymenostomatida
  - Family Parameciidae
  - Genus *Paramecium*
- Species *aurelia*, *bursaria*, and *caudatum*

All living organisms may be subdivided into three domains, the eukaryotes, which basically have a cell nucleus, and those that do not (the prokaryotes), which include bacteria and archaea. The domains are further subdivided into kingdoms. The kingdoms under Eukaryota include Animalia (all animals), Plantae (all plants), Protista (algae and protozoans; Figure 15.7), etc. So, for example, Animalia or the other kingdoms represents a vast and diverse assemblage of organisms that shares some common characteristics.

Now, to revisit the phytoplankton, the general groups mentioned at the beginning of this section were the blue-green algae, green algae, brown algae, and diatoms. The taxonomy of these groups is listed next. As indicated, blue-green "algae," or Cyanobacteria, are in a separate domain from the true algae and they are autotrophic bacteria. The "green" algae, "brown" algae, and diatoms are all eukaryotes and are representatives of separate divisions within the kingdom Protista.

- Domain Bacteria (prokaryotic)
- Kingdom Monera
  - Division Cyanobacteria
- Domain Eukarya (eukaryotic)

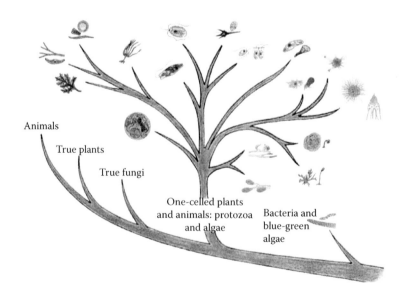

Animals

True plants

True fungi

One-celled plants
and animals: protozoa
and algae

Bacteria and
blue-green
algae

FIGURE 15.7 Family tree. (The organization is diagrammatic and may not reflect evolutionary associations, which are frequently updated and revised; Prepared by Robert G. Howells, Texas Parks and Wildlife Department, Ingram, Texas, 2003. Courtesy of the Texas Parks and Wildlife Department © 2003. With permission.)

- Kingdom Protista
  - Protozoans
  - Algae
  - Division Rhodophyta—red algae
  - Division Charophyta—desmids and stonewarts
  - Division Chrytophyta—chryptomonads
  - Division Euglenophyta—euglenoids
  - Division Phaeophyta—brown algae
  - Division Chrysophytes—chrosophytes
  - Division Haptophyte
  - Division Bacillariophyta—diatoms
  - Division Dinophyta—dinoflagellates
  - Division Chlorophyta—green algae
  - Plasmodial slime molds

The point is that these groups (blue-green, brown algae, green algae, and diatoms) do not represent individual organisms or groups, but rather vast and diverse assemblages of organisms including thousands of species, which do, however, share some common characteristics. Also, note that with the possible exception of species, each of the classification ranks (and the placement of organisms in those ranks) is subjective, and may vary between references. So, for example, in Reynold's (2006) *The Ecology of Phytoplankton*, rather than divisions, the major groups of eukaryotic algae are subdivided into phyla (e.g., phylum Bacillariophyta).

One interesting question is why are there so many different types of phytoplankton or so much species diversity? In a terrestrial environment, species diversity is often driven by the availability of a variety of specialized ecological niches (places organisms may live and roles they may take). However, for phytoplankton, there is water, light, and nutrients. That is, all plankton require, essentially, is the same nutrients, and all water appears to be the same, so how can any diversity be supported? This is what Hutchinson, in a 1961 paper, called the "paradox of the plankton" (Hutchinson 1961). So, what is the answer? As Hutchinson stated in that seminal paper "Perhaps it is, of course, possible that some people with greater insight might have seen further into the problem of the plankton ..., but for the moment I am content ... of presenting that problem to you."

15.3.1.1.1.1 Algae (Protista)

- Rhodophyta—Red Algae: Red algae are primarily (96%) marine organisms, and most seaweeds are members of this group. They are typically attached algae. None of the red algae is planktonic and few genera are found in freshwater (Wetzel 2001). They are red because of the presence of the pigments phycoerythrin and phycocyanin, also found in Cyanobacteria, which reflects red light and absorbs blue light and masks the other pigments (e.g., chlorophyll-a; Cole and Sheath 1990; Wilson 2000). Being able to absorb blue light is a competitive advantage since blue light penetrates to greater depths than the light of longer wavelengths. There are some freshwater species and Prescott (1962) lists some species found in the Great Lakes. However, most commonly, the freshwater forms of these algae are found in cold and fast-running streams.
- Charophyta—Stonewarts: These algae can be unicellular or multicellular and commonly occur in freshwater, primarily in slow-moving or standing water. These algae are commonly anchored on muddy or sandy substrates or hard limestone streambeds (Bold and Wynne 1985). The stonewarts commonly become encrusted with carbonates, hence their common name.
- Division Chrytophyta—Chryptomonads: Chryptomonads are typically very small, unicellular, mobile (with two flagella) phytoplankton found in almost any marine or freshwater environment, with the greatest diversity in temperate lakes (Suthers and Rissik 2009;

Likens 2010). This is an ecologically important group, providing a high nutrient food for zooplankton. Chryptomonads frequently dominate the phytoplankton assemblages of the Great Lakes. While populations may at times opportunistically increase with favorable conditions, aided by high production rates, they are typically year-round residents. They are able to grow at low light levels so that maximum population densities may occur at depth, resulting in deep chlorophyll maxima in some lakes. Diel vertical migrations (DVMs) may also occur (Likens 2010).

- Division Euglenophyta—Euglenoids: Euglenoids represent a large and diverse group of which only a few are planktonic (Figure 15.8). They are typically flagellated on one end, and most are colorless and nonphotosynthetic (Likens 2010) or photosynthetic and facultative heterotrophs. They are most often found in shallow water that is rich in organic matter (e.g., polluted lakes or farm pounds). Most euglenoids live in freshwaters (Suthers and Rissik 2009) with only a few marine species reported.
- Division Phaeophyta—Brown Algae: Brown algae are well-known, predominantly marine species, the largest of which is kelp. Other examples include seaweeds such as the genus *Sargassum*, which form floating mats and are the most prominent species in the area known as the Sargasso Sea, which is in the middle of the North Atlantic Ocean. The few freshwater forms are attached, not planktonic.
- Division Chrysophytes—Chrosophytes: These are commonly known as the golden-brown algae due to the presence of the pigment fucoxanthin in addition to chlorophyll. The groups are made up primarily of free-swimming, unicellular freshwater forms. Interestingly, while they are predominately phototrophs, in the absence of light or with the abundance of dissolved food nearly all of this group are facultative heterotrophs.
- Division Haptophyta—Golden Algae: These are also referred to as golden algae, which are flagellated (typically two slightly unequal flagella) and include some of the most common marine algae (Wehr and Sheath 2003). One of the well-known members of this group is *Prymnesium parvum*, which produces toxins that have resulted in massive fish kills in Texas fish hatcheries and reservoirs.
- Division Bacillariophyta—Diatoms: Diatoms are one of the more important, diverse, and abundant groups of phytoplankton. They are particularly dominant in cold, nutrient-rich waters. A distinguishing characteristic of diatoms is their cell wall (frustule) composed of silicon dioxide (Figure 15.9). Thus, in addition to other nutrients, diatoms require silica for growth. The frustules are composed of two parts (valves) that fit together, and diatoms are

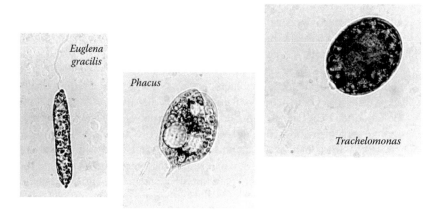

FIGURE 15.8 Examples of Euglenophyta. (Courtesy of Dr. Morgan L. Vis, Professor of Environmental & Plant Biology, Ohio University Algae Home Page http://www.ohio.edu/plantbio/vislab/algaeindex.htm.)

Examples of centric diatoms

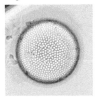

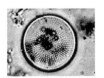

Coscinodiscus spp.

Actinocyclus spp.

Cyclotella spp.

Examples of pennate diatoms

Navicula spp.

Diploneis spp.

FIGURE 15.9 Examples of diatoms. (From NOAA's Great Lakes Water Life Photo Gallery.)

often distinguished based on the shape and structure of the valves. Two major groups include the centric diatoms (typically planktonic) with disk-shaped or cylindrical cells, and the pennate diatoms (typically benthic) that exhibit bilateral symmetry (elongated) (Figure 15.9).

- Division Dinophyta—Dinoflagellates: Dinoflagellates are primarily found in marine environments although they may become abundant in freshwaters as well; they are particularly abundant in oligotrophic lakes. One characteristic of this group is the presence of two flagella, which are long strands of protein groups that can be manipulated for movement. This allows them to move within the water column to areas where light and nutrients are optimal.
  - In marine waters, summer blooms may be so dense that the water appears golden or red, producing what is called the "red tide." In addition, in some of these blooms, the dinoflagellates produce a neurotoxin, which affects the muscle function in some organisms. Humans eating fish or shellfish contaminated with these neurotoxins may also be affected.
- Division Chlorophyta—Green Algae: Green algae are an abundant, diverse, and important group, and are considered the most closely related to higher plants. This group is almost totally freshwater in distribution (Wetzel 2001); and includes a great diversity of size, shape, and growth forms (single celled, colonial, filamentous, and flagellated), most of which are bright green, because chlorophyll is not masked by accessory pigments. They are almost exclusively phototrophic. Some of the more common forms in the Great Lakes are illustrated in Figure 15.10.

15.3.1.1.1.2 Bacteria (Domain: Bacteria, Kingdom: Monera, Division: Cyanobacteria) Bluegreen algae, or Cyanobacteria, as indicated previously, are not a true algae but rather an ancient group of bacteria. Cyanobacteria are prokaryotic organisms; they lack a nucleus and organelles (chloroplast, mitochondria). These are also a "problem child" of the phytoplankton community, and have been so for a long time. Cyanobacteria are believed to be the original "planet killer." Several billion years ago, free oxygen was rare in our atmosphere, so living organisms were mostly anaerobic. One of the first organisms to use water and light for photosynthesis was the Cyanobacteria, producing oxygen as a by-product, which was toxic to anaerobic organisms and favored the development of aerobic organisms. And here we are.

Cyanobacteria exhibit a number of features that make them very competitive, and undesirable. They are very productive, often dominating phytoplankton populations in lakes with high nutrient

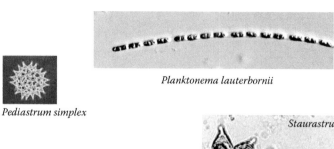

Planktonema lauterbornii

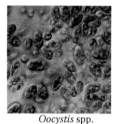

Pediastrum simplex

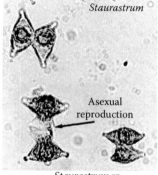

Oocystis spp.

Staurastrum sp.

FIGURE 15.10 Some of the more common groups of green algae in the Great Lakes. (From NOAA and Ohio University.)

inputs. They are able to fix nitrogen to form inorganic nitrogen, commonly a limiting nutrient, so while they can use inorganic forms if available, they can also make their own if necessary. They can also take up more phosphorus than immediately required for growth and store it for later use, known as luxury uptake. They have gas vacuoles, which enable them to regulate their buoyancy (Oliver and Walsby 1984). As a result, they can rise to the surface to compete for light and nutrients, creating unsightly scums that can be several feet thick (Smith 2001). In addition to occurring in large mats, they produce substances that cause strong tastes and odors, which cannot be removed by conventional water treatment, causing drinking water problems. When these large blooms die-off, their decay often results in hypoxia. They are also fairly inedible, so they are not a good food supply to support the food chain of lakes. Finally, as discussed in the following section, certain species may also release toxins into the water so that blooms of these, and comparable algae, are referred to as harmful algae blooms (HABs). A result of HABs is often lake closure, such as for the 2010 bloom in Grand Lake St. Marys, western Ohio (Figure 15.11), where reports indicated that people were suffering from "stomachaches, rashes and numbness" and that "one man, whose dog died after a swim in the lake, was hospitalized last week after he gave the dog a bath. Within days, the 43-year-old man began having trouble walking and lost feeling in his arms and feet."

15.3.1.1.2 HABs

While not a taxonomic grouping, there are a number of phytoplankton groups that, cause harm when they occur in excess (algal blooms); their blooms are classified as HABs. The harm may be due to populations being large enough to block light and to deplete oxygen. More commonly, HABs are focused on those phytoplankton that release toxic materials into waters, impacting human or ecosystem health. Many of the organisms responsible for toxic releases are found in estuarine or ocean environments (e.g., those causing "red tides").

In freshwater lakes, the primary organisms responsible for HABs are the Cyanobacteria. Cyanobacteria have a number of characteristics that make them problematic. For one, they may occur in large populations and are buoyant, so that they may form dense surface mats, like dense green paint, on the water surface up to several feet thick. Also, when they die, their decomposition results in hypoxia, taste, odor, and other problems. Cyanobacteria can produce a factor that inhibits

Algae bloom on Grand Lake St. Marys West Beach on June 23, 2010

NOTICE An algae bloom has made this area potentially unsafe for water contact. Avoid direct contact with visible surface scum.

Grand Lake St. Marys June 2010 algal bloom

West Beach recreational restrictions imposed on Grand Lake St. Marys users on June 23, 2010

FIGURE 15.11 Cyanobacterial algal bloom and lake in Grand Lake St. Marys in western Ohio, 2010. (From Davenport, T. and Drake, W., *LakeLine*, Fall, 41–46, 2011. Courtesy of the North American Lake Management Society. With permission.)

the growth of other algae (which may be more palatable to zooplankton and better support overall productivity). In addition, Cyanobacteria may produce toxins. The *Lyngbya* species produce a toxin that results in what is known as "swimmer's itch" or cyanobacterial dermatitis. Three of the most common and problematic organisms are in the genera *Anabaena*, *Aphanizomenon*, and *Microcystis*. For example, the Cyanobacteria of the genus *Microcystis* releases the hepatotoxin microcystin (Brittain et al. 2000; Vanderploeg et al. 2001), which is extremely toxic, causing death in livestock and human health problems. The toxicity of three blue-green algae toxins (anatoxin, microcystin, and aphantoxin) is compared to that of one of the deadly groups of mushrooms and the toxicity of cobra venom in Table 15.2.

The World Health Organization established a recommended limit of 1 μg L−1 for drinking water (Brittain et al. 2000; Vanderploeg et al. 2001), which can be exceeded many times in HABs. The

TABLE 15.2

Comparison of Toxicities of Three Cyanobacterial Toxins

Toxin	Source	LD50 (μg kg⁻¹)
Muscarine	*Amanita* mushroom	1100
Anatoxin	*Anabaena*	250
Microcystin	*Microcystis*	60
Neurotoxin	Cobra venom	20
Aphantoxin	Aphanizomenon	10

Source: Smith, V.H., Blue green algae in eutrophic fresh waters, *LakeLine*, Spring, 34–37, 2001.

Note: LD50 is the dose required to kill 50% of laboratory animals receiving the toxin.

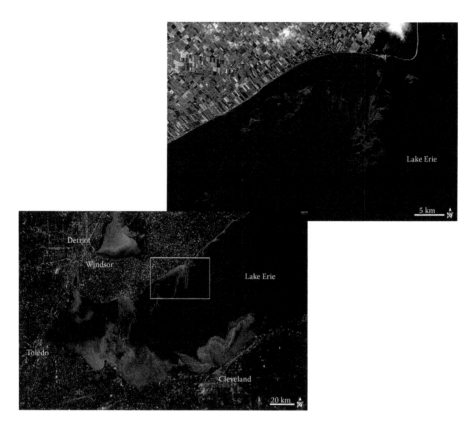

FIGURE 15.12 October 2011 satellite image of a toxic algae bloom (primarily *Microcystis*) in Lake Erie. (From NOAA Landsat-5 Imagery.)

Center for Disease Control (CDC) lists the possible human impacts from the consumption of these toxins, including stomach and intestinal illness; breathing difficulties; allergic responses; skin irritation; liver damage; and neurotoxic reactions, such as tingling fingers and toes. Figure 15.12 is an illustration of an extremely large HAB (primarily *Microcystis*) that occurred in Lake Erie during early October 2011 and was captured by satellite imagery.

Another toxic algae is the haptophyte, or golden algae, *P. parvum*. These algae are more common in brackish waters and can cause marine toxic algal blooms. While not considered a public health threat, they are believed to produce a number of toxins, collectively known as prymnesins, which include an ichthyotoxin (toxic to fish, Sallenave 2010) that ruptures gill membranes. In Texas, this alga became a major issue as a result of massive fish kills at Dundee State Fish Hatchery. The golden algae has caused fish kills in 33 Texas reservoirs and has invaded reservoirs and river systems in 15 other states, including Alabama, Arizona, Arkansas, California, Washington, Hawaii, New Mexico, Wyoming, North Carolina, South Carolina, Florida, and Georgia (Sager et al. 2008, cited by Sallenave 2010).

15.3.1.1.3 Methods for Enumeration

15.3.1.1.3.1 Direct Enumeration One method for quantifying phytoplankton is direct enumeration. This is a typically laborious process that initially involves collecting whole-water samples and then concentrating these samples (e.g., see Porter et al. [1993] for protocols used in the U.S. Geological Survey's National Water-Quality Assessment Program [NAWQA]). The samples are then counted via microscopy. Often, counting chambers are used to facilitate the process, such as the Sedgewick/Rafter counting cell (Figure 15.13).

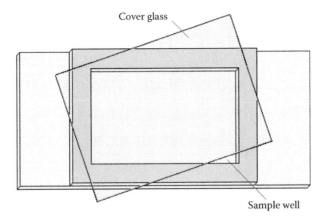

FIGURE 15.13 Sedgewick/Rafter counting cell. (From USAE, Analyzing plankton tows, pumped samples, and shallow water samples, Engineering Research and Development Centerweb site.)

Direct enumeration is usually prohibitively expensive and technically impossible for most monitoring programs. Further, in many lakes, a large portion of the algal biomass may be unidentifiable by most experts (subsequently these are appropriately called little round green things [LRGTs] and little round blue-green things [LRBGTs]).

Other methods, such as imaging flow cytometry, are also often used to assess phytoplankton biovolumes and functional group compositions.

15.3.1.1.3.2 Enumeration by Plant Pigments Phototrophs or photoautotrophs use light as an energy source for photosynthesis. They have certain pigments that absorb light, which raises the energy level of the pigment and provides the energy source. Similar to the Redfield ratios, the mass (or molar rations) for phytoplankton are approximately (Chapra 1997):

$$100 \text{ gD}:40 \text{ gC}:7200 \text{ mgN}:1000 \text{ mgP}:1000 \text{ mgA}$$

where gX = mass of element X (g) and mgY = mass of element Y (mg). The terms D, C, N, P, and A refer to dry weight, carbon, nitrogen, phosphorus, and chlorophyll-a, respectively. It should be noted that chlorophyll-a is the most variable of these quantities with a range of approximately 500–2000 mgA (Chapra 1997). Although the stoichiometry is variable, if chlorophyll can be measured then an estimate of the phytoplankton biomass can be obtained.

An advantage of using plant pigments, particularly chlorophyll (the green pigment), over other metrics used for enumeration (e.g., biomass, counting, etc.) is that it is relatively fast and inexpensive to measure. There are several forms of chlorophyll, but the most common pigment (and form used for enumeration) is chlorophyll-a. Since chlorophyll absorbs light, the amount of light absorbed can be used as an indication of the amount of chlorophyll present (e.g., using a spectrometer; Method 106-a in Standard Methods, e.g., APHA 1995). When chlorophyll-a absorbs light, it becomes excited and releases the light (fluoresces) at a lower wavelength. Therefore, an alternative method measures chlorophyll by fluorescence (Method 106b in Standard Methods). The wavelengths of light absorbed by chlorophyll-a peak at 665 nm (blue) and 465 nm (red), while chlorophyll-a fluoresces at 673 and 726 nm. The peak absorbance is near the peak of solar incident light, which makes chlorophyll-a an effective pigment (Figure 15.14).

There are a variety of other plant pigments that may participate in the process of photosynthesis, and the quantity (e.g., stoichiometry) and form of these pigments vary between different algal groups. These pigments are often used in the enumeration of these separate groups. For example, phycobilins are water-soluble pigments that occur in Cyanobacteria and Rhodophyta. These

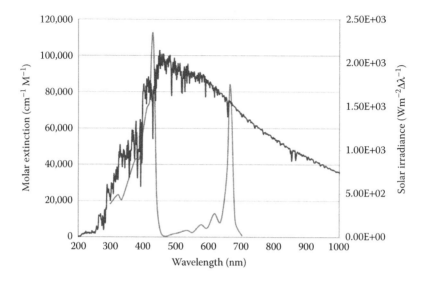

FIGURE 15.14 Absorbance spectra for chlorophyll versus solar energy on a mid-summer day.

pigments are often quantified using extraction and either fluorescence or spectrometry or by *in vivo* fluorometry. For example, probes may be used for the bluish pigment phycocyanin, which gives the Cyanobacteria their name, and/or the reddish pigment phycoerythrin, which gives the red algae their common name.

15.3.1.1.4 Size and Shape

The size and shape of phytoplankton vary enormously, and for a section title in his limnology textbook, Wetzel (2001) used "Small is Beautiful and Productive." Since all phytoplankton live in basically the same place, the photic zone of the water column, their size and shape may reflect their various adaptations to that environment.

One consequence of living in the photic zone is that any object will tend to settle out of it. Most phytoplankton are negatively buoyant, so they sink. So, logical questions may be why are they negatively buoyant and how can they persist? Vogel (2004) suggested that negative buoyancy may be an evolutionary adaptation to prevent phytoplankton from being trapped by surface tension. Regardless of why they sink, the distribution of phytoplankton is impacted by the rate of settling.

For quiescent waters, factors impacting the sinking velocity can be illustrated using Stoke's law:

$$v_s = \frac{2gr^2\left(\rho' - \rho\right)}{9\mu R}$$

where
   v_s is the sinking velocity
   g is the gravitational acceleration
   r is the radius of the sinking spherical particle
   ρ' is the density of the sinking particle
   μ is the viscosity of the fluid
   R is a form factor correction (with values of 0–1) for a shape other than spherical.

So, the sinking velocity is inversely proportional to the viscosity of the fluid, directly proportional to the difference in density with the fluid, and directly proportional to the square of the radius. Stoke's law was developed for the settling of a perfect sphere, where settling is independent from (or not influenced by) the settling of other particles, in a quiescent fluid. So, settling is also inversely proportional to the deviation of an object from a sphere (form resistance). Think of skydivers. If they are falling head down with their arms by their sides, frictional (form) resistance is minimized and their fall velocity is great (about 160–180 mph in the lower atmosphere). If the skydivers fall with their body perpendicular to the ground and with their arms stretched out, the terminal fall velocity is substantially decreased (to about 120 mph).

Similarly, phytoplankton may minimize their fall velocity by taking advantage of colder periods (e.g., spring or fall in contrast to summer), minimizing the density difference between their bodies and the water, and reducing their size. Since settling decreases nonlinearly with the square of the radius, the most effective method for reducing settling velocities is to reduce size. A common classification system by size is illustrated in Table 15.3. So, Cyanobacteria, which include some of the smallest phytoplankton, may have some competitive advantage over large phytoplankton in some cases.

The ratio of surface area to volume also increases as size decreases, which may facilitate the surface exchange of gases and nutrients as well as decrease settling rates. For picoplankton and nanoplankton, the transfer of material to and from cells is almost entirely by molecular diffusion (Wetzel 2001). Size may also impact the ease with which the phytoplankton may be eaten, such as by fish and zooplankton. The microplankton and macroplankton are not as vulnerable to predation (Wetzel 2001) but they are less efficient at nutrient uptake and more vulnerable to settling.

The size of phytoplankton may also impact the manner and accuracy of their enumeration. Sigee (2005), for example, suggested that in the past the ecological role of picoplankton and nanoplankton has been underestimated due to the difficulty of their detection and enumeration using conventional microscopy.

In addition to their size, phytoplankton can also decrease their rate of settling by varying their shape. Some of the potential strategies are to form groups or colonies, or to become asymmetrical where the form resistance decreases with increasing asymmetry.

In addition, many phytoplankton may actively influence settling velocities. Wetzel (2001), for example, indicated that microphytoplankton (Table 15.3) may be separated into two broad types of ecological strategies, those that are motile and those that are not motile. The motile forms include those with gas vacuoles, such as some Cyanobacteria. By regulating their gas vacuoles, these algae may become negatively buoyant and rise through the water column. Flagella may also be used for motility. The nonmotile forms may depend on the turbulence mixing of the water column to maintain their position within the photic zone. Other strategies include the accumulation of fats, the reduction of density by altering the cellular ion content, and the production of gelatinous sheaths

TABLE 15.3
Size of Phytoplankton

Category	Linear Size (Cell or Colony Diameter, λm)	Unicellular Organisms	Colonial Organisms
Picoplankton	0.2–2	Photosynthetic bacteria, blue-green algae; Chrysophyta	–
Nanoplankton	2–20	Blue-green algae, cryptophytes, small dinoflagellates and others	–
Microplankton	20–200	Dinoflagellates	Diatoms
Macroplankton	>200	–	Blue-green algae (Anabaena, Microcystis)

Source: Modified from Sigee, D.C., *Freshwater Microbiology: Biodiversity and Dynamic Interactions of Microorganisms in the Aquatic Environment*, Wiley, Chichester, UK, 2005.

(particularly Cyanobacteria and green algae), which are less dense and provide increased frictional resistance (Wetzel 2001).

15.3.1.1.5 Seasonal Succession and Spatial Distribution

Seasonal variations in phytoplankton concentrations occur due to a variety of factors, including light, water temperatures and density stratification, turbulent mixing, predation, nutrient distributions, and others, with the result that the species distribution of phytoplankton is not constant either temporally or spatially (Figure 15.15).

Seasonal variations often occur in types and population densities of phytoplankton. A general idealized trend in seasonal variations for temperate lakes is illustrated in Figure 15.16. Planktonic diatoms often produce spring and fall blooms in temperate lakes and oceans and summer blooms at higher altitudes. During spring with the onset of stratification, ample nutrients are trapped in the surface layer and ample light and low populations of herbivores (e.g., zooplankton). With relatively low temperatures, the density and viscosity of water are greater and there is still substantial vertical

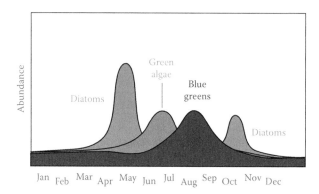

FIGURE 15.15 Typical seasonal succession patterns of phytoplankton populations. (From Water on the Web, Monitoring Minnesota lakes on the Internet and training water science technicians for the future—A national on-line curriculum using advanced technologies and real-time data, University of Minnesota–Duluth, Duluth, MN, 2004, Available at http://WaterOnTheWeb.org.)

FIGURE 15.16 Cyanobacteria (*Planktothrix rubescens*) on the surface of a Missouri lake. (From Lakes of Missouri Volunteer Program. With permission.)

mixing, so settling is minimized. Similarly, in the fall with fall turnover, waters are cooler, mixing is again high, and nutrient concentrations are high. Diatoms are capable of responding quickly and taking competitive advantage of such conditions, during which their populations "bloom." Following the spring bloom, nutrients become depleted, particularly the silica for diatoms, there is an increase in herbivores, and then the diatom populations are "bust." During these conditions, green algae often become dominant. Green algae are good competitors for nutrients and many have flagella or gas vacuoles to offset higher sinking rates. During late summer and early fall, the Cyanobacteria often dominate. In the early fall, herbivore populations are large, but Cyanobacteria are not very palatable. While nutrients are low, they can fix nitrogen. So, they bloom. Then, during winter with lower temperatures and increased vertical mixing, the population densities for all algae are typically low. However, winter productivity, such as under ice, may be significant in many systems.

Another more specific example of the complications in seasonal variations in spatial distributions is with fall blooms of the Cyanobacteria *Planktothrix rubescens*. This Cyanobacteria, like red algae, contains the pigment phycoerythrin, which reflects red light and absorbs blue light. So, blooms of this alga give water the color of deep red, rich wine (Sappington 2009; Figure 15.16). By absorbing blue light, which penetrates deeper into the water column than the light of longer wavelengths, these Cyanobacteria may commonly occur at greater depths than other phytoplankton. For example, Sappington (2009) discussed the distribution of these Cyanobacteria in Missouri lakes, where they may commonly occur in the metalimnetic zone during the summer months (Figure 15.17). As illustrated in Figure 15.17, during much of the year they may not occur at the depth of the first arrow (12 ft.) where there is still approximately 20% of surface light available and temperatures are near that of the surface. However, in the metalimnetic zone (36 ft.), where there is little or no light and the temperature is cooler, *P. rubescens* are at their greatest density. So, blooms, if they occur, do so

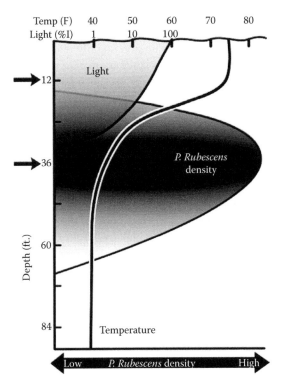

FIGURE 15.17 Illustration of the summer distribution of light, temperature, and the Cyanobacteria in a Missouri lake. (From Sappington, A., Red algae, *The Water Line*, 13, 1–2, 2009.)

near or in the metalimnion for much of the year. However, this phytoplankton also has gas vacuoles through which they can control their depth by regulating their buoyancy. In the fall, as temperatures cool and mixing occurs to a greater depth, *P. rubescens* blooms are more frequently reported in the surface waters (Figure 15.16). So, bloom density may not actually increase, but simply become more visible as this phytoplankton moves up in the water column.

In addition to seasonal variations in vertical positions in the water column, variations over diel (24 hour) periods in response to light are also well known, particularly for dinoflagellates. This movement is known as diel vertical migration. DVM is ubiquitous among plankton organisms (Hutchinson 1957), so it is common in both phytoplankton and zooplankton and must then provide some ecologic advantage. One benefit for phytoplankton, for example, may be increased exposure to light near the surface during the daytime and then increased nutrient concentrations in deeper waters at night (Eppley et al. 1968).

15.3.1.2 Zooplankton

15.3.1.2.1 Classification and Importance

Zooplankton comprise a very diverse group of organisms, which includes the primary consumers that graze on phytoplankton. Zooplankton provide the link between the primary producers and higher consumers. As such, they influence both the quantity and species composition of the phytoplankton community and of the secondary consumers. In addition, zooplankton are integral to the recycling of nutrients obtained from grazing on phytoplankton, which is important to both communities. For example, Axler et al. (1981) found that nitrogen regeneration via zooplankton excretion and microbial mineralization was critical for phytoplankton growth in the epilimnion of Castle Lake, California, since the levels of dissolved inorganic nitrogen were insufficient to maintain measured rates in the absence of regeneration.

As seen in the previous section, phytoplankton are typically taken to represent planktonic plants. However, phytoplankton are made up primarily of members of the kingdom Protista, which are not true plants, and also includes bacteria (the blue-green algae or Cyanobacteria), which are in a separate domain from plants. Similarly, zooplankton literally refers to planktonic animals ("zoo": animal). However, some of the important zooplankton are also members of the kingdom Protisa rather than Animalia. So, the zooplankton may be classified based on their location and feeding habits as well as by their phylogeny.

Planktonic animals are dominated by four different general groups (Wetzel 2001). They include members of the kingdom Protista (Protists) and three groups in the kingdom Animalia: the Rotifers (Phylum Rotifera), the Cladocerca, and the Copepods (both in the phylum Arthropoda). Note that Wetzel (2001) indicated that Protista is not a natural taxonomic group and is being abandoned.

15.3.1.2.1.1 Protists The majority of the protist zooplankton is protozoans or related forms. They include the flagellates, which move using flagella, and the ciliates, which move using cilia, with the flagellates being the most abundant. In addition to the planktonic forms, the flagellates include forms important for other reasons. For example, *Giardia* causes gastrointestinal trouble and *Trichomonas* causes venereal disease. Perhaps one of the best known of the protistan zooplankton is the Euglena, which include dinoflagellates (*Ceratium* and *Peridinium*), chrysomonads (*Dinobryon*, *Mallomonas*, and *Synura*), euglenids (*Euglena*), volvocids (*Volvox* and *Eudorina*), choanoflagellates (*Astrosiga*), and others (Wetzel 2001; Figure 15.18).

Wetzel (2001) indicated that while the actual biomass of protistan zooplankton is small in relation to other zooplankton, their generation rates are high. As a result, they are the most important microbial consumers and have a substantial impact on the utilization and cycling of nutrients and organic materials.

15.3.1.2.1.2 Rotifers Rotifers form an important group of soft-bodied invertebrates of which most are sessile (fixed or attached to some substrate, such as to macrophytes), but some important

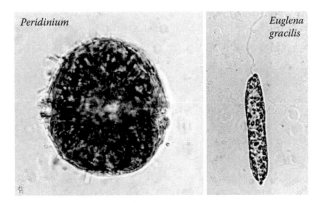

FIGURE 15.18   Examples of protistan zooplankton. (Images courtesy of Ohio University Algae Home Page.)

forms are planktonic. They are generally smaller than crustaceans and often have long spines to reduce settling. Rotifers have cilia, which function in locomotion but also aid in moving food toward the mouth of the zooplankton (see Figure 15.19). They exhibit a wide range of morphological variations and adaptations.

15.3.1.2.1.3   Cladocerans   Cladocerans (called waterfleas) are generally small (0.2–3.0 mm; Wetzel 2001), with a distinct head, a body covered with bivalve carapace, and a light-sensitive eye. They typically feed by filtering water to remove particulate organic matter, and the size of the particles eaten is a function of the setae. They are typically herbivorous, but there are some raptorial predators that grasp protozoans, rotifers, and small zooplankton with their forelegs. Cladocerans have hard exoskeletons and molt many times as they grow. Many species show changes in shape as they grow—characteristic of age, season, or environmental conditions (Balcer et al. 1984). They have large swimming appendages (second antennae; Wetzel 2001). Perhaps the best known of the cladocerans is the Daphnia (Figure 15.20).

15.3.1.2.1.4   Copepods   The copepods (class Cristacea) comprise of three groups: the suborders Calanoida, Cyclopoida, and Harpacticoida (Wetzel 2001). Of these, the Harpacticoids are primarily benthic with mouthparts for seizing and scraping, while the other two groups can be suspension feeders or raptorial predators. The body of a copepod consists of the anterior cephalothorax bearing

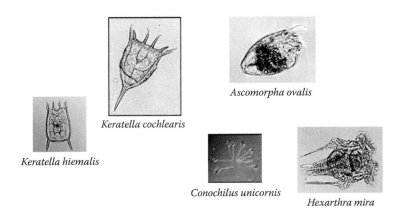

FIGURE 15.19   Examples of rotifers. (Courtesy of the NOAA Great Lakes Water Life Photo Gallery.)

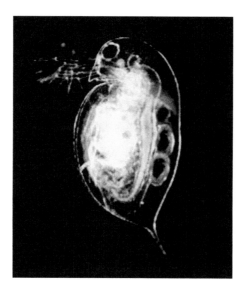

FIGURE 15.20   Daphnia. (Courtesy of Wikimedia Commons.)

five pairs of appendages, representing antennae and mouthparts, and the thorax with six pairs of swimming legs (Figure 15.21).

15.3.1.2.2   Seasonal and Spatial Variations

As with phytoplankton, zooplankton population densities and species distributions can vary seasonally and spatially. Variations may occur over seasons in part due to differences in the availability and quality of food. Increases in turbidity, flows, and other factors also impact seasonal abundance. Some copepods are able to survive unfavorable conditions (e.g., winter) by entering diapause, a resting state similar to hibernation (Balcer et al. 1984).

Zooplankton populations also vary horizontally. There is considerable "patchiness" of zooplankton in the pelagic zone, which affects fish feeding behavior (Wetzel 2001). In addition, Wetzel (2001) noted that the pelagic zooplankton exhibit an "avoidance of shore" so that the littoral and near littoral waters are virtually free of pelagic zooplankton.

As with some phytoplankton, vertical migrations of zooplankton over diel (24 hour) periods, or DVM, are well known. DVM is most conspicuous for the cladoceran zooplankton and to a lesser

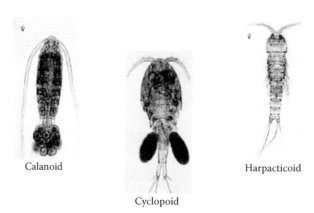

Calanoid

Harpacticoid

Cyclopoid

FIGURE 15.21   Examples of copepods. (Courtesy of the NOAA Great Lakes Water Life Photo Gallery.)

extent the copepods (Wetzel 2001). For zooplankton, most species migrate to the surface at night, and retreat to the bottom layers with the dawn.

15.3.2 LITTORAL ZONE

The organisms of the littoral zone are diverse and include a complex mixture of plants, animals, and microorganisms. Plants of the littoral zone include those that grow on some substrate, the periphyton, and the larger plants, the macrophytes.

15.3.2.1 Periphyton

Periphyton refers to a community of organisms that is associated with some surface, such as rocks, debris, macrophyte leaves, etc. Macrophytes, discussed in the next section, comprise the large ("macro") plants while periphytons are typically members of the small or microbiota. Microbiota, both plants and animals, grow on or attach to virtually any surface immersed or growing in water: living, dead, organic, or inorganic (Wetzel 2001).

The periphyton are largely composed of the same algal groups as the phytoplankton (e.g., diatoms, cyanophyta, and eugleonophyes). However, while the planktonic members of these algal groups are ecologically important, probably greater than 90% of all algal groups grow attached to a substrate. The attached algae often dominate the algal biomass of small streams and shallow lakes (Wetzel 2001).

15.3.2.2 Macrophytes

Macrophytes, also briefly introduced in Chapter 6, represent a diverse assemblage of plants that grow in the littoral zone of most lakes and reservoirs. Primary forms of macrophytes include emergents, floating-leaved and/or free floating, and submerged species (Westlake 1975).

Emergents: Emergent macrophytes are typically rooted in the lake bottom with the tops of the plant extending into the air. Thus, they can obtain nutrients from the sediments and water, oxygen and carbon dioxide from the air, and light from above the water surface (so they are not impacted by light attenuation in the water column). They must also be strongly rooted since they are subject to the wind and waves. The root and rhizome system (underground stems) exists under anaerobic conditions, as do the very young foliage for a brief period. Examples include cattails, bulrushes, pickerelweed, and others (Figure 15.22). Most emergent macrophytes are perennials (plants or plant

FIGURE 15.22 Examples of emergent macrophytes. (From Ohio Pond Management Plant Identification, Ohio Department of Natural Resources.)

parts living for longer than 1 year; LakeWatch 2007) and the emergent macrophytes share many similarities with terrestrial plants (Wetzel 2001).

Freely floating macrophytes: Freely floating macrophytes are a diverse group of plants, with the more elaborate plants consisting of aerial or floating leaves, a condensed stem, and submersed roots (Wetzel 2001). They can be very large, such as the water hyacinth or water lettuce, or very small surface-floating plants with few or no roots (e.g., duckweed). These plants generally grow in portions of lakes that are not subject to periodic drying, and in areas where currents are slow. Dense mats or stands may form, often for many species from spreading underground rhizomes (LakeWatch 2007), which often pose a problem for their control (Figure 15.23).

Submerged macrophytes: Submerged macrophytes are also a diverse group of plants that have submerged leaves, examples of which are illustrated in Figure 15.24. They are commonly rooted

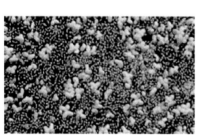
Duckweed and watermeal

Pond lilies

FIGURE 15.23 Examples of free-floating macrophytes. (From Ohio Pond Management Plant Identification, Ohio Department of Natural Resources.)

Water milfoil

Najas

Chara

Coontail

Elodea

FIGURE 15.24 Examples of submerged macrophytes. (From Ohio Pond Management Plant Identification, Ohio Department of Natural Resources.)

in the substrate so they can take nutrients from the water column and sediments. Some, such as the bladderworts, are not rooted. The submerged macrophytes are at some disadvantage to other forms, since light, oxygen, and carbon dioxide are not as available as they are at the water surface, but they have to invest less energy in structural support since that is provided by the water of which they are primarily (95%) composed (LakeWatch 2007).

Macrophytes are both influenced by, and they themselves influence, lake characteristics. Dense canopies may reduce the light availability to lower water levels by shading, so that productivity (and oxygen) may be limited to only the surface. Photosynthesis and respiration in dense stands can result in large diel variations in oxygen concentrations, as large as 12 mg L^{-1}, to occur in the surrounding waters (LakeWatch 2007). Conversely, they may also cause large variations in carbon dioxide concentrations, and can change the pH by 2–3 pH units during a 24-hour period (LakeWatch 2007). They may provide a substrate on which other organisms can grow (e.g., the microbiota), provide shelter and escape habitats for organisms, impact flows, impact nutrient cycling and sediment chemistry, and impact other limnological characteristics of lakes. Horne and Goldman (1994) indicated that emergent reeds and submerged macrophytes are the dominant primary producers and contribute the most biomass in small lakes.

The impact of macrophytes is proportional to their abundance and productivity, which are, in turn, impacted by lake characteristics (e.g., lake levels and the extent of the littoral zone). Excess biomass may eventually cause those littoral zones to be filled in, resulting in the succession of lakes to wetlands. Macrophytes are commonly used as indicators of the health of lakes and reservoirs. There may be too few macrophytes, or too few of the native or desirable species, as a result of toxicity, pollution, or other factors. Another common problem is "too much of a good thing," where macrophytes may be in excess, essentially choking some shallow reservoirs. As a result, macrophyte management is a large and expensive component of many lake management programs.

As indicated by the U.S. EPA's "Biological Indicators of Watershed Health," macrophytes are excellent indicators because they:

- Respond to nutrients, light, toxic contaminants, metals, herbicides, turbidity, water level change, and salt
- Are easily sampled through the use of transects or aerial photography
- Do not require laboratory analysis
- Are easily used for calculating simple abundance metrics
- Are integrators of environmental conditions

Macrophytes may be used as a component of a tiered assessment approach, as described by the U.S. EPA's (USEPA 1998) technical guidance on lake and reservoir bioassessment and biocriteria. Also, a number of ecological indices based on macrophytes are in use, particularly in Europe. An example is the free macrophyte index developed in Scotland (Free et al. 2007), which is based on the

- Maximum depth of colonization
- Mean depth of presence
- Percent relative frequency of *Chara* (for lakes with an alkalinity ≥ 100 mg L^{-1} $CaCO_3$)
- Percent relative frequency of *Elodeids* (which have grown profusely in some softwater lakes in Northwest Europe; Roelofs 1983)
- Plant trophic index
- Percent relative frequency of tolerant taxa

Another example is the U.K. macrophyte assessment system (LEAFPACS; UKTAG 2009). Penning et al. (2008a,b) discuss using macrophytes as indicators in European lakes. A disadvantage of macrophytes as an indicator is that macrophytes are often controlled (cutting, poisoning, etc.)

so they may not be representative. In addition, some criteria, such as the extent and percentage of cover, may not reflect the impacts due to species composition, such as that of invasive species.

15.3.2.3 Neuston

The neuston refers to organisms that live in the upper water surface film or surface microlayer. The term *neuston* was introduced by Naumann (1917) to designate the community of organisms associated with that surface microlayer. The neuston are typically subdivided into the epineuston, which are attached to the top of the surface, and the hyponeuston, which are attached to the bottom of the surface. They may also include some organisms that move over the surface film (Thorp and Covich 2010).

Organisms of the neuston include bacteria, protozoans, insects such as whirligig beetles and water striders, some spiders, and others. The neuston is known to accumulate nutrients and the base of the food chain is typically the layers of organic materials, such as lipids, lipoproteins, and polysaccharides (Bolsenda and Herdendorf 1993, Maier et al. 2009).

The neuston live on the water surface and are associated with the surface tension of water, so organisms such as Cyanobacteria, which rise to the surface because of their gas vacuoles, are excluded from this group. Some organisms spend only part of their life cycle in this surface microlayer.

15.3.2.4 Benthos

The benthos are the bottom dwellers. The characteristics of the benthos depend largely on the lake region (littoral zone, profundal) in which they live. In terms of the distribution of bottom-dwelling organisms, Cole (1979) further subdivides the benthos into the phytobenthos and the zoobenthos. The phytobenthos is comprised of the periphyton and macrophytes, which were discussed earlier, so this section will focus on the zoobenthos.

The zoobenthos represent a diverse assemblage of organisms, one of the more important of which is the insects (SWCSMH 2006):

> The freshwater world has more than 500,000 different species of insects. They occur in habitats that range from hot springs, discarded tin cans, temporary ponds, spring seeps, wetlands, rivers, lakes, to arctic and mountain pools. In fact, if water will stand for a few days, one or another of the ubiquitous chironomids will probably be the first to inhabit that water "... Take a walk down to the lake. It does not matter what time it is—there will be insect activity somewhere! (Narf 1997)

The type and biomass of zoobenthos depend on factors such as the substrate (e.g., gravel, sand, and silt), the organic content, etc., which impact their abundance and diversity, as illustrated in Table 15.4. The zoobenthos may be attached to the bottom, burrow in it, or be associated with aquatic plants.

TABLE 15.4

Abundance and Species Diversity of Aquatic Insects Found in Five Habitats (Characterized Mainly by Their Substrates) in a Quebec Stream

Habitat	Abundance (no. m^{-2})	No. of Species	Diversity = $(S - 1)/\log_e N$
Sand	920	61	1.96
Gravel	1300	82	2.31
Cobbles and pebbles	2130	76	2.02
Leaves	3480	92	2.40
Detritus (finely divided leaf material in pools and along stream margins)	5680	66	1.73

Source: SWCSMH, Chapter 1: Zoobenthos of Freshwaters—An Introduction, Soil & Water Conservation Society of Metro Halifax, 2006, Available at http://lakes.chebucto.org/ZOOBENTH/BENTHOS/i.html, accessed February 8, 2012.
Note: Values are annual averages.

Species diversity typically drops sharply in the profundal benthos as opposed to the littoral zoobenthos. This is particularly the case for eutrophic waters where the hypolimnion is usually anoxic. In those systems, the zoobenthos, where present, are often those forms indicative of highly polluted waters (such as some oligochaetes). For example, Lathrop (1992) demonstrated a decline in the zoobenthos of Lake Mendota, Wisconsin, from the early 1900s to the present day even though the extent and the duration of hypoxia have not substantially changed over that period. Lantrop suggested that the decline could be due to increased ammonia and sulfide concentrations resulting from increased eutrophication.

15.3.2.5 Fish

Fish are an ecologically and economically important component of the biota of lakes and reservoirs. Many lake management strategies for recreation are focused on optimizing the abundance of specific, economically important species, such as bass, lake trout, walleye, etc. These may be locally native species, but not necessarily so. For many reservoirs, which are not considered natural systems to begin with, the introduction of nonnative and often hatchery-raised sport fish is a common practice.

While the management of lakes has often focused on increasing angler harvest, fish are an integral component of lake ecosystems, playing important roles in energy flows, nutrient cycling, and maintaining community balance (Baker et al. 1993). Thus, fisheries management not only impacts the fish population, but it also impacts, and is impacted by, all other aspects of the lakes ecology.

15.3.2.6 Birds and Mammals

Birds and mammals are also important components of lake and reservoir ecosystems. Birds, for example, include those that are characteristic of open waters (e.g., ducks, geese, grebes, cormorants, kingfishers, terns, gulls, and pelicans); shoreline (e.g., stilts, greenshank, sandpipers, storks, ibises, spoonbill, herons, and egrets); contiguous meadows and grasslands; reed beds and other vegetation; and open air space above wetlands (e.g., rails, bitterns, coots, jacanas, moorhens, snipe, and painted snipe). These birds not only represent important components of the lake ecosystem, but they are also economically important (e.g., birding and hunting).

15.4 INVASIVE SPECIES

Invasive species are commonly taken as those nonnative species, which, when introduced to some new ecosystem, often become dominant to the detriment of the native species. However, a more general definition by the Nature Conservancy is that an invasive species is "any species not native to an ecosystem whose introduction does or is likely to cause economic or environmental harm or harm to human health." There are a wide variety of plant and animal species that by this definition are invasive, some of which are discussed in the following sections.

Invasive species are an enormous national problem. One response, motivated in part by the invasive zebra mussels, was the passage of the federal Nonindigenous Aquatic Nuisance Prevention and Control Act of 1990 amended by the National Invasive Species Act of 1996 that calls for the development of state and regional management plans to control aquatic nuisance species and for the secretary of the army to develop a program of research and technology development for the environmentally sound control of zebra mussels at public facilities. The purposes of the Nonindigenous Aquatic Nuisance Prevention and Control Act are to (Hanson and Sytsma 2001):

- Prevent unintentional introduction and dispersal of nonindigenous species into waters of the United States
- Coordinate federally conducted, funded, or authorized research, prevention, control, information dissemination, and other activities

- Develop and carry out environmentally sound control methods to prevent, monitor, and control unintentional introductions of nonindigenous species
- Understand and minimize the economic and ecological impacts of nonindigenous aquatic nuisance species that become established
- Establish a program of research and technology development and assistance to states in the management and removal of zebra mussels

There are a large number of invasive species in the United States and in other countries around the world (some of which originated in the United States and are invasive elsewhere). The following section provides an introduction to a few of these invasive species.

15.4.1 INVASIVE PLANTS

A variety of plants, including many macrophytes, are native to specific regions of the United States and can be problematic with excessive growth. However, a number of the really problematic species causing aquatic weed problems are introduced or invasive species. Examples of submerged invasive species include the watermilfoil and the curly-leaf pondweed that cause some of the worst problems in Florida (LakeWatch 2007) and other states. Other examples include those illustrated in Figure 15.25 and Table 15.5. As indicated, many of these species were introduced as either ornamental through the aquarium trade or ship ballast waters. These plants can grow rapidly to the water surface and form dense canopies.

Alligatorweed

Brazilian waterweed

Caulerpa, Mediterranean Clone

Eurasian watermilfoil

Hydrilla

Purple loosestrife

Water hyacinth

Water lettuce

Water spinach

FIGURE 15.25 Selected invasive macrophytes. (From USDA National Invasive Species Information Center, NISIC.)

TABLE 15.5
Selected Invasive Aquatic Species

Alligatorweed	*Alternanthera philoxeroides*	South America	First reported in Alabama in 1897	Probably introduced through ballast water
Brazilian waterweed	*Egeria densa*	South America	1893	Aquarium trade
Caulerpa, Mediterranean clone	*Caulerpa taxifolia*	Caribbean Sea; Indian Ocean	2000 (first infestation)	Aquarium trade
Common reed	*Phragmites australis*	Native to the United States, but the more invasive strains originated in Europe	Invasive European strains introduced during late 1800s	Possibly through ships' ballast
Curly pondweed	*Potamogeton crispus*	Eurasia, Africa, Australia	Mid-1800s	Possibly imported as an aquarium ornamental or introduced accidentally
Didymo	*Didymosphenia geminate*	Northern Europe and northern North America (Vancouver Island)	Was present in Canada in the late 1800s, but did not begin to cause problems until the early 1990s	Exact pathway unknown, but it spreads easily through contaminated boats and fishing gear
Eurasian water milfoil	*Myriophyllum spicatum*	Eurasia	Approx. 1900	Aquarium trade
Giant reed	*Arundo donax*	Thought to be native to eastern Asia	Early 1800s	Introduced for erosion control
Giant salvinia	*Salvinia molesta*	Brazil	1990s	Aquarium trade
Hydrilla	*Hydrilla verticillata*	Africa	1960	Aquarium trade
Melaleuca	*Melaleuca quinquenervia*	Australia	Early 1900s	Used as an ornamental and for erosion control
Purple loosestrife	*Lythrum salicaria*	Eurasia	Early 1800s	Through ships' ballast and as an ornamental
Water chestnut	*Trapa natans*	Eurasia	Late 1800s	Ornamental
Water hyacinth	*Eichhornia crassipes*	South America	1884	Ornamental
Water lettuce	*Pistia stratiotes*	Unknown	First described in Florida in 1765	Unknown
Water spinach	*Ipomoea aquatica*	Asia	1970s	Escaped ornamental

Source: Data from USDA National Invasive Species Information Center, NISIC.

15.4.2 INVASIVE ANIMAL SPECIES

15.4.2.1 Spiny Waterflea

The zooplankton spiny waterflea is a relatively recent invasive species, presently restricted primarily to the Great Lakes region (Figure 15.26). They were first discovered in Lake Huron in 1984 (Bur et al. 1986) and by the end of the 1980s, they were reported in the remaining Great Lakes. Currently, they are also found in a number of other lakes in the region.

The spiny waterflea have a single long tail with small spines along its length, hence its name (Figure 15.27). They may reproduce rapidly, and are often seen in gelatinous globs that collect on fishing lines, often clogging the eyelets of fishing rods, and on downrigger cables. They outcompete

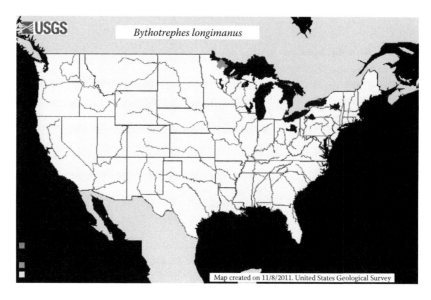

FIGURE 15.26 U.S. distribution of spiny waterflea. (From Fuller, P., Nico, L., Maynard, E., Larson, J., and Fusaro, A., USGS Nonindigenous Aquatic Species Database, Gainesville, FL, 2012, Available at http://nas.er.usgs.gov/queries/factsheet.aspx?SpeciesID = 836 Revision Date: 12/22/2011.)

FIGURE 15.27 Spiny waterflea. (From USGS; photograph by Kate Feil.)

native species, reduce food supplies for others, and have caused major changes in the zooplankton community structure of the Great Lakes (USEPA 2008).

15.4.2.2 Zebra Mussels

One of the most notorious of the invasive animal species is the zebra mussel (Figure 15.28). This freshwater mussel was originally from Asia but it presently occurs in most of Europe. Zebra mussels arrived in the United States in 1986 in the ballast water of a ship and were released into Lake St. Clair, Michigan (USACE 2001). Since their introduction into the Great Lakes, the mussels have spread rapidly, partially due to their ability to attach to vessels. They were first reported west of the continental divide in 2007, and are rapidly spreading (Figure 15.29).

Many of the problems associated with the relatively small (generally less than 4 cm long) zebra mussels are their explosive rates of growth and incredible population densities. Their maximum reported annual production, 29.8 g of dry tissue per square meter per year, is one of the highest recorded for freshwater or marine bivalves (USACE 2001). Biologists at Detroit Edison reported that zebra mussel densities on an intake screen climbed from 200 individuals per square meter (165 individuals per square yard) to 700,000 individuals per square meter (585,284 individuals per square yard) in 1989. A car submerged for 8 months in Lake Erie was 90% covered with mussels at an average density of 45,000 individuals per square meter (37,625 individuals per square yard). As many as 10,000 zebra mussels have been counted on a single freshwater mussel (USACE 2001).

FIGURE 15.28 Zebra mussels. (From USGS.)

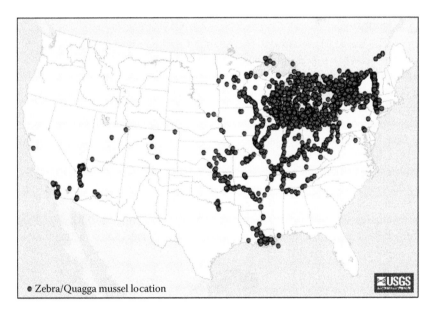

FIGURE 15.29 Zebra mussel distribution as of 2010. (From Benson, A.J., Zebra mussel sightings distribution, 2011, Retrieved 2/12/2012 from http://nas.er.usgs.gov/taxgroup/mollusks/zebramussel/zebramusseldistribution.aspx.)

One consequence of the incredible densities of the filter-feeding zebra mussels results from the quantity of water filtered (up to 1 gal. of water per day per mussel) and their highly efficient filtration mechanism. As a result, they remove a large fraction of the particulate materials from the water column (retaining nearly 100% of particles >1 μm; Jorgensen et al. 1984; Fanslow et al. 1995) reducing food supplies for other organisms.

A second consequence of their incredible densities is biofouling. Instead of burrowing into sands or silts like many other mussels, zebra mussels attach to a hard surface, any hard surface (rocks, logs, plants, other mussels, plastic, concrete, etc.). They can often clog water supply intakes

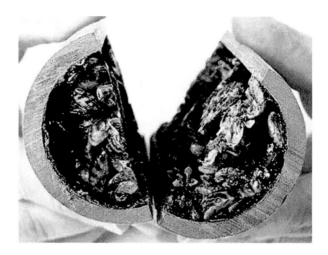

FIGURE 15.30   Pipe clogged by zebra mussels. (From USACE, 2001; photograph by Peter Yates.)

(Figure 15.30), and many power plants along Lake Erie now spend more than $250,000 each year on control (USACE 2001).

15.4.2.3   Sea Lamprey

The sea lamprey (Figure 15.31) is native to the Atlantic Coast and while generally marine, it ascends freshwater rivers to spawn. It is presently found throughout the Great Lakes (Figure 15.32). How it was introduced is somewhat controversial, with some indications that it was native to the Great Lakes. Its introduction is generally attributed to the Erie Canal, which allowed its introduction to Lake Ontario. It remained isolated from the other Great Lakes by the natural barrier imposed by Niagara Falls until the Welland Canal was opened in 1829 to bypass Niagara Falls and provide a navigable transportation route to Lake Erie from Lake Ontario (Aron and Smith 1971), which also allowed the sea lamprey access.

Sea lampreys are parasitic, feeding by attaching themselves to fish with their sucking mouth-parts, and causing tissue damage with their sharp teeth. This attached and parasitic feeding often results in the death of their prey, either directly from the loss of fluids and tissues or indirectly from secondary infection of the wound (Phillips et al. 1982; Fuller et al. 2012).

The overabundance of sea lampreys, combined with overfishing, pollution, and other factors contributed to the collapse of commercial fisheries in the Great Lakes in the 1940s and 1950s

FIGURE 15.31   Sea lamprey. (From NOAA's Great Lakes Environmental Research Laboratory.)

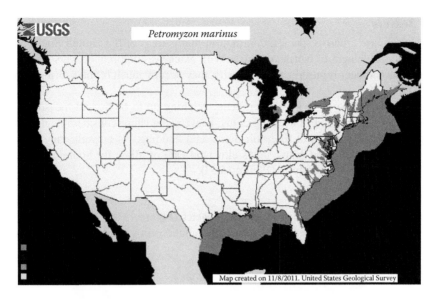

FIGURE 15.32 Sea lamprey distribution. (From Fuller, P., Nico, L., Maynard, E., Larson, J., and Fusaro, A., USGS Nonindigenous Aquatic Species Database, Gainesville, FL, 2012, Available at http://nas.er.usgs.gov/queries/factsheet.aspx?SpeciesID = 836 Revision Date: 12/22/2011.).

(Fuller et al. 2012). For example, lake trout catch in Lake Huron fell from 3.4 million pounds in 1937 to virtual failure in 1947, while in Lake Michigan the U.S. catch fell from 5.5 million pounds in 1946 to 402 pounds in 1953 (Scott and Crossman 1973; Fuller et al. 2012). Removing these large predators also allowed another nonindigenous species, the alewife, to invade. Alewifes are a problem species in many freshwater lakes, and the presence of the alewife could restructure a lake's food web, leaving less food for the native species. In the Great Lakes, alewife populations exploded, also contributing to large additional changes in species composition (Smith and Tibbles 1980).

The continued presence of sea lampreys in the Great Lakes has also impacted attempted restocking programs. As a result, a number of programs are in place that are designed to control lamprey populations, such as lampricides, barrier dams, and trapping. Traps are intended to remove the adult lamprey before they can spawn, while barrier dams prevent the adults from reaching spawning areas in tributaries to the Great Lakes. The lampricides typically target the larval sea lamprey, killing them before they can transform into their parasitic adult form. Although the number of sea lamprey in the Great Lakes has been reduced, they still kill substantial numbers of lake trout in some areas and thus are impeding the rebuilding of established populations (Schneider et al. 1996; Fuller et al. 2012).

15.4.2.4 Geese

Geese are not necessarily an invasive species, in that they are native. But, by the definition cited earlier by the Nature Conservancy, they could be considered so. For example, historically, Canada geese migrated through a number of flyways from northern nesting areas to the south. Increasingly though, populations of nonmigratory geese have taken up residence in or near many waterbodies, both creating a nuisance and contributing to water quality problems. Canada geese may be large (20 or more pounds) and live a relatively long time (20 or more years). A goose can convert about 4 lb. of grass per day to 3 lb. of fecal matter, which could be composed of about 76% carbon, 4.4% nitrogen, and 1.3% phosphorus (NHDES 2012). One old saying with considerable justification refers to speed or acting quickly being "like crap through a goose." So, for example, as a result, geese may contribute substantial quantities of coliform bacteria to a lake as well as nutrients and organic matter. Management strategies include removal (e.g., destroying), discouraging feeding, and changing landscape practices to reduce feeding and nesting habitats.

15.4.2.5   Beavers

Similarly, beavers are a native species, common in many areas of the United States and Canada before European settlement of North America. With extensive trapping, beavers were expatriated from many areas and later reintroduced. Currently, they occupy much of their former range (Baker and Hill 2003).

The beaver is North America's largest rodent, weighing up to 60 lb. and measuring 25–30 in. long. Due largely to their habit of creating dams, beavers can cause considerable damage, including timber damage, destruction of reservoir dams, and flooding. As a result, they are considered pest species in many areas and many states either provide or allow for bounties (e.g., Alabama, Arkansas, Louisiana, and Mississippi) (Figure 15.33).

15.4.2.6   Nutria

Unlike the beaver, the nutria is a nonindigenous species (Figure 15.34). Originally introduced for their fur, they have established localized breeding populations in a number of states (Figure 15.35).

FIGURE 15.33   Beaver. (From Ohio Department of Natural Resources.)

FIGURE 15.34   Nutria. (From Louisiana Department of Wildlife and Fisheries.)

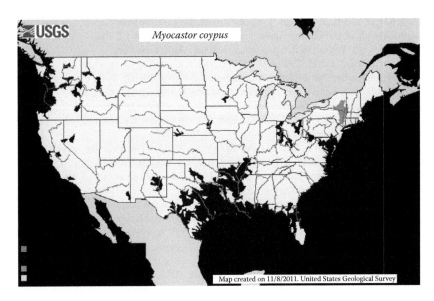

FIGURE 15.35 Nutria distributions in the United States. (From USGS; Fuller, P., Nico, L., Maynard, E., Larson, J., and Fusaro, A., USGS Nonindigenous Aquatic Species Database, Gainesville, FL, 2012, Available at http://nas.er.usgs.gov/queries/factsheet.aspx?SpeciesID = 836 Revision Date: 12/22/2011.)

They are a pest species in many areas, destroying the native aquatic vegetation and crops, and by burrowing into stream banks and dams.

Some of the wild nutria populations resulted from their escape from fur farms. However, in some areas, such as in Louisiana, nutria were released by state and federal agencies to provide a new fur resource and to control problems from other invasive species, such as the water hyacinth and the alligator weed. In Louisiana, by the late 1950s, there were an estimated 20 million nutrias in coastal areas, and the Louisiana legislature passed legislation providing a $0.25 bounty in 16 parishes (counties) for every nutria killed.

REFERENCES

APHA. 1995. *Standard Methods for the Examination of Water and Wastewater*, 19th and 20th ed. American Public Health Association, Washington, DC.

Aron, W.I. and S.H. Smith. 1971. Ship canals and aquatic ecosystems: Equilibrium has not been achieved since the Erie, Welland, and Suez canals were built. *Science* 174, 13–20.

Axler, R.P., G.W. Redfield, and C.R. Goldman. 1981. Productivity in a subalpine lake. *Ecology* 62 (2), 345–354.

Baker, J.P., H. Olem, C.S. Creaher, M.D. Marcus, and B.R. Parkhurst. 1993. Fish and fisheries management in lakes and reservoirs, EPA 841-R-93-002. Terrene Institute and U.S. Environmental Protection Agency, Washington, DC.

Baker, B.W. and E.P. Hill. 2003. Beaver (*Castor canadensis*). In G.A. Feldhamer, B.C. Thompson, and J.A. Chapman (eds), *Wild Mammals of North America: Biology, Management, and Conservation*, 2nd ed. The Johns Hopkins University Press, Baltimore, MD.

Balcer, M.D., N.L. Korda, and S.I. Dodson. 1984. *Zooplankton of the Great Lakes*. University of Wisconsin Press, Madison, WI.

Benson, A.J. 2011. Zebra mussel sightings distribution. Retrieved 2/12/2012 from http://nas.er.usgs.gov/taxgroup/mollusks/zebramussel/zebramusseldistribution.aspx.

Bold, H.C. and M.J. Wynne. 1985. *Introduction to the Algae*. Prentice-Hall, Englewood Cliffs, NJ.

Brittain, S.M., J. Wang, L. Babcock-Jackson, W.W. Carmichael, K.L. Rinehart, and D.A. Culver. 2000. Isolation and characterization of microcystins, cyclic heptapeptide hepatotoxins from a Lake Erie strain of *Microcystis aeruginosa*. *Journal of Great Lakes Research* 26, 241–249.

Bur, M.T., D.M. Klarer, and K.A. Krieger. 1986. First records of a European cladoceran, *Bythotrephes ceder-stroemi*, in Lakes Erie and Huron. *Journal of Great Lakes Research* 12, 144–146.

Chapra, S.C. 1997. *Surface Water Quality Modeling*. McGraw-Hill, New York.

Chapra, S.C., G.J. Pelletier, and H. Tao. 2007. QUAL2K: A modeling framework for simulating river and stream water quality, Version 2.07: Documentation and Users Manual. Tufts University, Medford, MA.

Cole, G.A. 1979. *Textbook of Limnology*, 2nd ed. C.V. Mosby Company, St. Louis, MO.

Cole, K.M. and R.G. Sheath (eds). 1990. *Biology of the Red Algae*. Cambridge University Press, Cambridge, UK.

Dalsgaard, T., B. Thamdrup, and D.E. Canfield. 2005. Anaerobic ammonium oxidation (anammox) in the marine environment. *Research in Microbiology* 156, 457–464.

Davenport, T. and W. Drake. 2011. Grand Lake St. Marys, Ohio—The case for source water protection: Nutrients and algae blooms. EPA Commentary, *LakeLine*, Fall 2011, 41–46.

Eppley, R.W., O. Holm-Hansen, and J.D.H. Strickland. 1968. Some observations on the vertical migration of dinoflagellates. *Journal of Phycology* 4, 333–340.

Fanslow, D.L., T.F. Nalepa, and G.A. Lang. 1995. Filtration rates of the zebra mussel (*Dreissena polymorpha*) on natural seston from Saginaw Bay, Lake Huron. *Journal Great Lakes Research* 21 (4), 489–500.

Free, G., R. Little, D. Tierney, K. Donnelly, and R. Caroni. 2007. A reference based typology and ecological assessment system for Irish lakes—preliminary investigations, ERTDI Report 57. University College, Dublin.

Fuller, P., L. Nico, E. Maynard, J. Larson, and A. Fusaro. 2012. USGS Nonindigenous Aquatic Species Database, Gainesville, FL. Available at: http://nas.er.usgs.gov/queries/factsheet.aspx?SpeciesID = 836 Revision. Date: 12/22/2011.

Hanson, E. and M. Sytsma. 2001. *Oregon Aquatic Nuisance Species Management Plan*. Portland State University, Portland, OR.

Horne, A.J. and C.R. Goldman. 1994. *Limnology*, 2nd ed. McGraw Hill, New York.

Hutchinson, G.E. 1957. *A Treatise on Limnology*, Vol. 2. Wiley, New York.

Hutchinson, G.E. 1961. The paradox of the plankton. *The American Naturalist* 95, 137–145.

Jorgensen, C.B., T. Kiorboe, F. Mohlenberg, and H.U. Riisgard. 1984. Ciliary and mucusnet filter feeding, with special references to fluid mechanical characteristics. *Marine Ecology Progress Series* 15, 383–392.

Kuypers, M.M.M., A.O. Sliekers, G. Lavik, M. Schmid, B.B. Jorgensen, and J.G. Kuenen. 2003. Anaerobic ammonium oxidation by anammox bacteria in the Black Sea. *Nature* 422, 608–611.

Kuypers, M.M.M., G. Lavik, D. Woebken, M. Schmid, B.M. Fuchs, and R. Amann. 2005. Massive nitrogen loss from the Benguela upwelling system through anaerobic ammonium oxidation. *PNAS* 102, 6478–6483.

LakeWatch. 2007. *A Beginner's Guide to Water Management: Aquatic Plants in Florida Lakes, Information Circular 111*. University of Florida, Gainesville, FL.

Lathrop, R.C. 1992. Decline in zoobenthos densities in the profundal sediments of Lake Mendota. *Hydrobiologia* 235–236 (1), 353–361.

Likens, G.E. 2010. *Encyclopedia of Inland Waters: Phytoplankton of Inland Waters*. Academic Press, San Diego, CA.

Maier, R.M., I.L. Pepper, and C.P. Gerba. 2009. *Environmental Microbiology*, 2nd ed. Academic Press, San Diego, CA.

Narf, R. 1997. Midges, bugs, whirligigs and others: The distribution of insects in Lake U-Name-It, *LakeLine*. 16–17, 57–62. North American Lake Management Society.

Naumann, E. 1917. Beiträge zur Kenntnis des Teichnannoplanktons. II. Über das Neuston des Süsswassers. *Biologisches Zentralblatt* 37, 98–106.

NHDES. 2012. Canada Geese facts and management options, Environmental Fact Sheet BB-53, New Hampshire Department of Environmental Services, Concord, NH.

Odum, E.P. 1971. *Fundamentals of Ecology*. Saunders, New York.

Oliver, R.L. and A.E. Walsby. 1984. Direct evidence for the role of light-mediated gas vesicle collapse in the buoyancy regulation of *Anabaena flos-aquae* (Cyanobacteria). *Limnology and Oceanography* 29, 879–886.

Penning, W.E., B. Dudley, M. Mjelde, S. Hellsten, J. Hanganu, A. Kolada, M. Van den Berg, G. Phillips, N. Willby, and F. Ecke. 2008a. Using aquatic macrophyte community indices to define the ecological status of European lakes. *Aquatic Ecology* 42, 253–264.

Penning, W.E., M. Mjelde, B. Dudley, S. Hellsten, J. Hanganu, A. Kolada, M. Van den Berg, et al. 2008b. Classifying aquatic macrophytes as indicators of eutrophication in European lakes. *Aquatic Ecology* 42, 237–251.

Phillips, G.L., W.D. Schmid, and J.C. Underhill. 1982. *Fishes of the Minnesota Region*. University of Minnesota Press, Minneapolis, MN.

Porter, S.D., T.F. Cuffney, M.E. Gurtz, and M.R. Meador. 1993. Methods for collecting algal samples as part of the national water quality assessment program. U.S. Geological Survey Open-File Report 93-409, Raleigh, NC.

Prescott, G.W. 1962. *Algae of the Western Great Lakes Area.* William C. Brown Company, Dubuque, IA.

Redfield, A.C., B.H. Ketchum, and F.A. Richards, 1963. The influence of organisms on the composition of seawater. In M.N. Hill (ed.), *The Sea*, Vol. 2. Wiley-Interscience, New York, pp. 27–46.

Reynolds, C.S. 2006. *The Ecology of Phytoplankton.* Cambridge University Press, Cambridge, UK.

Rissaned, A.J., M. Tiirola, and A. Ojala. 2011. Spatial and temporal variation in denitrification and in the denitrification community in a boreal lake. *Aquatic Microbial Ecology* 64, 27–40.

Roelofs, J.G.M. 1983. Impact of acidification and eutrophication on macrophyte communities in soft waters in The Netherlands I. Field observations. *Aquatic Botany* 17, 139–155.

Sager, D.R., A. Barkoh, D.L. Buzan, L.T. Fries, J.A. Glass, G.L. Kurten, J.J. Ralph, E.J. Singhurst, G.M. Southard, and E. Swanson. 2008. Toxic *prymnesium parvum*: A potential threat to U.S. reservoirs. In M.S. Allen, S. Sammons, and M.J. Macina (eds), *Balancing Fisheries Management and Water Uses for Impounded River Systems.* American Fisheries Society, Symposium 62, Bethesda, MD, pp. 261–273.

Sallenave, R. 2010. Toxic golden algae (*Prymnesium parvum*), Circular 647. Cooperative Extension Service, New Mexico State University, NM.

Sappington, A. 2009. Red algae, *The Water Line* 13(1), 1–2. (Newsletter of the Missouri Volunteer Program).

Saunders, D.L. and J. Kalff. 2001. Nitrogen retention in wetlands, lakes and rivers. *Hydrobiologia* 443, 205–212.

Schneider, C.P., R.W. Owens, R.A. Bergstedt, and R. O'Gorman. 1996. Predation by sea lamprey (*Petromyzon marinus*) on lake trout (*Salvelinus namaycush*) in southern Lake Ontario, 1982–1992. *Canadian Journal of Fisheries and Aquatic Sciences* 53 (9), 1921–1932.

Scott, W.B. and E.J. Crossman. 1973. Freshwater fishes of Canada. Fisheries Research Board of Canada, Bulletin 184. Ottawa.

Sigee, D.C. 2005. *Freshwater Microbiology: Biodiversity and Dynamic Interactions of Microorganisms in the Aquatic Environment.* Wiley, Chichester, UK.

Skulberg, O.M., G.A. Codd, and W.W. Carmichael. 1984. Toxic blue-green algae blooms in Europe: A growing problem. *Ambio* 13, 244–247.

Smith, V.H. 2001. *Blue Green Algae in Eutrophic Fresh Waters.* Spring, Lakeline. pp. 34–37.

Smith, B.R. and J.J. Tibbles. 1980. Sea lamprey (*Petromyzon marinus*) in Lakes Huron, Michigan, and Superior: History of invasion and control, 1936–78. *Canadian Journal of Fisheries and Aquatic Sciences* 37 (11), 1780–1801.

Suthers, I.M. and D. Rissik. 2009. *Plankton: A Guide to Their Ecology and Monitoring for Water Quality.* CSIRO Publishing, Collingwood, Australia.

SWCSMH. 2006. Chapter I: Zoobenthos of freshwaters—An introduction. Soil & Water Conservation Society of Metro Halifax from http://lakes.chebucto.org/ZOOBENTH/BENTHOS/i.html, accessed February 8, 2012.

Thorp, J.H. and A.P. Covich (eds). 2010. *Ecology and Classification of North American Freshwater Invertebrates*, 3rd ed. Academic Press, San Diego, CA.

USACE. 2001. Zebra mussels: Biology, ecology, and recommended control strategies, Technical Note ZMR-1-01. Zebra Mussel Research Program, U.S. Army Engineer Waterways Experiment Station, Vicksburg, MS.

UKTAG. 2009. UKTAG Lake Assessment Methods: Macrophyte and Phytobenthos, Macrophytes (Lake LEAFPACS). Water Framework Directive—United Kingdom Technical Advisory Group (WFD-UKTAG). Available at: http://www.wfduk.org/bio_assessment/.

USEPA. 1974. *Draft 316(a) Technical Guidance-Thermal Discharges.* U.S. Environmental Protection Agency, Washington, DC.

USEPA. 1998. Lake and reservoir bioassessment and biocriteria: Technical guidance document, EPA 841-B-98-007. U.S. Environmental protection Agency, Washington, DC.

USEPA. 2008. Predicting future introductions of nonindigenous species to the Great Lakes, EPA/600/R-08/066F. U.S. Environmental Protection Agency, National Center for Environmental Assessment, Washington, DC.

Vanderploeg, H.A., J.R. Liebig, W.W. Carmichael, M.A. Agy, T.H. Johengen, G.L. Fahnenstiel, and T.F. Nalepa. 2001. Zebra mussel (*Dreissena polymorpha*) selective filtration promoted toxic microcystis blooms in Saginaw Bay (Lake Huron) and Lake Erie. *Canadian Journal of Fisheries and Aquatic Sciences* 58, 1208–1221.

Vogel, S. 2004. Living in a physical world I: Two ways to move material. *Journal of Biosciences* 29, 391–397.

Water on the Web. 2004. Monitoring Minnesota lakes on the Internet and training water science technicians for the future—A national on-line curriculum using advanced technologies and real-time data. Available at: http://WaterOntheWeb.org. University of Minnesota–Duluth, Duluth, MN.

Wehr, J.D. and R.G. Sheath. 2003. *Freshwater Algae of North America: Ecology and Classification*. Academic Press, San Diego, CA.

Westlake, D.F. 1975. Macrophytes. In B.A. Whitton (ed.), *River Ecology*. California University Press, Berkeley, CA, pp. 106–128.

Wetzel, R.G. 2001. *Limnology: Lake and River Ecosystems*, 3rd ed. Academic Press, San Diego, CA.

Wilson, D. 2000. Rhodophyta. Red algae, Version 24 March 2000 (under construction). Available at: http://tolweb.org/Rhodophyta/2381/2000.03.24 in The Tree of Life Web Project, http://tolweb.org/.

16 Lake Production, Succession, and Eutrophication

16.1 PRIMARY AND SECONDARY PRODUCTIVITY

16.1.1 Primary Production

16.1.1.1 Introduction

Primary production refers to the rate at which new plant biomass is synthesized from inorganic materials and energy. This production forms the base of the food chain, supporting all other lake organisms. Since no transfer of energy is 100% efficient, the biomass that can be supported of organisms that feed on the primary producers must be less than the biomass of the organisms on which they feed. So, biomass or biomass productivity must decrease as one moves further up the food chain, often depicted graphically as an ecological pyramid. Or, in a quote usually attributed to G. Tyler Miller Jr., an American chemist: "Three hundred trout are needed to support one man for a year. The trout, in turn, must consume 90,000 frogs, that must consume 27 million grasshoppers that live off of 1,000 tons of grass." So, sufficient biomass is required to support a healthy food chain. On the other hand, excess primary productivity can also lead to problems (see e.g., Figure 16.1).

Primary production is controlled by a combination of physical (e.g., residence time, temperature, and light), chemical (e.g., nutrients), and biological (e.g., grazing) conditions. Primary production can be represented by Equation 16.1 (Stumm and Morgan 1996; Chapra et al. 2007; assuming for this example that ammonia is used as a substrate):

$$106CO_2 + 16NH_4^+ + HPO_4^{2-} + 106H_2O \underset{R}{\overset{P}{\rightleftharpoons}} C_{106}H_{263}O_{110}N_{16}P_1 + 106O_2 + 14H^+ \qquad (16.1)$$

where P refers to production and R to respiration. So, while plants may produce biomass in the presence of light, they also respire. The result produces diel variations in oxygen and carbon dioxide concentrations.

Gross primary productivity refers to the overall production of organic carbon, while net primary production refers to the difference between gross production and respiration. This may be based on individual plants, communities, or ecosystems. For example, an ecosystem's (or community) net productivity would be the difference between gross primary production (GPP) and respiration by all organisms: both autotrophs and heterotrophs.

Primary production (expressed as net or gross) refers to the rate of production of plant biomass, as opposed to the quantity of that biomass, or the standing crop. Although the two are related, they are different. The standing crop is a measure of the biomass at a point in time, but tells nothing about how long it took to attain that biomass or the factors that affected it. For example, there are many terrestrial ecosystems (e.g., some tropical forests) with high biomass, but with low rates of productivity. Conversely, in lake systems, the biomass of planktonic algae may be exceeded 100 times by annual production, so that the standing crop does not reflect annual production (Cole 1979).

In lakes, biomass or the standing crop is often directly related to the trophic state of lakes and the stage in the succession of lakes. For example, the widely used trophic state index (TSI) developed

FIGURE 16.1   Example of hypereutrophic system. (From National Eutrophication Monitoring Program—RQS, Department of Water Affairs and Forestry, South Africa.)

by Carlson (1977) is based primarily on algal biomass (as indicated by chlorophyll-a concentrations) and the relationships between biomass and nutrients or transparency to light of the water column.

16.1.1.2   Methods for Measuring Biomass or Standing Crop

The standing crop (or standing stock) refers to the mass of organic material that can be sampled at a particular time (Wetzel 2001). The biomass present is typically measured based on autotroph enumeration, weight, or pigment (e.g., chlorophyll-a) concentrations. The methods used will vary depending on the autotroph types, such as planktonic algae, periphyton, and macrophytes.

One common metric is ash-free dry weight (AFDW). First, the dry weight of organic material is determined by drying at a standardized temperature (e.g., 80°C–105°C). Then, the dry sample is burned in a furnace (at 550°C) so that only ash remains. The difference between the ash weight and the dry weight is referred to as the ash-free dry weight (APHA 1999), an approximation of the mass of organic materials. One difficulty is that without centrifugation or some other means of concentration, the ash-free dry mass (AFDM) concentrations from planktonic samples are near the detection level unless the sample is collected from a highly eutrophic stream or lake (Hambrook and Canova 2007). In addition, it is difficult to distinguish between the mass of algae, bacteria, detritus, and other small particles, so the use of AFDW measurements is often restricted to larger organisms (Wetzel 2001).

An alternative approach to measuring the AFDM of phytoplankton samples is to obtain a measure of carbon by an analysis of particulate organic carbon (POC). This typically involves filtration to obtain a particulate sample, an analysis for total POC followed by an analysis for particulate inorganic carbon, with the POC concentration determined by the difference. Some assumed stoichiometry may then be used (e.g., as in Equation 16.1) to relate the carbon to the dry (or wet) weight biomass.

Another commonly used metric is based on direct enumeration, counting, or volume determinations. The abundance, for phytoplankton, is commonly expressed as a biovolume of all individuals. Some of the techniques for enumeration were discussed in Chapter 15, and are discussed in detail in Vollenweider's (1969) *Manual on Methods for Measuring Primary Production in Aquatic Environments*. While commonly used, biovolumes are poor indicators of the biomass of macrophytes because of the variability in internal gas spaces (Wetzel 2001).

Cellular contents, such as chlorophyll-a, are also commonly used metrics for biomass. The biomass is estimated using some assumed relationship between biomass and chlorophyll or carbon and chlorophyll (the carbon/chlorophyll ratio). A number of cellular constituents have been employed to estimate biomass. However, with the exception of carbon, measures using other constituents are complicated by the extreme variability in the cellular composition (Wetzel 2001). While pigments such as chlorophyll also vary, they remain a commonly used metric for biomass as well as the trophic status of lakes and reservoirs.

16.1.1.3 Methods for Measuring Primary Production

Primary production is commonly estimated by the rates of change (e.g., rate of change in mass/time). For example, with reference to Equation 16.1, the rate can be estimated by either the rates of change of biomass, the consumption of carbon dioxide or nutrients, the production of oxygen, or other means. The method used is often specific to the types of autotrophs being sampled, such as phytoplankton, periphyton, and macrophytes.

One example is using changes in oxygen concentrations. For lakes and reservoirs, one method that has received wide attention is the light and dark bottle method (Wetzel 2001), which is used to estimate the rate of primary production by the rate of change in oxygen concentrations. Essentially, two standardized bottles are placed in a water column, where one can receive light (the light bottle) and the other cannot (the dark bottle). The oxygen concentrations are measured initially and then the bottles are incubated at specific depths in the water column. The change in oxygen concentrations is recorded. In the light bottle, both respiration and photosynthesis take place; so the change in oxygen is an indication of GPP. In the dark bottle, only respiration (R) takes place. So, the difference in oxygen concentrations between the two bottles reflects the net primary production (NPP = GPP − R).

Similarly, changes in carbon dioxide are used as indicators of production. However, the use of carbon dioxide is complicated by interactions between species of inorganic carbon, which provide a buffering capacity as indicated by their alkalinity. Alternatively, measurements of the rate of incorporation of the ^{14}C tracer into organic matter have been used as a measure of the rate of primary production (Wetzel 2001). The labeled carbon is added to water (such as in a bottle or an enclosure) and the quantity of organic material (e.g., of phytopklankton) is measured after incubation, providing a direct measure of the rate of production. The ^{14}C method is much more sensitive (by three orders of magnitude) than the light/dark bottle oxygen method (Wetzel 2001).

For periphyton and macrophytes, changes in the accrual of biomass may be measured. Other methods include changes in oxygen or carbon concentrations using methods similar to those used for phytoplankton, but in other types of enclosures. For example, apical portions of macrophyte shoots may be incubated in flasks. For *in situ* measurements of macrophyte productivity, Vollenweider (1969) recommends using clear Plexiglas cylinders.

16.1.2 SECONDARY PRODUCTION

Secondary production refers to the production by the consumers: both invertebrates and vertebrates. Secondary consumption is much more difficult to estimate accurately than is primary production (Wetzel 2001). The reasons include the following: trophic interactions are complex, the sizes between organisms and life stages vary greatly, organisms vary in the number of generations per year (voitinism), and they are mobile. For example, populations that are multivoltine have higher rates of production than those that are univoltine (Downing 1984).

Secondary production is based on changes in the number of animals, biomass, and growth rates. The rates may be for different types or communities of heterotrophs, such as the bacteria, zooplankton, and benthic communities or fish and may vary with time. For example, secondary production rates for benthic invertebrates are:

- Often lowest during summer
- Generally greater in running than standing water
- At least five to ten times greater for nonpredatory benthic organisms than for predacious benthic organisms
- Two to five times higher than the zooplankton in the zoobenthos of shallow lakes with low mean depths

Many of the methods for measuring secondary production are focused on specific species or groups of organisms. One group in particular that is a common focus of lake management is the production of those economically important species such as fish. So, from a management perspective, secondary production is used to estimate the capacity of a system to support fish (Naiman and Bilby 1998).

16.2 GEOLOGIC LAKE SUCCESSION

16.2.1 Definition

Ecological succession deals with the concept that natural systems tend toward, or transform themselves to, some final (relatively) or stable condition, and if disturbed or moved away from that condition, the systems are constantly undergoing change toward that condition. Natural succession then refers to the progression of ecological change along a continuum toward, or in the direction of, some climax condition.

Ecological succession from some "original" condition toward the climax community is often referred to as primary succession, such as from rock to soil to plant colonization, etc. Secondary succession occurs when there is some disturbance along that path, or some disturbance of an existing community.

Much of the early research on succession dealt with secondary forest succession, such as from some disturbed community. In September 1890, Henry David Thoreau (Figure 16.2) read a paper to the Middlesex Agricultural Society in Concord on "The succession of forest trees," which is perhaps the first account of how forest succession works. A common example is that of pine communities, which are often maintained by fire to which they are resistant. In the absence of fire or other disturbances, pines will eventually be replaced by hardwoods. The overstory of the hardwoods will eventually reduce light (by shading) such that the light-loving pines will no longer be able to grow. One issue is that with the climax hardwood community, as the understory is reduced, the overall biological productivity is reduced, even though there is substantial biomass. There are a number of examples, such as tropical forests, where high biomass does not equate to high productivity. So, the climax community is not necessarily the optimal community from some perspectives. While there are many important and valuable species that rely on old growth hardwoods, for others, such as many game species, the lack of an understory results in a biological desert. As a result, for example,

FIGURE 16.2   Henry David Thoreau.

Aldo Leopold wrote in his 1933 textbook titled *Game Management* that "game can be restored by the creative use of the same tools which have heretofore destroyed it-ax, plow, cow, fire, and gun."

Similarly, lakes can undergo a natural succession, the classical view of which is that lakes are temporary ecological systems, formed in some depression that will eventually fill in to become a terrestrial environment, or become extinct. The stages of succession that a lake will go through, and the rate of succession, are a function of the loads of water, nutrients and sediment that the lake receives. But, the natural process of succession will be for a lake to eventually form a pond, then a marsh, then a meadow, and eventually dry land (Figure 16.3).

As George E.P. Box wrote, "essentially, all models are wrong, but some are useful" (Box and Draper 1987). Similarly, the concept of lake aging is a useful tool, but that aging (succession) is not necessarily an inevitable and irreversible process.

16.2.2 Factors Affecting Succession

The origin of the concept of lake succession is often attributed to Jean André Deluc, who, in Volume 1 of his book *Geological Travels in Some Parts of France, Switzerland, and Germany*, published in 1810, described six stages in the transformation of a lake into a peaty meadowland. Deluc also indicated that the rate of succession is greatest on shallow shores, and the process of change is almost nonexistent on steep shores.

From Deluc's analyses, lakes with littoral zones (Figure 16.4) will have greater rates of succession than those with only a pelagic zone. Or, as lakes age and fill, the littoral zone will become

FIGURE 16.3   Stages of succession in a small lake. (From Kidfish, http://www.kidfish.bc.ca/.)

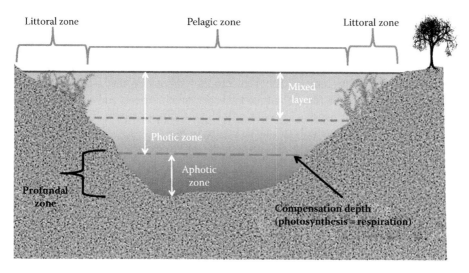

FIGURE 16.4 Lake zones.

increasingly dominant. Hutchinson (1975) noted that "zonation, therefore, is taken to be the spatial equivalence of succession in time, even in the absence of change."

16.2.3 LAKE ONTOGENY AND STAGES OF SUCCESSION

Lake ontogeny refers to the successional development of inland aquatic systems (Wetzel 2001). Ontology has multiple meanings, depending on the discipline to which the term is applied, such as to science or philosophy. One of its common scientific meanings refers to the growth of communities or organisms. One well-known use is in the theory of recapitulation, also called the biogenetic law or embryological parallelism—often expressed as "ontogeny recapitulates phylogeny." This hypothesis is that in developing from an embryo to an adult, animals undergo stages of development that have strong parallels with the development of their ancestors, thereby representing or resembling the stages of succession (or evolution).

Populations also undergo stages of growth, as illustrated by an idealized sigmoid (S-shaped) growth curve (Figure 16.5). Generally, when an area is first colonized, there is a period of acclimation when the population growth is slow, called the lag phase. Once acclimated, the rate of increase accelerates, with growth eventually becoming exponential. At some point, the population will approach the carrying capacity and as food or habitat becomes limiting, the rate of acceleration will decrease and the population will become stable, or will fluctuate around the carrying capacity in a dynamic equilibrium.

Hutchinson and Wollack (1940) stressed parallels between the early development of lake communities and the sigmoid growth phase of animal communities, such as the zooplankton in Linsley Pond, Connecticut. They implied that the apparent early developmental processes in lakes were dominated by colonization effects, the lag phase.

One of the most common schemes used to characterize lake ontogeny is one where lakes progressively develop through the following trophic (derived from Greek and meaning food or feeding) stages:

- Oligotrophic, which means "scant or lacking"
- Mesotrophic, which means "midrange"
- Eutrophic, which means "good or sufficient"
- Hypereutrophic, which means "over abundant" (Florida LakeWatch 2002).

The concept was introduced by Naumann (1919; Figure 16.6, cited by Wetzel 2001), who used oligotrophy and eutrophy to distinguish lakes based on their phytoplankton populations. Naumann's

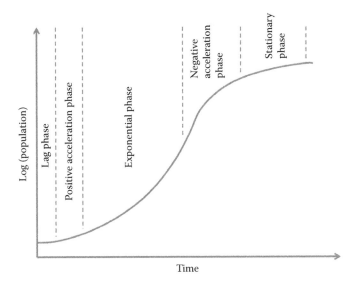

FIGURE 16.5   Stages of population growth.

FIGURE 16.6   Einar Naumann (1891–1934). Swedish limnologist and originator of the trophic state concept. (Courtesy of the Association for the Sciences of Limnology and Oceanography Image Library; the photographs were from an old ASLO slide collection submitted by W. Lewis; scanned by Dr. K.L. Schulz.)

oligotrophic lakes had few phytoplankton and were common in areas dominated by rocks, whereas eutrophic lakes contained abundant phytoplankton and occurred in fertile, nutrient-rich, lowland areas (Wetzel 2001). Hypereutrophy is the stage before senescing to a dystrophic stage and filling completely with sediment.

Naumann recognized that the "quantitative production of phytoplankton" on which his classification was based could be influenced by other factors, resulting in the classifications listed in Table 16.1 (Naumann 1929, cited in Carlson and Simpson 1996).

16.2.4 GENERAL CHARACTERISTICS AND TROPHIC STATUS

The differences in the trophic status of lakes are somewhat analogous to the differences between a "creek" and a "crick" described in Chapter 2. As characterized by Patrick McManus (1981), creeks "tend to be pristine," they "sparkle in the sunlight. Deer and poets sip from creeks, and images of eagles wheel upon the surface of their mirrored depths." Cricks, on the other hand, "shuffle through cow pastures, slog through beaver dams, gurgle through culverts, and ooze through barnyards."

Henry David Thoreau in *Walden* (1854, first published as *Walden; or, Life in the Woods*) stated that there is "nothing so fair, so pure, and at the same time so large, as a lake, perchance, lies on the surface of the earth." In the chapter on ponds, he described Walden Lake as "remarkable for its depth and purity as to merit a particular description. It is a clear and deep green well, half a mile long and a mile and three quarters in circumference, and contains about sixty-one and a half acres; a perennial spring in the midst of pine and oak woods, without any visible inlet or outlet except by the clouds and evaporation. The surrounding hills rise abruptly from the water to the height of forty to eighty feet, though on the southeast and east they attain to about one hundred and one hundred and fifty feet, respectively, within a quarter and a third of a mile. They are exclusively woodland." … "The hills which form its shores are so steep, and the woods on them were then so high, that, as you looked down from the west end, it had the appearance of an amphitheatre for some land of sylvan spectacle." Further, "the water is so transparent that the bottom can easily be discerned at the depth of twenty-five or thirty feet. Paddling over it, you may see, many feet beneath the surface, the schools of perch and shiners, perhaps only an inch long, yet the former easily distinguished by their transverse bars, and you think that they must be ascetic fish that find subsistence there."

Thoreau described the nearby "Flint's, or Sandy Pond, in Lincoln, our greatest lake and inland sea, lies about a mile east of Walden. It is much larger, being said to contain one hundred and ninety-seven acres, and is more fertile in fish; but it is comparatively shallow, and not remarkably pure."

TABLE 16.1
Trophic Lake Types Suggested by Naumann

Lake Type	Characteristics
Oligotrophy	Lakes with low production associated with low nitrogen and phosphorus
Eutrophy	Lake type with high production, associated with high nitrogen and phosphorus
Acidotrophy	Lake type with low production, with pH values less than 5.5
Alkalitrophy	Lake type with low production, associated with high calcium concentrations
Argillotrophy	Lake type with low production, associated with high clay turbidity
Siderotrophy	Lake type with low production, associated with high iron content
Dystrophy	Lake type with low production, associated with high humic color

Source: Carlson, R.E. and Simpson, J., *A Coordinator's Guide to Volunteer Lake Monitoring Methods*, North American Lake Management Society, Madison, WI, 1996.

TABLE 16.2

Selected Common Characteristics of Oligotrophic and Eutrophic Lakes

Characteristic	Oligotrophic	Eutrophic
Bathymetry	Steep shoreline, bottom	Shallow, moderate to low bottom slope
Nutrient and organic matter enrichment	Low	High
Phytoplankton growth	Little	High
Macrophyte growth	Low	Extensive
Plant biomass	Low	High
Productivity	Low	High
Sedimentation accumulation	Low	High
Light penetration (e.g., Secchi depth)	Deep	Shallow
Oxygen content	High throughout water column	Low to none in hypolimnion
Fisheries	Usually cold water, may be two layer	Warmwater fisheries

He then went on to make some rather disparaging remarks about the farmer, "whose farm abutted on this sky water, whose shores he has ruthlessly laid bare,"… "who never saw it, who never bathed in it, who never loved it, who never protected it, who never spoke a good word for it, nor thanked God that He had made it."

Thoreau's ponds illustrate many of the characteristics of, and controversies associated with, the trophic status of lakes. For example, are "lakes of crystal glass" the most desirable? To a poet perhaps, but to a fisherman? Also, what impact does man have on that trophic status, as did Thoreau's farmer? The topic of cultural eutrophication is discussed later, but first more comparisons of the characteristics of the trophic status of lakes.

Oligotrophic lakes, as with Thoreau's description of Walden Pond in 1854, can be thought of as crystal-clear waterbodies with typically rocky shores and steep shorelines. These lakes typically have low-nutrient concentrations and low productivity, so they remain clear. Since productivity is low, there is little decay, so the hypolimnion remains oxic. For example, extremely deep oligotrophic lakes, such as Lake Superior and Lake Tahoe, have hypolimnia that remain completely saturated with oxygen the entire year.

Eutrophic lakes, in contrast, are often relatively shallow, with larger contributing watersheds and higher nutrient loads and productivity. Often associated with higher productivity and solids loads, the light penetration is much reduced. Seasonal variations are greater and, due to the seasonally higher productivity and lower light penetration, the hypolimnion of many eutrophic lakes is void of oxygen for much of the year.

There are exceptions to each of the aforementioned trophic classifications, so that, for example, an anoxic hypolimnion does not necessarily indicate eutrophy. Some of the more common characteristics of oligotrophic and eutrophic lakes are summarized in Table 16.2. Mesotrophic lakes are intermediate between these two. One of the primary differences between the trophic statuses of lakes, as discussed in the following section, is due to their productivity and biomass.

16.3 EUTROPHICATION

16.3.1 EUTROPHICATION PARADIGM

The term *eutrophication* has been commonly used to characterize the natural process of the aging of lakes, or the successional changes in a lake's trophic status. The trophic state, as described by Naumann, refers to the amount of plant biomass in a body of water (Carlson and Simpson 1996) or plant productivity. Eutrophication is then taken as the natural progression of lakes from those with low plant biomass (or low productivity or both), Naumann's oligotrophic lakes, to those with high biomass (and high productivity or eutrophic). Note that there are many examples of systems with low productivity and high biomass, such as tropical forests, so the two terms are not synonymous.

One of the characteristics of succession, or the aging of lakes, is that productivity generally increases as age increases. As age increases and productivity increases, the biomass that is produced and dies increases, contributing to the increase in lake filling. This is in contrast to the successional paradigm for forest communities discussed earlier, where younger forests are inherently more productive than older forests (Ryan et al. 1997).

As with all such paradigms, the use of the term eutrophication or its definition is not universally accepted. In 1967, G. Evelyn Hutchinson, often called the father of modern limnology (cited in Nixon 1995), opened the proceedings of the First International Symposium on Eutrophication with the following observation: "it would be well at the beginning of this symposium to try to find out exactly what we are about to discuss."

Some of the alternative definitions of eutrophication are listed as follows:

- Eutrophication—"the process by which a body of water becomes enriched in dissolved nutrients (as phosphates) that stimulate the growth of aquatic plant life usually resulting in the depletion of dissolved oxygen." (Merriam-Webster Dictionary)
- Eutrophication—"Eutrophication is defined as an increase in the rate of supply of organic matter in an ecosystem." (Nixon 1995)
- Eutrophication—"The process by which a body of water acquires a high concentration of nutrients, especially phosphates and nitrates. These typically promote excessive growth of algae. As the algae die and decompose, high levels of organic matter and the decomposing organisms deplete the water of available oxygen, causing the death of other organisms, such as fish. Eutrophication is a natural, slow-aging process for a water body, but human activity greatly speeds up the process." (Art 1993)
- Eutrophication—"The term 'eutrophic' means well-nourished; thus, 'eutrophication' refers to natural or artificial addition of nutrients to bodies of water and to the effects of the added nutrients…. When the effects are undesirable, eutrophication may be considered a form of pollution." (National Academy of Sciences 1969)
- Eutrophication—"The enrichment of bodies of fresh water by inorganic plant nutrients (e.g. nitrate, phosphate). It may occur naturally but can also be the result of human activity (cultural eutrophication from fertilizer runoff and sewage discharge) and is particularly evident in slow-moving rivers and shallow lakes … Increased sediment deposition can eventually raise the level of the lake or river bed, allowing land plants to colonize the edges, and eventually converting the area to dry land." (Lawrence et al. 1998)
- Eutrophic—"Waters, soils, or habitats that are high in nutrients; in aquatic systems, associated with wide swings in dissolved oxygen concentrations and frequent algal blooms." (Committee on Environment and Natural Resources 2000)

16.3.2 Cultural Eutrophication

As with Thoreau's description of Walden Pond and nearby Flint's pond, the characteristics of the drainage basin (e.g., the amount of nutrients available to the lake system) and the mean depth of the lake or reservoir are the primary factors controlling eutrophication (Horne and Goldman 1994). As also with the farmer on Thoreau's Flint's pond, activities that increase the amount of nutrients and sediments in the runoff from the drainage basin, accelerate the eutrophication process. This accelerated eutrophication due to human-related activities is referred to as cultural eutrophication.

The cause of cultural eutrophication, or increased biological productivity, is most frequently due to enhanced nutrient availability, or nutrient loads. The term *nutrient enrichment* is in many, if not most, instances an alternative term for eutrophication (USEPA 2000). There are a variety of activities that can lead to this increase in nutrient loads, including urban or agricultural development and stormwater runoff, point sources such as wastewater discharges, livestock, pets, fertilizers used on lawns, golf courses, croplands, etc. While sediments are important, the U.S. Environmental Protection Agency (U.S. EPA) identified excess nutrients as a major reason for impaired water quality in the nation's waters, indicating that "more lake acres are affected by nutrients than any other pollutant or stressor.... States reported that excess nutrients pollute 3.8 million lake acres (which equals 22% of the assessed lake acres and 50% of the impaired lake acres)" (USEPA 2002). As a result, the U.S. EPA directed that all states and authorized tribes develop numeric criteria for nutrients to protect the designated uses of the nation's waters. The purpose of numeric nutrient criteria is to address cultural eutrophication (waters enriched with nutrients because of human activities) and guidance for criteria development is provided in the U.S. EPA's (2000) "Nutrient criteria technical guidance manual: Lakes and reservoirs."

While reservoirs are not natural waterbodies, their ontogeny is similar to that of natural lakes, but it is more rapid (Kimmell and Groeger 1986). The ontogeny of reservoirs can also be accelerated by cultural eutrophication.

Reservoirs commonly become more productive soon after filling, partially as a result of the leaching of nutrients from flooded soils and the decomposition of inundated vegetation (Kennedy and Walker 1990), although ontogeny does not always lead to reservoir eutrophication (Kimmel and Groeger 1986; Hall et al. 1999). If the external load remains constant after the initial upsurge period, the reservoir productivity will decline.

Reservoirs are also impacted by relatively rapid (in comparison to lakes) sediment accumulation. As discussed in Chapter 10 (see Table 10.2), reservoirs typically have higher drainage to surface area ratios (usually greater than 10:1), higher variations in water levels, greater changes in shoreline (due to changes in water levels and shoreline erosion), and usually have larger nutrient loadings, all of which contribute to sedimentation and aging. Seasonal water-level fluctuations are important components of reservoir ecosystems, affecting nutrient dynamics, underwater light climate, plankton dynamics, and littoral development (Kimmel et al. 1990). The rate of sedimentation is related to the trapping efficiency of specific reservoirs, but a regional approach to estimating sediment damages in lakes and reservoirs indicated that 0.22% of the nation's water storage capacity is lost annually (Crowder 1987). Crowder also indicated that of this, an average of 24% is due to soil erosion on cropland. Crowder also indicated that in the central United States, the greatest water storage capacity losses were due to deposited sediment originating from cropland. However, Graf et al. (2010) indicated that in western reservoirs, the rates of sedimentation have often been overestimated, and sustainability has been underestimated.

16.4 METRICS FOR EUTROPHICATION

16.4.1 Single Metric Methods

A variety of methods have been developed based on single metrics related to lake productivity. Some of the more commonly used metrics include:

- Phosphorus loading
- Phosphorus concentrations
- Nitrogen loading
- Nitrogen concentrations
- Algal productivity
- Algal biomass
- Chlorophyll concentrations
- Secchi depth
- Hypolimnetic oxygen
- Typology

All of these metrics may be related to biomass or productivity as a useful metric for eutrophication. For example, Vollenweider (1968) demonstrated the importance of phosphorus loading as an indicator of lake production, and these and similar relationships have been widely used in lake management. Similarly, while phosphorus is commonly thought of as the limiting nutrient for freshwater lakes, nitrogen loadings are important for some lakes, so they are related to production and biomass. Chlorophyll can be related to biomass and, since light is essential for growth, so can the Secchi depth. Algal biomass and productivity may also be measured directly, but that measurement is commonly difficult, particularly in computing variations in annual averages. Each of these has advantages and disadvantages as discussed by Carlson and Simpson (1996). The single metric that will be discussed here is related to oxygen consumption. Carlson and Simpson suggested that some limnologists would use oxygen consumption as the sole indicator of a trophic state, but that it would be better to represent a result of the lake's trophic state than an indicator of that state.

A potential impact of eutrophication is a reduction of the oxygen concentration in the hypolimnion. That is, material from the surface falls to the hypolimnion and is decomposed, which reduces oxygen concentrations. The greater the productivity is (e.g., for a eutrophic system), the greater the resulting hypolimnetic oxygen demands will be. The larger the hypolimnion is, the greater will be its capacity to absorb these losses related to oxygen depletion.

Hypolimnetic oxygen depletion has been used to estimate the trophic status of lakes, usually based on the hypolimnetic oxygen deficit, or the difference between the oxygen concentrations and the saturation concentrations (Thienemann 1926, 1928). The rate of hypolimnetic oxygen consumption (increase in deficit) is impacted by, among other factors, the productivity in surface waters and the morphometry of the hypolimnion. To eliminate the impact of morphometry, Strom (1931) and Hutchinson (1938) proposed that the demand be expressed as an areal rate, the areal hypolimnetic oxygen deficit (AHOD g O_2 m^{-2} day^{-1}). Other factors that influence the rate include the temperature and the presence of dissolved organic compounds. Dissolved organic compounds also contribute to the depletion of oxygen, even in lakes of low productivity, and thus have become one of the defining characteristics of dystrophy (Thienemann 1921; Naumann 1932).

The AHOD is normally calculated as the slope of the linear plot of the product of mean depth and mean hypolimnetic DO against time. Mortimer (1941) proposed limits of 0.25 g m^{-2} day^{-1} for the upper limit of oligotrophy and 0.55 g m^{-2} day^{-1} for the lower limit of eutrophy.

Since this is only applicable to stratified lakes, other metrics may be used, such as net dissolved oxygen (Porcella et al. 1980). Another oxygen-related index is the anoxic factor (AF) represented in Equation 16.2:

$$AF = \frac{\sum t_a A_s}{A_o} \tag{16.2}$$

where
t_a is the days of detectable anoxic conditions
A_s is the sediment area
A_o is the surface area (Nurnberg 1995)

So, the AF describes the relationship of the lake bottom area exposed to anoxia, which is often more useful in assessments of benthic organisms than the AHOD.

16.4.2 Forsberg and Ryding's Criteria

Forsberg and Ryding (1980) developed criteria for classifying lakes into trophic states based on four water chemistry parameters (total chlorophyll, total phosphorus [TP], total nitrogen [TN], and water clarity). These criteria were based on studies of a series of Swedish lakes; however, the criteria

TABLE 16.3

Forsberg and Ryding Criteria Lake Trophic Status

	Oligotrophic	Mesotrophic	Eutrophic	Hypereutrophic
Total chlorophyll ($\mu g\ L^{-1}$)	<3	3–7	7–40	>40
Total phosphorus ($\mu g\ L^{-1}$)	<15	15–25	25–100	>100
Total nitrogen ($\mu g\ L^{-1}$)	<400	400–600	600–1500	>1500
Water clarity (ft.)	>13	8–13	3–8	<3

Source: Forsberg, C. and Ryding, S.O., *Archiv Für Hydrobiologie*, 89, 189–207, 1980.

have been found to work well in other areas. For example, Florida LakeWatch (coordinated through the University of Florida) has adapted these criteria to track the trophic status of Florida lakes (Table 16.3).

16.4.3 TROPHIC STATE

Other common metrics of trophic status based on nutrients and chlorophyll include the trophic state classification by the Organization for Economic Development and Co-Operation (OECD 1982) and that for tropical lakes by Salas and Martino (1991). The OECD (1982) classification has been internationally accepted for nearly 30 years, including acceptance by the U.S. EPA (USEPA 2000, 2010). Table 16.4 shows these trophic classes, associated TP chlorophyll-a values, and consensus assessment of use impairment.

16.4.4 CARLSON'S TSI

One of the most widely used metrics for estimating the trophic status of lakes was developed by Carlson (1977) and is referred to as Carlson's TSI. Carlson indicated that the factors impacting the trophic status of lakes are multidimensional, and cannot be adequately characterized by a single metric. He developed an index based on data on algal biomass (as indicated by chlorophyll-a concentrations) from primarily northern lakes and developed relationships between biomass and transparency (as indicated by the Secchi depth) and the TP. He then based his classification of the trophic status on the doubling of biomass. A 10-unit change in the index then represents a doubling or

TABLE 16.4

OECD and Salas and Martino Trophic Categories

Trophic Category	OECD (1982)			Salas and Martino (1991)	
	TP (mg L^{-1})	Chlorophyll-a ($\mu g\ L^{-1}$)	Use Impairment	TP (mg L^{-1})	Use Impairment
Ultraoligotrophic	<0.004	<1.0			Low
Oligotrophic	<0.010	<2.5	Little	<0.028	Low
Mesotrophic	0.010–0.035	2.5–8	Variable	0.028–0.070	Variable
Eutrophic	0.035–0.10	8–25	Great	>0.070	High
Hypertrophic	>0.10	>25			Very high

Source: OECD, *Eutrophication of Waters: Monitoring, Assessment and Control*, Organization for Economic Development and Co-Operation, Paris, 1982; Salas, H.J. and Martino, P., *Water Research*, 25(3), 341–350, 1991; USEPA, *Technical Support Document for U.S. EPA's Final Rule for Numeric Criteria for Nitrogen/Phosphorus Pollution in Florida's Inland Surface Fresh Waters*, U.S. Environmental Protection Agency, Washington, DC, 2010.

halving of algal biomass (chlorophyll-a). The simplified forms of the equations derived by Carlson (1977) are given as follows (from Carlson and Simpson 1996):

$$TSI(SD) = 60 - 14.41 \ln(SD)$$

$$TSI(CHL) = 9.81 \ln(CHL) + 30.6$$

$$TSI(TP) = 14.42 \ln(TP) + 4.15$$

where
  SD is the Secchi depth (meters)
  CHL is chlorophyll-a concentrations (micrograms per liter)
  TP is total phosphorus (micrograms per liter)

Carlson created the interpretation scheme in Table 16.5 for the computed TSI (Carlson 1977).

Note that the values in Table 16.5 are interrelated, based on Carlson's data analyses, so that the TSI can be based on any one of the metrics (SD, TP, or CHL) and is not additive and should not be averaged (Carlson and Simpson 1996) since it is predicated on the idea that it is predicting algal biomass. Carlson and Simpson (1996) suggested using chlorophyll as the primary index for trophic state classification where data for chlorophyll and phosphorus are available. Deviations of the Secchi depth and TP indices from the chlorophyll index could then be used to infer additional information about the functioning of the lake. Carlson (1977) also emphasizes that these relationships are indications of trophic status, not water quality. Carlson points out a common misconception that eutrophic is often taken as synonymous with poor water quality. Water quality is a subjective determination while the trophic status is a "neutral" determination; basically, "it is what it is."

A limitation to Carlson's method is that it was developed for northern lakes, so its applicability to southern lakes is limited. The following section describes a modification of the TSI method that

TABLE 16.5
Carlson's Trophic State Index

TSI	CHL (µg L⁻¹)	SD (m)	TP (µg L⁻¹)	Attributes
<30	<0.95	>8	<6	Classical *oligotrophy*: Clear water, oxygen throughout the year in the hypolimnion, salmonid fisheries in deep lakes.
30–40	0.95–2.6	8–4	6–12	Deeper lakes still exhibit classical oligotrophy, but some shallower lakes will become anoxic in the hypolimnion during the summer.
40–50	2.6–7.3	4–2	12–24	Water moderately clear, but increasing probability of anoxia in hypolimnion during summer.
50–60	7.3–20	2–1	24–48	Lower boundary of classical *eutrophy*: Potential for decreased transparency, anoxic hypolimnia during the summer and macrophyte growth, warmwater fisheries only.
60–70	20–56	0.5–1	48–96	Dominance of blue–green algae, algal scums probable, extensive macrophyte problems.
70–80	56–155	0.25–0.5	96–192	Heavy algal blooms possible throughout the summer, dense macrophyte beds, but extent limited by light penetration. Often would be classified as *hypereutrophic*.
>80	>155	<0.25	192–384	Algal scums, summer fish kills, few macrophytes, dominance of rough fish.

Source: Carlson, R.E. and Simpson, J., *A Coordinator's Guide to Volunteer Lake Monitoring Methods*, North American Lake Management Society, Madison, WI, 1996.

the state of Florida uses for evaluating lakes in that state. Additionally, the method is not applicable to macrophyte-dominated lakes.

16.4.5 FLORIDA'S TROPHIC STATE INDEX

The Florida Department of Environmental Protection (FDEP) has commonly used a modified version of Carlson's TSI to manage lakes throughout the state (FDEP 2009). The Florida TSI has been used, for example, to determine lakes that are impaired (e.g., for 305(b) and 303(d) assessment). As with Carlson's TSI, the Florida TSI is not a direct measure of any single lake characteristic, but combines TN, TP, and chlorophyll-a on a common scale that can also be related to trophic state categories.

Carlson's TSI was modified by Florida since Florida lakes do not share some of the characteristics of northern lakes, such as an oxygenated hypolimnia and cool/cold-water fisheries (FDEP 2009). Florida's high frequency of dark or stained waters also precluded use of the Secchi depth as a measure of trophic state. Florida based the modified TSI on studies by Salas and Martino (1991, see Section 16.4.3). Salas and Martino analyzed tropical and subtropical lakes of the Americas (including lakes in north Florida) to identify trophic states corresponding to TP concentrations, concluding that tropical-subtropical TP concentrations were higher than the equivalent temperate concentrations by a factor of approximately 2. The FDEP (2009) determined that the subtropical lakes of Florida were more similar to the tropical and subtropical lakes analyzed by Salas and Martino (1991) than to the lakes analyzed by the OECD (1982), resulting in the development of a modified Florida TSI. That modified TSI was based on replacing the Secchi depth, used in Carlson's TSI, with TN; including chlorophyll-a (Salas and Martino only used TP); and based on an analysis of data from 313 Florida lakes. The Florida TSI is computed from (FDEP 2009):

$$\text{Florida TSI} = \frac{(\text{CHL TSI} + \text{Nutrient TSI})}{2} \tag{16.3}$$

where

CHL TSI $= 16.8 + 14.4 \times \ln$ (Chlorophyll-a) and

For TN/TP > 30 (phosphorus limited)

Nutrient TSI $= 10 \times [2.36 \times \ln(\text{TP} \times 1000) - 2.38]$

For TN/TP < 10 (nitrogen limited)

Nutrient TSI $= 10 \times [5.96 + 2.15 \times \ln(\text{TN} + 0.0001)]$

For 10 < TN/TP < 30 (colimited)

Nutrient TSI $= (\text{TP\_TSI} + \text{TN\_TSI}) / 2$, where

TN_TSI $= 56 + [19.8 \times \ln(\text{TN})]$

TP_TSI $= [18.6 \times \ln(\text{TP} \times 1000)] - 18.4$

The relationship between the values of chlorophyll-a, TP, and TN are tabulated in Tables 16.6 and 16.7.

The values for the TSI were adjusted so that a chlorophyll-a concentration of 20 μg L^{-1} was equal to a TSI value of 60, used to represent "fair" lakes, while lakes above 70 were assessed as "poor" as tabulated in Table 16.7.

TABLE 16.6
Relationship between Chlorophyll a, Total Phosphorus, and Total Nitrogen, as Described by Florida's TSI

Trophic State Index	Chlorophyll-a ($\mu g \, L^{-1}$)	Total Phosphorus (mg L^{-1})	Total Nitrogen (mg L^{-1})
0	0.3	0.003	0.06
10	0.6	0.005	0.1
20	1.3	0.009	0.16
30	2.5	0.01	0.27
40	5	0.02	0.45
50	10	0.04	0.7
60	20	0.07	1.2
70	40	0.12	2
80	80	0.2	3.4
90	160	0.34	5.6
100	320	0.58	9.3

Source: FDEP (Florida Department of Environmental Protection), *Technical Support Document: Development of Numeric Nutrient Criteria for Florida Lakes and Streams*, Florida Department of Environmental Protection, Tallahassee, FL, 2009.

TABLE 16.7
Warmwater Florida TSI Categories

TSI	Category	TP ($\mu g \, L^{-1}$)	Chlorophyll-a
40	Oligotrophic	21.3	5
50	Mesotrophic	39.6	10
70	Eutrophic	118.7	40

Source: Salas, H.J. and Martino, P., *Water Research*, 25, 341–350, 1991.

16.4.6 MACROPHYTES AND TROPHIC STATUS

The TSI methods described by Carlson (1977) and FDEP (2009) do not consider macrophyte-dominated lakes. Carlson and Simpson (1996) indicated that a potential solution is to use relationships based on the TP of a lake, to include the TP both in the water column and in the plant biomass (Canfield et al. 1983), so that the trophic state could include both macrophytes and algae, and have internally consistent units.

REFERENCES

American Public Health Association. 1999. *Standard Methods for the Examination of Water and Wastewater.* American Public Health Association, American Water Works Association, and Water Pollution Control Federation, 20th edn, Washington, DC.

Art, H.W. (ed.). 1993. *Eutrophication. A Dictionary of Ecology and Environmental Science*, 1st ed. Henry Holt and Company, New York.

Box, G.E.P. and N.R. Draper. 1987. *Empirical Model-Building and Response Surfaces*, Wiley, New York.

Canfield, Jr. D.E., K.A. Langeland, M.J. Maceina, W.T. Haller, J.V. Shireman, and J.R. Jones. 1983. Trophic state classification of lakes with aquatic macrophytes. *Canadian Journal of Fisheries and Aquatic Sciences* 40, 1713–1718.

Carlson, R.E. 1977. A trophic state index for lakes. *Limnology and Oceanography* 22, 361–369.

Carlson, R.E. and J. Simpson. 1996. *A Coordinator's Guide to Volunteer Lake Monitoring Methods.* North American Lake Management Society, Madison, WI.

Chapra, S.C., G.J. Pelletier, and H. Tao. 2007. QUAL2K: A modeling framework for simulating river and stream water quality, Version 2.07: Documentation and Users Manual. Civil and Environmental Engineering Dept., Tufts University, Medford, MA.

Cole, G.A. 1979. *Textbook of Limnology*, 2nd ed. C.V. Mosby, St. Louis, MO.

Committee on Environment and Natural Resources. 2000. Integrated assessment of hypoxia in the northern Gulf of Mexico. National Science and Technology Council, Washington, DC.

Crowder, B. 1987. Economic costs of reservoir sedimentation: A regional approach to estimating cropland erosion damages. *Journal of Soil and Water Conservation* 42, 3.

Downing, J.A. 1984. Assessment of secondary production: The first step. In J.A. Downing, and F.H. Rigler (eds), *A Manual on Methods for the Assessment of Secondary Productivity in Fresh Waters.* Blackwell Scientific Publications, Oxford, UK, p.1–18.

FDEP (Florida Department of Environmental Protection). 2009. *Technical Support Document: Development of Numeric Nutrient Criteria for Florida Lakes and Streams.* June. Florida Department of Environmental Protection, Tallahassee, FL. Accessed September 6, 2010 from http://www.dep.state.fl.us/water/wqssp/nutrients/docs/tsd_nutrient_crit.docx.

Florida LakeWatch. 2002. Trophic state: A waterbody's ability to support plants, fish, and wildlife. Department of Fisheries and Aquatic Sciences, University of Florida, Institute of Food and Agricultural Sciences, Gainesville, FL.

Forsberg, C. and S.O. Ryding. 1980. Eutrophication parameters and trophic state indices in 30 Swedish waste-receiving lakes. *Archiv Für Hydrobiologie* 89, 189–207.

Graf, W.L., E. Wohl, T. Sinha, and J.L. Sabo. 2010. Sedimentation and sustainability of western American reservoirs. *Water Resources Research* 46, W12535.

Hall, R.I., P.R. Leavitt, R. Quinlan, A.S. Dixit, and J.P. Smol. 1999. Effects of agriculture, urbanization, and climate on water quality in the northern Great Plains. *Limnology and Oceanography* 44, 739–756.

Hambrook Berkman, J.A. and M.G. Canova. 2007. Algal biomass indicators (ver. 1.0). U.S. Geological Survey techniques of water-resources investigations. Accessed 7 February 2012 from http://pubs.water.usgs.gov/twri9A/.

Horne, A.J. and C.R. Goldman. 1994. *Limnology*, 2nd ed. McGraw-Hill, New York.

Hutchinson, G.F. 1938. On the relation between the oxygen deficit and the productivity and typology of lakes. *Internationale Revue Gesamten Hydrobiologie und Hydrogrographie* 36, 336.

Hutchinson, G.E. and A. Wollack. 1940. Studies on Connecticut lake sediments. Chemical analysis of a core from Linsley Pond. *American Journal of Science* 238, 493–517.

Hutchinson, G.E. 1975. *A Treatise on Limnology,* v. 1, Part 1. Geography and Physics of Lakes. Wiley, New York.

Kennedy, R.H. and W.W. Walker. 1990. Reservoir nutrient dynamics. In K.W. Thornton, B.L. Kimmel, and F.E. Payne (eds), *Reservoir Limnology: Ecological Perspectives.* Wiley, New York, pp. 109–131.

Kimmel, B.L. and A.W. Groeger. 1986. Limnological and ecological changes associated with reservoir aging. In G.E. Hall and M.J. van den Avyle (eds), *Reservoir Fisheries Management: Strategies for the 80's.* Reservoir Committee, Bethesda, MD, pp. 103–109.

Kimmel, B.L., O.T. Lind, and L.J. Paulson. 1990. Reservoir primary production. In K.W. Thornton, B.L. Kimmel, and F.E. Payne (eds), *Reservoir Limnology: Ecological Perspectives.* Wiley, New York, pp. 133–193.

Lawrence, E., A.R.W. Jackson, and J.M. Jackson. 1998. Eutrophication. In *Longman Dictionary of Environmental Science*, Addison Wesley, London, UK, pp. 144–145.

McManus, P.F. 1981. *A Fine and Pleasant Misery.* Henry Holt & Co., New York.

Mortimer, C.H. 1941. The exchange of dissolved substances between mud and water in lakes. Parts I and II. *Journal of Ecology* 29, 280–329.

Naiman, R.J. and R.E. Bilby (eds). 1998. *River Ecology and Management: Lessons from the Pacific Coastal Ecoregion.* Springer, New York.

National Academy of Sciences. 1969. Introduction, summary, and recommendations, in eutrophication—Causes, consequences, correctives. National Academy of Sciences, Washington, DC, pp. 3–4.

Naumann, E. 1919. Nagrasynpunkterangaendelimnoplanktonsokologi med sarskildhansyn till fytoplankton. *Svensk Botanisk Tidskrift Utgifven af Svenska Botaniska Foreningen* 13, 129–163.

Naumann, E. 1929. The scope and chief problems of regional limnology. *Internationale Revue Gesamten Hydrobiologie* 21, 423.

Naumann, E. 1932. Grundzuge der regionalen limnologie. *Die Binnengewässer* 11, 1–176.

Nixon, S.W. 1995. Coastal marine eutrophication: A definition, social causes and future concerns. *OPHELIA* 41, 199–219.

Nurnberg, G.K. 1995. Quantifying anoxia in lakes. *Limnology and Oceanography* 40 (6), 1100–1111.

OECD. 1982. *Eutrophication of Waters. Monitoring, Assessment and Control.* Organization for Economic Development and Co-Operation, Paris.

Porcella, D.B., S.A. Peterson, and D.P. Larson. 1980. Index to evaluate lake restoration. *Journal of Environmental Engineering American Society of Civil Engineers* 106, 1151–1169.

Ryan, M.G., D. Binkley, and J.H. Fownes. 1997. Age-related decline in forest productivity: Patterns and process. *Advances in Ecological Research* 27, 213–262.

Salas, H.J. and P. Martino. 1991. A simplified phosphorus trophic state model for warm-water tropical lakes. *Water Research* 25 (3), 341–350.

Stumm, W. and J.J. Morgan. 1996. *Aquatic Chemistry.* Wiley-Interscience, New York.

Strom, K.M. 1931. Ferovatn: A physiographic and biological study of a mountain lake. *Archiv für Hydrobiologie*, 22, 491–536.

Thienemann, A. 1921. Seetypen. *Naturwissenschaften* 18, 1–3.

Thienemann, A. 1926. Der NahrungskreislaufimWasscr. *Verhandlungen der Deutschen Zoologischen Gesellschaft* 2, 29–79.

Thienemann, A. 1928. Der SauerstoffimeutrophenImdoligotrophen Seen. *Die Binncngcwaesser* 4.

Thoreau, H.D. 1854. *Walden; or, Life in the Woods.* Ticknor and Fields, Boston, MA.

USEPA. 2000. Nutrient criteria technical guidance manual: Lakes and reservoirs, EPA-822-B-00-001. U.S. Environmental Protection Agency, Office of Water, Washington, DC.

USEPA. 2002. National Water Quality Inventory 2000 Report, August 2002. EPA-841-R-02-001. Office of Water, Washington, DC. Accessed July 3, 2007 from www.epa.gov/305b.

USEPA. 2010. *Technical Support Document for U.S. EPA's Final Rule for Numeric Criteria for Nitrogen/ Phosphorus Pollution in Florida's Inland Surface Fresh Waters*, U.S. Environmental Protection Agency, Washington, DC.

Vollenweider, R.A. 1968. Scientific fundamentals of the eutrophication of lakes and flowing waters, with particular reference to nitrogen and phosphorus as factors in eutrophication, Rep. Organization for Economic Cooperation and Development, DAS/CSI/68.27, Paris.

Vollenweider, R.A. 1969. *Manual on Methods for Measuring Primary Production in Aquatic Environments*, IBP Handbook 12, Blackwell Scientific Publications, Oxford, UK.

Wetzel, R.G. 2001. *Limnology: Lake and River Ecosystems*, 3rd ed. Elsevier, San Diego, CA.

17 Restoration and Management of Lakes and Reservoirs

17.1 LAKE MANAGEMENT AND RESTORATION

Restoration typically deals with restoring the social, economic, and ecological functions of some degraded system, such as a lake or a reservoir, while management could be considered as actions either to prevent the necessity of restoration or to maintain those functions at a desirable level. For both restoration and management a holistic approach is required, since there is rarely any single "magic bullet" that can solve or prevent the problem, and often the source of the problem may be removed from the lake or reservoir proper, such as in the watershed.

There is a fairly extensive amount of literature and guidance on lake management and restoration, including the North American Lake Management Society's (NALMS) manual *Managing Lakes and Reservoirs* (Holdren et al. 2001), which provides a good organizational approach for any lake management or restoration effort.

The first component of the lake management or restoration process is developing a fundamental understanding of the ecological concepts associated with lake and reservoir ecosystems, including the lake and its watershed and in-lake processes. A number of these concepts were discussed in earlier chapters of this text.

The second component of the lake management or restoration process is to "have a plan." The NALMS (Holdren et al. 2001) manual emphasizes the need for citizen (or stakeholder) involvement in the process and discusses in detail the following generic steps of the process:

- Clarify goals
- Gather information
- Conceptualize alternatives
- Make formal decisions
- Define measurable objectives
- Implement
- Evaluate
- Repeat the process

NALMS (Holdren et al. 2001) provides, for example, a hierarchical approach to conceptualizing alternatives, as illustrated in Figure 17.1 for a recreational lake suffering winterkills.

In repeating the processes, Holdren et al. (2001) indicated that "the end is not the end," since things change, from societal goals to scientific understanding. This is often referred to as adaptive management.

The next steps in the lake management or restoration process, as described by NALMS (Holdren et al. 2001), are problem identification followed by the implementation of methods to predict lake water quality. Today, this prediction is most commonly accomplished using predictive watershed, hydrodynamic, and water quality models, which are used to provide the mechanistic link between management alternatives and predicted impacts. Some of the common problems in lakes and reservoirs, such as sedimentation, controlling algae and nuisance plants (e.g., Figure 17.2), and managing fisheries, and their potential solutions are introduced in the following sections of this chapter.

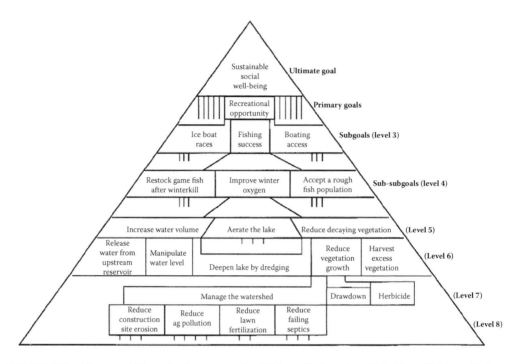

FIGURE 17.1 Means-end hierarchy for a recreational lake suffering winterkill. (From Holdren, C., Jones, W., and Taggart, J., *Managing Lakes and Reservoirs*, North American Lake Management Society and Terrene Institute, in cooperation with the Office of Water Assessment and Watershed Protection, U.S. Environmental Protection Agency, Madison, WI, 2001.)

FIGURE 17.2 Aquatic herbicide treatment. (Courtesy of U.S. Army Corps of Engineers.)

17.2 CLEAN WATER ACT

In other chapters, the role of the 1972 Federal Water Pollution Control Act, now known as the Clean Water Act (CWA), was introduced. This act established water quality as a national priority. A specific objective of the CWA was to "restore and maintain the chemical, physical, and biological integrity of the nation's waters," which, of course, include lakes and reservoirs.

The CWA also included a number of sections and provisions targeted at lakes. Under Section 314, the Clean Lakes program was established to fund and provide technical assistance to states for the restoration of publicly owned lakes. The initial focus of the Section 314 grants program was on research, the development of lake restoration techniques, and the evaluation of lake conditions. The specified evaluation of lake conditions by individual states included:

- A classification of the eutrophic condition of the state's publicly owned lakes
- A description of the methods for controlling sources of pollution of those lakes
- A description of the methods and procedures to restore the quality of those lakes
- A list of those publicly owned lakes known to be impaired, which are then to be included in the state's annual 305(b) report

The evaluations were stipulations for the receipt of funds under this program. In the early years, there was substantial funding for this program, with in excess of $145 million distributed. However, although reauthorized in 2000 (as part of the Estuaries and Clean Water Act of 2000), no funds have been allocated for this program since 1994, so that it presently exists only on paper.

Section 314 funding, when it was available, provided funding for a variety of data collection efforts. Data are critical for all stages of lake management and restoration, and usually long-term data are required in order to establish trends, both spatial and temporal. In addition to trend data, data are required for predictions, relating trends (effects) to causes. For example, trend data may include periodic water quality data at specific horizontal and vertical stations within a lake. Predictive models also require flows and loadings (based on flows and concentrations) for all sources impacting the lake. Data are expensive, and finding sources of funding for data collection and implementing and maintaining data collection efforts are an essential component of lake management and restoration.

In the absence of 314 funding, some federal funding is provided under Section 319 of the CWA. Section 319 is targeted at nonpoint source control, such as urban and agricultural runoff. While nonpoint source pollution is important for lakes and reservoirs, this program is not targeted toward lakes, so that lakes and reservoirs are only one waterbody type of interest. While some federal funding is periodically available under other programs (e.g., U.S. EPA 2008 Watershed Survey) or legislation targeted to specific waterbodies (e.g., Lake Tahoe), funding for lake restoration remains problematic and often relies solely on local or state sources.

17.3 CONTROL AND MANAGEMENT OF SEDIMENTATION

One of the major issues impacting many lakes and reservoirs is sedimentation, not only because of its impact on water quality but also because of its impact on life expectancy. For example, the classical view of lake succession is that lakes are temporary ecological systems, formed in some depression that will eventually fill in to become a terrestrial environment, or become extinct. That timescale varies enormously for natural lakes, if it is applicable at all.

For reservoirs, the rates of sedimentation are generally greater since they are designed to store water from the flow of the rivers and streams that they dam. They also store much of the sediments contained in those flows (acting as sediment traps). Traditionally, most reservoirs are considered to be sediment traps so that, ultimately, they may be filled in unless the sediments are removed the dams are destroyed by floods or are removed.

The control and management of sedimentation in reservoirs include both external source control and the control of lake shorelines. The erosion of lake shorelines not only impacts sedimentation, but it also impacts many other ecological characteristics of lakes. The U.S. Environmental Protection Agency's (EPA) 2010 National Lakes Assessment reported that lakes with poor lakeshore habitats are three times more likely to be in poor overall biological condition than lakes with good quality shorelines. The EPA maintains a website as a clearinghouse for information on lakeshore protection and restoration.

Since the sediment directly impacts the storage capacity and purpose of reservoir projects, quantification (e.g., through sedimentation surveys) and control of that sedimentation are an integral component of reservoir design and management. For example, sediment management plans are components of most reservoir operations. The design of many reservoirs is based on 50 or 100 years of sediment storage, thereby establishing the design "life of the reservoir" (Morris et al. 2008).

The rate of sedimentation is related to the trapping efficiency of specific reservoirs, but a regional approach to estimating sediment damages in lakes and reservoirs indicated that 0.2% of the nation's water storage capacity is lost annually (Crowder 1987; White 2001). Crowder also indicated that of this, an average of 24% is due to soil erosion on cropland, with the greatest losses occurring in the central United States.

Morris et al. (2008) divided sediment management in reservoirs into five categories:

1. Sediment yield reduction
2. Sediment storage
3. Sediment routing
4. Sediment removal
5. Sediment focusing

Each of these categories is briefly introduced in the following sections and is discussed in greater detail in Morris and Fan (1998) and Morris et al. (2008).

17.3.1 SEDIMENT YIELD REDUCTION

Sediment yield reduction refers to the reduction of sediment loads (or yields) to the reservoir. Loads to the reservoir are from the reservoir's watershed, due to the point and nonpoint sources from the watershed, and from the reservoir's shoreline.

Figure 17.3 illustrates shoreline erosion, which can contribute to reservoir sedimentation. Management and restoration efforts for lakes and reservoirs may include a variety of methods to

Eroded material
forming a shelf

Recently eroded material that has not
moved further into the reservoir

FIGURE 17.3 Examples of shoreline erosion from a USBR reservoir. (From Ferrari, R.L. and Collins, K., *Erosion and Sedimentation Manual*, U.S. Department of the Interior Bureau of Reclamation, Sedimentation and River Hydraulics Group, Denver, CO, 2006. Photograph by S. Nuanes.)

protect shorelines. McComas (2003), in his *Lake and Pond Management Guidebook*, described a variety of shoreline protection and landscaping methods, including:

- Native landscaping and upland buffers
- Wave breaks for shoreline protection
- Biostabilization of the lakeshore
- Structural landscape protection (e.g., riprap, gabions, and retaining walls)
- Aquascaping, working with plants and woody debris in shallow water

There has been an extensive amount of literature and guidance on identifying and controlling erosion and the consequent sediment loads from watersheds since that erosion often has multiple negative impacts. Much of that literature, while not specifically focused on sediment yields to reservoirs, is still applicable. But, as indicated by Morris et al. (2008), while erosion control to reduce sediment is widely recommended to prolong a reservoir's function, it is difficult to implement successfully.

17.3.2 SEDIMENT STORAGE

Sediment storage simply refers to allowing for sufficient or increased storage to accommodate sediment inflows and subsequent sedimentation. One commonly used method is to simply raise the dam. Garbrecht and Garbrecht (2004; described by Morris et al. 2008) provided an example of successively raising the Marib diversion over the period from 940 BC to AD 570 to accommodate increased sediment upstream of the dam and an increase in land levels downstream due to diverted irrigation water.

17.3.3 SEDIMENT ROUTING

Sediment routing methods pass sediment loads around or through storage areas or structures while minimizing deposition (Morris et al. 2008). Some of the pass-through methods include (Morris and Fan 1998):

- Seasonal drawdown: where the reservoir storage is seasonally decreased to allow flows representing high sediment loads to pass through the reservoir with reduced rates of sedimentation (lower volumes, higher velocities, and lower retention time), and the reservoir is refilled with wet season flows with lower sediment loads.
- Flood drawdown by hydrograph prediction: where instead of a seasonal drawdown, the drawdown is timed to allow the passage of the leading edge or rising limb of a flood hydrograph where sediment loads are highest and the reservoir is then refilled during the flood recession of the hydrograph.
- Flood drawdown by rule curve: A similar tactic for small reservoirs based on the reservoir's rule curve.
- Venting turbid density currents: In cases where a dense, sediment-laden plume "short-circuits" the reservoir (passes through it without substantial mixing), the density current may be intentionally vented through the outlet structure.

Some of the sediment bypass methods include (Morris and Fan 1998):

- Sediment bypass for instream reservoirs: where channels or tunnels are constructed to pass sediment-laden flows around the instream reservoir.
- Off-stream reservoirs: where flows with high sediment loads are diverted to an off-channel storage area and only relatively clear flows pass to the storage reservoir.

17.3.4 SEDIMENT REMOVAL

Sediment is most commonly removed by dredging or hydraulic flushing. Dredging is commonly used, albeit an often expensive practice. Dredging in the absence of other controls is also often impractical, since the sediment problem just reoccurs. Scouring typically involves opening bottom outlets in order to empty the reservoir and allow streamflows to scour sediments from the bottom as they pass through the reservoir (Morris and Fan 1998; Morris et al. 2008).

17.3.5 SEDIMENT FOCUSING

Focusing refers to using hydraulic techniques to redistribute sediments within a reservoir (Morris and Fan 1998). Sedimentation is typically not uniform in reservoirs with, for example, more coarse materials falling out in the transition zone while cohesive sediments may extend further into the reservoir. As a result, shoals may form. Also, sedimentation may be most problematic around structures. As a result, a more focused removal is desirable via dredging, or in some cases a focused redistribution of sediments such as to deeper parts of the reservoir.

17.4 CONTROL OF ALGAE

Excesses of algae can cause a wide variety of problems in lakes and reservoirs, depending on the timing, duration, and magnitude of the algal concentrations. Blooms of planktonic algae or periphyton may be unsightly, so they are simply perceived as a problem; they can cause taste and odor problems; they can consist of harmful or toxic algae (harmful algae blooms); they can result in hypoxia; and they can cause a wide variety of other problems. Conversely, these are at the bottom of the food chain so they are the necessary "grass" for the lake food chain. Maintaining acceptable algal concentrations (and determining what is "acceptable") is then a necessary component of lake management.

One of the most commonly used strategies to control algae is to control those nutrients that control plant growth, principally nitrogen and phosphorus. These strategies may involve the addition of nutrients to promote algal growth where nutrients are (and hence productivity is) limited. A more common problem is excess nutrients, and management strategies are directed toward preventing excess nutrients from entering the lake or reservoir or reducing nutrient concentrations within the lake or reservoir.

17.4.1 NUTRIENT LOAD REDUCTION (EXTERNAL)

One commonly used strategy to control lake algal concentrations is the control or management of the source of those loads, the nutrient loads coming to the lake or reservoir through point and nonpoint sources. The methods typically involve, first, the application of methods for predicting the impacts of nutrient loads, such as using mathematical models (e.g., Chapra 1997 and others) of the watersheds, tributary waterbodies, and the lake or reservoir, to first quantify the load sources and establish acceptable limits. Then, methods for the control of nutrients, such as watershed management, improved wastewater treatment, and the application of best management practices (BMPs), are developed and implemented, often as part of watershed management plans. Examples include the extensive efforts to reduce phosphorus loads to the Great Lakes in the 1970s, which led to the removal of phosphorus from detergents as well as changes in agricultural practices and improvements to wastewater treatment plants. The problems extant in the Great Lakes in the 1970s (e.g., eutrophication and hypoxia) still remain common today in parts of the Great Lakes and elsewhere.

Currently, there is a great deal of interest in quantifying nutrient loads and in nutrient load reductions for all waterbodies. Quantifying loads and the impacts of those loads from all sources (point and nonpoint) are a component of all total maximum daily loads (TMDLs) studies. Quantifying nutrient loads and impacts is also a necessary component of the recent and/or ongoing development of nutrient standards by states. The advantage to lake management and restoration is that the TMDLs and nutrient standards provide a link between, and regulatory control of, external sources and in-lake quality.

17.4.2 INTERNAL MANAGEMENT

A variety of techniques can be used for the internal management and control of nuisance algae in lakes, a number of which are listed in the following sections and described by NALMS (e.g., 1990, 2001; Table 17.1). It should be noted that many of these techniques are not effective over long periods unless they are implemented in conjunction with the management of external loads.

17.4.3 PHOSPHORUS INACTIVATION

Phosphorus inactivation is typically intended to remove or reduce stored phosphorus in lakes by adding salts of aluminum, iron, or calcium to the lakes to complex with and inactivate phosphorus. Of these, aluminum is the most common and is a popular lake management tool (Cooke et al. 2005).

Aluminum salts, such as aluminum sulfate (alum), may be added to the water column or sediments or both. In the water column, alum may precipitate the formation of flocs (e.g., $Al(OH)_3$ at a typical pH of 6–8; Cooke et al. 2005) that bind phosphorus and remove it via settling. As they settle, the flocs also remove other particulates (including phytoplankton), so that the water column transparency increases as the flocs settle. The flocs on the sediment surface, or those added to the sediments, form a "blanket" that binds and inhibits the release of phosphorus from the sediments (NALMS 1990; Holdren et al. 2001). One consideration is that the addition of alum tends to reduce the pH (the formation of AlOH), which affects both its efficiency (greatest floc formation occurs between a pH of 6–8) and its toxicity below $pH = 6$. For lakes with low alkalinities, it may be necessary to add a buffer, such as sodium aluminate, along with the alum to act as a buffer and maintain a stable pH (Cooke et al. 2005).

TABLE 17.1

Comparison of Lake Management Techniques for the Control of Nuisance Algae

	Short-Term Effect	Long-Term Effect	Cost	Chance of Negative Effects
Phosphorus inactivation	E	E	G	L
Dredging	F	E	P	F
Dilution	G	G	F	L
Flushing	F	F	F	L
Artificial circulation	G	?	G	F
Hypolimnetic aeration	F	?	G	F
Sediment oxidation	G	E	F	?
Algicides	G	P	G	H
Food chain manipulation	?	?	E	?
Rough fish removal	G	P	E	?
Hypolimnetic withdrawal	G	G	G	F

Source: NALMS, *The Lakes and Reservoir Restoration Guidance Manual*, 2nd ed., North American Lake Management Society, U.S. Environmental Protection Agency, Washington, DC, 1990.

Note: E = Excellent, G = Good, H = High, F = Fair, P = Poor, L = Low, ? = not listed.

17.4.4 Dredging

The dredging of lake sediments physically removes sediments and the internal nutrient source they represent by removing the organics stored in sediments that can contribute to nutrient releases and oxygen demands as they decay (sediment diagenesis). Dredging also removes the stored nutrients. Removal via dredging may have long-term benefits if it is conducted in conjunction with the control of external sources. Some of the issues involved in dredging include costs, short-term impacts on benthic biota, and the disposal of potentially contaminated sediments. Examples of dredging approaches are illustrated in Figure 17.4.

17.4.5 Dilution and Flushing

Some of the same methods described in the previous section for sediment routing (bypass or pass-through) may also be applicable to the management of sediments. Nutrient-poor waters may also be diverted to lakes to aid in diluting and flushing nutrients. NALMS (1990) indicated that there are relatively few documented cases of the effective use of dilution and flushing, primarily due to the lack of availability or the expense involved in developing a suitable water source of sufficient volume. Well-documented cases include Green Lake in Seattle and Moses Lake in Grant County, Washington. For Moses Lake, low-nutrient water from the Columbia River was diverted through the lake at a rate of about three times the normal rate on an annual basis, to achieve a reduction of nutrient concentrations on the order of 50%. For Green Lake, water from the municipal water source was added, which resulted in a 90% decline in chlorophyll concentrations (Welch 1981).

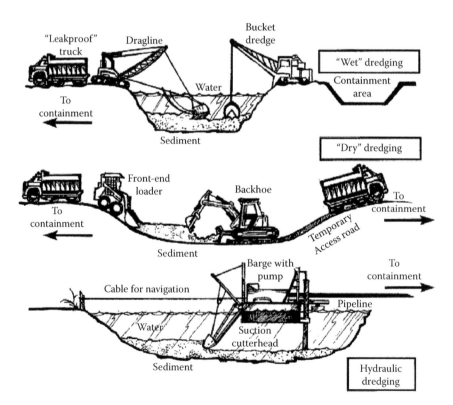

FIGURE 17.4 Examples of dredging approaches. (From Holdren, C., Jones, W., and Taggart, J., *Managing Lakes and Reservoirs*, North American Lake Management Society and Terrene Institute, in cooperation with the Office of Water Assessment and Watershed Protection, U.S. Environmental Protection Agency, Madison, WI, 2001.)

17.4.6 Artificial Circulation

Artificial circulation typically involves methods to destratify a lake or reservoir, or to remove vertical stratification. The primary benefit is to improve the overall water column dissolved oxygen concentrations, but there may be improvements in algal concentrations too. The most common method is air injection, although other methods have been used, such as axial flow pumps, like a big floating fan, to "push" the water down into the epilimnion. Some of the alternative methods are illustrated in Figure 17.5 and are described by Holdren et al. (2001).

17.4.7 Oxidation

Similar to aeration, oxidation is used to introduce oxygen to the hypolimnion and sediments, where, under oxic conditions, nutrient concentrations and release rates from sediments would be reduced. Unlike aeration when used for artificial circulation, hypolimnetic oxygen injection is typically designed not to destratify the lake or reservoir. Hypolimnetic oxidation is a commonly used technique to improve the quality of reservoir releases and reduce nutrient releases by sediments. There are a variety of methods to introduce oxygen, including diffusers on the bottom of the lake or reservoir or sidestream oxygenation, where hypolimnetic water is pumped onto the shore, injected with oxygen, and then discharged back into the hypolimnion (Fast et al. 1975). Beutel and Horne (1999) reviewed the effects of hypolimnetic oxygenation on lake and reservoir water quality. NALMS (1990) indicated that there is little documentation of its successful use to control nuisance algae. However, there is evidence that sediment oxygenation successfully reduces nutrient releases and Cooke et al. (2005) indicated that it may be, albeit more expensive, an effective alternative to alum treatment and dredging.

17.4.8 Algicides

Algicides kill algae, and the most commonly used are the toxic forms of copper, with copper sulfate ($CuSO_4$) being the most basic and common form (Holdren et al. 2001). Copper commonly kills many

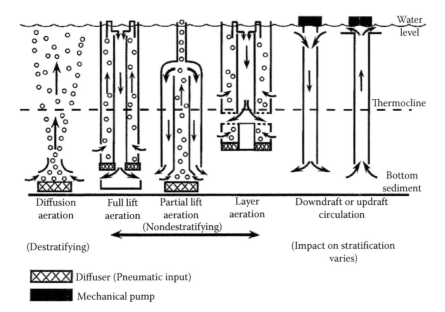

FIGURE 17.5 Aeration methods for artificial circulation. (From Holdren, C., Jones, W., and Taggart, J., *Managing Lakes and Reservoirs*, North American Lake Management Society and Terrene Institute, in cooperation with the Office of Water Assessment and Watershed Protection, U.S. Environmental Protection Agency, Madison, WI, 2001. With permission.)

green and blue-green algae and most diatoms. While it may be effective in the short term, there are major negative impacts to nontarget organisms and an additional concern is the contamination of sediments (Cooke et al. 2005). Holdren et al. (2001) reported that 58 years of copper sulfate treatment of several Minnesota lakes led to the following:

- Depleted dissolved oxygen
- Increased internal nutrient cycline
- Occasional fish kills
- Accumulated copper in sediments
- Increased tolerance to copper by some nuisance algae
- Negative impacts on fish and zooplankton

This led Hanson and Stefan (1984) to conclude that a short-term control method had been traded for the long-term degradation of lakes. Cooke et al. (2005) indicated that a search for a viable alternative algicide to copper with fewer negative effects has, to date, been unsuccessful.

17.4.9   FOOD CHAIN MANIPULATION OR BIOMANIPULATION

The term *biomanipulation* was coined by Sharpio (1979, cited in Cooke et al. 2005) to represent "a series of manipulations of the biota of lakes and their habitats to facilitate … reduction of algal biomass, and in particular, of blue-greens." Biomanipulation usually involves predator–prey relationships. An example would be altering the structure of the fish community. The first example, cited by Cooke et al. (2005), was the study by Caird (1945) where the introduction of largemouth bass reduced phytoplankton biomass. An approach is to reduce the dominance of fish such as blue gills and perch that feed on zooplankton, which, in turn, feed on algae, and replace them with bass, pike, or walleye (Holdren et al. 2001). This reduces the pressure on the zooplankton community, which then consumes greater quantities of algae. Cooke et al. (2005) summarize case histories of a variety of biomanipulation projects.

17.4.10   ROUGH FISH REMOVAL

Rough fish include shad, carp, buffalo, carpsuckers, white perch, and other "undesirable" fish and their removal can aid in reducing algal concentrations in two ways. First, these fish typically feed on algae on the lake bottom, stirring up sediments and releasing nutrients into the water column. Secondly, some of these fish, such as the gizzard shad, eat zooplankton. As with biomanipulation, removing these fish may result in a larger population of zooplankton, which will consume greater amounts of algae and provide increased food resources for juvenile, more "desirable" game fish species.

17.4.11   HYPOLIMNETIC WITHDRAWAL

Hypolimnetic withdrawal refers to removing nutrient-rich and oxygen-depleted waters from the hypolimnion of lakes and reservoirs. This also reduces the residence time of water in the hypolimnion, which can aid in reducing deoxygenation and internal nutrient loadings from sediments. Cooke et al. (2005) evaluated reports of the use of hypolimnetic withdrawal in 21 lakes, 15 of which were located in Europe and 2 in the United States (Lake Waramaug, CT, and Lake Ballinger, WA). He found that after long periods (in excess of 5 years) there were documented substantial reductions in epilimnetic phosphorus, while the case of the reduction of hypolimnetic hypoxia was not as strong.

17.5 AQUATIC PLANT MANAGEMENT

Aquatic plants refer to the macrophytes or the "larger" (macro) vascular plants. These include emergent, submersed (submerged aquatic vegetation [SAV]), and free-floating forms. As with algae, these organisms form the base of the food chain. In addition, they provide a substrate on which other organisms can grow (e.g., the microbiota), provide shelter and an escape habitat for organisms, impact flows, impact nutrient cycling and sediment chemistry, and impact other limnological characteristics of lakes. Horne and Goldman (1994) indicated that emergent reeds and submerged macrophytes are the dominant primary producers and contribute the most biomass in small lakes.

Also, as with algae, the excessive growth of macrophytes can cause a wide variety of problems, such as limiting swimming and other recreational activities, affecting navigation, shading other species, increasing sedimentation, causing oxygen depletion, and other impacts. In many southern lakes and reservoirs in the littoral zone, they can completely cover the lakes surface during the summer months. Many of the real "problem children" of the macrophytes are also invasive species, such as the Eurasian water milfoil, which infest a wide variety of lakes and reservoirs and can grow at incredible rates.

The methods for the control or, in the case of invasive species, the eradication of macrophytes have traditionally consisted of either cutting or herbicide usage. However, there are a wide variety of alternative methods in common use, such as those listed and briefly discussed next.

- Prevention
- Benthic barriers
- Dredging
- Mechanical removal
- Water level control
- Herbicide usage
- Biological controls

17.5.1 PREVENTION

The prevention of the spread of aquatic plants, particularly invasive species, is a common and cost-effective part of weed control programs in many states. It is cost effective since it focuses on the initial phase of invasion when costs are least (Figure 17.6) and it produces longer-lasting results than those

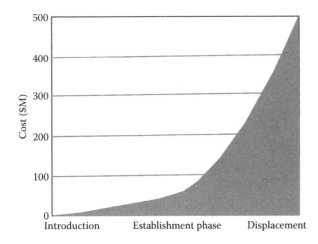

FIGURE 17.6 Comparison of invasion phase versus management costs. (After Chilton, E.W., Aquatic vegetation management in Texas: A guidance document, Texas Parks and Wildlife, Inland Fisheries, San Marcos, TX, 2004.)

directed at managing problematic vegetation. Chilton (2004) listed three components of prevention programs: monitoring and rapid response, education, and law enforcement. One example of a successful program implemented by a number of states is the "Clean Boat/Clean Waters" program whereby education and volunteer efforts are used to clean boats and prevent the transport of noxious plants, such as hydrilla, that can commonly become attached to boats and trailers (Figure 17.7).

17.5.2 BENTHIC BARRIERS

Benthic barriers (illustrated in Figure 17.8) block light and provide a surface through which plants grow, so they are effective at controlling plant growth. These barriers may be natural, such as clay, silt, and gravel, or they may consist of a variety of synthetic materials (e.g., polyethylene, polypropylene, fiberglass, and nylon sheets; Holdren et al. 2001). The Washington Department of Ecology, Aquatic Plant Management Program, lists the following advantages and disadvantages of benthic barriers:

- Advantages
  - Installation of a bottom screen creates an immediate open area of water.
  - Bottom screens are easily installed around docks and in swimming areas.
  - Properly installed bottom screens can control up to 100% of aquatic plants.
  - Screen materials are readily available and can be installed by homeowners or divers.

- Disadvantages
  - Because bottom screens reduce habitats by covering the sediment, they are only suitable for localized control.
  - For safety and performance reasons, bottom screens must be regularly inspected and maintained.
  - Harvesters, rotovators, fishing gear, propeller backwash, or boat anchors may damage or dislodge bottom screens.
  - Improperly anchored bottom screens may create safety hazards for boaters and swimmers.
  - Swimmers may be injured by poorly maintained anchors that are used to pin bottom screens to the sediment.

Wisconsin Clean
Boats/Clean Waters
Program

Watercraft check points

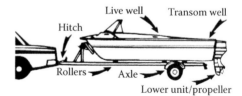

Michigan Clean Boats/Clean
Waters Program

FIGURE 17.7 Examples of state "Clean Boat/Clean Waters" programs (sources include the Save Black Lake Coalition and the Wisconsin and Michigan Clean Boats/Clean Waters programs).

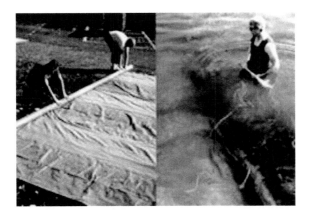

FIGURE 17.8 Benthic barrier. (From ERDC, Bottom barriers, USACE Environmental Research and Development Center, Noxious and Nuisance Plant Management Information System, 2005. Available at http://el.erdc.usace.army.mil/pmis/MechanicalControls/MechanicalControlInfo.aspx?mechID=2.)

- Some bottom screens are difficult to anchor on deep muck sediments.
- Bottom screens interfere with fish spawning and bottom-dwelling animals.
- Without regular maintenance, aquatic plants may quickly colonize the bottom screen.

17.5.3 Dredging

Dredging is most commonly used to restore lakes that have been filled in with sediments, to deepen them, to remove sediment nutrients and organics, and for purposes other than aquatic plant management. However, dredging may be effective for restructuring a lake or deepening it to reduce the littoral zone and hence reduce plant growth.

An additional benefit to dredging is the physical removal of the aquatic plants, including roots, tubers, and turions (buds along the leafy stem) that are found in or settle to the bottom of a lake and start new plants. In some cases, the removal of all plant parts, surfaces, and subsurfaces via dredging may be the only feasible alternative to controlling noxious plants.

An example of effective targeted dredging is diver dredging, where a diver uses a suction or a pump system to physically remove plants, both surface and subsurface portions, such as hydrilla (Figure 17.9). This method is most effective for removing pioneering infestations of submersed invasive plant species.

FIGURE 17.9 Diver-assisted dredging. (From ERDC, Diver dredge, USACE Environmental Research and Development Center, Noxious and Nuisance Plant Management Information System, 2005. Available at http://el.erdc.usace.army.mil/pmis/MechanicalControls/MechanicalControlInfo.aspx?mechID=2.)

17.5.4 MECHANICAL REMOVAL

Mechanical removal can include (Holdren et al. 2001):

- Hand pulling
- Cutting without collection
- Harvesting with collection
- Rototilling
- Hydroraking

Hand pulling, as the name applies, is the removal of plants by hand. It can be effective particularly for removing pioneer infestations and can be coupled with diver dredging to increase efficiency.

Cutting without collection can be likened to cutting grass without a bagger. It is a temporary measure, since the grass will regrow. In addition, the cut biomass is left in the water, which may result in increased nutrient additions and oxygen depletions.

Harvesting is cutting and removal of the material, and includes traditional barge-type harvesters with both vertical and horizontal cutting blades, and a conveyor belt that gathers the cut material for later offloading or shredding (Chilton 2004; Figure 17.10).

Rototilling (also called rotovation, Figure 17.11) is equivalent to tilling a garden (Holdren et al. 2001), where a "tiller" is lowered into the water to till the sediments and the plants are typically removed using a rake. Hydroraking is similar to a "floating backhoe" (Holdren et al. 2001), which is used to remove roots, sediments, and other debris.

17.5.5 WATER LEVEL CONTROL

Water level drawdown is a commonly used practice in multipurpose lakes with an outlet control, where the plants, and in particular their root systems, are exposed to conditions (hot, dry, freezing, etc.) for a period sufficient to kill them (Cooke et al. 2005). The effectiveness of this practice varies, and Holdren et al. (2001) and Cooke et al. (2005) listed the response of a variety of aquatic plants to drawdown. Drawdown is effective for a number of species; however, for a number of other species, biomass increased in response to drawdown. For example, hydrilla and alligator weed were rarely controlled by drawdown (Cooke et al. 2005) and since it is tolerant to a wide range of conditions, drawdown may actually give hydrilla a competitive advantage over desirable native plants (Chilton

FIGURE 17.10 Example of mechanical harvesters. (From ERDC, Harvesting, USACE Environmental Research and Development Center, Noxious and Nuisance Plant Management Information System, 2005. Available at http://el.erdc.usace.army.mil/pmis/MechanicalControls/MechanicalControlInfo.aspx?mechID=10.)

Hand removal

Rotovation

FIGURE 17.11 Examples of mechanical removal. ((From Jacksonville District, Invasive species management, n.d. Available at http://www.saj.usace.army.mil/Missions/Environmental/InvasiveSpecies/Management.aspx.)

2004). Winter drawdowns are most effective where the roots are exposed to periods of freezing and thawing, while drawdown is generally ineffective for control in moist and mild climates where seepage keeps lake sediments moist (Cooke et al. 2005).

17.5.6 HERBICIDES

Herbicides have commonly been used to either kill macrophytes or interrupt their growth processes (Figure 17.12). Herbicides are typically applied to (Cooke et al. 2005) (1) eradicate exotic species, (2) change plant community composition, or (3) treat excessive vegetative growth in direct or high-use areas. As with the algicide copper sulfate, some of the herbicides have short-term benefits but may introduce long-term harm. An example provided by Cooke et al. (2005) was the prolonged use of sodium arsenate, where records of the Wisconsin Department of Natural Resources between 1950 and when its use was discontinued in Wisconsin in 1970 indicated that a total of 798,799 kg was added to 167 lakes (Lueschow 1972). Sodium arsenate was later determined to be a highly persistent hazardous material and according to the U.S. EPA TEACH summary (last revised in 2007), it is a known human carcinogen (cancer-causing agent).

A variety of herbicides are available and approved for use by the U.S. EPA for the control of aquatic plants. These include (Cooke et al. 2005; Getsinger et al. 2011) 2,4-D, diquat, endothall, fluridone, gylphosphate, and triclopyr. These herbicides vary by their mode of application and action. Generally, chemicals with a similar chemical structure have similar actions, and control similar

FIGURE 17.12 Example of a herbicide application. (Courtesy of U.S. Army Corps of Engineers.)

plant species. Herbicides can be "contact" or "systemic" (Table 17.2). Contact herbicides act immediately on the tissues contacted but do not affect the areas not contacted, such as root crowns, roots, or rhizomes. Systemic herbicides are translocated throughout the plant, causing the mortality of the entire plant. Herbicides can also be broad spectrum or selective. Examples of use recommendations for these herbicides are provided in Table 17.3.

17.5.7 Biological Controls

Biological controls usually focus on plant-eating organisms, such as the grass carp. In addition, varieties of organisms have been introduced that impact the growth or reproduction of specific noxious host plants. The goal, of course, is to identify biological controls that are economical, relative, for example, to mechanical or chemical controls, that provide long-term benefits, and are targeted at specific noxious plants, while not harming native species. The goal, typically, of biological controls is not to eradicate the noxious species, but to obtain a dynamic equilibrium between the control and the plant, at an acceptable level of plant biomass (Cooke et al. 2005). Some examples of biocontrols are briefly introduced in the following sections.

17.5.7.1 Triploid Grass Carp (*Ctenopharyngodon idella*)

Grass carp, also called white amur, are plant-eating fish native to Asia and were introduced into the United States to control aquatic vegetation. They were first introduced in 1963 to aquaculture facilities in Auburn, Alabama, and Stuttgart, Arkansas. They may grow rapidly, reaching weights of up to 80–100 pounds, and they may eat up to three times their body weight per day, with submerged plants such as hydrilla typically being their preferred food (Chilton 2004).

The grass carp introduced to control vegetation are most commonly triploid grass carp, which are sterile, so the goal is, typically, to introduce only sterile carp to a confined area (e.g., lakes with no or a controlled inlet or outlet) so they may not escape and ensure that they do not reproduce.

TABLE 17.2
Examples of Approved Herbicides

Compound	Formulation	Half-Life in Water (Aerobic Conditions) (Days)	Mode of Action
Contact			
Diquat	Liquid	1–7	Disrupts plant-cell membrane integrity
Endothall	Liquid or granular	4–7	Inactivates plant protein synthesis
Systemic			
2,4-D	BEE salt DMA liquid	7–48	Selective plant-growth regulator
Fluridone	Liquid or granular	20–90	Disrupts carotenoid synthesis
Glyphosate	Liquid	14 days; used over but not in water	Disrupts synthesis of amino acids
Triclopyr	Liquid	1–10	Selective plant-growth regulator

Source: Data from Getsinger, K.D., Poovey, A.G., Glomski, L., Slade, J.G., and Richardson, R.J., Utilization of herbicide concentration/exposure time relationships for controlling submersed invasive plants on Lake Gaston, Virginia/North Carolina, U.S. Army Corps of Engineers, Aquatic Plant Control Research Program, Vicksburg, MS, 2011; Cooke, G.D., Welch, E.B., Peterson, S.A., and Nichols, S.A., *Restoration and Management of Lakes and Reservoirs*, 3rd ed., CRC Press, New York, 2005.

TABLE 17.3

Usage Suggestions for Selected Aquatic Herbicides

Compound	Exposure Time	Advantages	Disadvantages	Systems Where Used Effectively	Plant Species Response
2,4-D	Intermediate (18–72 hours)[a]	Inexpensive, systemic	Nontarget plants may be affected	Lakes and slow-flow areas	Selective to broadleaves, acts in 5–7 days up to 2 weeks
Diquat	Short (12–36 hours)[b]	Rapid action, limited drift	Does not affect underground portions	Shoreline, localized treatments, higher exchange rate areas	Broad spectrum, acts in 7 days
Endothall	Short (12–36 hours)[c]	Rapid action, limited drift	Does not affect underground portions	Shoreline, localized treatments, higher exchange rate areas	Broad spectrum, acts in 7–14 days
Fluridone	Very long (30–60 days)[d]	Very low dosage required, few label restrictions, systemic	Very long contact period	Small lakes, slow-flowing systems	Broad spectrum, acts in 30–90 days
Glyphosate	Not applicable	Widely used, few label restrictions, systemic	Very slow action, no submersed control	Emergent and floating-leaved plants only	Broad spectrum, acts in 7–10 days, up to 4 weeks
Triclopyr	Intermediate (12–60 hours)[e]	Selective, systemic	Can injure other nearby broadleaved species	Lakes and slow-flow areas	Selective to broadleaves, acts in 5–7 days, up to 2 weeks

Source: ERDC, Herbicide overview, Use suggestions for U.S. Environmental Protection Agency-approved aquatic herbi-cides, U.S. Army Corps of Engineers, Environmental Research and Development Center, Aquatic Plant Information System, APIS, n.d. Available at http://el.erdc.usace.army.mil/apis/HerbicideInformation/HerbicideInfoMain.aspx.

[a] Green and Westerdahl (1990)
[b] Westerdahl (1987)
[c] Netherland et al. (1991)
[d] Netherland (1992)
[e] Netherland and Getsinger (1992)

However, these measures including sterilization and confinement are not always effective, as cur-rently grass carp have been recorded from 45 states; with breeding populations in a number of states in the Mississippi River basin and elsewhere (Figure 17.13).

17.5.7.2 Alligatorweed Flea Beetles (*Agasicles hygrophila*)

The alligatorweed flea beetle was introduced into the United States for the control of the invasive alligator weed (*Alternanthera philoxeroides*). Alligator weed is an invasive species in the United States, native to South America. It is an emergent or rooted floating plant that invades aquatic areas and adjoining uplands, growing in thick interwoven mats, sometimes completely covering ponds, lakes, or canals (Cooke et al. 2005). Alligator weed displaces native vegetation and wildlife habitats, clogs waterways, restricts the oxygen levels of water, increases sedimentation, interferes with irriga-tion, and prevents drainage.

Alligatorweed flea beetles are native to Argentina and were first introduced into North and South California in 1964 by the U.S. Army Corps of Engineers to control the invasive alligator weed

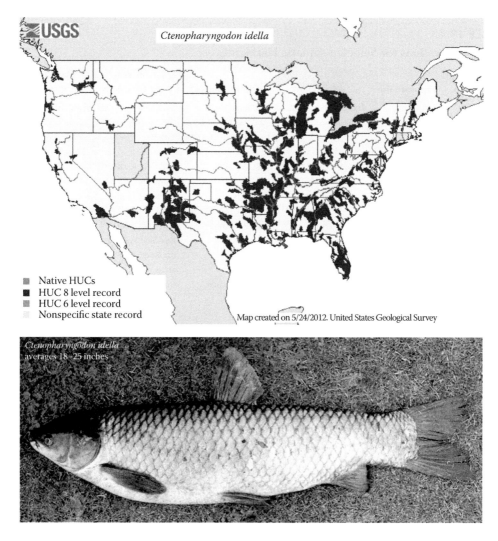

FIGURE 17.13   U.S. distribution of grass carp from the USGS Nonindigenous Aquatic Species center and a photograph of the triploid grass carp. (Courtesy of New York Department of Environmental Conservation.)

(Center et al. 1998). Their larvae burrow into the hollow stem of the alligator weed and feed on the leaves, while the adults feed primarily on the leaves. They have been so successful that alligator weed is currently only a nuisance in local areas (Cooke et al. 2005).

17.5.7.3   Water Hyacinth Weevils (*Neochetina eichhorniae* and *N. bruchii*)

Water hyacinth weevils (Figure 17.14) are native to Central and South America, and were introduced to control water hyacinths. A water hyacinth is a free-floating invasive aquatic plant, native to South America and was first introduced as an ornamental plant into the United States in 1884 at the Cotton States Exposition in New Orleans. The two species of weevils (Figure 17.15) were introduced into the United States in the 1970s to help control water hyacinths (Chilton 2004). These weevils do not kill water hyacinths quickly, but cause stunted plant growth with less flowering and a reduced competitive ability against native plants. Past studies have demonstrated that it may take from 3 to 6 years before significant impact, ultimately leading to decreases in the water hyacinth populations (Grodowitz et al. 2000).

FIGURE 17.14 Alligator weed and alligatorweed flea beetle. (From Buckingham, G.R., *Biological Control of Invasive Plants in the Eastern United States*, United States Department of Agriculture Forest Service. Forest Health Technology Enterprise Team, Morgantown, WV, 2002.)

17.5.7.4 Water Lettuce Weevils (*Neohydronomous affinis*)

Water lettuce is a free-floating invasive plant with high vegetative reproduction, allowing water lettuce to form dense floating mats of vegetation (Figure 17.16). The impacts of severe infestations include preventing light penetration, reducing oxygen levels, increasing siltation, reducing fish habitats, impeding navigation, reducing suitable fish spawning habitats, and restricting water flow and boating traffic. Water lettuce is believed to be native to Africa or South America, and was identified in North America as early as the late 1700s.

Water lettuce weevils are native to Central and South America and were introduced into the United States in the 1980s to help control water lettuce in Florida. Water lettuce weevils have proven very effective and where they have become established, nearly complete control is often achieved in 18–24 months (Chilton 2004).

17.5.7.5 Salvinia Weevils (*Cyrtobagous salviniae*)

The giant salvinia is an aquatic fern with floating leaves that form chains that run together to form thick mats on the surface of the water. The plant is native to South America and was first found in the United States in the 1990s. This plant has extremely rapid rates of growth, estimated under ideal conditions to double every 36–53 hours. The giant salvinia weevil (Figure 17.17), originally occurring in southeastern Brazil, Bolivia, Paraguay, and northern Argentina, was introduced to control this invasive species (Mudge and Harms 2012). These weevils have been very successful in other areas of the world in controlling giant salvinia, but have not proven comparably successful in the

FIGURE 17.15   Water hyacinth (a) (From USDA Agricultural Research Service Invasive Weeds Research Unit) and water hyacinth weevils (*Neochetina* sp.) (b,c). (b) Adults of *N. eichhorniae* (top right) and *N. bruchi* (top left) and an example of feeding damage produced by both adults. (c) Grub-like *Neochetina* larvae and damage within the plant crown (top right) and within the leaf petioles (bottom). (b,c: From Grodowitz, M.J., Freedman, J.E., Jones, H., Jeffers, L., Lopez, C.F., and Nibling, F., Status of waterhyacinth/hydrilla infestations and associated biological control agents in Lower Rio Grande Valley cooperating irrigation districts, U.S. Army Corps of Engineers, Vicksburg, MS, 2000.)

United States to date, and the conditions for their effectiveness are not totally understood (Chilton 2004).

17.6   FISH AND FISHERIES MANAGEMENT

Often, as long as the lake or reservoir is not "unsightly or odorous," the perception by the public as to the health of a lake or a reservoir is based to a large degree on its fishery, where fishing is one of the more common and widespread uses of lakes and reservoirs. Fisheries management is then a critical component of lake and reservoir management.

The overall goal of fisheries management is to "produce sustainable biological, social, and economic benefits from renewable aquatic resources" (Lackey 2005). Historically, the target of fisheries management was to maximize sustainable populations of some specific target fish of recreational or commercial value, such as largemouth bass, walleye, or others. Management goals may include (Baker et al. 1993):

- Maximizing yield
- Maximizing catch rates per unit of effort

FIGURE 17.16 (a) Water lettuce weevil. (*Neohydronomus affinis*) (Courtesy of U.S. Army Corps of Engineers Aquatic Plant Control Research Program.) (b) Water lettuce plants. (Photo from Forest and Kim Starr, U.S. Geological Survey, bugwood.org.)

- Maximizing "bites per hour"
- Maximizing the number and size of trophy fish
- Providing fishing opportunities for a particular fish species or strain that is unique and desirable
- Providing fishing opportunities for "wild" fish
- Maximizing the diversity of fishing opportunities
- Providing a relatively pristine, natural, and undisturbed environment for fishing
- Maximizing the ease and convenience of fishing
- Providing edible fish, with little or no health risk from bioaccumulated toxic substances

The maximization of sustainable yields of specific species is based on specific habitat requirements, reproductive patterns and requirements, and feeding habits and requirements. Baker et al. (1993) discuss each of these as well as management concerns for a number of target fish species (including largemouth, smallmouth, and spotted bass; sunfish and crappie; striped and white bass; bullheads and catfish; walleye and yellow perch; pike, pickerel, and muskellunge; trout and salmon; and threadfin shad, gizzard shad, and alewife).

As described by Lackey (2005), to support that "fish" it is necessary to manage the "fishery," which includes the management of the aquatic biota, the aquatic habitat, and the human users of these renewable resources, each of which influences how the fishery performs. So, managing sediments, algae, and aquatic plants, discussed previously, is also critical to fisheries management, since sedimentation can impact habitats and algae and aquatic plants prove the base of the food chain. In addition, aquatic plants provide habitat and cover (such as escape cover). Currently, fish management is based on species and habitat protection (Lackey 2005), based on the principles of lake ecology as well as fish biology and population dynamics (Baker et al. 1993).

Fisheries management strategies may differ between large lakes and reservoirs. For example, reservoirs may provide more opportunities for management, since they have outlet controls that may

Giant salvinia weevil (*Cyrtobagous salviniae*)
on giant salvinia

Tertiary growth stage of giant salvinia

Dense infestation of giant salvinia in a
Louisiana lake

FIGURE 17.17　Giant salvinia and giant salvinia weevil. (From Mudge, C.R. and Harms, N.E., Development of an integrated pest management approach for controlling giant salvinia using herbicides and insects, U.S. Army Corps of Engineers Aquatic Plant Control Research Program, Vicksburg, MS, 2012.)

allow, for example, water level management. There are often more constraints as well, since large reservoirs may be operated for multiple purposes (water supply, hydropower, flood control, etc.) that may impose operations (such as spring drawdown) not optimal to fisheries management. Another management strategy in many reservoirs, since they are not "natural" systems, is to introduce and maintain yields of nonnative fish species to provide enhanced recreational or commercial opportunities. A specific concern for reservoirs is their impact on and management of diadromous fisheries, where, for example, dams provide barriers to migration.

REFERENCES

Baker, J.P., H. Olem, C.S. Creager, M.D. Marcus, and B.R. Parkhurst. 1993. Fish and fisheries management in lakes and reservoirs, Technical Supplement to Lake and Reservoir Restoration Guidance Manual, EPA-841-R-93-002. Terrene Institute and U.S. Environmental Protection Agency, Washington, DC.

Beutel, M.W. and A.J. Horne. 1999. A review of effects of hypolimnetic oxygenation on lake and reservoir water quality. *Lake and Reservoir Management* 15 (4), 285–297.

Buckingham, G.R. 2002. Alligatorweed. In F.V. Driesche, B. Blossey, M. Hoodle, S. Lyon, and R. Reardon (eds), *Biological Control of Invasive Plants in the Eastern United States*. United States Department of Agriculture Forest Service. Forest Health Technology Enterprise Team, Morgantown, WV, pp. 5–15.

Caird, J.M. 1945. Algae growth greatly reduced after stocking pond with fish. *Water Works Engineering* 98, 240.

Center, T.D., D.L. Sutton, V.A. Ramey, and K.A. Langeland. 1998. Other methods of aquatic plant management. In K.A. Langeland (ed.), *Training Manual for Aquatic Herbicide Applicators*. University of Florida, Gainesville, FL, pp. 63–70.

Chapra, S.C. 1997. *Surface Water-Quality Modeling*. McGraw-Hill, New York.

Chilton, E.W. 2004. Aquatic vegetation management in Texas: A guidance document, PWD PL T3200-1066. Texas Parks and Wildlife, Inland Fisheries, San Marcos, TX.

Cooke, G.D., E.B. Welch, S.A. Peterson, and S.A. Nichols. 2005. *Restoration and Management of Lakes and Reservoirs*, 3rd ed. CRC Press, New York.

Crowder, B. 1987. Economic costs of reservoir sedimentation: A regional approach to estimating cropland erosion damages. *Journal of Soil and Water Conservation* 42, 3.

ERDC. 2005. Bottom barriers. USACE Environmental Research and Development Center Noxious and Nuisance Plant Management Information System, available at: http://el.erdc.usace.army.mil/pmis/MechanicalControls/MechanicalControlInfo.aspx?mechID=2.

ERDC. n.d. Herbicide overview. Use suggestions for US Environmental Protection Agency-approved aquatic herbicides. U.S. Army Corps of Engineers, Environmental Research and Development Center, Aquatic Plant Information System, APIS. Available at http://el.erdc.usace.army.mil/apis/HerbicideInformation/HerbicideInfoMain.aspx.

Fast, A.W., W.J. Overholtz, and R.A. Tubb. 1975. Hypolimnetic oxygenation using liquid oxygen. *Water Resources Research* 11 (2), 294–299.

Ferrari, R.L. and K. Collins. 2006. Reservoir survey and data analysis, Chapter 9, *Erosion and Sedimentation Manual*. U.S. Department of the Interior Bureau of Reclamation, Sedimentation and River Hydraulics Group, Denver, CO.

Garbrecht, J.D. and G.K.N. Garbrecht. 2004. Siltation behind dams in antiquity. *ASCE Conference Proceedings* 140, 6.

Getsinger, K.D., A.G. Poovey, L. Glomski, J.G. Slade, and R.J. Richardson. 2011. Utilization of herbicide concentration/exposure time relationships for controlling submersed invasive plants on Lake Gaston, Virginia/North Carolina, ERDC/EL TR-11-5. U.S. Army Corps of Engineers, Aquatic Plant Control Research Program, Vicksburg, MS.

Green, W.R. and H.E. Westerdahl. 1990. Response of Eurasian watermilfoil to 2,4-D concentrations and exposure times. *Journal of Aquatic Plant Management* 28, 27–32.

Grodowitz, M.J., J.E. Freedman, H. Jones, L. Jeffers, C.F. Lopez, and F. Nibling. 2000. Status of waterhyacinth/hydrilla infestations and associated biological control agents in Lower Rio Grande Valley cooperating irrigation districts, ERDC/EL SR-00-11. U.S. Army Corps of Engineers, Aquatic Plant Control Research Program, Vicksburg, MS.

Hanson, M.J. and H.G. Stefan. 1984. Side effects of 58 years of copper sulfate treatment in the Fairmont Lakes, Minnesota. *Water Resources Bulletin* 20, 889–900.

Holdren, C., W. Jones, and J. Taggart. 2001. Managing lakes and reservoirs. North American Lake Management Society and Terrene Institute, in cooperation with the Office of Water Assessment and Watershed Protection, U.S. Environmental Protection Agency, Madison, WI.

Horne, A.J. and C.R. Goldman. 1994. *Limnology*, 2nd ed. McGraw Hill, New York.

Jacksonville District. n.d. Invasive species management. Available at: http://www.saj.usace.army.mil/Missions/Environmental/InvasiveSpecies/Management.aspx.

Lackey, R.T. 2005. Fisheries: History, science, and management. In H.L. Jay and J. Keeley (eds), *Water Encyclopedia: Surface and Agricultural Water*. John Wiley and Sons, New York, pp. 121–129.

Lueschow, L.A. 1972. Biology and control of selected aquatic nuisances in recreational water. Technical Bulletin No. 57, Wisconsin Department of Natural Resources, Madison, WI.

McComas, S. 2003. *Lake and Pond Management Guidebook*. Lewis Publishers, Boca Raton, FL.

Morris, G.L. and J. Fan. 1998. *Reservoir Sedimentation Handbook*. McGraw-Hill, New York.

Morris, G.L., G. Annandale, and R. Hotchkiss. 2008. Reservoir sedimentation. In M. Garcia (ed.), *Sedimentation Engineering: Processes, Measurements, Modeling, and Practice*. American Society of Civil Engineers, Reston, VA, pp. 579–612.

Mudge, C.R. and N.E. Harms. 2012. Development of an integrated pest management approach for controlling giant salvinia using herbicides and insects, Volume A-12-1. U.S. Army Corps of Engineers Aquatic Plant Control Research Program, Vicksburg, MS.

NALMS. 1990. *The Lakes and Reservoir Restoration Guidance Manual*, 2nd ed. EPA-440/4-90-006. North American Lake Management Society, U.S. Environmental Protection Agency, Washington, DC.

Netherland, M.D. 1992. Herbicide concentration/exposure time relationships for Eurasian watermilfoil and hydrilla. Proceedings, 26th Annual Meeting, Aquatic Plant Control Research Program. Miscellaneous Paper A-92-2, U.S. Army Engineer Waterways Experiment Station, Vicksburg, MS, pp. 79–85.

Netherland, M.D. and K.D. Getsinger. 1992. Efficacy of triclopyr on Eurasian watermilfoil: Concentration and exposure time effects. *Journal of Aquatic Plant Management* 30, 1–5.

Netherland, M.D., W.R. Green, and K.D. Getsinger. 1991. Endothall concentration and exposure time relationships for the control of Eurasian watermilfoil and hydrilla. *Journal of Aquatic Plant Management* 29, 61–67.

Sharpio, J. 1979. The need for more biology in lake restoration. In *Lake Restoration*, U.S. Environmental Protection Agency National Conference on Lake Restoration. U.S. EPA, Washington, DC, 440/5-79-001, pp. 161–167.

U.S. EPA. 2008. CWNS 2008 report to Congress. U.S. Environmental Protection Agency, Office of Water Management, Washington, DC.

Welch, E.B. 1981. The dilution/flushing technique in lake restoration, EPA-600/3-81-016. U.S. Environmental Protection Agency, Washington, DC.

Westerdahl, H.E. 1987. Herbicide concentration/exposure time relationships. Proceedings, 21st Annual Meeting, Aquatic Plant Control Research Program. Miscellaneous Paper A-89-1, U.S. Army Engineer Waterways Experiment Station, Vicksburg, MS, pp. 169–172.

White, R. 2001. *Evacuation of Sediments from Reservoirs*. Thomas Telford, London.

18 Dam Tailwaters

18.1 INTRODUCTION AND ISSUES

Tailwaters (Figure 18.1) begin immediately downstream of some hydraulic structure, such as a dam. As such, the spatial extent of the tailwater is easy to define, at least at its beginning. The conveyance areas (channel and structures) below the dam are often called the tailrace. The downstream extent of the tailwaters is more problematic, and is taken as that zone influenced by the structure and the outflows from that structure. That zone can reach far downstream of the dam.

Since tailwaters are directly impacted by the upstream structure, they may be considered a unique type of environmental system (WES 1998) with physical, chemical, and biological characteristics directly impacted by the operation of the dam. The characteristics of the tailwater are largely influenced by the dam, the associated hydraulic structures, and how much, when, and from where water is released from. By their nature and operation, dams often modify or eliminate the natural flow hydrograph in the tailwater and downstream river reach.

The tailwater inflow hydrograph (the outflow hydrograph from the dam) varies with the operation and design purpose of the dam. Operational purposes include hydropower, water supply, flood storage, and other uses, and, most commonly, large reservoirs are operated for multiple uses.

The purpose of a flood storage facility is to attenuate the flood inflow to the reservoir, so that the inflow hydrograph to the tailwater would have a lower peak but a longer duration than the reservoir inflow hydrograph. An example of releases from a flood storage reservoir is illustrated in Figure 18.2. The timing and flows of the releases would depend on the timing and the magnitude of the flood events, which would vary between locations. The impact for the tailwater and the downstream river is to (Richter and Thomas 2007):

- Eliminate small floods, that is, 2- to 10-year recurrence interval floods.
- Eliminate all but the most extreme large floods (>50 year).
- Introduce artificially long, high-flow pulses following flood peaks.

For a storage reservoir, the tailwater inflow hydrograph may be relatively constant. One impact of these facilities, particularly those with a large storage capacity, is to rearrange the seasonal patterns of water flows. In storage facilities, wet-season flows are stored for release in the dry season to support water supply users. When demands are low, flows may be limited or eliminated. When demands are high, release flows may be unnaturally high (Richter and Thomas 2007).

Run-of-the-river hydropower facilities may have only a limited impact on natural variations in river discharge. For a peaking hydropower facility, the inflow may be limited to seepage during nongeneration and then rapidly increased to some peak during periods of generation, as illustrated by Figure 18.2.

The inflows to the tailwaters may also be from and over spillways, from mid-level or bottom release structures, and through turbines or a variety of types of conduits (see Chapter 10). The location and the magnitude of the flows, and the characteristics of the reservoir at the level(s) from which the water was withdrawn, impact not only the hydraulic characteristics of the flow but also the flow water quality. For example, as previously discussed, the bottom waters of many reservoirs are hypoxic during much of the year, so that bottom releases are low in dissolved oxygen (DO), often

FIGURE 18.1   Center Hill Tailwater, TN. (Courtesy of U.S. Army Corps of Engineers.)

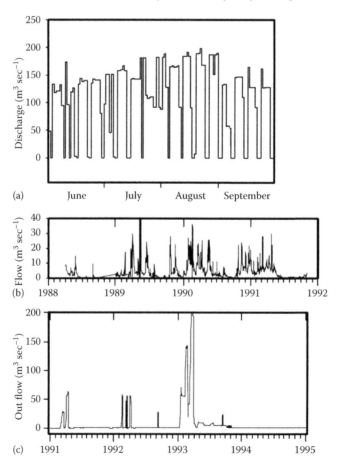

FIGURE 18.2   Tailwater inflow hydrographs from a peaking hydropower operation (a), a flood control project in New York (b), and a flood control project in Arizona (c). (From WES, *The Handbook on Water Quality Enhancement Techniques for Reservoirs and Tailwaters*, U.S. Army Corps of Engineers, Waterways Experiment Station, Vicksburg, 1998.)

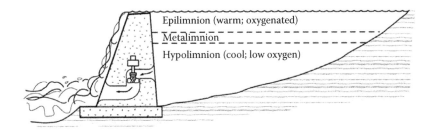

FIGURE 18.3 Illustration of hydropower release. (From Peterson, M.J., Cada, G.F., Sale, M.J., and Eddlemon, G.K., Regulatory approaches for addressing dissolved oxygen concerns at hydropower facilities, U.S. Department of Energy, Idaho Falls, ID, 2003.)

high in nutrients (e.g., ammonia and dissolved phosphorus), and high in reduced materials (reduced iron and manganese, sulfides, etc.; Figure 18.3).

There are a variety of other issues related to tailwater systems, including:

- Release magnitude and designated uses (minimum flows/maximum flows, pulsed flows)
- Erosion, scour, and bank sloughing
- Gas bubble disease
- Entrainment and impingement (e.g., from pumpback operations)
- Release temperature and water quality
- Habitat loss for existing or desirable species
- Changes from cold-water to warmwater species (or inversely, changes from warmwater to cold-water species)
- Blockage of fish passage, or loss of spawning or necessary habitats

Many of the same characteristics of tailwaters that may be judged from some perspectives as undesirable often make them popular and productive from other perspectives. For example, the dam may provide a barrier to migrating fish, and may also provide conditions unfavorable to the predam river biota. On the other hand, dam tailwater often provides very productive managed fisheries. For example, the high nutrient and organic concentrations of reservoir releases often result in high productivity in the reservoir tailwaters. Also, due to the presence and operation of dams, tailwaters often have less seasonally variable conditions than natural streams, so there is often more food available for longer periods of the year than there is in natural systems. As a result, tailwaters are often some of the most popular recreational fisheries available. Since tailwaters are not considered natural systems, often the fisheries may include introduced rather than, or in addition to, native species. Examples include popular hybrid striped bass tailwater fisheries. In some areas, bottom releases of cold hypolimnetic waters change the tailwater areas from a warmwater system (such as prior to the construction of the dam) to a cold-water environment (such as with bottom releases), allowing for highly popular and productive trout fisheries below many dams in the south, where such a fishery would not be possible without the dam. The presence of these fisheries has resulted in a highly successful and productive recreational industry below many dams.

18.2 DAM RELEASES AND IMPACTS

18.2.1 Release Magnitude and Designated Uses

One of the major influences on the physical, chemical, and biological characteristics of tailwaters is the magnitude and the timing of releases from a dam. These releases may be highly variable, as illustrated by Figure 18.2.

As discussed in Chapter 10, the operation of a large dam is often dictated by the purpose for which it was constructed, as specified in the reservoir operations manual developed when the reservoir was constructed and in the reservoir rule curve, which provides target lake level elevations. In many cases, these operational rules and rule curves were developed without considerations to the tailwaters. As such, the reservoir may be precluded from operating in a manner consistent with the optimal management of the tailwater. An example is reservoir releases. If the reservoir is under drought management, there may be no allowance for minimum releases, an operation that may be catastrophic to the downstream fishery.

18.2.2 Erosion and Scour

One of the issues below dams, particularly for peaking hydropower operations, is the erosion scour that results from the high, and highly variable, flows and from the dam itself. The presence of the dam alters the natural sediment transport characteristics of the predammed river by acting as a sediment trap and reducing sediment loads. Sediment-starved rivers are prone to channel bed and bank erosion, channel incision, coarsened bed material, and reduced habitat heterogeneity (Collins and Dunne 1989; Kondolf 1997). Within a short distance from the dam, increased erosion and bank instability contribute sediments that often block or limit the use of preferred habitat and spawning areas downstream (Gore et al. 1990).

18.2.3 Gas Bubble Disease

Gas bubble disease occurs when fish are exposed to highly or supersaturated gases in water. As discussed in Chapter 14, gas saturation occurs when equilibrium conditions exist between the gas phase and the liquid phase, such as with water and the overlying atmosphere. The solubility, or the liquid concentration at which the liquid is saturated with the gas, varies with the gas characteristics and the gas partial pressure as well as with the temperature, as illustrated in Figure 18.4 for oxygen and nitrogen at one atmosphere of pressure. Also illustrated is the vapor pressure of water, or the

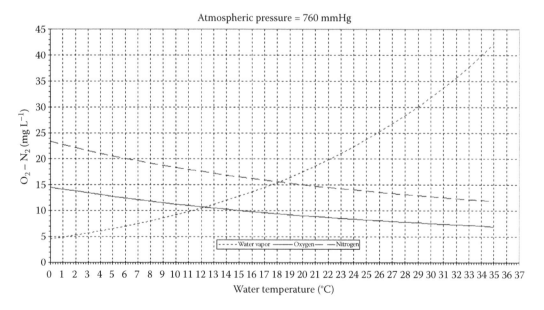

FIGURE 18.4 Solubility curves for oxygen and nitrogen and water vapor pressure at one atmosphere pressure. (From Fidler, L.E. and Miller, S.B., British Columbia water quality guidelines for dissolved gas supersaturation, prepared by Aspen Applied Sciences Ltd. for BC Ministry of Environment, Canada Department of Fisheries and Oceans, Environment Canada, Canada, 1994.)

equilibrium pressure exerted by the water just above the water surface (when the air is saturated with water), as a function of temperature. At the water surface, the total gas pressure would be the sum of the atmospheric pressure (the sum of the partial pressure of all gases) and the water vapor pressure. The gas saturation is also strongly a function of pressure, so the greater the pressure is, the greater the saturation concentration will be. Conversely, lower pressures, such as would occur in the Rocky Mountains versus sea level, would result in lower saturation concentrations. In water, the hydrostatic pressure increases with depth such that gas supersaturation decreases (or solubility increases) by about 10% per meter of increase in water depth due to hydrostatic pressure.

Engineers are familiar with what happens with cavitation in pipe systems, which occurs when the pressure head in the pipe (e.g., due to an increase in elevation and head loss in valves) falls below the sum of the atmospheric and vapor pressures. Under reduced pressure, the saturation concentration of all gases or the total gas solubility drops and the liquid becomes supersaturated. As a result, the liquid degasses and bubbles form, which can cause excessive wear or damage to pipes and pipe parts. Similarly, the bends in human divers is also well known; it occurs when divers rise through the water column and as the pressure decreases, bubbles form in the body.

In fish exposed to changes in pressure (changes in gas saturation concentrations and gas release), bubbles may form in the gills, fins, eyes, and other body parts. The impacts can vary, but the bubbles can block the blood flow (form emboli or an embolism), causing overinflation of the swim bladder, subdermal emphysema on the body surfaces, bulging of the eye (exophthalma), and other impacts (Figures 18.5 and 18.6). The impacts can range from minor distress to death and can vary with the type of fish, life stage, exposure time, and other factors.

The U.S. water quality criteria guideline for the protection of aquatic life from gas bubble disease is that total dissolved gas (TDG) concentrations should not exceed 110% of the saturation value for gases at the existing atmospheric and hydrostatic pressures (USEPA 1976).

Any condition that may result in rapid changes in gas solubility may result in gas bubble disease. These include rapid changes in temperatures or pressures. The rapid heating of cold saturated waters such as in power plant or hydropower releases may also result in bubble release and gas bubble disease. For reservoir tailwaters, one common and problematic cause is due to the spill of waters. As water falls over a spillway, it entrains air, and when pressures increase by the impact and as the water plunges into a stilling basin, the TDG concentration increases. This is a transient condition, so that soon afterward, as pressures decrease, the total gas solubility decreases and the water will degas, releasing the excess gas into solution.

The areas below spillways attract fish, and this congregation of fish creates a problem for many reservoirs when they are exposed to excess TDG, particularly those with salmonids. For example,

FIGURE 18.5 Close-up of left eye of sockeye salmon afflicted with gas bubble disease. (From USEPA National Environmental Research Center; Photo by Gene Daniels, 1970.)

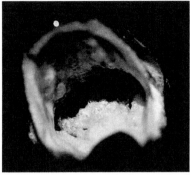

Subdermal emphysema in the mouth of
a rainbow trout

Subdermal emphysema on the head of
a rainbow trout

Severe exopthalmia in a juvenile
rainbow trout

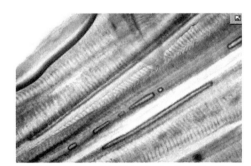

Intracorporeal bubbles in the lamella of
a rainbow trout

FIGURE 18.6 Examples of the impacts of gas bubble disease in rainbow trout. (From Fidler, L.E. and Miller, S.B., British Columbia water quality guidelines for dissolved gas supersaturation, Canada Department of Fisheries and Oceans, Environment Canada, Canada, 1994.)

the states of Oregon and Washington have both listed basically the entire lower Columbia River as being impaired due to TDG exceeding their state standards (McGrath et al. 2006). That is, the states have listed multiple reaches as being impaired due to TDG on their federal Clean Water Act 303(d). A TDG total maximum daily load (TMDL) was developed for the lower Columbia River in 2002.

The spill of water over the spillway can be voluntary or not. Uncontrolled releases may occur during high-flow events. Spillway release may also occur due to dam maintenance (such as the maintenance of turbines for hydropower facilities). In some cases, voluntary spills also occur, such as to facilitate the downstream movement of migratory fish, such as salmon.

A variety of methods have been developed to reduce or mitigate the impact of gas bubble disease in reservoir tailwaters. These include (Richardson and Baca 1974):

- Reducing the spill volume and modifying spillways to reduce the degree of air entrainment to stilling basin depths.
- Increasing the water flow through powerhouses, thereby reducing the volume passing over the spillway.
- Using spillway deflectors to reduce plunging.
- Transporting salmonids around the dams.

A variety of other alternatives have been developed, such as low water ports, spillway bypasses, and others, many of which are described in feasibility and abatement plans, such as in the "Alternatives for TDG abatement at Grand Coulee Dam, feasibility design report" (Frizell and Cohen 2000).

18.2.4 Entrainment and Impingement in Pumpback

Peaking hydropower facilities have the advantage that they can generate during peak demand periods. The energy that they generate is a function of the elevation head and the flows that they can pass through turbines. Therefore, the water in the reservoir can be viewed as stored energy.

Since the value of that energy varies with the demand, such as over the course of a day, the stored energy is essentially moved around to make the most effective use of it when the demand (hence economic value) is greatest, called load balancing. One load balancing strategy is to store releases in the reservoir tailwaters and then pump that water back into the reservoir when demands are low (pumpback). This allows for the "reuse" of that water (energy) during periods of peak demand.

The interest in pumpback operations, also referred to as pumped-hydro energy storage (PHES), is increasing in the United States and around the world, with an additional 76 GW PHES capacity worldwide expected by 2014. In the United States, the Federal Energy Regulatory Commission (FERC) has granted 32 preliminary permits as of April 5, 2010 (Ingram 2009; Yang and Jackson 2011), to 25 licensees who are interested in developing new PHES facilities.

One of the issues impacting tailwater storage and pumpback is the entrainment of fish during the pumpback operations. That is, when the turbines are reversed and water is pumped back into the upper reservoir, fish may be literally sucked into the turbine and killed by the blades. An example of an entrainment controversy impacting the operation of a pumpback facility is the Richard B. Russell (RBR) Reservoir between Georgia and South Carolina. The RBR Dam is a peaking hydropower facility constructed with four conventional generation units and four reversible pump-turbines, which began operation in 1985. The pump-turbines were planned for use in generating power and then to be reversed and used as pumps to move water from the downstream J. Strom Thurmond Reservoir back to the RBR Dam during periods of low power. The controversy was related to the estimated entrainment of fish, and subsequent fish kills, during pumpback. As a result, in 1988, the South Carolina Department of Wildlife and Marine Resources joined by the South Carolina, Georgia, and National Wildlife Federations, filed an injunction to stop the pumpback operations. It was not until 2002 that the injunction was lifted by the Federal District Court in Charleston, South Carolina, and the four reversible pump-turbines were brought on-line for commercial generation. This followed a series of studies and design alterations to mitigate or reduce fish mortality, including a sound repulsion system, bar screens at the pumping intakes, and other measures.

18.2.5 Release Water Quality

The water quality of tailwaters is largely influenced by the water quality characteristics of the upstream reservoir at the level (depth) from which the water is withdrawn. The depth from which water is released from a stratified reservoir can be considered to be one of the most important factors determining the composition and abundance of tailwater biota (Gore et al. 1990). Water quality parameters of concern are usually

- Temperature
- Particulate organic matter (POM)
- DO
- Metals (e.g., iron and manganese)
- Methane
- Sulfides
- Nutrients

The temperatures immediately below dams and the daily and seasonal variations in those temperatures are primarily determined by the depth of the withdrawal and variations in seasonal stratification. Those temperatures largely impact tailwater chemical and biotic characteristics. Reservoirs often vertically stratify during the summer months, and the release temperature is a function of the layers(s) from which the water is withdrawn. For example, a surface or bottom release may, if there is sufficient energy and a lack of stratification, result in water being withdrawn from the entire vertical profile. Alternatively, with less energy or a greater degree of stratification, the withdrawal may be confined to a relatively thin layer of the reservoir (either surface or bottom). Some reservoirs are also equipped with selective withdrawal structures, so they can select the layers from which water is withdrawn. These are all important in meeting downstream temperature objectives.

The temperature objectives will vary with the location and the fishery. For example, to maintain the tailwater trout fishery below Buford Dam in Georgia, a peaking hydropower facility, the Georgia target criteria are for temperatures below an instantaneous maximum of 22°C and below 20°C as a 5-day average. Representative daily cycles in release temperatures and oxygen concentrations from Buford Dam are illustrated in Figure 18.7. For the state of Oregon, the water-temperature standard for the lower Columbia River is a 7-day average maximum temperature of 20°C, while the state of Washington regulations state that the 1-day maximum should not exceed 20°C as a result of human activities.

The magnitude of the flows from a dam also influences the downstream extent of the dam's impact. For example, the existence of trout fisheries in the tailwaters of many southern reservoirs is dependent on the release of cold waters from the dam. The greater the magnitude of the releases, often the further downstream the waters would remain at within the suitable temperature envelope and support a viable fishery.

For mid-level or bottom releases, the concentrations of many water quality constituents are directly influenced by the seasonal progression of hypoxia in the hypolimnion of many reservoirs (Gore et al. 1990). Subsequently, concentrations are often much higher than would normally occur in natural streams.

18.3   REGULATORY ISSUES

18.3.1   License Conditions

One of the regulatory issues discussed in Chapter 10 is that reservoirs typically have an approved design purpose that is translated into specific operation rules. These rules may often leave little flexibility in the way that the reservoir can be operated for other purposes not included in the design purpose, such as may be the case for the optimal management of tailwaters and downstream water uses.

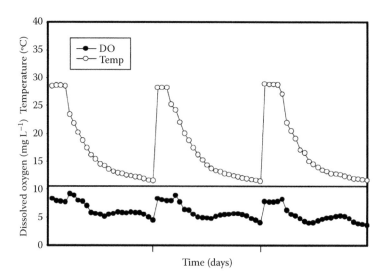

FIGURE 18.7   Representative daily cycles in release temperatures and oxygen concentrations from Buford Dam. (From WES, *The Handbook on Water Quality Enhancement Techniques for Reservoirs and Tailwaters*, U.S. Army Corps of Engineers, Waterways Experiment Station, Vicksburg, MS, 1998.)

Similarly, for hydropower facilities, a license is required from the FERC. As of July 2012, the FERC database includes over 1000 of such permits that have been issued, beginning with Escondido in 1924 to the Jennings Randolph project licensed in March 2012. Each of these is a 50-year license. A variety of regulatory issues are associated with these licenses, as discussed in Martin et al. (2007). Some of the license requirements include:

- 401 certification: Under Section 401 of the Clean Water Act, certification (by the state or an appropriate tribal authority) is a condition of any federal license that authorizes the construction or operation of a project. The certification is that the project will not result in a violation of water quality standards.
- Section 18 of the Federal Power Act (Chapter 12 of "Federal Regulation and Development of Power") provides the requirement that fishways may be required as prescribed by the secretary of the interior or the secretary of commerce.
- The Endangered Species Act requires federal agencies to ensure that their actions are not likely to jeopardize the continued existence of federally listed threatened and endangered species, or result in the destruction or adverse modification of their designated critical habitats.

Each of these requirements (and the results of environmental assessments or environmental impact studies) may result in the specification of certain conditions for the license, impacting, for example, construction, maintenance, and operation. As an example, some of the license requirements for the Franklin Hydroelectric Project on the Little Tennessee River (No. 2603 for Duke Energy Carolinas, LLC issued September 2011) are:

- To protect the aquatic resources in Lake Emory and the Little Tennessee River, the license requires Duke to: (1) operate the project in a run-of-river mode; (2) monitor project operations as described in its Lake Level and Flow Management Plan (Addendum No. 1 to the certification); and (3) revise its Sediment Management Plan (Addendum No. 3 to the certification) to include (a) provisions, when dredging, to minimize the impact of the project drawdown and sediment removal on environmental resources; (b) provisions for evaluating downstream effects on aquatic resources; (c) best management practices for sediment management; and (d) a schedule for implementation.
- To protect the aquatic resources in the Little Tennessee River, the license requires Duke to release a minimum flow of 309 cfs (the September median flow) into the project tailrace during reservoir maintenance or drawdown and refill periods.
- To enhance wood duck nesting at the project, the license requires Duke to install and maintain wood duck nesting boxes.
- For the continued protection and maintenance of Virginia spiraea and its habitat in the project area, the license requires Duke to prepare a Virginia spiraea management plan.

Many of the specified license requirements may be protective of fish, wildlife, cultural resources, and recreation. In some cases though, the license requirements may restrict, by statement or omission, the operation of the facility, limiting the capability of the hydropower operators to alter releases (such as, for example, to benefit downstream uses) because of license conditions (Peterson et al. 2003).

18.3.2 WATER QUALITY: POINT OR NONPOINT SOURCE?

One of the interesting questions for dams and dam releases is whether or not they should be considered point sources, and, as such, subject to permitting requirements under the National Pollutant Discharge Elimination System (NPDES permits) or for a "load allocation" as part of TMDL studies (under

Section 303(d) of the Clean Water Act). To help resolve this and other issues, the U.S. Environmental Protection Agency (U.S. EPA) assembled a federal advisory commission (FAC) to review and make recommendations to the agency. In their report on the TMDL program (USEPA 1998), the FAC identified existing large dams as an "extremely difficult problem" since they create a physical structure or physical modification that would be impossible or virtually impossible to remove. The FAC's recommendation was that the U.S. EPA require states to include waters impaired (not meeting water quality standards) wholly or partially by the existence of dams, or their operations, on their 303(d) list (Part 1, thereby requiring TMDL studies to quantify allowable loads to alleviate the impairment; Martin and Kennedy 2000). The U.S. EPA instead chose to categorize the water quality impairments due to dam releases as resulting from "pollution" rather than a pollutant, that is, solely resulting from hydromodification, and excluded the impact of dams for the portion of the list of impaired waters (Part 1) requiring a TMDL. That is, the U.S. EPA stated that it "does not believe that TMDLs should be the solution to problems substantially caused by hydromodification" (*Federal Register*, Vol. 64, No. 162). That categorization has been challenged, and the courts to date have agreed that, essentially, regulation as a point or a nonpoint source, requires the "addition" of a pollutant and that a dam does not result in the "discharge of a pollutant" as defined by the Clean Water Act, so that, for example, a reduction in oxygen due to a dam release does not constitute an "addition."

However, there have been a number of TMDL studies by states on tailwater quality, such as those related to low DO concentrations. In 2000, for example, the Georgia Environmental Protection Division (GAEPD) issued a multibasin TMDL report for "Total maximum daily load (TMDL) development for low dissolved oxygen below dams," in which they listed eight tailwaters as being impaired and argued that "since the impairment is caused by the low DO levels in the dams' release waters and is not due to a specific pollutant, the TMDL is equal to the dams' release waters meeting the appropriate DO criterion for the downstream waters' designated use." Other more recent examples include:

- TMDLs for dissolved oxygen for White River below Bull Shoals Dam and North Fork River below Norfork Dam (reaches 11010003-002U and 11010006-001); Arkansas Department of Environmental Quality (prepared by FTN Associates, Ltd.)
- TMDL for Lake Taneycomo in Taney County, Missouri; Missouri Department of Natural Resources (submitted November 15, 2010; approved December 30, 2010)

Other examples of water quality concerns for tailwaters were related to temperature and gas supersaturation (TDG and gas bubble disease). Like oxygen, it could easily be argued that gas supersaturation may not be considered due to the "addition" of a pollutant, but due to the physics associated with water spilling, such as over a spillway. However, TMDLs have been completed, which address TDG in the mainstream Columbia River from the Canadian border to its mouth at the Pacific Ocean in addition to specific locations (ODEQ and WADEQ 2002), such as below Lake Pend Oreille, Idaho. In those cases, the allocation was based on a change in concentration (or temperature "delta") rather than a specific loading.

It has been argued that regardless of whether the discharge from dams could be viewed as a point or a nonpoint source, resulting in the "addition" of pollutants, the water quality below dams is subject to meeting water quality standards and to a variety of regulatory controls. These controls have and will continue to impact the design and operation of dams.

18.3.3 RELEASE FLOWS

Dams by their design alter the flow regime in the dam tailwaters. Changes may be pulses in flows, such as due to hydropower facilities, or, in some cases, the near absence of release flows. That is, in some cases, the releases from dams may be primarily limited to seepage through the dam rather than intentional releases. That, of course, may have dramatic impacts on the chemical and biological characteristics of the tailwaters.

The flow releases from dams are often determined by the project purpose and the license requirements. Flows are presently not subject to the Clean Water Act or water quality standards, although they directly impact quality and can be used to mitigate water quality concerns. Recall that the goal of the Clean Water Act included "physical" as well as chemical and biological integrity.

In the absence of the consideration of the downstream impacts, the flows or the absence of flows can have a devastating impact on the tailwater biota, and the related economy. As indicated in Section 18.3.1, there may be license conditions that impose minimum flows. But, where there are no such impositions, the lack of a specification to maintain a release may impact the ability to later include a release in the licensed operation, and the licenses are in effect for periods of 50 years.

One illustrative example of impacts and what may often be required to implement minimum flows is the White River Basin in Arkansas and Missouri, consisting of five U.S. Army Corps of Engineers (the Corps) lakes operated primarily for hydropower and flood control (USACE 2008; Figure 18.8) and constructed during the period of 1940–1970. The reservoirs altered the tailwaters from a warmwater to a cold-water system. Trout were introduced into the tailwaters, resulting in what is today a world-class trout fisheries and a major recreational industry below the dams. The fisheries extend from 5 mi. below Norfolk Lake (to its confluence with the Bull Shoals tailwater) to 89 mi. of the White River below Bull Shoals (USACE 2008). The dams are operated by the Southwestern Power Administration (SWPA).

A reasonably typical release schedule and water quality profile for Bull Shoals Dam is illustrated in Figure 18.9. This case demonstrates a metalimnetic oxygen maxima and low DO conditions near the bottom. The reservoir hypolimnion commonly becomes hypoxic during the summer months (see DO TMDL in Section 18.3.2). Also demonstrated are the relatively cool hypolimnetic temperatures and the daily variation in turbine discharges.

Previous authorizations for the five projects contained no provisions for minimum releases. In the absence of power generation, flows are restricted to turbine and dam leakages (about 210 cfs for Bull Shoals and 75 cfs for Norfork; USACE 2008). Of course, without flows, the downstream

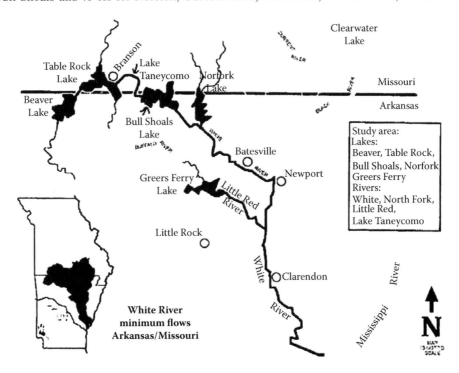

FIGURE 18.8 Projects in the White River Basin, Arkansas and Missouri. (From USACE, White River Basin, Arkansas, minimum flows project report, U.S. Army Corps of Engineers, Little Rock, AR, 2008.)

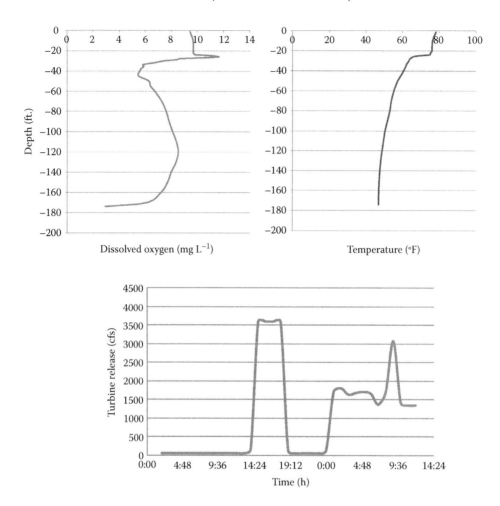

FIGURE 18.9 Temperature and oxygen vertical profiles near the dam and turbine releases for Bull Shoals Lake for July 11–12, 2012. (Data from USACOE Little Rock District Water Management online project data reports, Available at http://www.swl-wc.usace.army.mil/.)

fisheries would not exist. So, in times of water shortages, or drought conditions, the lack of water to support the fisheries became a major problem. The USACE (2008) indicated that worst-case conditions were 3-day weekends during the summer months with minimum hydropower generation, when the pools in the river became isolated by shoals and prevented trout from seeking refuge. But, with the absence of a designated project purpose, the reservoir could/would not be operated to support and protect the downstream fishery.

Decades of concern and negotiation led to authorized studies (in Section 375 of the 1999 Water Resources Development Act), including environmental impact studies under the National Environmental Policy Act (NEPA), and then authorization (in the 2006 Energy and Water Development Appropriations Act; Public Law 109–103; USACE 2008) that led to the establishment of reallocations of water storage for minimum flows. For example, under the authorized plan for the Bull Shoals project, 5 ft. of storage for minimum flows was to be reallocated from the flood control pool with provisions to provide a portion of the reallocated storage for the hydropower's use to maintain the yield of the current hydropower storage. In addition, for Bull Shoals, additional releases to maintain the minimum flow were to be implemented by generating power with one of the main units at a low rate.

Since the change in storage impacted the economics of power generation, the 2006 act required a determination of compensation necessary for losses of income from hydropower. SPAW (2009) estimated a federal hydropower loss of $109,920,200, and a loss to the nonfederal project of $41,319,400. Those losses were to be mitigated, in part, by a relief from the federal project debt and compensation for FERC licensing (USACE 2008). This example illustrates some of the complexities and costs that may be associated with the reauthorization of projects.

18.4 METHODS TO IMPROVE TAILWATER QUALITY

18.4.1 RELEASE MANAGEMENT

Release management could involve the timing, magnitude, and duration of releases. The target could be either to meet some tailwater flow or water quality objective or both.

Particularly during periods of stratification and hypolimnetic DO depletion, the magnitude and location of the withdrawals can have large impacts on the release temperatures and water quality. Some reservoirs have the capability of withdrawing from multiple levels (see selective withdrawal, Chapter 10) so that they can manage the zone from which water is withdrawn from the reservoir in order to manage the water quality of those releases (e.g., as a function of density and quality vertical stratification). Changes in operations such as start-up procedures for turbines can also influence the vertical zones of withdrawal from the reservoir and manage water quality.

The timing and duration of the release flow also directly impact tailwaters. Minimum flows, in addition to other benefits, may be used to flush waters from the tailwaters, preventing stagnation and the resulting oxygen problems. Minimum flows may have many other benefits for tailwater fisheries and downstream users. Flow pulsing, separate from hydropower or flood release, is also a tool used to manage tailwater quality and biota, such as flushing streambeds, scouring vegetation, etc. Spillage over structures may also be used to increase oxygen concentrations. The spillages, however, may also increase problems associated with TDGs and reduce the hydropower capacity.

18.4.1.1 Aerating, Venting Turbines

Turbine venting is a method that is used to increase the oxygen concentration of reservoir releases from hydropower facilities. Turbine venting introduces or injects oxygen into the flow as it passes through the turbines and is often a relatively economical alternative. The forced injection of oxygen into turbines has been used to increase the oxygen concentrations of reservoir releases. Perhaps a more common practice is autoventing. In autoventing, areas of low pressure are used to draw or force oxygen into the water. Hubs or baffles are often used to reduce the pressure, thereby increasing the aspiration of oxygen, as illustrated in Figure 18.10 for Norris Dam, Tennessee (including graphical comparisons indicating improvements in tailwater oxygen concentrations). However, techniques such as autoventing turbines are very site-specific and outcomes will vary considerably (USEPA 2006).

18.4.1.2 Gated Conduits

Gated conduits are also used to increase the release of oxygen concentrations. Gated conduits are hydraulic structures that divert the flow of water under the dam and are designed to create turbulent mixing to enhance oxygen transfer (USEPA 2006). A configuration is illustrated in Figure 18.11, where a structure such as a sluice gate is used to create high downstream velocities with reduced pressures, which are then vented to increase the oxygen concentrations and downstream pressures (e.g., to prevent cavitation).

18.4.1.3 Oxygen or Air Injection Systems

These systems introduce air or oxygen into the upstream reservoir to improve the oxygen concentrations of releases. They include surface-water pumps, like big fans, that push or pump oxygen-rich air into the withdrawal zone or inject oxygen directly. Oxygen injection systems usually take liquid oxygen, convert it into a gas, and pump it through diffusers to the reservoir hypolimnion or directly into the withdrawal waters prior to them entering the reservoir intake structure.

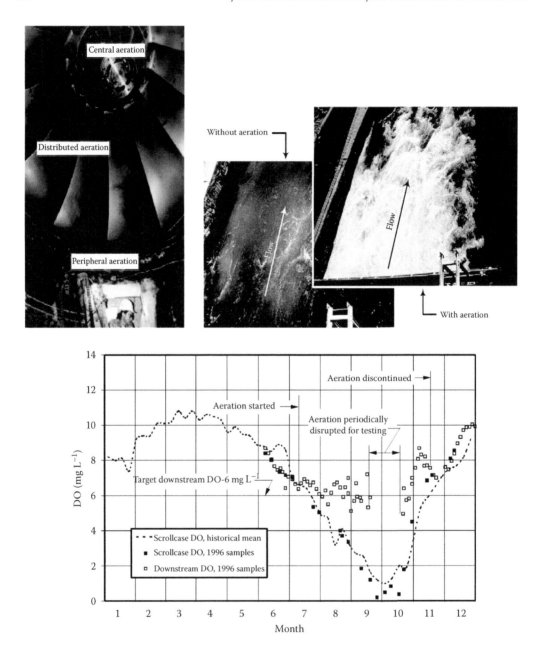

FIGURE 18.10 Autoventing turbine for Norris Dam, Tennessee Valley Authority, and an illustration of a bubble plume and DO concentrations with and without reaeration. (From Hopping, P., Patrick, M., Thomas, B., and Joseph, C., Update on development of auto-venting turbine technology, Tennessee Valley Authority, Norris, TN, nd.)

18.4.1.4   Weirs

Small dams and weirs may be built below the main dam of some reservoirs to partially mitigate their impact. One example is a regulation dam, which can be used to create a pool below the reservoir. Regulation pools are often used to smooth out or attenuate the dam flows, such as the pulses from peaking hydropower facilities. Downstream weirs are also used for reaerating the releases, such as using aerating weirs. Aerating weirs are passive systems that typically induce turbulent mixing and increased reaeration using waterfalls. Here, the spillage and pressure difference is usually not enough

FIGURE 18.11 Illustration of a high-pressure gated conduit.

to cause TDG problems. Examples of aerating weirs include linear, labyrinth, and infuser weirs. A labyrinth weir is shaped like a series of "Ws" in the channel. The infuser weir may have an equivalent waterfall length to the labyrinth weir using a slotted decking or grates to create a series of waterfalls, but it is more compact (Hauser and Brock nd). Examples of the three weir types are illustrated in Figure 18.12, and a photograph of a labyrinth weir on the Brazos River, Texas is shown in Figure 18.13.

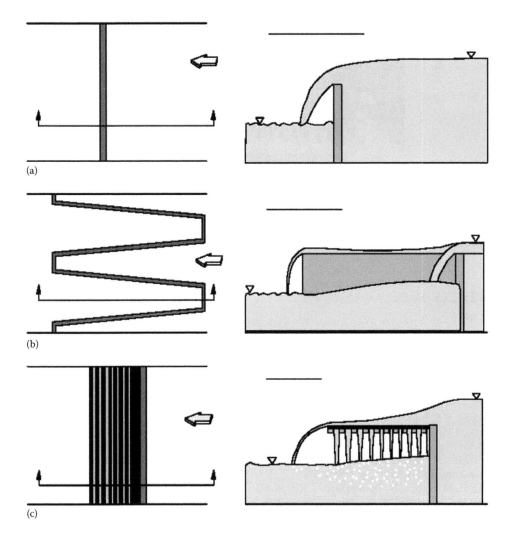

FIGURE 18.12 Comparison of conventional linear (a), labyrinth (b), and diffusing weirs (c) used for aeration. (From WES, *The Handbook on Water Quality Enhancement Techniques for Reservoirs and Tailwaters*, U.S. Army Corps of Engineers, Waterways Experiment Station, Vicksburg, MS, 1998.)

FIGURE 18.13   Labyrinth weir on Brazos River. (From Waco Texas Water Services.)

REFERENCES

Collins, B.D. and T. Dunne. 1989. Gravel transport, gravel harvesting, and channel-bed degradation in rivers draining the southern Olympic mountains, Washington, USA. *Environmental Geology and Water Sciences* 13, 213–224.

Fidler, L.E. and S.B. Miller. 1994. British Columbia water quality guidelines for dissolved gas supersaturation. Prepared by Aspen Applied Sciences Ltd. for BC Ministry of Environment, Canada Department of Fisheries and Oceans, Environment Canada, Canada.

Frizell, K.H. and E. Cohen. 2000. Alternatives for TDG abatement at Grand Coulee Dam, feasibility design report. U.S. Bureau of Reclamation, Pacific Northwest Region, Denver, CO.

Gore, J.A., J.M. Nestler, and J.B. Layzer. 1990. Habitat factors in tailwaters with emphasis on peaking hydropower, Technical Report EL-90-2. U.S. Army Engineer Waterways Experiment Station, Vicksburg, MS.

Hauser, G.E. and W.G. Brock. nd. The aerating infuser for increasing dissolved oxygen concentrations. Tennessee Valley Authority, Norris, TN.

Hopping, P., M. Patrick, B. Thomas, and C. Joseph. nd. Update on development of auto-venting turbine technology. Tennessee Valley Authority, Norris, TN.

Ingram, E. 2009. Pumped storage development activity snapshots. *Hydro Review* 17 (6), 12–25.

Kondolf, G.M. 1997. Hungry water: Effects of dams and gravel mining on river channels. *Environmental Management* 21, 533–551.

Martin, J.L. and R.H. Kennedy. 2000. Total maximum daily loads (TMDLs): A perspective, ERDC WQTN-MI-071. U.S. Army Engineering Research and Development Center, Vicksburg, MS.

Martin, J.L., H. Craig, and A.B. Jeffery. 2007. Energy production, reservoir and river water quality: Regulatory framework. In J.L. Martin, J. Edinger, J.A. Gordon, and J. Higgins (eds), *Energy Production and Reservoir Water Quality*. American Society of Civil Engineers, Reston, VA. pp. 2-1–2-30.

McGrath, K.E., E.M. Dawley, and D.R. Geist. 2006. Total dissolved gas effects on fishes of the Lower Columbia River, PNNL-15525. Battelle Pacific Northwest National Laboratory, Richland, VA.

ODEQ and WADEQ. 2002. Total maximum daily load (TMDL) for Lower Columbia River total dissolved gas. Prepared jointly by the Oregon Department of Environmental Quality and the Washington State Department of Ecology.

Peterson, M.J., G.F. Cada, M.J. Sale, and G.K. Eddlemon. 2003. Regulatory approaches for addressing dissolved oxygen concerns at hydropower facilities, DOE/ID-11071. U.S. Department of Energy, Idaho Falls, ID.

Richardson, G.C. and R. Baca. 1974. Round table discussion: Physics of dissolved gases and engineering solutions. In D.H. Fickeisen and M.J. Schneider (eds), *Gas Bubble Disease*, proceedings of a workshop held at Richland, Washington.

Richter, B.D. and G.A. Thomas. 2007. Restoring environmental flows by modifying dam operations. *Ecology and Society* 12 (1), 12.

SPAW. 2009. Final determination, White River minimum flows study determination of offset to the federal hydropower purpose and impacts on non-federal project. Southwestern Power Administration.

USACE. 2008. White River Basin, Arkansas, minimum flows project report. U.S. Army Corps of Engineers, Little Rock, AR.

USEPA. 1976. Quality criteria for water, EPA 440-9-76-023. U.S. Environmental Protection Agency, Washington, DC.

USEPA. 1998. Report of the federal advisory committee on the total maximum daily load (TMDL) program, EPA 100-R-98-006. National Advisory Council for Environmental Policy and Technology.

USEPA. 2006. National management measures to control nonpoint source pollution from hydromodification. Office of Wetlands, Oceans and Watersheds, U.S. Environmental Protection Agency, Office of Water, Washington, DC.

WES. 1998. *The Handbook on Water Quality Enhancement Techniques for Reservoirs and Tailwaters*. U.S. Army Corps of Engineers, Waterways Experiment Station, Vicksburg, MS.

Yang, C.-J. and R.B. Jackson. 2011. Opportunities and barriers to pumped-hydro energy storage in the United States. *Renewable and Sustainable Energy Reviews* 15, 839–844.

19 Freshwater Wetlands
An Introduction

"In the end, we will conserve only what we love, and we love only what we understand."

Baba Dioum, Senegal

19.1 INTRODUCTION

Wetlands (see e.g., Figure 19.1) are typically transitional systems where land meets water. They are defined by the following attributes:

- Hydrology: wetlands are permanently or periodically flooded or saturated with water.
- Soils: wetlands have hydric soils that are frequently or periodically saturated with water.
- Vegetation: wetlands support hydrophytes or plant species that are adapted to wet environments.

As discussed in Chapter 16, wetlands could be considered an intermediate stage in the succession of lakes toward some climax condition. One classical view is that lakes are temporary ecological systems that are formed in a depression that will gradually fill in to become a terrestrial environment, or become extinct. One of the intermediate stages in that succession is a wetland. However, while wetlands may be transitional zones between aquatic and terrestrial environments, they may also stand alone, be permanent or temporary, natural or man-made systems of considerable ecological and hydrological importance. They are known to be valuable ecological systems that provide a wide variety of ecological services, ranging from flood control, to providing habitats, to nutrient control.

Note that wetlands are very diverse systems and, as such, they are difficult to characterize and identify. One of the problems for wetlands is in determining whether they are subject to protection by law, such as under the Clean Water Act (CWA). Therefore, unlike other chapters, much of this chapter will focus on the legal rather than the scientific definition of wetlands and how that definition has changed due to both the historical shifts in the nation's environmental ethics and the winds and vagaries of national policy and politics.

19.1.1 DEFINITION

First, what is a wetland? One of the issues associated with wetland management and protection is defining specifically what wetlands are and how to identify them. Definitions vary, often depending on the specific purpose of that definition, such as research, general habitat classification, natural resource inventories, and environmental regulations (Tiner 1999). For example, the definition used to allow the U.S. Fish and Wildlife Service to inventory the nation's wetlands, and to determine changes in that inventory over time, may vary with the definition used by regulatory agencies to protect those wetlands.

Definitions may also vary with the technical background of the scientist composing those definitions (Tiner 1999; Lefor and Kennard 1977). For example, a botanist's definition may emphasize plants while a soil scientist would focus on the soil properties and a hydrologist would focus on the water table.

FIGURE 19.1   Example of a wetland. (From the cover page of Dahl, T.E., Status and trends of wetlands in the conterminous United States 1986 to 1997. U.S. Fish and Wildlife Service, Washington, DC, 2000.)

Some of the available definitions by federal agencies are provided in Table 19.1. In many cases, individual states or other organizations also have specific definitions, which, in some cases, are in conflict.

19.2   WETLAND TYPES

One aspect of defining a wetland is to develop a wetlands classification system. Conversely, the classification of wetlands, as discussed later, emphasizes the difficulty in arriving at a general definition.

There are a variety of types and classification systems of wetlands. The wetland types or classifications often depend upon the specific purpose of that classification, such as for research or regulatory purposes or to delineate wetlands from other terrestrial or aquatic systems. The characteristics of wetlands are also a foundation for their protection and management. For example, bogs and fens are often characterized as difficult or impossible to replace, so they may receive greater protection than other wetland types.

19.2.1   BOGS AND FENS

Bogs are found on saturated, low-nutrient, acidic, peat soils that only support low shrubs, herbs, and a few tree species on a mat of sphagnum moss. While their productivity is low, bogs have substantial peat accumulation (>40 cm) and accumulations of partially decayed vegetation. Additionally, they have high water tables and contain acid-loving vegetation. Bogs are stagnant systems, with no

TABLE 19.1
Examples of Wetland Definitions

Source	Definition
USEPA Office of Wetlands, Oceans, and Watersheds	"Wetlands are areas where water covers the soil, or is present either at or near the surface of the soil all year or for varying periods of time during the year, including during the growing season."
U.S. Fish and Wildlife Service nonregulatory definition (Cowardin et al. 1979)	"Wetlands are lands transitional between terrestrial and an aquatic system where the water table is usually at or near the surface or the land is covered by shallow water." For the purposes of this classification, wetlands must have one or more of the following three attributes: (1) at least periodically, the land supports predominantly hydrophytes; (2) the substrate is predominantly undrained hydric soil; and (3) the substrate is nonsoil and is saturated with water or covered by shallow water at some time during the growing season of each year.
U.S. Army Corps of Engineers (33 CFR 328.3) U.S. Environmental Protection Agency (40 CFR 230.3)	"Wetlands are those areas that are inundated or saturated by surface or groundwater at a frequency and duration sufficient to support, and that under normal circumstances do support, a prevalence of vegetation typically adapted for life in saturated soil conditions. Wetlands generally include swamps, marshes, bogs, and similar areas."
U.S. Soil Conservation Service (National Food Security Act manual [SCS 1988]) (The act is commonly known as the Swampbuster)	"Wetlands are defined as areas that have a predominance of hydric soils and that are inundated or saturated by surface or ground water at a frequency and duration sufficient to support, and under normal circumstances do support, a prevalence of hydrophytic vegetation typically adapted for life in saturated soil conditions, except lands in Alaska identified as having high potential for agricultural development and a predominance of permafrost soils."

significant inflows or outflows of water so that most of their moisture is accumulated from rainfall. Similarly, fens are a type of peatland, but they are slightly more alkaline with a higher species diversity.

One plant commonly associated with bogs is the carnivorous pitcher plant (see Figure 19.2). While bogs are low in nutrients, they typically have ample light and moisture. The pitcher plant overcomes the low nutrients and low productivity of bogs by capturing insects in their leaves and digesting them in deep, slippery pools filled with digestive enzymes, with the aid of symbiotic bacteria. A variety of other carnivorous plants commonly occur in bogs and fens and as many as 13 species of carnivorous plants have been found in a single bog (Folkerts 1982).

One of the problems with bogs and fens is that they are difficult or impossible to replace. As will be discussed later, mitigation is commonly used to compensate for wetlands that have been damaged or destroyed. However, in "Compensatory Mitigation for Losses of Aquatic Resources (33 C.F.R. Part 332)," bogs, fens, springs, streams, and Atlantic white cedar swamps are identified as "difficult to replace resources" (Gardner 2011), strongly encouraging the avoidance of activities impacting these wetland types.

19.2.2 Nontidal (Inland or Freshwater) Marshes

Freshwater marshes are wetlands that are frequently or continually inundated with water, and are characterized by emergent soft-stemmed vegetation, as opposed to trees and shrubs that characterize swamps. Marshes constitute one of the most commonly occurring natural wetlands. Man-made marshes, constructed wetlands, are also becoming increasingly used as a best management practice (BMP) to control flooding or for nutrient removal (e.g., ASCE 1992). There is a wide variety of freshwater marshes, including playa lakes, prairie potholes, and wet meadows or prairies (Figure 19.3).

FIGURE 19.2 Northern bogs. (From USEPA, Learn the issues: Learn about water, Available at: http://www2.epa.gov/learn-issues/learn-about-water, 2013.)

FIGURE 19.3 Freshwater marshes. (From USEPA, Learn the issues: Learn about water, Available at: http://www2.epa.gov/learn-issues/learn-about-water, 2013.)

19.2.2.1 Prairie Potholes

Prairie potholes, often also called "sloughs," are most common in the Great Plains of North America, and, in particular, the Upper Midwest (e.g., North Dakota, South Dakota, Wisconsin, Iowa, and Minnesota; Figure 19.4). They were primarily formed by glaciation and the subsequent melting of buried blocks of ice, leaving a great number of shallow depressions, often round like a "pot," which formed these wetlands. The water sources are primarily either direct precipitation or snowmelt. Some prairie pothole marshes are ephemeral or temporary, while others are seasonal or semipermanent. Often, the ephemeral wetlands are not wet long enough to support hydric vegetation (Stewart and Kantrud 1969). While some are large, most prairie potholes are less than 1 acre in size. Although these wetlands are small, they are at least historically common. They provide a variety of ecological services but are most often noted for their importance to migratory waterfowl.

19.2.2.2 Playas

Playa lakes are similar to prairie potholes in that they are usually shallow, round depressions in semiarid regions such as in West Texas, Oklahoma, New Mexico, Colorado, and Kansas. Unlike potholes, playa lakes are most commonly formed by wind action rather than glaciation, or, in some cases, by dissolution or land subsidence. As with some potholes, they are ephemeral, characterized by short, infrequent, and unpredictable water availability, and are usually shallow with sparse vegetation. Some playa lakes are "wet" only once every few years and then often for short periods of time. Historically, playas have served as the main water source on the plains, such as for American Indians and nineteenth-century settlers (Graves 2006). These wetlands are also extremely important to waterfowl. But since they are only periodically and often infrequently "wet," waterfowl often rely on a large network of playas, selecting those that are sufficiently wet (Robinson and Oring 1996; Robinson and Warnock 1996) rather than single or specific wetlands. Comanche, and later hide hunters, knew that bison drank from the playas and often hunted them at water holes.

FIGURE 19.4 Prairie pothole regions of North America. (From Euliss, Jr. N.H., Mushet, D.M., and Wrubleski, D.A., In: D.P. Batzer, R.B. Rader, and S.A. Wissinger (eds), *Invertebrates in Freshwater Wetlands of North America: Ecology and Management*, John Wiley & Sons, New York, p. 471, 1999.)

19.2.2.3 Vernal Pools

Vernal pools are seasonal, often isolated, depressions, usually containing water for only a few months of the year. They may fill with the rising water table during the fall and winter or during the spring snowmelt and rain, and are commonly dry during late summer and fall and may be dry for sustained periods during drought conditions. The dry period eliminates fish and many predators, such as bullfrogs, which take more than a year to develop from tadpole to adult. While only periodically "wet," these pools also provide a critical habitat for a variety of plants and animals that are adapted to reproduce in temporary wetlands (Brown and Jung 2005). These include a variety of amphibians, some of which breed exclusively in vernal pools, and other organisms such as the fairy shrimp that rely on these pools during their entire life cycle.

19.2.2.4 Wet Meadows

Wet meadows occur in poorly drained areas such as shallow basins, low-lying farmland, and between shallow marshes and upland areas. They often look like grassland or a fallow field except that they are dominated by water-loving grasses and sedges. Generally, they are completely covered with vegetation and only periodically and temporarily have standing water. Wet meadow soils are commonly saturated and mucky. Wet meadows are common throughout the United States in areas ranging from mountain valleys to Florida. In Florida, they occupy large areas called "flats," which are extensive in the northwest Everglades (McPherson et al. 1976; Figure 19.5).

An example is the saw grass marshes common in Florida where they make up about 70% of the remaining Everglades (McPherson et al. 1976). Saw grass (*Cladium jamaicense*) is not a "true" grass, but a member of the sedge family, characterized by sharp teeth along the edges of each blade. Saw grass marshes are usually flooded with water for most of the year and the period over which they are flooded determines the growth of the saw grass, with taller, thicker stands of saw grass corresponding to more prolonged and deeper inundation.

19.2.2.5 Swamps

Swamps are wetlands dominated by woody plants, which include forested swamps, dominated by trees, or brush swamps (Figure 19.6). Forested swamps commonly occur in the floodplains of rivers, and are typically saturated for most of the year and periodically inundated. Common trees in the northern United States include the red maple and pin oak (*Quercus palustris*). In the south, the overcup oak (*Q. lyrata*) and cypress provide the characteristics of many forested swamps, more

FIGURE 19.5 Saw grass marsh. (From McPherson, B.F., Hendrix, C.Y., Klein, H., and Tyus, H.M., The environment of South Florida, a summary report, Geological Survey Professional Paper 1011, 1976.)

Forested swamps

Shrub swamps

FIGURE 19.6 Examples of swamps. (From the National Wetlands Research Center.)

frequently called bottomland hardwood swamps. In the Northwest, the dominant trees include willows (*Salix* spp.) and western hemlock (*Tsuga* sp.). Brush swamps are commonly dominated by shrubs, such as the buttonbush or smooth alder and are frequently found near or associated with forested swamps.

19.3 WETLAND IDENTIFICATION AND CLASSIFICATION

The types of wetlands previously discussed represent a very diverse assemblage of ecological systems. This diversity makes it difficult to develop a universally accepted definition of wetlands, and such a definition will vary depending on the purpose of that definition (e.g., scientific study or regulatory purpose). Similarly, a wide variety of classification systems has been developed for various purposes, such as the scientific study of wetlands, to establish trends in wetlands, or to identify systems as wetlands in order to preserve and protect them (regulatory classification).

One example of a wetland identification system is that developed for identifying whether a system is subject to protection under Section 404 of the CWA. Under Section 404, discussed in detail in the following sections as it relates to wetlands, a permit is required from the U.S. Army Corps of Engineers (the Corps) for the discharge of dredged or fill material into any waters of the United States, including wetlands. Therefore, the Corps has to have criteria that it can use in this process to identify if a particular system is a "water of the United States," and then whether that system is a wetland (a "jurisdictional" determination). The methods developed to delineate a wetland were published in the "Corps of Engineers wetland delineation manual" in 1987 (USACE 1987). In that manual, the Corps specified the criteria to be used to establish whether or not a system is a wetland, subject to protection under the CWA. The criteria were based on the presence of three factors:

1. Hydrophytic vegetation
2. Wetland hydrology
3. Hydric soils

The manual also stressed the need for "sound professional judgment, providing latitude to demonstrate whether an area is a wetland or not based on a holistic and careful consideration of evidence

for all three parameters" ("U.S. Army Corps of Engineers wetland delineation manual modifications and clarifications"; USACE 1992).

Today, it is mandatory to use the Corps 1997 manual in the 404 permitting process. However, one
of the issues concerning its use relates to regional differences in technical and policy considerations
impacting wetland delineation. For example, technical issues could include whether wetland criteria
are appropriate in a particular region, and whether indicators used to identify wetlands in the field
are sensitive to regional variations in environmental conditions (National Research Council; NRC
1995). These issues resulted in the Corps developing supplements to the 1997 manual for the following regions (Figure 19.7):

- Alaska
- Arid West
- Atlantic and Gulf Coastal Plain
- Caribbean Islands
- Eastern Mountains and Piedmont
- Great Plains

FIGURE 19.7 Corps of Engineers wetlands delineation manual and regional supplements.

- Hawaii and Pacific Islands
- Midwest Region
- Northcentral and Northeast
- Western Mountains, Valleys, and Coast

The methods described in the Corps manuals, while not the only wetland delineation methods, are of particular importance in determining whether the protection of a wetland is under the jurisdiction of the Corps. While the Corps administers the 404 program, the U.S. Environmental Protection Agency (U.S. EPA) has the ultimate authority to determine the geographic scope of the waters of the United States subject to jurisdiction under the CWA, including the 404 program.

Another important federal act impacting wetlands is the Food Security Act (FSA), the wetland conservation (Swampbuster) provisions of which will be discussed later in this chapter. This act is the regulatory responsibility of the U.S. Department of Agriculture (USDA), acting through the Natural Resources Conservation Service (NRCS). For Swampbuster implementation, the NRCS has the ultimate authority to determine the geographic extent of wetlands. Based on memorandums of agreement, the NRCS is also responsible for making wetland delineations for implementing CWA regulations (e.g., 404 permits) on agricultural lands. The NRCS delineations were originally based on methods described in the National Food Security Act manual (NRCS 1987).

Note that the methods used by federal agencies for delineation, or wetland determination, have been subject to the winds and vagaries of politics and policy. Methods for the delineation of wetlands subject to federal jurisdiction under the CWA and FSA were described in both the Corps 1997 manual (USACE 1987) and the National Food Security Act manual (NRCS 1987). In addition, the U.S. EPA had its own delineation manual, the 1988 Environmental Protection Agency Wetland Identification and Delineation manual (Sipple 1988). Since the methods for delineation (and the resulting delineations) differed, the agencies agreed to develop a joint manual, resulting in the 1989 Interagency Wetland Delineation manual (Interagency 1989). However, this manual was criticized by some in that it supposedly broadened the definition of a wetland, and therefore increased the acreage of wetlands subject to protection under these two laws. The response by the agencies was a 1991 revision, the 1991 Interagency Wetland Delineation manual. However, subsequently, the Corps was directed by law to go back to using its original 1987 manual. In addition, the NRCS is required to use the Corps 1987 manual when conducting wetland determinations for the purposes of the FSA on nonagricultural lands and agricultural lands with undisturbed native vegetation, and it currently uses the hydrology wetland indicators from the Corps 1987 manual for conducting wetland determinations on agricultural land for the purposes of the FSA.

19.4 WETLAND TRENDS: HISTORICAL

While the importance of wetlands and their protection is now commonly recognized, this was not always the case. In much of U.S. history, wetlands have been considered an impediment rather than a valuable resource, and something to be removed rather than to be protected. An example of attitudes toward wetlands is William Byrds description of the "Dismal Swamp," an area that at that time formed a disputed boundary between the colonies of Virginia and North Carolina, as a "horrible desert [where] the foul damps ascend without ceasing, corrupt the air and render it unfit for respiration." That area, still so named, now includes a protected state recreation area and the Dismal Swamp National Wildlife Refuge. In 1900, the U.S. Supreme Court proclaimed that wetlands were "the cause of malarial and malignant features" and that "the police power is never more legitimately exercised than in removing such nuisances" [*Leovy v. United States*, 177 U.S. 621, 636 (1900)]. In their chapter in *National Water Summary on Wetland Resources*, Dahl and Allord (1996) discussed the changing attitudes toward wetlands during U.S. history, and wetland losses (Figure 19.8), which are briefly summarized next.

FIGURE 19.8   States with notable wetland loss, 1780s to mid-1980s. (From Dahl, T.E., Status and trends of wetlands in the conterminous United States 1986 to 1997. U.S. Department of the Interior, Fish and Wildlife Service, Washington, DC, 1990; From Dahl, T.E. and Allord, G.J., *National Water Summary on Wetland Resources*, U.S. Geological Survey, Washington, DC, 1996.)

During the period from the early 1600s to the 1800s, a period of colonial development, relatively few records existed to document the extent and change in wetlands. It was not until 1785 that the Land Ordinance Act established the U.S. Public Land Survey, which required that in surveying land, some quantitative information was available on the extent of wetlands. During this period of extensive agricultural development, wetlands were considered an impediment to development, and were often converted to farmland, particularly in the most southern of the original colonies (Dahl and Allord 1996).

The period from 1800 to 1860 was a period of growth and westward expansion. This period included the Louisiana Purchase (1803) and the annexation of Texas (1845). It also included extensive agricultural development and the drainage of wetlands. The Swamp Land Acts in 1849, later modified in 1850, granted all swamp and overflow lands to states for "reclamation," first in Louisiana and then in 12 additional states (Shaw and Fredine 1956; Dahl and Allord 1996), establishing the prevailing tone and policy for land use trends for the next century.

During 1860–1900, westward migration continued and, with technological advances, the conversion of wetlands to farmlands rapidly increased. This rapid conversion continued during the 1900s to 1950, and by the 1920s about 70% of the original wetland acreage had been modified by levees, drainage, and water diversion projects (Frayer et al. 1989, cited in Dahl and Allord 1996). The development of mechanical tractors allowed large areas of wetlands and prairie potholes to be drained. By the 1930s, large portions of the Florida Everglades had been drained. Flooding in Florida prompted the Corps to build the Central and Southern Florida Project for flood control, which resulted in very substantial losses of wetlands. This period also included the proliferation of organized drainage districts throughout the country that coordinated efforts to remove surface water from wetlands (Wooten and Jones 1955, cited in Dahl and Allord 1996). During the 1930s, the U.S. government basically provided free engineering services to farmers to drain wetlands and, by the 1940s, it shared the cost of drainage projects (Burwell and Sugden 1964, cited in Dahl and Allord 1996).

In addition to the dewatering of large portions of the Florida Everglades, one other particularly notable effort that resulted in enormous losses of wetlands was that targeted at the control of the flooding of the Mississippi River. The control of the mighty Mississippi remains a problem today, as noted by Mark Twain in his *Life on the Mississippi*:

> One who knows the Mississippi will promptly aver … that ten thousand River Commissions, with the mines of the world at their back, cannot tame the lawless stream, cannot curb it or confine it, cannot say to it Go here or Go there, and make it obey; cannot save a shore that it has sentenced.

The attempted control of the Mississippi River was initiated early in U.S. history, such as the initial construction of a man-made levee system by Bienville, the founder of the city of New Orleans, in 1717. The responsibility of the Mississippi levee system was ultimately delegated to the Corps and since 1882, the Corps in conjunction with the Mississippi River Commission has maintained and extended the levee system. The levee system, while providing flood control, also isolated the bottomland hardwood wetlands that were common along the Mississippi River valley. More than 75% of the historic bottomland hardwood wetlands in the lower Mississippi valley have been lost (Dahl 1990; Dahl and Allord 1996). As stated frequently elsewhere in this text, it is often common that "one generation's solution is the next generation's problem," or for engineers "one generation's solution is the next generation's job." For example, currently, there are a variety of wetland restoration activities along the Mississippi River corridor and elsewhere for the purposes of flood control, nutrient management, habitat enhancement, and other beneficial uses.

By the 1960s, a variety of political, financial, and institutional incentives to drain or destroy wetlands was in place (Dahl and Allord 1996). During this time, an increasing awareness of the values of wetlands resulted in increasing efforts at their protection and restoration. Federal policies have shifted to eliminate many of the incentives for the destruction of wetlands, and new laws, such as the CWA in 1972 and the Emergency Wetland Resources Act of 1986, have further protected and curtailed wetland losses. In 1974, the U.S. Fish and Wildlife Service initiated an inventory of the nation's wetlands. The 1986 act also required that the Fish and Wildlife Service update the initial wetlands status and trends information every 10 years and provide a report to Congress.

- The "Status and trends of wetlands and deep-water habitats in the conterminous United States, 1950's to 1970's" (Frayer et al. 1983) reported: "Total acreage of wetlands and deep-water habitats in the 48 conterminous United States in the 1950's was 179.5 million acres. In the 1970's it was 171.9 million acres, a net loss of 7.6 million acres. Average annual net loss for the 20-year period was 380 thousand acres."
- The "Status and trends of wetlands in the conterminous United States, mid-1970's to mid-1980's" (Dahl and Johnson 1991) reported that: "there were an estimated 105.9 million acres of wetlands in tile conterminous United States in the mid-1970's. In the mid-1980's, there were 103.3 million acres of wetlands. This translates into a net loss of over 2.6 million acres over the study period. Freshwater wetlands experienced 98.0 percent of the losses that occurred during the study period. By the mid-1980s, an estimated 97.8 million acres of freshwater wetlands and 5.5 million acres of estuarine (coastal) wetlands remained."
- The "Status and trends of wetlands in the conterminous United States 1986 to 1997" (Dahl 2000) reported that: "An estimated 105.5 million acres of wetlands remained in the conterminous United States in 1997. Between 1986 and 1997, the net loss of wetlands was 644,000 acres. The annual loss rate during this period was 58,500 acres, which represents an eighty percent reduction in the average annual rate of wetland loss as compared to the last wetlands status and trends report" (Dahl and Johnson 1991).
- The "Status and trends of wetlands in the conterminous United States 1998 to 2004" (Dahl 2006) reported that: "there were an estimated 107.7 million acres of wetlands in the conterminous United States in 2004." "Wetland area increased by an average 32,000 acres annually" and that "agricultural conservation programs were responsible for most of the gross wetland restoration." Additionally, "despite the net gains realized from restoration and

creation projects, human induced wetland losses continued to affect the trends of fresh-water vegetated wetlands—especially freshwater emergent marshes which declined by an estimated 142,570 acres."

- The "Status and trends of wetlands in the conterminous United States 2004 to 2009" (Dahl 2011) reported that: "an estimated 110.1 million acres of wetlands in the conterminous United States in 2009" and while "the difference in the national estimates of wetland acreage between 2004 and 2009 was not statistically significant. Wetland area declined by an estimated 62,300 acres (25,200 ha) between 2004 and 2009."

So, in over 50 years of studies by the Fish and Wildlife Service, some improvements have been noted, or at least reductions in loss rates. However, there have been a number of regulatory changes and judicial rulings, as will be discussed in the following sections, which have accelerated losses in recent years.

19.5 WETLANDS AND THE CWA

19.5.1 FEDERAL STATUTES AND ADMINISTRATIVE LAW: REGULATIONS AND GUIDANCE

Before discussing the CWA and federal law, let us make a brief divergence to discuss administrative law and procedures defined by the Administrative Procedures Act. A simplified version of the process is illustrated in Figure 19.9. Federal laws are, of course, acts of Congress, which are typically assigned to agencies for implementation. That is, the agencies are mandated to take some action, such as promulgations or regulations, to enforce the intent of the statute. Administrative law deals with the decision making of the administrative units of government, such as those agencies involved in the development and implementation of those regulations.

Since these agencies are tasked with the implementation of the law, it is instructive to know how they receive their powers. While part of the executive branch, the agencies are typically created by Congress. An exception is the U.S. EPA, which was created by executive order (signed by President Nixon). Regardless, the U.S. EPA and other agencies receive their authority to act, and their funding, from Congress. For example, an agency cannot generate or obtain funding for its own use, such as from fines or fees, since that would diminish Congress's oversight and any funding so generated must be directly added to the federal treasury.

Once a federal agency is mandated to take some action based on a federal statute, such as the CWA, that action typically involves developing regulations and rules (rulemaking), based on the intent of Congress, to enforce that statute. The regulations are drafted by the agency. The public is then notified and the proposed regulation is posted in the *Federal Register* (published daily and presently available online). The agency receives and responds to public comments after which it

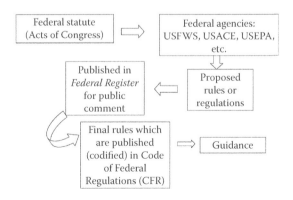

FIGURE 19.9 Federal laws and regulations.

will either withdraw the proposed regulation or post the final regulation, along with the agency's response to public comments, in the *Federal Register* together with the date when the regulation will become effective. On that date, the regulation has the force of law (Gardner 1990, 2011) and is eventually published in the Code of Federal Regulations.

The agency can also, and commonly does, affect guidance or interpretative rules that describe how the agency plans on interpreting and implementing the statutes or regulations. These guidance documents do not have the force of law and the agency is not required to, and commonly does not, post the guidance in the *Federal Register* for public review before being implemented. However, as described by Gardner (2011), "many of the rules governing wetlands are found in guidance documents; these are where the real details are."

19.5.1.1 Federal Law and the Courts

The federal law of the United States consists of the U.S. Constitution, laws enacted by Congress, and decisions of the U.S. Supreme Court and other federal courts. So, for example, after regulations that have the force of law are implemented by agencies, based on interpretation of the federal statute (e.g., the CWA), issues can be brought to court and the U.S. Supreme Court or other federal courts may agree or disagree with that interpretation.

It is not easy to bring a suit against a federal agency to challenge a regulation. Gardner (2011) indicates that the two chief requirements are the requirements of standing and ripeness. Standing generally means that the plaintiff must demonstrate that there is sufficient connection to and harm from the regulation to support his or her participation in the case. That often makes it difficult for third parties, such as environmental groups, to challenge federal regulations. Gardner (2011) indicates that establishing standing typically requires the establishment of "injury in fact" or harm; that there is "causation" so that the harm is traceable to the regulation (in this case); and "redressability" so that the court's decision would have some practical effect. The "ripeness" requirement simply means that the court may decide that the time is not yet right or "ripe" to hear the case.

Challenging a regulation in court typically means challenging the agency's interpretation of the federal statute. One of the most important principles in administrative law is called the *Chevron* deference, established by the U.S. Supreme Court in *Chevron U.S., Inc. v. Natural Resources Defense Council, Inc.* 467 U.S. 837 (1984). In that case, th e issue was raised as to how the courts should treat agency interpretations of statutes that mandated that agency to take some action. The U.S. Supreme Court held that the courts should defer to the agency's interpretations of such statutes unless they are unreasonable, after which this principle is referred to as the *Chevron* deference. There are two parts to the *Chevron* deference (Gardner 2011). The first is to determine whether Congress addressed the issue in plain language, and if so then that intent must be followed. If the intent was not clear, then the court must determine if the agency's interpretation is reasonable. If the latter is the case, then the court may provide their interpretation of the intent of Congress and render a decision.

If the case is heard by the U.S. Supreme Court or other federal court, the decision of that court becomes federal law. As will be discussed shortly, a number of such decisions have dramatically altered how wetlands in the United States have been defined under law and protected (or not).

19.5.1.2 "The Waters of the United States Saga" Part 1: CWA Section 404

A number of sections of the CWA have been discussed elsewhere, but in this section we will focus on Section 404 and related sections. Section 404 of the CWA establishes a program to regulate the discharge of dredged or fill material into "navigable waters." We will return to the dredging and fill issue shortly, but first, given the topic of this chapter, a question is: what is a navigable water and does it include wetlands?

The traditional definition of a navigable water is that it is "navigable" for the purposes of commerce. The Corps has been responsible for maintaining navigable waterways since the 1890s under the Rivers and Harbors Act and it has a wealth of experience, regulations, and procedures for defining what is navigable and what is not. So, when first mandated with the responsibility for implementing Section 404 of the CWA, the first approach was to limit the Corps regulatory activities to those navigable waters.

However, under CWA Section 502, navigable waters are broadly defined as "the waters of the United States, including territorial seas." So, the question then becomes, what are "waters of the United States," and does that include wetlands, and if so which ones? That has become the subject of considerable administrative law.

The Corps original regulations, which narrowly restricted the implementation of Section 404 to navigable waterways, were invalidated by the federal court in *Natural Resources Defense Council v. Callaway* (Gardner 2011). As a result, the Corps Regulatory program revised its regulations in the Code of Federal Regulations (CFR)—Title 33: Navigation and Navigable Waters, Chapter II: Corps of Engineers, Part 328: Definition of Waters of The United States. A few of the definitions of "waters of the United States" in this section include the traditional definition:

- All waters which are currently used, or were used in the past, or may be susceptible to use in interstate or foreign commerce, including all waters which are subject to the ebb and flow of the tide (3(a)(1))

and additional definitions including:

- All interstate waters including interstate wetlands (3(a)(2))
- All other waters such as intrastate lakes, rivers, streams (including intermittent streams), mudflats, sand flats, wetlands, sloughs, prairie potholes, wet meadows, playa lakes, or natural ponds, the use, degradation or destruction of which could affect interstate or foreign commerce including any such waters (3(a)(3))
- All impoundments of waters otherwise defined as waters of the United States under the definition (3(a)(4))
- Tributaries of waters identified in paragraphs (a) (1) through (4) of this section (3(a)(5))
- The territorial seas (3(a)(6))
- Wetlands adjacent to waters (other than waters that are themselves wetlands) identified in paragraphs (a) (1) through (6) of this section (3(a)(7))

The regulation went on to define wetlands as "those areas that are inundated or saturated by surface or ground water at a frequency and duration sufficient to support, and that under normal circumstances do support, a prevalence of vegetation typically adapted for life in saturated soil conditions. Wetlands generally include swamps, marshes, bogs, and similar areas."

Based on this revised regulation, the Corps also issued guidance documents such as the Wetland Delineation manual (TR Y-87-1, USACE 1987) as well as a series of regional supplements. One change of course in the revised regulations was that they now protected wetlands directly or indirectly connected to navigable waters as well as isolated wetlands as long as they "are currently used, or were used in the past, or may be susceptible to use in interstate or foreign commerce." Of course, that introduces the further question of how isolated wetlands may be connected to (have nexus with) interstate commerce.

19.5.1.3 "The Waters of the United States Saga" Part 2: The Saga Continues

Of course, the aforementioned regulation did not make some people happy, as described by Gardner (2011). One result was a series of lawsuits that challenged and changed the interpretation of what the intent of Congress was in defining "waters of the United States."

19.5.1.3.1 *United States v. Riverside Bayview Homes*

In this case, a housing developer, Riverside Bayview Homes, began developing a property in Macomb County, Michigan, near Lake St. Clair, a traditional navigable waterway. The developer began filling an 80-acre mash and did not seek a 404 permit from the Corps. The case moved through the court system and was argued in the U.S. Supreme Court in 1986. One of the major issues was whether not allowing the development constituted "taking" just as if the government had seized the property. The court ruled that "neither the imposition of the permit requirement itself nor the denial of a permit necessarily constitutes a taking."

An additional issue was whether the Corps had regulatory authority over "adjacent" wetlands or waters adjacent to but not regularly flooded by a navigable water (Gardner 2011). In this case, the court ruled that the Corps' jurisdiction over the adjacent wetlands was reasonable since Congress broadly defined "waters" covered by the CWA and that adjacent wetlands play an important role in "protecting and enhancing water quality." One of the issues then is, can adjacent be taken to include isolated wetlands? (Bueschen 1997). The NRC (1995) suggested that there is little justification for legislation and judicial decisions that hold isolated wetlands as less significant than adjacent wetlands.

One of the components of the Corps regulations is that to be protected, an isolated wetland must have a substantial connection to interstate commerce, as discussed by Bueschen (1997). Gardner (2011) indicated that when the Corps attorney was asked under what constitutional authority did the Corps regulate isolated wetlands, her response was "the Commerce clause," stating that the presence of migratory birds at an isolated wetland could provide a sufficient connection to interstate commerce. Gardner (2011) indicated that the transcript of the oral argument reported parenthetically that there was general laughter in response to the concept that migratory birds could create a nexus to interstate commerce. But, following the decision, the Corps announced that it would regulate activities in isolated waters (including wetlands) that "are or would be used as habitat by other migratory birds that cross state lines" (Gardner 2011), which became known as the Migratory Bird Rule.

19.5.1.3.2   Solid Waste Agency of Northern Cook County v. Army Corps of Engineers

The Solid Waste Agency of Northern Cook County (SWANCC) was a consortium of 23 towns and villages in the Chicago area that managed solid wastes on a regional basis. They wanted to develop a landfill for nonhazardous solid waste on a 533-acre parcel that had in the past been used for sand and gravel mining. But, the sand and gravel pits had become a series of lakes and ponds of ideal bird habitat, with 121 species reported, and home to the second largest blue heron rookery in northeastern Illinois (Gardner 2011). These ponds and lakes were isolated waterbodies, not connected or associated with a navigable water.

SWANCC sought a Section 404 permit from the Corps, but the Corps determined that the area, while not wetlands, qualified as "waters of the United States" pursuant to the Migratory Bird Rule and denied the permit. SWANCC filed suit against the Corps, which ultimately went to the U.S. Supreme Court for a decision.

In a split decision (5–4), the U.S. Supreme Court decision in January 2001 struck down the application of the Migratory Bird Rule to protect isolated waters. The rationale of the decision though was broader, appearing to preclude federal assertion of Section 404 jurisdiction over isolated waters on any basis. The court stated that "In order to rule for [the Corps], we would have to hold that the jurisdiction of the Corps extends to ponds that are not adjacent to open water. But we conclude that the text of the statute will not allow this."

The impact of this split decision was huge. The decision represented a major reinterpretation of the CWA by emphasizing the importance of "navigability" in the definition of "waters of the United States" and implying that federal protection required a "significant nexus" to navigable waters (or that the protection should lie with the states). As such, it effectively removed isolated waters (including wetlands) from protection under the CWA, estimated to make up 40%–60% of all freshwater wetlands (Kusler 2004, cited by Gardner 2011). The impact was illustrated in "Lawful Loss," an article published in *Texas Parks and Wildlife* magazine (Harvey 2006), which stated that 3000 acres of coastal wetlands formerly protected under the CWA in the Houston area are now becoming housing subdivisions with names such as "Mar Bella" and "Tuscan Lakes" and went on to state that if the Corps' Galveston district's "interpretation was in place in Florida, most of the Everglades would be considered isolated and subject to filling."

Riverside Bayview Homes confirmed the Corps' jurisdiction on wetlands adjacent to navigable waters, while SWANCC indicated that the Corps had no jurisdiction over isolated wetlands, at least using the Migratory Bird Rule (Gardner 2011). But, what about other wetlands, such as those adjacent to tributaries to navigable waters (Gardner 2011)? The issue then became one of determining what a "significant nexus" was in order to establish protection under the CWA (Downing et al. 2003).

19.5.1.3.3 Rapanos v. United States

Enter John Rapanos (with his wife and several wholly owned companies) who wanted to construct a shopping center near Midland, Michigan, on a site that was adjacent to tributaries that eventually entered into navigable waterways that were from 11 to 20 miles away. The Michigan Department of Natural Resources told Mr. Rapanos that he would need a permit to fill wetlands on the property. Mr. Rapanos then hired an environmental consultant who reported the same, concluding that 48–58 acres of the site were wetlands. "Rapanos threatened to 'destroy' Dr. Goff if he did not destroy the wetland report, and refused to pay Dr. Goff unless and until he complied" (records of *John A. Rapanos, et ux., et al. v. United States, June Carabell et al. v. United States Army Corps of Engineers et al.*, Supreme Court of the United States argued February 21, 2006, decided June 19, 2006). Both civil and criminal charges were brought, and in the criminal trial he was found guilty of violating the CWA. The civil trial eventually went to the U.S. Supreme Court (records of *John A. Rapanos, et ux., et al. v. United States, June Carabell et al. v. United States Army Corps of Engineers et al.*, Supreme Court of the United States argued February 21, 2006, decided June 19, 2006; see USEPA Region 5 [2006] for a summary).

In this case, the court (June 19, 2006 decision) was split 4–1–4, with Justice Kennedy, unlike in *SWANCC*, failing to break the tie. It was his opinion, with which the other justices apparently did not agree, that the appropriate implementation of the intent of Congress in the formulation of the CWA was to require a "significant nexus" of the wetlands to traditional navigable waters (TNWs). He further stated that the "nexus required must be assessed in terms of the Act's goals and purposes." Congress enacted the law to "restore and maintain the chemical, physical, and biological integrity of the Nation's waters," and it pursued that objective by restricting dumping and filling in "waters of the United States."

The result of the lack of a decision was to not provide precise guidance to the lower courts or agencies on how to clearly interpret wetland protection under the CWA. As indicated by Gardner (2011), one option for concise guidance would have been to limit federal protection to only traditional navigable waterways, but that option was rejected by Justice Kennedy and four dissenters in *Rapanos*. Another option was to define the nexus as being a hydrological connection to a navigable water, but that too was rejected by Justice Kennedy and the other four dissenters. Strangely, while there was no majority opinion, it was the single opinion of Justice Kennedy that had the greatest influence on subsequent wetland protection (Gardner 2011). The impact of lack of a decision is illustrated by the 2006 article (Beardsley 2006) in *Scientific American* titled *The End of the Everglades? Supreme Court Case Jeopardizes 90 percent of U.S. Wetlands*.

19.5.1.4 So What Are "Waters of the United States" as They Apply to Wetlands Now?

In the absence of a clear legal definition of wetlands protected by the CWA, agencies proposed what they could provide "fuzzy and ambiguous guidance." That guidance is then "subject to interpretation," meaning that individual agencies or organizations, or even districts within the Corps and ultimately the courts may interpret it differently. The proposed new guidance was posted by the U.S. EPA and the Corps in the *Federal Register* on May 2, 2011, after which it received about 230,000 public comments. In the proposed new guidance, wetlands are basically divided into three areas (see Table 19.2).

The agencies are still working on the final rulemaking (as of May 2012) "and the saga (and confusion) continues."

TABLE 19.2

Proposed 2011 Guidance on What Constitutes "Waters of the United States"

Type 1: Those that are absolutely protected include:

Traditional navigable waters

Interstate waters

Wetlands adjacent to either traditional navigable waters or interstate waters

Nonnavigable tributaries to traditional navigable waters that are relatively permanent, meaning that they contain water at least seasonally

Wetlands that directly abut relatively permanent waters

Type 2: Those that may be, or where a "fact-specific" analysis determines that they have a "significant nexus" to a traditional navigable water or interstate water:

Tributaries to traditional navigable waters or interstate waters

Wetlands adjacent to jurisdictional tributaries to traditional navigable waters or interstate waters

Waters that fall under the "other waters" category of the regulations. The guidance divides these waters into two categories, those that are physically proximate to other jurisdictional waters and those that are not, and discusses how each category should be evaluated

Type 3: And finally, those that definitely (generally?) do not:

Wet areas that are not tributaries or open waters and do not meet the agencies' regulatory definition of "wetlands"

Waters excluded from coverage under the CWA by existing regulations

Waters that lack a "significant nexus" where one is required for a water to be protected by the CWA

Artificially irrigated areas that would revert to upland should irrigation cease

Artificial lakes or ponds that are created by excavating and/or diking dry land and are used exclusively for such purposes as stock watering, irrigation, settling basins, or rice growing

Artificial reflecting pools or swimming pools created by excavating and/or diking dry land

Small ornamental waters created by excavating and/or diking dry land for primarily aesthetic reasons

Water-filled depressions created incidental to construction activity

Groundwater drained through subsurface drainage systems

19.5.2 Section 404 Permitting and Wetland Mitigation

19.5.2.1 Jurisdictional?

Section 404 of the CWA regulates the discharge of dredged material, the placement of fill material, or the excavation within "waters of the United States" and it authorizes the secretary of the army, through the chief of engineers, to issue permits for such actions. One of the questions addressed in the preceding section was what are "waters of the United States"? While it is not clear what they are, if you plan on doing one of the aforementioned actions regulated by Section 404 in a "water of the United States," a permit is required, which is usually obtained from the Corps division and district in which the water occurs (Figure 19.10).

Back to "waters of the United States," the determination of which is usually made by the Corps district offices. For those waters that are "absolutely protected" (see Table 19.2), such as TNWs or interstate waters, that determination is relatively easy. The Corps district offices typically make a "jurisdictional determination" for waters, such as under the River and Harbors Act of 1899 (Section 10) definition that navigable waterways are those that "are presently used, or have been used in the past, or may be susceptible for use to transport interstate or foreign commerce," and post a list of the TNWs where permits are required on their website. For other waters, the Corps may make a delineation verification and/or a jurisdictional determination. The verification identifies or approves the boundaries of a wetland or waterbody, based on a preliminary identification by the Corps or some other agency or organization, while the jurisdictional determination defines the regulatory status of wetlands and waterbodies. Based on *Rapanos* and other decisions, the

FIGURE 19.10   U.S. Army Corps of Engineers locations.

Corps developed the 2007 *Jurisdictional Determination Form Instructional Guidebook*, which is widely used to establish jurisdictional procedures, an example of which is illustrated in Figure 19.11. While the Corps is responsible for the process and for making the decision on whether a particular waterbody is a "water of the United States" (jurisdictional) or not (nonjurisdictional), the U.S. EPA has the final authority and is provided an opportunity to agree with, or to reverse, the decision.

Currently, a number of activities are exempt from 404 requirements. They include, for example, normal farming, silviculture (forestry), or ranching practices that are part of an established, ongoing operation and maintenance of structures, such as dikes, dams, levees, breakwaters, and causeways.

19.5.2.2   404 Permits

For an activity requiring a permit, there are generally three forms of permits that the Corps may issue: general permits, letters of permission, and standard individual permits. The general permits may be regional or nationwide.

Nationwide general permits authorize activities across the country. These general permits are updated every 5 years following public notice in the *Federal Register* (latest notice in February 2012). The proposal to reissue general permits was published in the *Federal Register* in February 2011 and received approximately 26,600 comments. According to the Corps' "Nationwide Permit Reissuance" (February 15, 2012), there are currently 49 nationwide permits (two new ones were added in the revision related to renewable energy) and the nationwide permits annually authorize about 40,000 reported (to the district offices) projects as well as about 30,000 projects that do not require reporting to the district offices. For the reported activities, a preconstruction notification to the Corps district is required, while for the nonreporting activities only

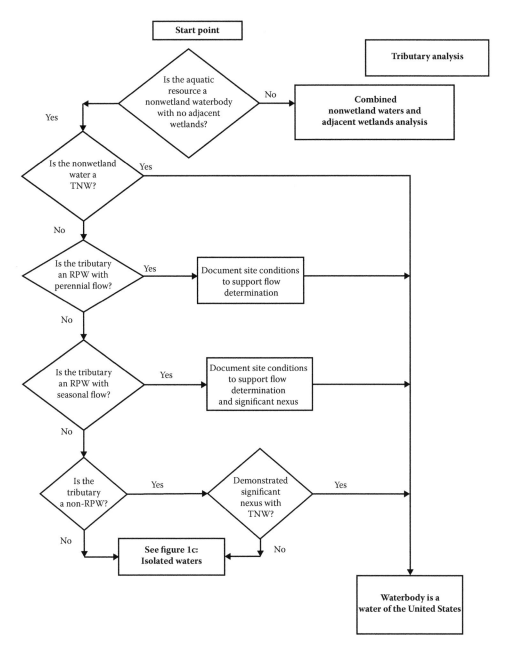

FIGURE 19.11 Example of waterbody determinations (TNW = traditional navigable waterway; RPW = relatively permanent waters; figure 1c can be found in USACE 2007). (From USACE, *Jurisdictional Determination Form Instructional Guidebook*, U.S. Army Corps of Engineers, 2007.)

a "letter to file" may be required providing justification for the authorization of the project under the nationwide permit.

"Letters of permission" is also an abbreviated permit process, where the Corps notifies other agencies but does not have to publish a public notice. These abbreviated permits are restricted to projects that have "minor" impacts with little opposition.

For everything else involving dredge or fill operations in jurisdictional waters, individual permits are required. The permits are issued by the Corps office, usually after a preapplication consultation,

a public notice and comment period on the permit application, and the preparation of permit decision documents. While issuing the permits is within the purview of the Corps, there are also coordination requirements with other federal and state agencies to meet the requirements of the CWA,
the National Environmental Policy Act, the Endangered Species Act, and other federal statutes. In
addition, a state 401 certification is required for a permit to be issued. These permits are based on
Section 401 of the CWA, which grants certifying authority for federal permits, which is basically
a state certifying that the activity is not expected to result in a violation of state water quality standards or criteria. Finally, the U.S. EPA (under Section 404(c)) has veto authority if it determines that
a project may result in significant environmental damage.

In some cases, the Corps may also issue "after the fact" permits. Basically, someone or some
organization can complete a dredge or fill project that should have required a permit and then
request a permit after the fact. If the permit is denied, then the person or organization would have
to perform restoration.

For waters that are "jurisdictional," permits such as 404 permits are required before they can be
dredged or filled. Copeland (2010), based on information from the Corps, indicated that the Corps
evaluates more that 85,000 permits annually of which only 0.3% are denied. Of the permits issued,
over 90% were based on general permits (regional or national), so that only 9% were required to
go through the more detailed (in terms of both documentation and review) individual permit process (Copeland 2010). While the U.S. EPA does have veto authority, as of 2010 that authority has
only been exercised 13 times, one of the most notable of which was the 2008 veto of the Yazoo
Backwater Area Pumps Project—Issaquena County, Mississippi (USEPA 2012).

19.5.2.3 Avoid–Minimize–Mitigate

The most common procedure used by the Corps is based on what could be called an avoid–
minimize–compensate (or mitigate) policy (Gardner 2011), as described in a 1990 memorandum of
agreement between the Department of the Army and the U.S. EPA. That is, the first priority is to
avoid sensitive areas that are, for example, difficult or impossible to restore, such as fens and bogs.
The second priority is to minimize impacts where an impact is unavoidable. The third priority is
then to compensate for those unavoidable impacts. That is, the permittee must agree to restore,
create, enhance, or preserve other wetlands to offset or compensate for those unavoidable impacts
(Gardner 2011), known as compensatory mitigation.

Mitigation is a common practice in some environmental areas, but it is unusual in the context of
the CWA other than Section 404. That is, in general, the quality of "waters of the United States" are
equally protected and one cannot basically write off one waterbody in favor of another, even if, in so
doing, one improves some other waterbody of similar or even greater value (such as in the national
pollutant discharge elimination system (NPDES) permitting process). There may be different water
quality standards for different waterbodies, based on their beneficial uses, but all waters must meet
quality standards. Another practice is effluent trading, which is when a discharger cleans up some
source other than his or her own as long as the total load requirements to that specific waterbody
are met; however, effluent trading is not comparable to compensatory mitigation since it applies to
the same waterbody.

Section 404 of the CWA deals with permitted impacts, such as dredging, and stipulates that those
permitted impacts on "aquatic resources" must be mitigated. While most commonly applied to
wetlands, "aquatic resources" also include streams, although stream mitigation has not received as
much attention (Lave et al. 2008; Doyle and Shields 2012). Regardless, common practice has been to
require compensatory mitigation as part of the 404 permitting process. That mitigation has resulted
in a reasonably large environmental industry centered on wetland mitigation. There are basically
three types of mitigation in common practice: site-specific compensatory mitigation, in-lieu-fee
mitigation, and mitigation banking.

Site-specific compensatory mitigation is based on a specific permit and site to compensate for
unavoidable damage resulting from a project. The mitigation may be at the particular site or located

elsewhere, generally within the same watershed. Most commonly, some environmental group or firm is hired to perform the mitigation, but the responsibility (and liability) still lies with the permittee. The mitigation activities could involve:

- Creating a new wetland
- Restoring a former wetland
- Enhancing a degraded wetland
- Preserving an existing wetland

and the mitigation is a condition of the permit.

In-lieu-fee mitigation involves the payment of a fee, which is placed in a pooled fund. The idea is simply to pool money from a series of small projects to fund a larger one by a third party. By providing funds, the permittee is relieved of his or her mitigation responsibility, and the responsibility then shifts to the third party.

Mitigation banks are also a form of third-party mitigation and are generally larger-scale mitigation projects constructed specifically to sell mitigation credits to others. These are usually created by for-profit organizations, but they may also be created by nonprofit or government agencies. For example, some of the earliest mitigation banks were State Departments of Transportation (DOT), such as the Montana DOT (Gardner 2011), where the Montana DOT entered into an agreement to establish wetlands before its road projects and then "banked" these credits for later use. Similar to in-lieu-fee mitigation, the purchase of these credits can be used to remove the liability for mitigation from the permittee, and the responsibility shifts to the management of the mitigation bank.

The process of mitigation and the evolution of the mitigation process is described in detail in Gardner's (2011) *Lawyers, Swamps and Money*. Until relatively recently, one of the problems was that the responsibility for the success (or failure) of the mitigation was largely that of the permittee, for site-specific mitigation.

As described in Gardner (2011), if an agency is failing to enforce laws to which it is assigned responsibility, then private citizens or organizations can sue to force the agency to do so. For example, the enforcement of Section 303(d) of the CWA requiring the establishment of total maximum daily loads largely occurred following lawsuits against the U.S. EPA and most states, as a result of which they were required to do so. But, can an individual or organization sue because permit requirements have not been met? That seems to depend on the type of permit. For example, in Section 505 of the CWA, Congress specifically provided for citizen suits for the violation of permits under Section 402, such as for the violation of water quality standards. However, there is no such language for Section 404 and, as described by Gardner (2011), the courts have decided that this was intentional. Therefore, if the Corps does not choose to enforce the 404 permit compensatory mitigation requirements, the public cannot sue because permit requirements have not been met. Another issue, as raised by Gardner (2011), is whether in third-party mitigations, such as in-lieu-fee mitigation or mitigation banks, responsibility equates with liability. His response is, somewhat surprisingly, "probably not" since the third party is not actually "engaging in an unauthorized discharge." The impact was that until recently, in the absence of regulatory oversight, the success rate for mitigating wetlands was somewhat less than desirable. As a result, a series of critical reviews of the compensatory mitigation process was undertaken by organizations such as the NRC (2001), the Government Accounting Office (GAO 2005), and others. In 2008, the U.S. EPA and the Corps issued a new rule (*Federal Register* April 10, 2008) that (from USEPA and USACE 2006) (Figure 19.12):

- Emphasizes that the process of selecting a location for compensation sites should be driven by assessments of watershed needs and how specific wetland restoration and protection projects can best address those needs

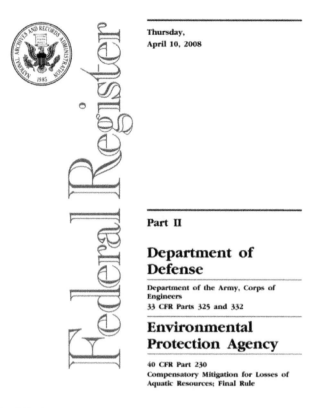

FIGURE 19.12   *Federal Register.*

- Requires measurable and enforceable ecological performance standards for all types of compensation so that project success can be evaluated
- Requires regular monitoring to document that compensation sites achieve ecological performance standards
- Clearly specifies the components of a complete compensation plan based on the principles of aquatic ecosystem science
- Emphasizes the use of science-based assessment procedures to evaluate the extent of potential water resource impacts and the success of compensation measures

The new rule retained and reformed compensatory mitigation and also retained in-lieu-fee and mitigation banks as third-party mitigation approaches, but placed new rules on them. For example, in-lieu-fee mitigations must meet the same performance standards as mitigation banks, but they are restricted to government agencies and not-for-profit organizations.

19.5.2.4   State Involvement

The Corps is the federal permitting authority for Section 404 of the CWA. However, the CWA provided a mechanism for state/tribal and federal cooperation in the 404 program, allowing, for example, a state to "administer its own individual and general permit program" in place of the federal dredge and fill permit program. To date though, only Michigan and New Jersey have elected to do so (ASWM 2011).

One section of the CWA where states do maintain control is through Section 401. Under CWA Section 401, states have the authority to ensure that federal agencies will not issue permits or licenses that violate the water quality standards, or other applicable authorities, of a state or tribe through a process known as water quality certification. That is, before the Corps can issue a 404 permit, the state has to issue a Section 401 certification.

19.6 SWAMPBUSTER PROVISIONS

Another major federal legislation impacting wetland protection is the FSA of 1985 and 1990 (the Farm Bills), the wetland conservation (Swampbuster) provisions. Unlike the 404 permit, which provides legal protection, this provision provides only economic incentives to protect wetlands. It does not provide funds for those who protect wetlands; rather it withholds funds from those who do not. That is, farm program benefits can be withheld from any person who plants an agricultural commodity on a converted wetland that was converted by drainage, dredging, leveling, or any other means after December 23, 1985, or converts a wetland for the purpose of or to make agricultural commodity production possible after November 28, 1990.

The Swampbuster program is implemented by the USDA NRCS. While most normal agricultural activities are exempt from 404 permit requirements, certification is required to maintain eligibility to USDA programs. Typically, a participant in a USDA program is required to obtain a "certified wetland determination" from the NRCS prior to planned activities in order to verify that those activities do not impact USDA program eligibility.

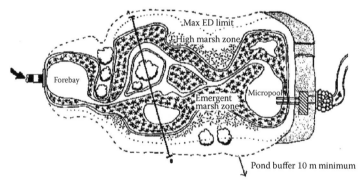

(a) Schematic design of a shallow marsh system
(adapted from Schueler, 1992)

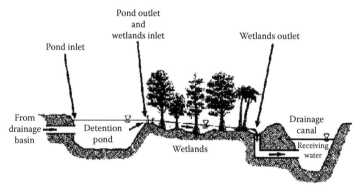

(b) Movement of water through a detention pond-wetlands system
(Martin and Smoot, 1986)

FIGURE 19.13 (a,b) Design schematics for a stormwater wetland BMP. (From U.S. Federal Highway Administration. Fact sheet—Wetlands and shallow marsh systems, *Stormwater Best Management Practices in an Ultra-Urban Setting: Selection and Monitoring.* Available at: http://environment.fhwa.dot.gov/ecosystems/ultraurb/index.asp.)

19.7 WETLAND RESTORATION AND CONSTRUCTION

A national goal for wetlands has been "no net loss" and achieving that goal has involved both wetland construction and wetland restoration (national policy for "no net loss of wetlands" established in 1989 by President George H.W. Bush). Restoration activities are defined by the Society of Wetland Scientists (2000) as "actions taken in a converted or degraded natural wetland that result in the reestablishment of ecological processes, functions, and biotic/abiotic linkages and lead to a persistent, resilient system integrated within its landscape." The focus is on the return of a degraded wetland or a former wetland to its preexisting, naturally functioning condition, or a condition as close to that as possible. (Interagency Workgroup 2003) Wetland construction for mitigation commonly has the goal of providing the same ecological function as the wetland that is being replaced. As introduced previously, wetland mitigation has resulted in an industry in the United States that revolves around the construction of wetlands for mitigation.

In addition to wetlands constructed solely for mitigation, constructed wetlands have evolved as a commonly used BMP for stormwater runoff and for the treatment of wastewater. As a result, there are a wide variety of books and manuals for the construction of wetlands and a wide number of organizations and private companies involved in their construction. One difference from "natural" wetlands is that, since they are considered artificial systems, nonnative plants such as those more tolerant of high nutrient loads are often used in their design.

Constructed wetlands are popular BMPs for stormwater treatment and most stormwater manuals include their design criteria. Stormwater wetland construction guidance is provided not only by environmental agencies but also by Departments of Transportation and local and county governments (see, e.g., ASCE 1992; USEPA 2000). Figure 19.13 illustrates an example of a design schematic from the Federal Highway Administrations web-based "Stormwater best management practices in an ultra-urban setting: Selection and monitoring."

Stormwater wetlands have the advantage that in addition to reducing peak flows and removing/storing pollutants, they can be aesthetically pleasing and support healthy populations of plants and wildlife. Stormwater wetlands, also commonly called stormwater treatment areas (STAs), have become particularly popular in the southeastern United States and often these systems have become so successful at attracting wildlife that they are often very popular birding and photography destinations.

REFERENCES

ASCE. 1992. Design and construction of urban stormwater management systems. The Urban Water Resources Research Council of the American Society of Civil Engineers (ASCE) and the Water Environment Federation. American Society of Civil Engineers, New York.

ASWM. 2011. Clean Water Act Section 404 program assumption: A handbook for states and tribes. Association of State Wetland Managers Inc. and the Environmental Council of the States. Windham, ME.

Beardsley, S. 2006. The end of the Everglades? *Scientific American*, May 22.

Brown, L.J. and R.E. Jung. 2005. An introduction to mid-Atlantic seasonal pools, EPA-903-B-05-001. U.S. Environmental Protection Agency.

Bueschen, E. 1997. Comment: Do isolated wetlands substantially affect interstate commerce? *The American University Law Review* 46, 931.

Burwell, R.W. and L.G. Sugden. 1964. Potholes—Going, going… In J.P. Linduska (ed.), *Waterfowl Tomorrow*. U.S. Fish and Wildlife Service, Washington, DC, pp. 369–380.

Copeland, C. 2010. Wetlands: An overview of issues. Congressional Research Service, 7-5700, www.crs.gov, RL33483.

Cowardin, L.M., V. Carter, F.C. Golet, and E.T. LaRoe. 1979. Classification of wetlands and deepwater habitats of the United States. U.S. Department of the Interior, Fish and Wildlife Service, Washington, DC.

Dahl, T.E. 1990. Wetlands losses in the United States 1780s to 1980s. U.S. Department of the Interior, Fish and Wildlife Service, Washington, DC.

Dahl, T.E. 2000. Status and trends of wetlands in the conterminous United States 1986 to 1997. U.S. Department of the Interior, Fish and Wildlife Service, Washington, DC.

Dahl, T.E. 2006. Status and trends of wetlands in the conterminous United States 1998 to 2004. U.S. Department of the Interior, Fish and Wildlife Service, Washington, DC.

Dahl, T.E. 2011. Status and trends of wetlands in the conterminous United States 2004 to 2009. U.S. Department of the Interior, Fish and Wildlife Service, Washington, DC.

Dahl, T.E. and G.J. Allord. 1996. History of wetlands in the conterminous United States. In J.D. Fretwell, J.S. Williams, and P.J. Redman (eds), *National Water Summary on Wetland Resources*. Water-Supply Paper 2425. U.S. Geological Survey, Washington, DC, pp. 19–26.

Dahl, T.E. and C.E. Johnson. 1991. Wetlands—Status and trends in the conterminous United States, mid-1970s to mid-1980s. U.S. Fish and Wildlife Service, Washington, DC.

Downing, D.M., C. Winer, and L.D. Wood. 2003. Navigating through Clean Water Act jurisdiction: A legal review. *Wetlands* 23, 3.

Doyle, M.W. and F.D. Shields. 2012. Compensatory mitigation for streams under the Clean Water Act: Reassessing science and redirecting policy. *Journal of the American Water Resources Association (JAWRA)* 48 (3), 494–509.

Euliss, Jr. N.H., D.M. Mushet, and D.A. Wrubleski. 1999. Wetlands of the prairie pothole region: Invertebrate species composition, ecology, and management. In: D.P. Batzer, R.B. Rader, and S.A. Wissinger (eds), *Invertebrates in Freshwater Wetlands of North America: Ecology and Management*. John Wiley & Sons, New York, pp. 471–514. Available at: http://www.npwrc.usgs.gov/resource/wetlands/pothole/index.htm.

Folkerts, G.W. 1982. The Gulf Coast pitcher plant bogs. *American Scientist* 70, 260–267.

Frayer, W.E., T.J. Monahan, D.C. Bowden, and F.A. Graybill. 1983. Status and trends of wetlands and deep-water habitats in the conterminous United States, 1950s to 1970s. U.S. Fish and Wildlife Service, Washington, DC.

Frayer, W.E., D.D. Peters, and H.R. Pywell. 1989. Wetlands of the California Central Valley: Status and trends, 1939 to mid-1980s. U.S. Fish and Wildlife Service, Portland, OR.

GAO. 2005. Corps of Engineers does not have an effective oversight approach to ensure that compensatory mitigation is occurring. Wetlands Protection, Government Accounting Office, GAO-05-898, September, 2005.

Gardner, R.C. 1990. The Army-EPA mitigation agreement: No retreat from wetlands protection. *Environmental Law Reporter* 20, 10337–10344.

Gardner, R.C. 2011. *Lawyers, Swamps, and Money: U.S. Wetland Law, Policy and Politics*. Island Press, Washington, DC.

Graves, R.A. 2006. Playas in peril, Special issue on The State of the Wetlands, *Texas Parks & Wildlife*, July, 2006.

Harvey, T. 2006. Lawful loss, Special issue on The State of the Wetlands, *Texas Parks & Wildlife*, July, 2006.

Interagency. 1989. Federal manual for identifying and delineating jurisdictional wetlands. An Interagency Cooperative Publication. U.S. Fish and Wildlife Service, U.S. Environmental Protection Agency, U.S. Army Corps of Engineers, and USDA Soil Conservation Service, January.

Interagency Workgroup. 2003. An introduction and user's guide to wetland restoration, creation, and enhancement. Interagency Workgroup on Wetland Restoration: National Oceanic and Atmospheric Administration, Environmental Protection Agency, Army Corps of Engineers, Fish and Wildlife Service, and Natural Resources Conservation Service.

Kusler, J. 2004. The SWANCC decision: State regulation of wetlands to fill the gap. Association of State Wetland Managers, Berne, NY, March, pp. 6–8.

Lave, R., M.M. Robertson, and M.W. Doyle. 2008. Why you should pay attention to stream mitigation banking. *Ecological Restoration* 26 (4), 287–289.

Lefor, M.W. and W.C. Kennard. 1977. Inland wetland definitions: Storrs, Conn., University of Connecticut., Institute of Resources, Report 28.

McPherson, B.F., C.Y. Hendrix, H. Klein, and H.M. Tyus. 1976. The environment of South Florida, a summary report. Geological Survey Professional Paper 1011.

NRC. 1995. Wetlands: Characteristics and boundaries. National Research Council, 156, 166–167.

NRC. 2001. Compensation for wetland losses under the Clean Water Act. National Research Council, National Academic Press, Washington, DC.

NRCS. 1987. *NFSAM—National Food Security Act Manual*, 5th ed. M180, USDA Natural Resources Conservation Service.

Robinson, J.A. and L.W. Oring, 1996. Long-distance movements by American avocets and black-necked stilts. *Journal of Field Ornithology* 67, 307–320.

Robinson, J.A. and S.E. Warnock. 1996. The staging paradigm and wetland conservation in arid environments: Shorebirds and wetlands of the North American Great Basin. *International Wader Studies* 9, 37–44.

SCS. 1988. *National Food Security Act Manual*, 4th ed. USDA Soil Conservation Service, February 11, 1988.

Shaw, S.P. and C.G. Fredine. 1956. Wetlands of the United States—Their extent and their value to waterfowl and other wildlife. Circular 39, U.S. Fish and Wildlife Service, Washington, DC.

Sipple, W.S. 1988. Wetland identification and delineation manual. Volume I. Rationale, wetland parameters, and overview of jurisdictional approach. Revised interim final report. U.S. Environmental Protection Agency, Office of Wetlands Protection, Washington, DC.

Society of Wetland Scientists. 2000. Position paper on the definition of wetland restoration, Society of Wetland Scientists.

Stewart, R.E. and H.A. Kantrud. 1969. Proposed classification of potholes in the glaciated prairie region. In *Small Water Areas in the Prairie Pothole Region-Transactions of a Seminar*, Ser. 6, Canadian Wildlife Service Rept., Ottawa, Canada, pp. 57–69.

Tiner, R.W. 1999. Technical aspects of wetlands: Wetland definitions and classifications in the United States. In United States Geological Survey Water Supply Paper 2425.

USACE. 1987. Corps of Engineers wetlands delineation manual. Wetlands Research program technical report Y-87-1. U.S. Army Corps of Engineers, Vicksburg, MS.

USACE. 1992. U.S. Army Corps of Engineers wetlands delineation manual modifications and clarifications, CECW-OR 2/20/1992. U.S. Army Corps of Engineers.

USACE. 2007. *Jurisdictional Determination Form Instructional Guidebook*. U.S. Army Corps of Engineers. Available at: http://www.usace.army.mil/Portals/2/docs/civilworks/regulatory/cwa_guide/jd_guidebook_051207final.pdf.

USEPA. 2000. Manual: Constructed wetlands treatment of municipal wastewaters. United States Environmental Protection Agency Office of Research and Development Cincinnati, Ohio 45268 EPA/625/R-99/010 September 2000.

USEPA. 2012. Clean Water Act Section 404(c) veto authority. United States Environmental Protection Agency. Available at: http://water.epa.gov/type/wetlands/outreach/.

USEPA. 2013. Learn the issues: Learn about water. Available at: http://www2.epa.gov/learn-issues/learn-about-water.

USEPA Region 5. 2006. The landmark case, *United States v. John A. Rapanos, et al.*, (E.D. MI.), Resolved, Region 5 Enforcement and Compliance, U.S. Environmental Protection Agency.

USEPA and USACE. 2006. Proposed wetland conservation rule, Fact Sheet. U.S. Environmental Protection Agency and U.S. Army Corps of Engineers.

Wooten, H.H. and L.A. Jones. 1955. The history of our drainage enterprises. In *The Yearbook of Agriculture, 1955*, 84th Congress, 1st Session, House Document no. 32. U.S. Department of Agriculture, Washington, DC, pp. 478–498.

Index

Printed and bound by CPI Group (UK) Ltd, Croydon, CR0 4YY

18/10/2024

01776231-0004